Respiration in Archaea and Bacteria

Advances in Photosynthesis and Respiration

VOLUME 16

The scope of our series, beginning with volume 11, reflects the concept that photosynthesis and respiration are intertwined with respect to both the protein complexes involved and to the entire bioenergetic machinery of all life. *Advances in Photosynthesis and Respiration* is a book series that provides a comprehensive and state-of-the-art account of research in photosynthesis and respiration. Photosynthesis is the process by which higher plants, algae, and certain species of bacteria transform and store solar energy in the form of energy-rich organic molecules. These compounds are in turn used as the energy source for all growth and reproduction in these and almost all other organisms. As such, virtually all life on the planet ultimately depends on photosynthetic energy conversion. Respiration, which occurs in mitochondrial and bacterial membranes, utilizes energy present in organic molecules to fuel a wide range of metabolic reactions critical for cell growth and development. In addition, many photosynthetic organisms engage in energetically wasteful photorespiration that begins in the chloroplast with an oxygenation reaction catalyzed by the same enzyme responsible for capturing carbon dioxide in photosynthesis. This series of books spans topics from physics to agronomy and medicine, from femtosecond processes to season long production, from the photophysics of reaction centers, through the electrochemistry of intermediate electron transfer, to the physiology of whole orgamisms, and from X-ray christallography of proteins to the morphology or organelles and intact organisms. The goal of the series is to offer beginning researchers, advanced undergraduate students, graduate students, and even research specialists, a comprehensive, up-to-date picture of the remarkable advances across the full scope of research on photosynthesis, respiration and related processes.

The titles published in this series are listed at the end of this volume and those of forthcoming volumes on the back cover.

Respiration in Archaea and Bacteria

Diversity of Prokaryotic Respiratory Systems

Edited by

Davide Zannoni
Universita di Bologna,
Italy

A C.I.P. Catalogue record for this book is available from the Library of Congress.

ISBN 1-4020-2002-3 (HB)
ISBN 1-4020-3163-7 (e-book)

Published by Springer,
P.O. Box 17, 3300 AA Dordrecht, The Netherlands.

Sold and distributed in North, Central and South America
by Springer,
101 Philip Drive, Norwell, MA 02061, U.S.A.

In all other countries, sold and distributed
by Springer,
P.O. Box 322, 3300 AH Dordrecht, The Netherlands.

The camera ready text was prepared by
Lawrence A. Orr, Center for the Study of Early Events in Photosynthesis,
Arizona State University, Tempe, Arizona 85287-1604, U.S.A.

Printed on acid-free paper

Printed in the Netherlands.

Obituary for Achim Kröger
(1937-2002)

The scientific community lost Achim Kröger, one of the pioneers in the research of the biological function of quinones in cells. Achim Kröger died unexpectedly at the age of 65 in the midst of his research activities. He studied Chemistry at the University of Marburg and expressing a keen interest in biochemistry he joined my group in 1962 for his graduate work. After his PhD he stayed as a post doc and moved to Munich where he worked until 1980 when he assumed the position of an associate professor at the University of Marburg. In 1985, he followed the offer of a full professorship of microbiology at the University of Frankfurt.

When Achim Kröger started his thesis work, there was a big question mark about the role of ubiquinone in the respiratory chain. Kinetic studies of Britton Chance's group and some reconstitution work in a few other laboratories were in conflict with a proposed role of ubiquinone in the main pathway of the electron transport. By using new sensitive spectrophotometric and chemical methods, Achim Kröger was able to determine the in- and out-flux of electrons to and from ubiquinone in mitochondria and found that ubiquinone fully complied with a central electron transfer role. He studied the impact of oxidative phosphorylation on the redox state of ubiquinone in various types of mitochondria under the influence of specific substrates, also demonstrating the participation of ubiquinone in reversed electron transfer. On this basis the role of ubiquinone was developed and shown in further experiments to be a flexible pool of collecting and redistributing electrons.

Achim Kröger then embarked on the still unknown role of menaquinone in anaerobic bacteria where he established its electron transfer role analogous to ubiquinone. By elucidating the electron transfer system in anaerobic bacteria from formiate to fumarate, he demonstrated the role of menaquinone in the vectorial H^+ transfer without the complications of the mitochondrial 'Q-cycle.' He cloned and characterized the components of this elementary energy transducing system even to the stage of getting the crystal structure.

Achim Kröger performed his research with the greatest and uncompromising profoundness. His contributions are milestones in the role of quinones in electron transfer and vectorial H^+ transport. He leaves a great void in the biological sciences.

Achim Kröger was a man of family and had three children. He was a person who enjoyed social contacts and heated discussions which were always conducted with a great sense of humor. He was a profound connoisseur of classical music and it is no surprise that his son Ralf became a professional musician. His other son Nils followed the main occupation of his father and became an accomplished biochemist at the University of Regensburg. His loss is mourned by all who knew and cherished his wonderful personality

Martin Klingenberg
(University of Munich, Germany)

Editorial

Advances in Photosynthesis and Respiration

Volume 16: Respiration in Archaea and Bacteria: Diversity of Prokaryotic Respiratory Systems

I am extremely delighted to announce the publication of the second book on 'Respiration' since the Series *Advances in Photosynthesis* was enlarged to become *Advances in Photosynthesis and Respiration*. Volume 16, edited by Davide Zannoni, deals specifically with *Respiration in Archaea and Bacteria: Diversity of Prokaryotic Respiratory Systems*. The earlier volume 15, also edited by Davide Zannoni, was titled: Respiration in Archaea and Bacteria: *Diversity of Prokaryotic Electron Transport Carriers*. Volume 16 is a sequel to the fifteen volumes in the *Advances in Photosynthesis and Respiration* (AIPH) series.

Published Volumes

(1) *Molecular Biology of Cyanobacteria* (Donald A. Bryant, editor, 1994);
(2) *Anoxygenic Photosynthetic Bacteria* (Robert E. Blankenship, Michael T. Madigan and Carl E. Bauer, editors, 1995);
(3) *Biophysical Techniques in Photosynthesis* (Jan Amesz and Arnold J. Hoff, editors, 1996);
(4) *Oxygenic Photosynthesis: The Light Reactions* (Donald R. Ort and Charles F. Yocum, editors, 1996);
(5) *Photosynthesis and the Environment* (Neil R. Baker, editor, 1996);
(6) *Lipids in Photosynthesis: Structure, Function and Genetics* (Paul-André Siegenthaler and Norio Murata, editors, 1998);
(7) *The Molecular Biology of Chloroplasts and Mitochondria in Chlamydomonas* (Jean David Rochaix, Michel Goldschmidt-Clermont and Sabeeha Merchant, editors, 1998);
(8) *The Photochemistry of Carotenoids* (Harry A. Frank, Andrew J. Young, George Britton and Richard J. Cogdell, editors, 1999);
(9) *Photosynthesis: Physiology and Metabolism* (Richard C. Leegood, Thomas D. Sharkey and Susanne von Caemmerer, editors, 2000);
(10) *Photosynthesis: Photobiochemistry and Photobiophysics* (Bacon Ke, author, 2001);
(11) *Regulation of Photosynthesis* (Eva-Mari Aro and Bertil Andersson, editors, 2001);
(12) *Photosynthetic Nitrogen Assimilation and Associated Carbon and Respiratory Metabolism* (Christine Foyer and Graham Noctor, editors, 2002).
(13) *Light Harvesting Antennas* (Beverley Green and William Parson, editors, 2003).
(14) *Photosynthesis in Algae* (Anthony Larkum, Susan Douglas and John Raven, editors, 2003)
(15) *Respiration in Archaea and Bacteria: Diversity of Prokaryotic Electron Transport Carriers* (Davide Zannoni, editor, 2004)
(17) *Plant Mitochondria: From Genome to Function* (David A. Day, Harvey Millar and James Whelan, editors, 2004)
(19) *Chlorophyll a Fluorescence: A Signature of Photosynthesis* (George Papageorgiou and Govindjee, editors);

See <http://www.wkap.nl/series.htm/AIPH> for further information and to order these books. Please note that the members of the International Society of Photosynthesis Research, ISPR (<http: //www. Photosynthesis research.org>) receive special discounts.

Respiration in Archaea and Bacteria: Diversity of Prokaryotic Respiratory Systems

As I mentioned in my Editorial for volume 15 (subtitled: *Diversity of Prokaryotic Electron Transport Carriers*), I find it quite exciting to write this Editorial sitting in Room 669 Morrill Hall of the University of Illinois at Urbana-Champaign, since just three floors down is the office and the laboratory of Carl

Woese. Woese, along with others, brought before us the discovery of the third form of life, Archaea (other two forms of life being Bacteria and Eukarya). For his discovery, Carl Woese was recognized by the Award of the coveted Crafoord Prize for Biosciences by the King of Sweden on September 24, 2003.

As explained in the 'Preface' of volume 15 by Davide Zannoni, it is not by chance that the title of this book (operationally divided into two volumes) recognizes the distinct features of Archaea from those of Bacteria as far as the enormous variety of respiratory systems is concerned. As volume 15 dealt beautifully with the question — how complex is a respiratory bacterial complex — the current volume 16 is conceived to deal with the question — how diverse is the diversity in bacterial respiration? This metabolic process probably arose under an anaerobic atmosphere between three and four million years ago (Mya) with a successive multiple-lateral phylogenetic distribution in many lineages and specialized niches (oxic and anoxic). Microbiologists consider the so called 'aerobic respiration,' which also occurs in cellular organelles such as mitochondria of plants and animals, a specialized way to transduce chemical energy. As a matter of fact, dioxygen reduction by membrane bound oxidases has been the most successful way for both bacteria and eukaryotes to grow, after the advent of an aerobic atmosphere generated by small, but efficient, oxygen evolving bioreactors — the cyanobacteria. However, bacterial respiration is not simply restricted to describe the function of membrane bound redox enzymes but it plays a fundamental role in several biogeochemical cycles of important nutrient elements (e.g. carbon, nitrogen, sulfur and iron) under both aerobic and anaerobic habitat.

The book entitled *Respiration in Archaea and Bacteria*, edited by Davide Zannoni, one of the leading authorities in this field, is a unique piece of work in that it emphasizes the integration of several disciplines (evolutionary biology, biochemistry, microbiology, molecular genetics and bioenergetics) with the goal to define in details the complexity of the respiratory chains in the microbial world.

The current volume 16 of *Advances in Photosynthesis and Respiration* contains a series of exciting chapters. It starts with a fantastic overview on the respiratory chains and bioenergetics of Archaea (Chapter 1) and it continues with a chapter analyzing the aerobic respiration in Gram-positive bacteria (Chapter 2). Chapters 3 and 4 illustrate the most recent advances on the function and structure of the respiratory chains in the pathogenic genera Helicobacter and Campylobacter and in acetic acid bacteria, respectively, while Chapter 5 deals with the energetic demand of respiration in nitrogen fixing bacteria (genera Rhizobium) having a membrane-bound oxidase with an extraordinary oxygen affinity (encoded by the gene cluster *fix NOQP*). In this respect, I remember that this latter type of oxidase was described 30 years ago by D. Zannoni in the facultative phototroph, *Rhodobacter capsulatus*. On the other hand, it is now established that the genera Rhizobium (obligate-aerobe) and Rhodobacter (phototroph-anaerobe) are phylogenetically closer than previously thought on the basis of their apparent metabolic differences (they both belong to the α subdivision of the Proteobacteria).

In the light of the contrasting or converging features reflecting the type of habitat in which the various bacterial species exist, Zannoni invited colleagues and friends to write six chapters having important microbiological and ecological relevance, namely: aerobic respiration of ammonium, methane, iron, sulfur and hydrogen (Chapters 6, 7, 9, 10, 11, respectively) and anaerobic respiration of nitrous compounds (Chapter 8). The book ends with two chapters dedicated to facultative phototrophs: Chapter 12 deals with respiration in cyanobacteria, the microorganisms unanimously accepted as the 'creators' of the present aerobic atmosphere while Chapter 13 covers the intriguing topic of the interaction (structural and functional) between photosynthesis and respiration in facultative phototrophs, the microrganisms which have been closest to Zannoni's heart for almost three decades!

For further details on the book, I refer the reader to the 'Table of Contents' of *Respiration in Archaea and Bacteria: Diversity of Prokaryotic Respiratory Systems*, as well as the 'Preface' by Davide Zannoni.

In the end, I quote two influential plant physiologists: F. F. Blackman (an Englishman) and P. Parija (an Indian) (1928) about respiration 'Of all protoplasmic functions, the one which is, by tradition, most closely linked with our conception of vitality is the function for which the name of respiration has been accepted.' Thus, I am looking forward to further discussions on respiration of other systems e.g., 'Plant Respiration' in our Series (*Advances in Photosynthesis and Respiration*). It is ironic that in early days, one of the major efforts of some plant physiologists was to get rid of bacterial respiration in their tissues!

The Scope of the Series

Advances in Photosynthesis and Respiration is a book series that provides, at regular intervals, a comprehensive and state-of-the-art account of research in various areas of photosynthesis and respiration. Photosynthesis is the process by which higher plants, algae, and certain species of bacteria transform and store solar energy in the form of energy-rich organic molecules. These compounds are in turn used as the energy source for all growth and reproduction in these and almost all other organisms. As such, virtually all life on the planet ultimately depends on photosynthetic energy conversion. Respiration, which occurs in mitochondria and in bacterial membranes, utilizes energy present in organic molecules to fuel a wide range of metabolic reactions critical for cell growth and development. In addition, many photosynthetic organisms engage in energetically wasteful photorespiration that begins in the chloroplast with an oxygenation reaction catalyzed by the same enzyme responsible for capturing carbon dioxide in photosynthesis. This series of books spans topics from physics to agronomy and medicine, from femtosecond (10^{-15} s) processes to season long production, from the photophysics of reaction centers, through the electrochemistry of intermediate electron transfer, to the physiology of whole organisms, and from X-ray crystallography of proteins to the morphology of organelles and intact organisms. The intent of the series is to offer beginning researchers, advanced undergraduate students, graduate students, and even research specialists, a comprehensive, up-to-date picture of the remarkable advances across the full scope of research on bioenergetics and carbon metabolism.

Future Books

The readers of the current series are encouraged to watch for the publication of the forthcoming books:

(1) *Chlorophylls and Bacteriochlorophylls: Biochemistry, Biophysics and Biological Function* (Editors: Bernhard Grimm, Robert J. Porra, Wolfhart Rüdiger and Hugo Scheer);
(2) *The Water/Plastoquinone Oxido-reductase in Photosynthesis* (Editors: Thomas J. Wydrzynski and Kimiyuki Satoh);
(3) *Plant Respiration* (Editors: Miquel Ribas-Carbo and Hans Lambers);
(4) *The Plastocyanin/Ferredoxin Oxidoreductase in Oxygenic Photosynthesis* (Editor: John Golbeck);
(5) *Photoprotection, Photoinhibition, Gene Regulation and Environment* (Editors: Barbara Demmig-Adams, William W. Adams III and Autar Mattoo);
(6) *Photosynthesis: A Comprehensive Treatise; Biochemistry, Biophysics and Molecular Biology*, 2 volumes (Editors: Julian Eaton-Rye and Baishnab Tripathy)
(7) *The Structure and Function of Plastids* (Editors: Kenneth Hoober and Robert Wise); and
(8) *Discoveries in Photosynthesis Research* (Editors: Govindjee, Howard Gest, J. Thomas Beatty and John F. Allen);

In addition to these contracted books, we are interested in publishing several other books. Topics under consideration are: Molecular Biology of Stress in Plants; Global Aspects of Photosynthesis and Respiration; Protein Complexes of Photosynthesis and Respiration; Biochemistry and Biophysics of Respiration; Protonation and ATP Synthesis; Functional Genomics and Proteonomics; The Cytochromes; Laboratory Methods for Studying Leaves and Whole Plants; and C-3 and C-4 Plants.

Readers are requested to send their suggestions for these and future volumes (topics, names of future editors, and of future authors) to me by E-mail (gov@uiuc.edu) or fax (1-217-244-7246).

In view of the interdisciplinary character of research in photosynthesis and respiration, it is my earnest hope that this series of books will be used in educating students and researchers not only in Plant Sciences, Molecular and Cell Biology, Integrative Biology, Biotechnology, Agricultural Sciences, Microbiology, Biochemistry, and Biophysics, but also in Bioengineering, Chemistry, and Physics.

I take this opportunity to thank Davide Zannoni for his timely and prompt editorial work; as noted in the Editorial of volume 15, he is one of the most efficient editors I have encountered in many years. Respiration in *Archaea and Bacteria: Diversity of Prokaryotic Respiratory Systems* was possible because of all the authors of volume 16: without their authoritative chapters, there will be no book. I owe Larry Orr special thanks for his friendly and wonderful work as always. Thanks are also due to Jacco Flipsen, and Noeline Gibson (both of Springer/Kluwer Academic Publishers), and Jeff Haas (Director of Information Technology, Life Sciences, University of Illinois) for their support. My wife Rajni Govindjee deserves my special praise for being a role model for my life and well- being. Our daughter Anita Govindjee and her husband Morten Christiansen provided facilities at the time this book was being prepared for publication.

August 26, 2004

Govindjee
Series Editor, *Advances in Photosynthesis and Respiration*
University of Illinois at Urbana-Champaign
Department of Plant Biology
505 South Goodwin Avenue
Urbana, IL 61801-3707, USA
E-mail: gov@uiuc.edu;
URL: http://www.life.uiuc.edu/govindjee

Govindjee

Govindjee is Professor Emeritus of Biochemistry, Biophysics and Plant Biology at the University of Illinois at Urbana-Champaign (UIUC), Illinois, USA. He received his Ph.D. in Biophysics from the University of Illinois at Urbana-Champaign in 1960, with a thesis on the 'Action Spectra of the Emerson Enhancement Effect in Algae', under Eugene Rabinowitch. From 1960-1961, he served as a United States Public Health (USPH) Postdoctoral Fellow; from 1961-1965, as Assistant Professor of Botany; from 1965-1969 as Associate Professor of Biophysics and Botany; and from 1969-1999 as Professor of Biophysics and Plant Biology, all at the UIUC. In 1999, he became Professor Emeritus of Biochemistry, Biophysics and Plant Biology at UIUC. Julian Eaton-Rye, Prasanna Mohanty, George Papageorgiou, Alan Stemler, Thomas Wydrzynski, Jin Xiong, Chunhe Xu and Barbara Zilinskas are among his more than 20 PhD students. His honors include: Fellow of the American Association of Advancement of Science (1976); Distinguished Lecturer of the School of Life Sciences, UIUC (1978); President of the American Society of Photobiology (1980-1981); and Fulbright Senior Lecturer (1996-1997). Govindjee's research has focused on the function of 'Photosystem II' (water-plastoquinone oxido-reductase), particularly on the primary photochemistry; role of bicarbonate in the electron and proton transport; thermoluminescence, delayed and prompt fluorescence (particularly lifetimes), and their use in understanding electron transport and photoprotection against excess light. He has coauthored 'Photosynthesis' (1969); and has edited (or co-edited) 'Bioenergetics of Photosynthesis' (1975); 'Photosynthesis' (in 2 volumes, 1982);'Light Emission by Plants and Bacteria' (1986), among other books. He is a member of the American Society of Plant Biology, American Society for Photobiology, Biophysical Society of America, and the International Society of Photosynthesis Research (ISPR). Govindjee's scientific interest, now, includes Fluorescence Lifetime Imaging Microscopy (FLIM) (with Robert Clegg) and regulation of excitation energy transfer in oscillating light (with Lada Nedbal). In addition, Govindjee is interested in 'History of Photosynthesis Research', and in 'Photosynthesis Education'. His personal background appears in Volume 13 (edited by B. Green and W. Parson); and contributions to photosynthesis and fluorescence in algae in Volume 14 (A. Larkum, J. Raven and S. Douglas, editors) of the *Advances in Photosynthesis and Respiration* (AIPH). He serves as the Series Editor of AIPH, and as the 'Historical Corner' Editor of 'Photosynthesis Research'. For further information, see his web page at: http://www.life.uiuc.edu/govindjee.

Contents

Preface

During the past three decades, the microbial physiology has been significantly changed by the advent of molecular biology. Indeed, the way to approach the degree of metabolic sophistication of prokaryotes, reflecting the type of habitat in which they exist, was greatly enhanced by the possibility to study gene expression and regulation. The creation of the prokaryotic Archaean domain, mainly based on the pioneering molecular studies of Carl Woese at the University of Illinois in the early 1970s, not only has modified the bacterial taxonomy but resulted in a tremendous rise in efforts aiming to define the so called 'microbial diversity.'

In this book, entitled *Respiration in Archaea and Bacteria*, I have attempted to provide an extensive coverage of one of the most ancient biochemical processes: respiration. The book has been operationally divided into two volumes of the series *Advances in Photosynthesis and Respiration* (Govindjee, Ed.). The first volume has recently been published (Vol. 15) and is sub-titled *Diversity of Prokaryotic Electron Transport Carriers*. It includes a group of chapters covering all molecular, functional and structural aspects of the membrane-bound bacterial redox complexes and respiratory co-factors such as pyrrolo-quinoline quinones and hemoglobins. Volume 15 also includes a chapter illustrating the evolution of respiration in the prokaryotic world along with two chapters on topics such as cytochrome *c* biogenesis and membrane oxidases as redox sensors regulating the expression of photosynthetic genes in facultative phototrophs.

The current volume, Vol. 16 of the series, is subtitled *Diversity of Prokaryotic Respiratory Systems*, and contains 13 chapters. This book is conceived as a comprehensive account of respiratory systems in selected genera of the domain Bacteria along with an extensive chapter (Chapter 1) on the redox chains and bioenergetics of extremophiles belonging to the Archaean domain. The present volume contains six chapters dealing with metabolic features having important microbiological and ecological relevance such as the use of ammonium (Chapter 6), methane (Chapter 7), iron (Chapter 9), sulfur (Chapter 10) and hydrogen (Chapter 11) as respiratory substrates or nitrous compounds in denitrification processes (Chapter 8). The remaining chapters are dedicated to respiration of selected groups of bacteria such as Gram positives (Chapter 2), the pathogenic microaerophilic genera *Helicobacter* and *Campylobacter* (Chapter 3), acetic acid bacteria (Chapter 4), nitrogen fixing symbionts and free-living bacteria (Chapter 5), oxygenic phototrophs (Cyanobacteria) (Chapter 12), and anoxygenic (purple non-sulfur) phototrophs (Chapter 13).

In view of the extensive on-going research on the functional annotation of microbial genomes revealing new insights into respiratory metabolism, I'm conscious that a few of the topics selected in this book might require some reappraisal within a year or so. To partially overcome this problem, I had recommended to all the contributing authors to make a serious effort to describe not only what we know at present but also try to indicate those aspects deserving our future attention. I hope that these two volumes will be a long-term source of information for Ph.D. students, researchers and undergraduates from disciplines such as microbiology, biochemistry, chemistry and ecology, studying basic sciences, medicine and agriculture.

As explained in the 'Preface' of the preceding Volume 15, there are many persons that I would like to thank since they have been crucial for the completion of this work. Primarily, I wish to remember Professor A. Kröger, who tragically passed away a few months after the conclusion of his chapter (see his obituary written by Professor M. Klingenberg); secondly, I like to thank all the contributing authors and the series Editor of the *Advances in Photosynthesis and Respiration*, Govindjee, who provided me encouragement and useful suggestions to complete a project of such a magnitude. In the end, I'm glad to thank Larry Orr for his friendly and wonderful assistance in producing the page layout of the two volumes.

This work is dedicated to my family and my parents.

Davide Zannoni
Department of Biology,
University of Bologna,
40126-Bo-Italy
davide.zannoni@unibo.it

Davide Zannoni

Davide Zannoni is Professor of General Microbiology in the Department of Biology at the University of Bologna, Italy. He received the 'Laurea' in Biological Sciences from the University of Bologna in 1973 with a thesis on the bioenergetics of the facultative phototroph *Rhodobacter (Rb.) capsulatus*. In 1974 he reported (FEBS Letters 48:152-155) the first example of a high-potential *b*-type heme cytochrome oxidase (today recognized as cbb_3 type oxidase). From November 1977 to November 1978, he was a research fellow of the North Atlantic Treaty Organization (NATO) at the St. Louis School of Medicine, Department of Biochemistry, St. Louis MO, USA, under the supervision of Professor Barry L. Marrs. In 1979, he was appointed Lecturer in Plant Biochemistry at the University of Bologna and promoted to Associate Professor in 1981. As a research fellow of the European Molecular Biology Organization (EMBO) in 1981, 1983 and 1991, he visited different European laboratories (Department of Biochemistry and Microbiology, St. Andrews University, St. Andrews Scotland U.K.; Départment de Biologie Cellulaire et Moléculaire, CNRS URA, CEA Saclay, Gif sur Yvette, France; Department of Microbiology, University of Göttingen, Göttingen, Germany) to investigate the structure of electron transport redox complexes in various genera and species of facultative phototrophs and aerobes. The role played by D. Zannoni on the discovery and description of the membrane-anchored cytochrome c_y of *Rb. capsulatus* was published in 2003 (Photosynth Res 76:127). Zannoni's scientific interest, now, includes a research line on the bioenergetics of microbial remediation of metals and metalloids. In addition, D. Zannoni is presently Chairman of the Italian Society for General Microbiology and Microbial Biotechnologies (2004-2007).

Color Plates

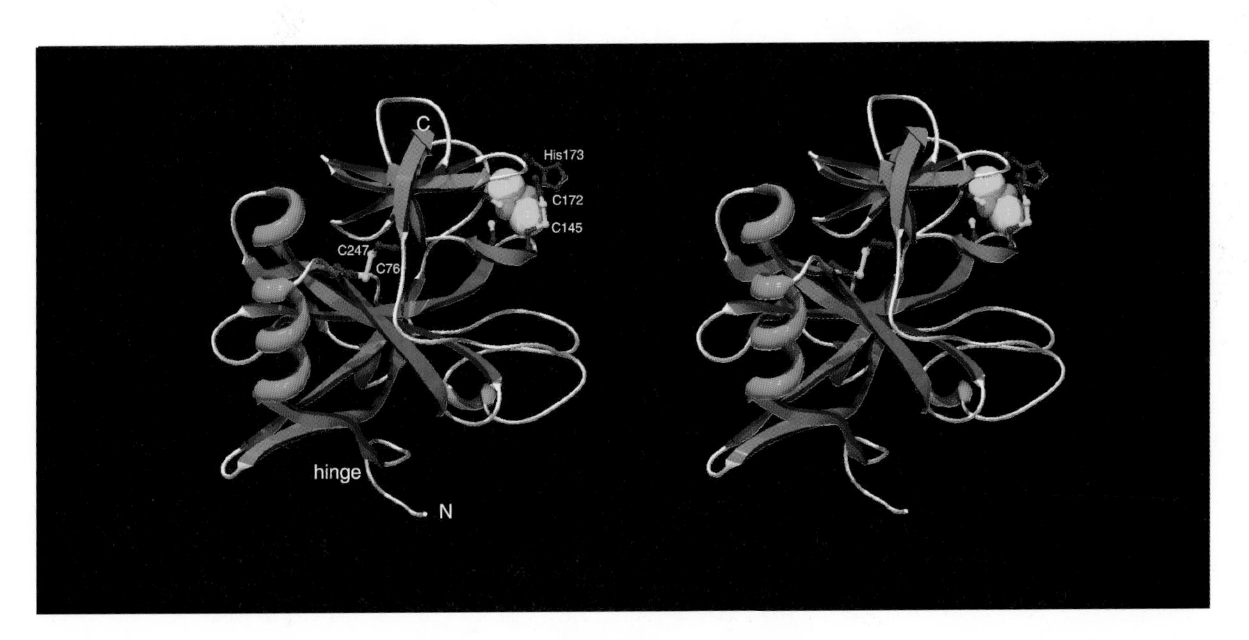

Color Plate 1. Stereo representation of the SoxF Rieske protein from *S. acidocaldarius* structure (view with crossed eyes). The structure was resolved with 1.1 Å. The additional disulfide bridge (C247-76) and the fixation of the C-terminus onto an additional β-sheet are presumably important for the extreme thermostability. See Chapter 1 for details.

Davide Zannoni (ed): Respiration in Archaea and Bacteria. Vol 2: Diversity of Prokaryotic Respiratory Systems, pp. CP-1– CP-2.

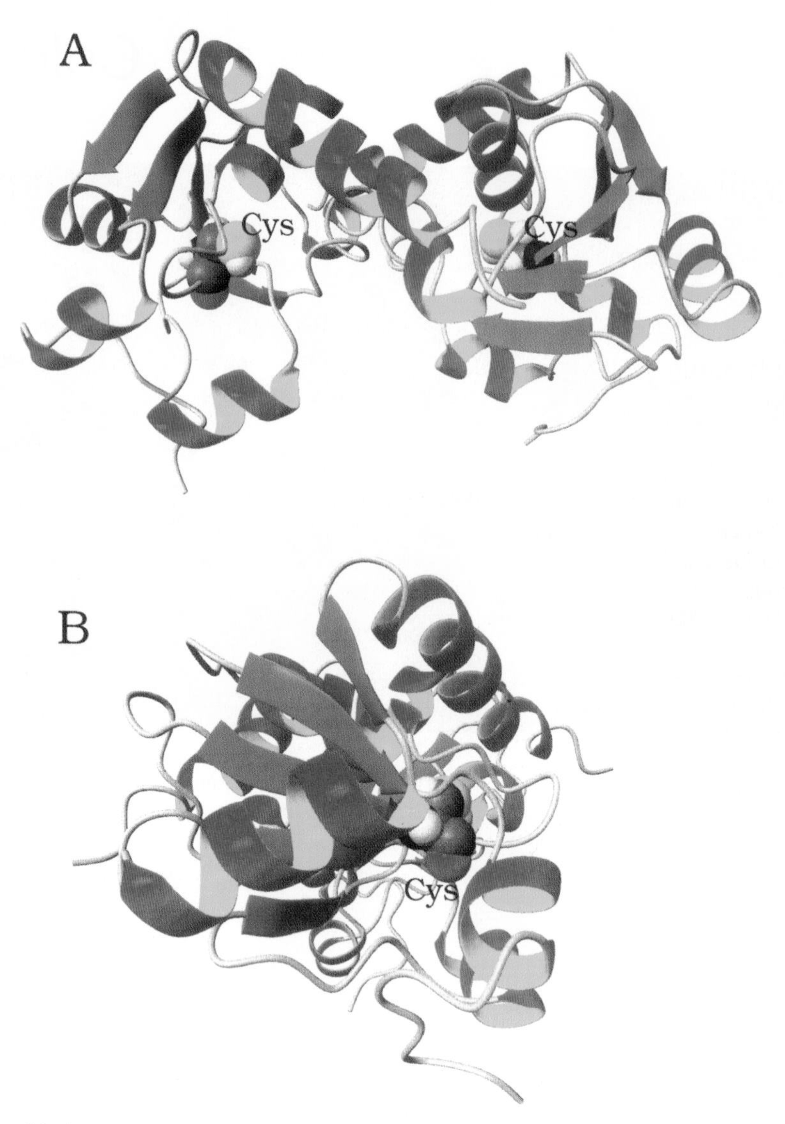

Color plate 2. Structure of Sud. Ribbon representations of Sud dimer drawn with program MOLMOL (Koradi et al., 1996). The A and B views are related by a 90° rotation around the symmetry axis. The cysteine residues were depicted using a CPK model (spheres with WdV radius and standard colours: red for O, yellow for S, blue for N, black for C and grey for H). (See Chapter 10)

Chapter 1

Respiratory Chains in Archaea: From Minimal Systems to Supercomplexes

Günter Schäfer*
Institute of Biochemistry, University of Luebeck, D-23538 Lübeck, Germany

Summary

Oxygen respiration originated in evolution within the prokaryotic kingdom and was acquired by eukaryotes later via endosymbiosis. Prokaryotes are divided into the domains of *Bacteria* and *Archaea*. The latter represent the so called third phylogenetic domain and are not only characterized by their unusual extremophilic life styles, but also by a variety of molecular peculiarities including their bioenergetic systems. Though the principle of chemiosmotic energy conservation holds also for Archaea, some primary energy transducing reactions are restricted to this domain, as for example methanogenesis and rhodopsin dependent phototrophic energetics. Whereas the majority of known Archaea can thrive anaerobically, also a significant number of species is capable of obligate or facultative aerobic growth. The present chapter highlights the essential and archetypical differences with respect to orthodox pro- and eukaryotic respiratory systems. In particular the following aspects

*Email: guenter.schaefer@biochem.uni-leubeck.de

Davide Zannoni (ed): Respiration in Archaea and Bacteria. Vol 2: Diversity of Prokaryotic Respiratory Systems, pp. 1–33.

are addressed: 1. Several archaeal genomes have been completely analyzed, but protein chemical verification of the genomic data on electron transport systems lacks behind; 2. According to present knowledge, aerobic Archaea can oxidize NADH by a type-II NADH dehydrogenase while an equivalent to complex-I has not been verified; 3. In several reactions the function of NAD is replaced by ferredoxins; 4. Other auxiliary cofactors are small copper proteins which may replace the function of *c*-type cytochromes, as well as unusual thio-pheno-benzoquinones which act as membrane residing pools for reducing equivalents. Further, most *a*-type cytochromes contain an archetypical heme-A_S which also replaces heme-*b* in several respiratory complexes of thermoacidophilic Archaea; 5. Archaeal complex-II equivalents contain modified S3 FeS-clusters as well as unusual cysteine-rich motifs in their small subunits which host a novel [2Fe2S] cluster and, in addition, can be devoid of their typical membrane anchors; 6. A novel type of Rieske/cytochrome complex, equivalent to respiratory complex-III, has been detected in thermoacidophilic Archaea; 7. Minimal respiratory chains can consist simply of a quinol oxidase and are contrasted by supercomplexes comprising a genetic and structural fusion of a whole respiratory chain into one unique protein assembly; 8. A novel regulatory adaptation to acidic environments is proposed based on the unique properties of a Cu_A containing terminal oxidase from an extreme thermoacidophile; 9. A revised view on the structural basis for proton pumping capabilities of terminal oxidases is provided. 10. The Conclusions paragraph addresses the necessary experimental approaches required to investigate the unsolved questions on archaeal respiratory systems.

I. Introduction

Archaea are representing the so called third domain of life, forming a distinct branch of the universal phylogenetic tree based on the evolutionary distances of 16s ribosomal RNA. This novel domain was identified first by C. Woese (Woese and Fox, 1977; Woese et al., 1990; Wheelis et al., 1992; Doolittle, 1996) named as Archaea (formerly *archaebacteria*) illustrating the evolutionary distance and fundamental differences of lifestyles. Despite the large evolutionary distance to Bacteria and Eukarya, Archaea cannot be considered as living fossils; indeed the archaeal domain reveals the shortest distance to a hypothetical common ancestor preceding the split into the three major domains. Initially, a few genera of methanogens and halobacteria along with a restricted number of thermoacidophilic sulfur metabolizing microbes were identified to belong to the archaeal domain. In the past ten years, however, the large diversity of archaeal organisms became evident (Boone et al., 1993; Barns et al., 1994, 1996; Hershberger et al., 1996; Takai and Sako, 1999; Smith, 2001) with a continuously increasing number of newly detected species. Archaea stand out also phenotypically because most of them are inhabiting unusual environments and therefore are real extremophiles. Especially the latter features as, for example to grow optimally at temperatures close to the boiling point of water, at extremely acidic or alkaline pHs, or at almost saturating NaCl concentration, or a combination of these properties, challenged both the basic research on the molecular aspects of life under extreme conditions and the biotechnological applications of these organisms. For a detailed discussion on taxonomy, metabolism, membrane structure and molecular biology of Archaea the reader is referred to more specialized reviews (Danson, 1993; Schönheit and Schäfer, 1995; Blöchl et al., 1996; Stetter, 1996, 1999; Danson and Hough, 1998).

The bioenergetic systems of Archaea have been recently reviewed extensively (Schäfer et al., 1999). Clearly two metabolic aspects are restricted to the Archaea domain — besides other molecular properties like membrane structure and lipid composition— namely: (a) the use of light driven rhodopsin-systems in cellular energetics, and (b) the reductive fermentation of low molecular carbon compounds as the sole cellular energy source to form methane. In contrast, the basic principles of aerobic or anaerobic respirations appear to be shared between Archaea, Bacteria and Eucarya (mitochondria) at least to a certain extent. Nevertheless, fundamental differences have been discovered and a merely partial homology between bacterial, mitochondrial and archaeal respiratory complexes have raised the discussion whether or not the formerly anaerobic Archaea might have gained their capability to cope with oxygen from extensive lateral gene transfer. Indeed, a frequent exchange of

Abbreviations: Fdx – ferredoxin; HIPIP – high-potential iron-sulfur protein; HQNO – hydroxyquinoline-N-oxide; IFP – iron flavoprotein; NDH-1 – type-1 NADH dehydrogenase; NDH-II – type-II NaDH dehydrogenase; QFR – quinol:fumarate reductase; SDH – succinate dedydrogenase; SQR – succinate:quinone reductase; TFR – thiol-fumarate reductase

genetic material between primordial organisms has been assumed to have occurred prior to branching into the presently existing kingdoms of life (Woese, 1998). On the other hand, significant molecular data support the idea of an origin of oxygen reducing heme/copper oxidoreductases (Castresana and Saraste, 1995) (see also Chapter 1 by Castresana, Vol. 1) as preceding the evolutionary advent of oxygenic photosynthesis, which incidentally is completely absent from the archaeal domain.

In this chapter the different types of anaerobic respirations are not discussed; emphasis is put on the respiratory systems of Archaea using oxygen reduction compulsory or facultatively as an energy source. Though this issue has been included in a rather recent review (Schäfer et al., 1999) novel results make necessary a revision of the previously proposed models. In particular, the question whether or not the unusual archaeal terminal oxidases are proton pumps, as well as the issue whether Archaea do have a redox complex equivalent to the orthodox respiratory complex-III are addressed below. Also new structural data on the function and composition of electron transport super-complexes and high resolution structures of intermediate electron carriers are presented.

II. Aerobic Archaea and Components of Electron Transport

Aerobic Archaea with widely different lifestyles have been found. Besides organotrophic organisms litho-autotrophic species exist and, moreover, obligate as well as facultative aerobes are among these. An exemplary compilation is contained in a previous review (Schäfer et al., 1999). Therein those Archaea were listed which can be grown as laboratory cultures. One has to realize however, that a much higher number of species exists which have only been identified by RNA extraction and analysis of soil and water samples taken from a large variety of environments (Barns et al., 1996). Unfortunately some species are restricted to symbiotic communities and therefore they remain inaccessible to growth as pure cultures in vitro. At least one has to accept that only a relatively small number of aerobic Archaea can be grown in sufficient amounts to allow extensive biochemical and bioenergetic studies. Nevertheless, respiratory systems could be studied from halobacteria (actually the first archaeal genus investigated bioenergetically), from hyperthermophilic, acidophilic, alkaliphilic, or neutrophilic genera including CO_2 fixating sulfur metabolizers as well as organotrophic species. The best investigated species are *Halobacterium (Hb.) salinarum*, *Haloferax (Hf.) volcanii* and *H. mediteranei*, *Natronomonas (N.) pharaonis*, *Thermoplasma (Tp.) acidophilum*, *Sulfolobus acidocaldarius*, *S. solfataricus*, S. *tokodaii* strain 7, *S. metallicus*, and *Acidianus (Ac.) ambivalens*. In addition, a number of respiratory systems can be deduced from fully sequenced genomes as for example from *Aeropyrum (Ae.) pernix* (Kawarabayashi, 1999), or *Pyrobaculum (Pb.) aerophilum* (Fitz-Gibbon et al., 1997, 2002). From the latter and others which are difficult to cultivate a number of redox proteins related to bioenergetics has been obtained by heterologous expression of single genes (Janssen et al., 1997; Schmidt et al., 1997; Henninger et al., 1999; Janssen et al., 2001; Komorowski and Schäfer, 2001).

Generally, the experimental approach to Archaeal respiratory components was characterized by a search for analogies to components of known—for example mitochondrial—respiratory chains, i.e. the presence of complex I to IV as classical electron transport and proton pumping devices. However, whereas spectroscopy of intact and solubilized plasma membranes clearly revealed the occurrence of *a*- and *b*-type cytochromes as well as of cyanide sensitive terminal oxidases of the aa_3-type, *c*-type cytochromes were often absent, e.g. *Sulfolobales*. Instead, unusual bands in reduced-oxidized difference spectra indicated novel intermediate electron carriers like the cytochrome-a_{587} of *Sulfolobus acidocaldarius*. Moreover, archaeal respiration appeared in most cases insensitive to rotenone, piericidine derivatives, antimycin-A, or stigmatellin which are all specific inhibitors of 'orthodox' respiratory complexes (reviewed in Schäfer et al., 1999). This insensitivity largely hampered the use of routine approaches for the resolution of electron transport systems especially in view of the finding that Archaea, similarly to many species of Bacteria, can possess multiple terminal oxidases and branched respiratory pathways. Therefore investigation of the distinct properties of archaeal respiratory systems is still a challenge and many details remain to be resolved. In this context the identification of proton pumping sites is an important aspect. One has to remember that thermoacidophilic Archaea are maintaining an unusually large proton gradient across their plasma membrane of up to ≥ 4 (Moll and Schäfer, 1988; Lübben and Schäfer, 1989; Van de Vossenberg et al., 1998). Though respiration coupled active proton extrusion has been shown (Moll and Schäfer, 1988) the identification of individual proton pumps and especially of their mechanism is still insufficient.

Table 1 summarizes aerobic Archaea which have been investigated in culture. Among these several thermophiles and hyperthermophiles as well as extremely acidophiles were identified; however, the full genome sequences have been determined in a few species and even less has been done with regard to the protein structure of their respiratory systems. Among the Archaea domain the halobacterial members belong to the major group of Euryarchaeota and they can use also the light-energy employing various bacteriorhodopsin-based systems. In the dark, however, they have to rely on oxygen respiration. In addition they are mesophilic. *Haloferax* species can grow also with nitrate as terminal electron acceptor. The same holds for the thermoextremophilic crenarchaeon *Pyrobaculum aerophilum*, which is a facultative aerobe. Other facultative aerobes include species of the *Acidianus* genus which was described as strict chemolithoautotroph similarly to the species *Sulfolobus metallicus* and *Metallosphera sedula*. All other listed Archaea can grow organotrophically, and of these the *Sulfolobales* and the halophiles are obligate aerobes.

Respiratory components which have been spectroscopically identified or were characterized protein chemically from aerobic Archaea are also briefly summarized in Table 1. Specialized descriptions of archaeal cytochromes and of the hosted hemes have been published elsewhere (Lübben and Morand, 1994; Lübben, 1995). An interesting deviation from known *a*-type cytochromes is the archaeal heme-A_S which has been described first in *Sulfolobales*. This heme is present in terminal aa_3 oxidases as well as in their auxiliary *a*-type cytochromes like Cyt a_{587}. A number of components not included in the table were only tentatively assigned as respiratory electron carriers in various archaeal genome projects by sequence similarities. These include *c*-type cytochromes and constituents of complex I-type NADH dehydrogenases. In most cases, however, it is yet unknown whether or not the respective genes are actually expressed and/or leading to functional complexes. On the other hand, protein chemical isolation and characterization of several functional complexes revealed that unusual associations of redox components have been detected and that even supercomplexes exist comprising almost an entire respiratory chain as it will be discussed below.

An obvious fact is the canonical finding of succinate:quinone-reductases (SQR) either by activity or by genome analysis. This complex — in aerobes considered to represent complex-II of respiratory chains—reflects the importance of the reversible equilibrium reaction between fumarate and succinate as part of either an aerobically or anaerobically functioning citric acid cycle. Actually, the genes (or operons) of both, SQRs and quinol:fumarate-reductases (QFRs) originated presumably from the same ancestral genes, thus underlining the importance of this reaction in the evolution of intermediary metabolism. Likewise, NADH oxidizing activities have been frequently observed in archaeal membrane preparations (Wakagi and Oshima, 1987) and a certain number of genes has been attributed to NADH dehydrogenases in the above listed genome projects. This is in contrast, however, to the lack of protein chemical proof for the presence of typical proton translocating NADH:quinone-reductases as known from mitochondria and bacteria (Weiss et al., 1991; Friedrich, 1998), an issue to be discussed separately in detail.

III. Auxiliary Redox Factors

In general, membrane residing electron transport complexes of respiratory systems require both soluble cytosolic hydrogen donors and redox mediators. Commonly, the former are pyridine nucleotides while various quinones and cytochromes of *c*-type are coupling the electron flow between primary donor complexes and the terminal oxidases. In archaea, particular deviations from this classical picture have been discovered. Especially the use of ferredoxin instead of NAD as primary electron acceptor is a typical archaeal property of several metabolic reactions as for example 2-oxoacid dehydrogenases, aldehyde dehydrogenases or even glyceraldehyde dehydrogenase (Kerscher et al., 1982; George et al., 1992; Mukund and Adams, 1993; Bock et al., 1996; Zhang et al., 1996). It has also been assumed that NADH: ferredoxin-reductases may play a role as hydrogen donors to the respiratory chain (Iwasaki et al., 1993). Another is the function of blue copper proteins presumably replacing *c*-type cytochromes (Scharf and Engelhard, 1993; Komorowski and Schäfer, 2001). High potential FeS-proteins (HIPIP) as described in the respiratory chain of the marine bacterium

Table 1. Comprehensive list of obligate and facultative aerobic archaea.

Species	growth T_{max} °C	pH range	genome sequenced	characterized or isolated respiratory complexes / enzymes
Acidianus ambivalens	95	1–4	*in peparation*	SQR; aa_3-quinol-oxidase; Fdx; NDH-II
Acidianus brierleyi	75	1.5–4		
Acidianus infernus	95	1.5–4		
Aeropyrum pernix	95	n	+	
Halobacterium saccharovorum	m	n		
Halobacterium salinarum	m	n	+ *sp.*GRB	SQR; Cyt*b*; Cyt*c*;Rieske FeS; *c*-oxidase; Fdx; NDH-II
Haloferax mediteranei	m	n		
Haloferax volcanii	m	n	+	cytochrome aa_3
Haloferax denitrificans	m	n		
Metallosphera sedula	80	1–4.5		
Natronomonas pharaonis	45	7.9–9.5		SQR; Halocyanin; Cyt*c*; Cyt*b; c*-oxidase.
Picrophilus oshimae	60	0.5–2.5		
Picrophilus toridus	60	0.5–2.5		
Pyrobaculum aerophilum	104	5.8–9	+	SQR*; Rieske FeS; heme/Cu-oxidase*
Sulfolobus acidocaldarius	85	2–4.5	+ DSM639	SQR; SoxABCD; SoxM; Cytb_{558};Fdx; Sulfocyanin; SoxLN; NDH-II
Sulfolobus islandicus	75–85	3–5		
Sulfolobus metallicus	75	1–4		NDH (type?); SQR; cytb_{562}; cyta_{586}; Cyta$_{600}$; Rieske FeS; HIPIP(?);Fdx
Sulfolobus shibatae	86	3–5		
Sulfolobus solfataricus	87	3–5	+ P2	SQR; heme/Cu-quinol/*c*-oxidase*; Cyt*b*; Cyt *a*;
Sulfolobus tokodaii	80	3–5	+ strain 7	SQR; Cyt*b*; Cyt*a*; Rieske-FeS;heme/Cu-quinol-oxidase; NDH-II;Fdx
Thermoplasma acidophilum	65	0.5–4	+	SQR; Cyt*b*; heme/Cu terminal-oxidase*
Thermoplasma volcanium	65	0.5–4		

m = mesophilic, n = neutrophilic. Respiratory electron transport components listed in the right-hand column were either determined as enzymatic activity of membranes, or were studied with enriched/purified compounds and complexes, respectively. In some cases they were only predicted from genetic data, as identified by asterisks. SQR = succinate:quinone reductase; NDH-II = NADH dehydrogenase type-II; Fdx = ferredoxins; HIPIP = high-potential iron-sulfur protein; ? indicates exact type unknown.

Rhodothermus marinus (Pereira et al., 1999) might assume a similar role also in archaea.

A. Ferredoxins

In sulfur metabolizing archaea an abundance of ferredoxins has been observed (Teixeira et al., 1995; Iwasaki and Oshima, 2001). Their high thermostability (Aono et al., 1989) may reflect an advantage over NAD as a primary hydrogen-transducing coenzyme in hyperthermophiles. They have the advantage to provide reducing equivalents at potentials negative enough to drive hydrogenase-catalyzed reactions ($E_0 = -450$ mV), or $NAD(P)^+$ reduction ($E_0 = -320$ mV). Also from the alkaliphilic *Natronomonas pharaonis* a ferredoxin with $E_0 = -342$ mV has been purified (Scharf et al., 1997). The primary structures of several archaeal ferredoxins are known (Kerscher et al., 1982; Minami et al., 1985; Pfeifer et al., 1993; Janssen et al., 2001) and recent genetic and sequencing data suggest that in *Sulfolobales* multiple genes may encode highly similar ferredoxins. A peculiarity first demonstrated from the 3D-structure of crystallized *Sulfolobus* ferredoxin (Fujii et al., 1996) is the presence of Zn^{2+} in addition to the [3Fe-4S] and [4Fe-4S] clusters. The alignment of 9 sequences from thermophilic Archaea reveals that the presence of a Zn^{2+} chelating domain is the prevailing case; only two of the sequences are devoid of this Zn binding motif (Fig. 1). The role of zinc has been emphasized as a thermostability mediating factor (Iwasaki et al., 1997; Kojoh et al., 1999; Iwasaki and Oshima, 2001). This is in contradiction, however, to other observations and it can not be used as a general rule. For example, the two ferredoxins from *Acidianus ambivalens* have been shown to exhibit almost identical thermostability and thermal unfolding characteristics (Wittung-Stafshede et al., 2000; Janssen et al., 2001); however, only one of these contains the bindings site for Zn^{2+} as depicted from Fig. 1. Actually, for the Zn-free ferredoxin from *A. ambivalens* a melting temperature of 108 °C has been extrapolated from GdHCl-induced unfolding isotherms at various temperatures. Figure 2 shows the EPR spectrum of the dithionite reduced ferredoxin and the EPR redox titration yielding a half oxidation reduction potential of –235 mV which is very close to that of pyridine nucleotides. Also *S. metallicus* expresses two ferredoxins, with and without Zn, exhibiting practically identical thermostability (Gomes et al., 2002). As an alternative hypothesis for the role of Zn^{2+} we have proposed that such ferredoxins might differ in their specificity against various electron donors or acceptors, respectively (Janssen et al., 2001); a proposal which requires further investigation. Another open issue is the physiological significance of a reversible redox linked protonation equilibrium which we determined with the purified ferredoxin from *Sulfolobus acidocaldarius* (Breton et al., 1995). Also from *Sulfolobus* a ferredoxin reoxidizing iron-flavoprotein (IFP) has been described ($E_0 = -57$ mV) to be involved in an electron transport pathway ferredoxin → IFP → FMN → X, where X represents an unknown acceptor putatively mediating electron entry into the respiratory chain (Iwasaki et al., 1993).

Whereas the ferredoxins of thermoacidopihilic archaea (Fig. 1) are clustering within a very homologous group, *halobacterial* ferredoxins (Ng et al., 2000; Pfeiffer and Oesterhelt, 2001) form a more distant cluster with low homology to those from *Sulfolobales*.

B. Quinones

Archaeal respiratory chains do not contain ubiquinone or derivatives thereof. A comprehensive analysis on archaeal quinones was first performed with acidophilic Archaea (Collins and Langworthy, 1983). These were found to contain unusual benzo-[*b*]-thiopheno-4,7-quinones as for example in *S. acidocaldarius* (Nicolaus et al., 1992). Their structure and properties have been extensively discussed in a preceding review (Schäfer et al., 1999). Unusual is the occurrence of tricycloquinone which is present only in traces, however (Nicolaus et al., 1992; Trincone et al., 1992). The most abundant compound appears to be caldariella-quinone, first detected in *Caldariella acidophila* which, interestingly, was not definitely classified as an archaeon but is an extreme thermophile (De Rosa et al., 1977). It is the only archaeal quinone of which exact electrochemical parameters have been determined (Anemüller and Schäfer, 1990) and which can be isolated from the genuine membranes in sufficient amounts for use after reduction as a substrate for several archaeal quinol oxidases. In *Sulfolobus* about 60% of the total quinone pool is formed by caldariella-quinone.

Thermoplasma quinone was isolated from *Thermoplasma acidophilum* (Shimada et al., 2001) and has close similarity to menaquinones. The latter are the major quinones of halophilic archaea, predominantly MK-8 with an unsaturated isoprenoid

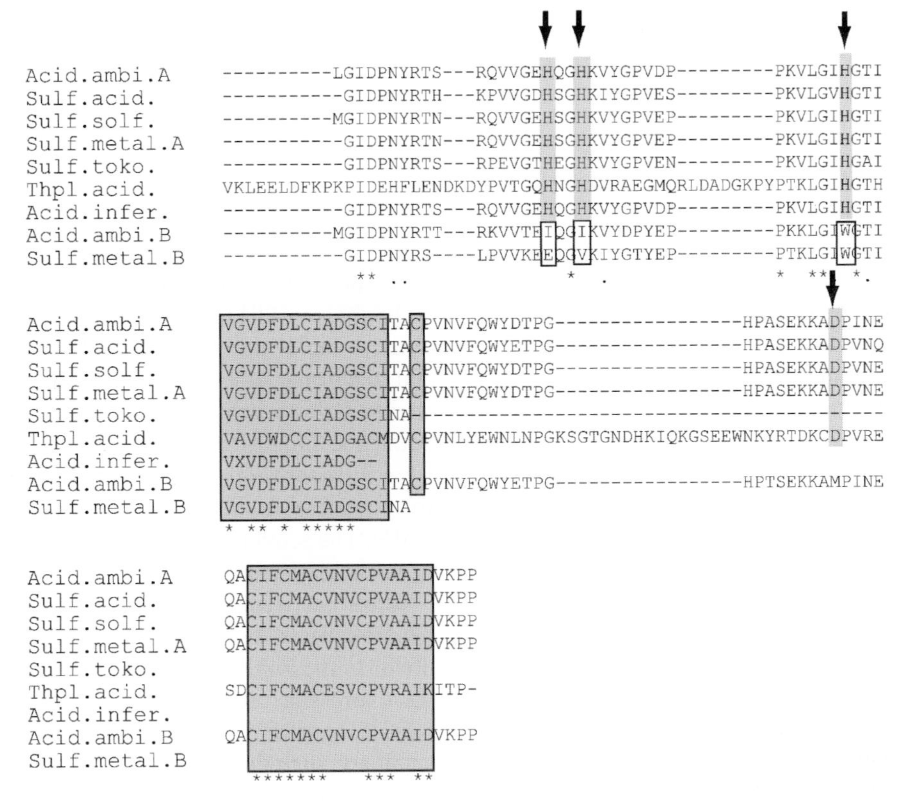

Fig. 1. Alignment of ferredoxin sequences from *Archaea*. Shaded plus boxed = binding domains for the two Fe-clusters; shaded plus arrows = metal liganding residues in Zn containing ferredoxins; boxed open = replacements in the corresponding Zn free ferredoxins. Sequence data were taken from gene banks. Abbreviations denote: Acid. ambi, *A. ambivalens*; Sulf. acid, *S. acidocaldarius*; Sulf. toko, *S. tokodaii*; Sulf. metal, *S. metallicus*; Thpl. acid, *Thermoplasma acidophilum*; Aci. inf, *A. infernus*; those species expressing both the Zn-containing and the Zn-free ferredoxins, are labeled with A and B, respectively.

side chain (Collins and Langworthy, 1983; Collins, 1985). In contrast, the quinones of *Sulfolobus* and hyperthermophilic anaerobic Archaea exhibit fully reduced side chains.

C. Blue Cu Proteins

In plants and bacteria a large number of small blue copper proteins has been found which play an important role in electron transfer reactions (Sykes, 1990). Usually these are soluble proteins with molecular masses from 10–20 kDa which fold into stacked β-sheets with one Cu-ion ligated in a trigonal equatorial coordination. The typical Cu-ligands include 2 histidines and a cysteine in the equatorial plain, and additionally one axially bound methionine (type-I Cu protein).

The first archaeal blue copper protein was detected in *Natronomonas pharaonis* (Scharf and Engelhard, 1993) and named 'halocyanin.' Its UV/vis-, CD-, and EPR spectral properties are in good agreement with typical type-I Cu-proteins and are partially resembling those of plastocyanin. In contrast to soluble members of this family it could be removed from the plasma membrane only by mild detergents. The primary structure indicates indeed a lipid anchor for membrane fixation (Mattar et al., 1994) of the protein and a total molecular mass of 17.2 kDa. The actual mass of the mature protein, 15.45 kDa, indicates post-translational processing as found with other prokaryotic lipoproteins. Spectro-electrochemical studies in the UV- and IR region revealed an oxido-reduction midpoint potential of E_m = +183 mV at neutral pH (Brischwein et al., 1993). However, the E_m changed dramatically with pH (from +333 mV at pH 4.0 to +113 mV at pH 10.0) indicating a reversible protonation coupled to the redox equilibrium. FTIR spectra revealed significant structural alterations along with the redox process, and suggested the participation of the histidine ligands in the protonation equilibrium as well as the function of a carboxyl (Asp, or Glu) as a possible fifth ligand. The pH dependence of FTIR

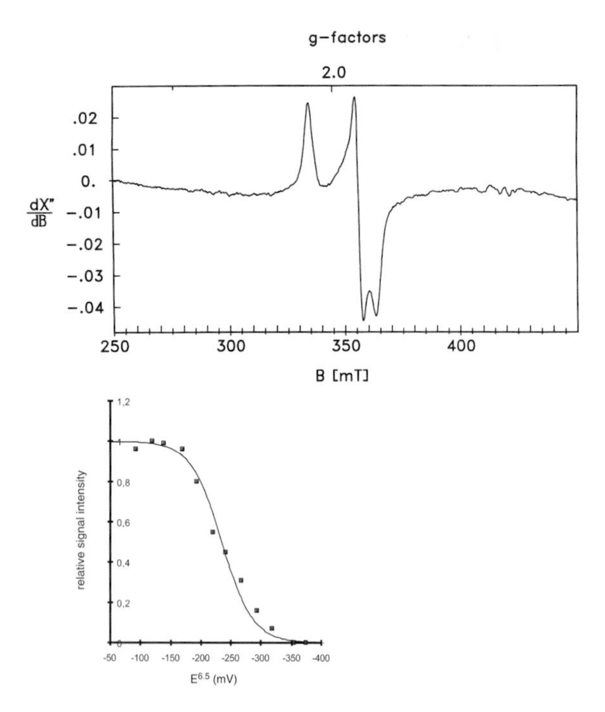

Fig. 2. EPR spectrum of the recombinantly produced Zn-free 7-iron ferredoxin from *S. acidocaldarius*. The spectrum of the dithionite reduced form is shown together with a redox titration indicating a midpoint potential around –250 mV. Data from (Janssen et al., 2001).

spectra and midpoint potential were not reflected in the Resonance-Raman spectra of the protein (Hildebrandt et al., 1994). Thus, it was concluded that the redox potential is not exclusively controlled by the geometry of the coordination sphere; the redox potential falls into the range of *c*-type cytochromes and the membrane residing halocyanine might assume a similar function in the respiratory system of *N. pharaonis*.

The completion of the genome from *H. salinarum* (strain R1) revealed genetic evidence for at least seven small blue copper proteins with homology to halocyanin from *N. pharaonis*. One of these exhibits two copper binding sites and might have arisen from gene duplication. Some are carrying a membrane anchor (Pfeiffer and Oesterhelt, 2001). Also in *H. salinarum* blue copper proteins might be functioning in replacement for cytochrome *c*. An azurin has been annotated in the genome of *Ae. pernix* (Kawarabayashi, 1999).

Another archaeal blue Cu-protein was initially predicted from genetic data of *S. acidocaldarius* (Castresana et al., 1995). The *soxE* gene within a large gene cluster encoding the SoxM supercomplex (see below) exhibits the signature of a typical Cu-binding protein of type-I and was named 'sulfocyanin.' Though transcription of the gene had been shown it could not be identified protein chemically until lately. It was thought to be absent from the isolated SoxM complex, thus causing its catalytic inactivity (Schäfer et al., 1999). Recently we succeeded to express this protein in *E. coli* cells and to study its properties in detail (Komorowski, 2001; Komorowski and Schäfer, 2001). From the amino acid sequence a native molecular mass of 20.4 kDa is calculated, and according to hydropathy plots the protein reveals a typical membrane anchor at its N-terminus. Sequence similarities can be derived to rusticyanin from *T. ferrooxidans* which was taken as a scaffold for tentative homology modeling (Komorowski, 2001). The sequence has only low homology to any of the halocyanins.

Nevertheless, neither rusticyanin nor any other blue Cu-protein was able to replace Sulfocyanin as electron donor to the SoxM respiratory complex. Moreover, the expression in *E. coli* failed with the genuine *Sulfolobus* gene; for efficient production at least in small amounts the gene had to be fully synthesized according to the optimum codon usage of *E. coli*. For simplification of handling the recombinant protein was truncated of the membrane anchor and further investigated as Δ(2,33)-sulfocyanin. With that in hand its fundamental properties could be determined.

Figure 3 shows the vis-spectrum in the oxidized state together with an EPR spectrum clearly exhibiting the typical Cu-hyperfine splitting of a type-I copper ligation; $g_{\|} = 2.22$, $g_{\perp} = 2.05$ and the hyperfine splitting is small ($A_{\|} = 93 \times 10^{-4}$ cm^{-1}). It appears nearly axial. The CD spectrum shows a minimum at 218 nm and the relative β-ratio is calculated as 55% in line with most other blue copper proteins. Remarkably, the absorption spectrum remains unchanged even on exposure to temperatures above 80 °C reflecting the high thermostability of the sulfocyanin fold.

Sulfocyanin can be reversibly reduced with ascorbate, dithionite, or reduced cytochrome *c*. Reduction results in the loss of the 592 nm absorption band. An equilibrium titration of reduced cytochrome *c* with oxidized sulfocyanin allowed the determination of the equilibrium constant at pH 6.5, as 2.35, equivalent to a half-reduction potential of E_0= +300 mV. Similar to halocyanin this potential suggests a cytochrome-*c* like function of sulfocyanin. The pH dependence of the redox potential could not be determined with the applied method though it appears of significant

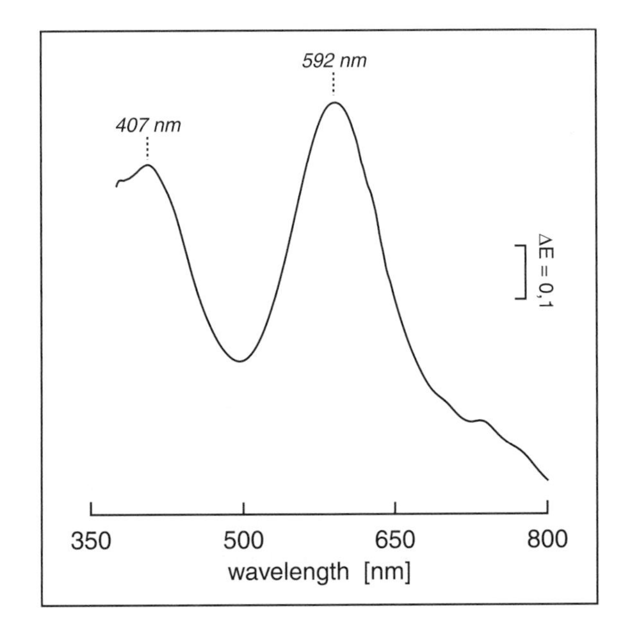

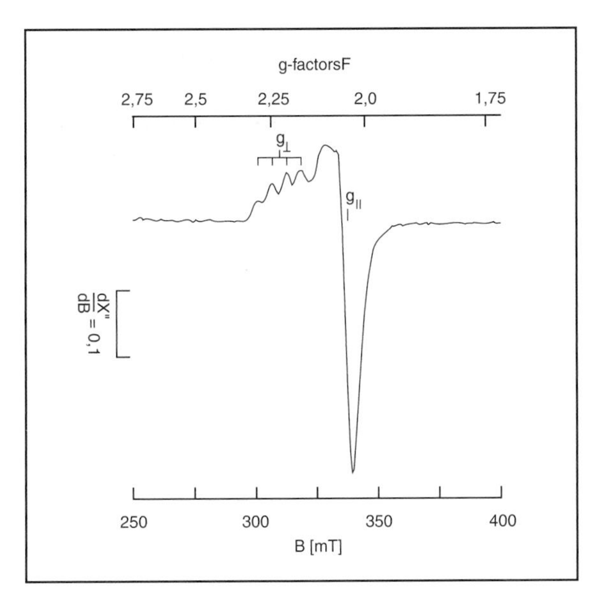

Fig. 3. Optical (top) and EPR (bottom) spectra of the oxidized blue copper protein sulfocyanin from S. *acidocaldarius*. Spectra were produced with the truncated (Δ2,33) form of the recombinant protein. The data are taken with permission from L. Komorowski (Dissertation, University of Luebeck, 2001) and (Komorowski and Schäfer, 2001).

importance because it is assumed to be exposed to the outer surface of the plasma membrane.

In membrane denaturating gels, sulfocyanin shows a molecular mass (~30 kDa) which is likely to result from glycosylation or lipid association (Komorowski and Schäfer, 2001) although none of these possibilities has been experimentally verified so far. Actually, the sequence of sulfocyanin reveals a number of possible glycosylation domains. For comparison, an high degree of glycosylation has been shown for cytochrome-b_{558}, another electron carrier exposed to the outer surface of the *Sulfolobus* plasma membrane (Hettmann et al., 1998; Zähringer et al., 2000).

D. Hemoproteins

Many thermoacidophilic archaea were found to lack *c*-type cytochromes. Besides blue copper proteins also novel hemoproteins might take on the function of cytochrome *c* as a linker between larger electron transport complexes. At least two candidates come into question. One is the as yet functionally unassigned cytochrome-$b_{558/562}$ from *S. acidocaldarius*. This highly glycosylated, membrane bound monoheme cytochrome is facing the outer surface of the plasma membrane (Hettmann et al., 1998). Previously it was discussed that it could be involved in redox reactions of a hypothetical periplasmatic metabolism. More likely, however, cytochrome-$b_{558/562}$ may function as a membrane-fixed intermediate electron carrier with a movable head. In fact, by EPR studies it has been demonstrated that the heme bearing extrinsic domain of this cytochrome can assume two distinct limit orientations in the native membrane (Schoepp-Cothenet et al., 2001) of *S. acidocaldarius*. The implications for the function of a bc_1-analogous novel complex is discussed below.

Another candidate might be the mono-heme cytochrome-b_{562} described as member of a respiratory supercomplex in *Sulfolobus* sp. strain 7 (Iwasaki et al., 1995a; Iwasaki and Oshima, 1996), now classified as *Sulfolobus tokodaii* (Kawarabayashi, 2001). As it is a mono-heme cytochrome, a cytochrome *b*-like function in a proton transducing Q-cycle as originally proposed appears unconceivable.

IV. Is Complex-I Present in Archaea?

The proton pumping NADH dehydrogenase of mitochondria and many aerobic bacteria was previously postulated as missing in archaea (Schäfer et al., 1994, 1999; Schäfer, 1996). After the full resolution of several archaeal genomes the above question has to be taken up again. Actually, the first reports on respiratory properties of halobacteria and thermoacidophilic *Sulfolobales* demonstrated the presence of NADH oxidizing activities in their membranes (Lanyi, 1969, 1972; Oshima, 1987; Wakagi and Wakao et al., 1987). However, later studies using EPR spectroscopy and site-specific inhibitors failed to demonstrate the presence of any of the iron-sulfur signals typical for energy transducing orthodox type-I NADH dehydrogenases (Complex-I); furthermore, it has been shown that archaeal NADH oxidizing systems are insensitive to rotenone, amytal, or piericidines. It is noteworthy, that hybridization attempts, made in our laboratory, with archaeal DNA using NDH-I directed probes, failed as well (unpublished results). The present-day situation is unchanged and therefore one has to assume that indeed a proton pumping complex-I is absent from archaeal respiratory systems.

This may appear unlikely in view of the positive hits of tentative gene assignments in recently sequenced archaeal genomes. However, in none of these cases the protein chemical or functional equivalent of a complex-I has been demonstrated. In that context one has to keep in mind that a surprising structural and topological relation, or even homology, exists between components of complex-I and hydrogenases (Albracht and Hedderich, 2000; Friedrich, 2001). Moreover, only single homologous genes have been assigned but in no case the presence of a full set of (up to 14; Friedrich, 2001) genes for completion of a functional complex-I have been identified. In evolution the modular association of initially separate membrane redox-components and peripheral dehydrogenase modules with transport modules may have given rise to the formation of both, hydrogenases and NADH dehydrogenases. Also, the typical operon structure of bacterial complex-I genes is not verified in Archaea. Thus, the above statement stands as long as the function of the tentatively assigned genes has not been shown.

A hint to the function of proteins encoded by the newly detected genes may be derived from our knowledge on the modular design of complex-I (Friedrich, 2001). Three modules are essential of which the two membrane integral fragments contain the iron-sulfur hosting device and coupling the redox process to proton translocation, and the module transferring electrons to a quinone. Homologues to those have been identified for example in *H. salinarum* (Ng et al., 2000; Pfeiffer and Oesterhelt, 2001). What generally is missing is the peripheral module receiving electrons from NADH. Thus, it may be proposed as a working hypothesis that this module could be replaced in the respective archaea by a different redox device using other donors than NADH. Ferredoxins might be suitable candidates. The abundance of ferredoxins in archaea should challenge further experimental investigation on this issue, especially because iron-sulfur centers can link the electron flow from the substrate oxidizing module to the membrane core of the whole machinery also in the known complex-I architectures.

In contrast, the ability of aerobic archaea to oxidize NADH by type-II dehydrogenases without coupled proton translocation has been repeatedly demonstrated and some of the respective enzymes could be characterized. This implies, however, that archaeal respiratory chains in general are lacking one possible energy coupling site.

An early report describes the isolation of an NADH:acceptor oxidoreductase from *Sulfolobus* sp. strain 7 (now *S. tokodaii*) (Wakao et al., 1987). The water-soluble dimeric enzyme (95 kDa) contains two FAD/molecule and may represent the peripheral component of a larger membrane bound type-I NADH dehydrogenase, i.e. without coupling the reaction to proton pumping. This preparation had virtually no activity with caldariella quinone as hydrogen acceptor. A similar enzyme (76 kDa) was first described from *Acidianus ambivalens* (Gomes and Teixeira, 1998) and has initially been considered to act as an NADH: ferredoxin oxidoreductase. In this latter organism, the same laboratory reported recently the identification of a new type-II NADH dehydrogenase which was reconstituted in liposomes to form an artificial respiratory chain together with the terminal aa_3-type quinol oxidase (Gomes et al., 2001b). The enzyme (47kDa) has no iron-sulfur clusters, it reacts rapidly with caldariella-quinone (>200 nmol/min/mg protein) and it can transfer electrons also to synthetic quinones with a K_M for NADH of ~6 μM. Notably, this in vitro system resembles the previously postulated minimal respiratory chain of primordial aerobic organisms (Schäfer et al., 1996a). 4-Hydroxyquinoline-N-oxide (HQNO) is an inhibitor of type-II NADH dehydrogenases. The NADH oxidizing activity of membranes from *Halobacterium salinarum* has been shown to be due to a type-II NADH dehydrogenase which is sensitive to this inhibitor (Sreeramulu et al., 1998). In contrast to *Sulfolobus*, the membrane bound enzyme is capable to transfer electrons to cytochrome *c* as an artificial acceptor.

In conclusion, to the best of our knowledge aerobic archaea possess only type-II NADH dehydrogenases which not linked to proton pumping. A peculiarity is the existence of a $H_2O + H_2O_2$ forming NADH oxidase in the obligate anaerobe *Pyrococcus furiosus* (Ward et al., 2001) which presumably functions as an oxygen scavenger.

V. Archaeal Complex-II Equivalents (SQR/ QFR)

Reversible succinate dehydrogenase (SDH) activities have been ubiquitously detected in organisms from the three domains of life. They represent constituents either of respiratory complex II in aerobes, or of fumarate reductases complexes in anaerobes. Archaeal succinate:quinone oxidoreductases have been comparatively analyzed extensively in a recent review (Schäfer et al., 2002), therefore only the most notable peculiarities of Archaeal SQRs will be discussed here. These are the presence of unusual iron-sulfur clusters in subunit B of succinate dehydrogenases from thermoacidophilic cren-archaeota, the occurrence of an unusual cysteine-rich sequence motif in subunit C, and the lack of typical membrane anchors or membrane spanning helices in the small subunits. These observations, taken together, justify an assignment as a novel type of succinate dehydrogenases.

The operon structure of Archaeal SDHs usually follows the order of genes for the four subunits *sdhA, sdhB, sdhC, sdhD*, with exception of *Natronomonas pharaonis* resembling the *E. coli* order C,D,A,B, and *H. salinarum* with *sdhD* preceding *sdhC*.

Polypeptide A is the central substrate- and primary acceptor binding large subunit containing the canonical FAD in all SQR/ QFR complexes. Polypeptide B contains 3 FeS clusters as a link to further electron acceptors or donors. The small polypeptides C and D usually serve as membrane anchors for the complex and in some cases contain functionally important histidine residues. When present these serve as heme-liganding residues for *b*-type cytochromes buried between the membrane spanning α-helices. Such heme-binding sites were identified in complex II from *T. acidophilum, H. salinarum, P. aerophilum, N. pharaonis*, but are missing from the hyperthermo-acidophilic archaeal species *Sulfolobus* and *Acidianus*.

The recently achieved high resolution 3D-structure of QFR from *Wolinella succinogenes* (Lancaster et al., 1999) can serve as a very general and basic structural model also for complex II due to their common evolutionary origin and the over-all similarity of both membrane residing redox systems (see also Chapter 4 by Lancaster, Vol. 1).

A. The FeS Clusters

From 13 Archaeal species with SQR activity or genetic evidence for a complex II only 5 partially or entirely purified preparations have been described (Moll, 1991; Iwasaki et al., 1995b; Janssen et al., 1997; Scharf et al., 1997; Lemos et al., 2001). In vivo the above mentioned Archaeal quinones serve as a terminal electron acceptor, and also the reduction of *b*-cytochromes by succinate has been shown very early in Archaeal membranes (Hallberg and Hederstedt, 1981, Anemüller et al., 1985). Surprisingly a quinone reductase activity was not preserved in all purified preparations as for example from *S. acidocaldarius* (Moll, 1991).

In several Archaeal membranes SQR is so abundant that the typical iron-sulfur centers S1, S2, and S3 could be studied by EPR spectroscopy already in the intact membrane (Anemüller et al., 1995; Gomes et al., 1998). Specific Archaeal deviations from the usual cluster structure were identified in *S. acidocaldarius* (Janssen et al., 1997) and *A. ambivalens* (Gomes et al., 1999; Lemos et al., 2001). Generally it was assumed that subunit B hosts 1 [2Fe-2S], 1 [4Fe-4S], and 1 [3Fe-4S] cluster. As shown by the sequence alignment of Fig. 4, in a number of cases including the related thiol-fumarate reductase from *Methanococcus jannashii* an additional cysteine was found at the binding domain of cluster S3. In fact, EPR spectra of SQRs from *S. acidocaldarius*, and *A. ambivalens* revealed the presence of an additional [4Fe-4S] cluster replacing the classical S3 cluster. This altered motif is also found in *M. thermoautotrophicum*, as well as in bacterial fumarate reductase subunits FrdB from *Aquifex aeolicus*, *Synechocystis* sp., and *Campylobacter jejuni* (Schäfer et al., 2002). For further details the reader is referred to the specialized review (Schäfer et al., 2002).

B. A Novel Sequence Motif

Another irregularity as compared to 'classical' complex II is the appearance of an unusual 10-cysteine motif in subunit C of SQRs from thermoacidophilic *Sulfolobales*. It has no equivalent in any other SQR or QFR but has been found also in polypeptides B of thiol-fumarate reductases (TFR) from methanogenic Archaea and also in a small number of completely

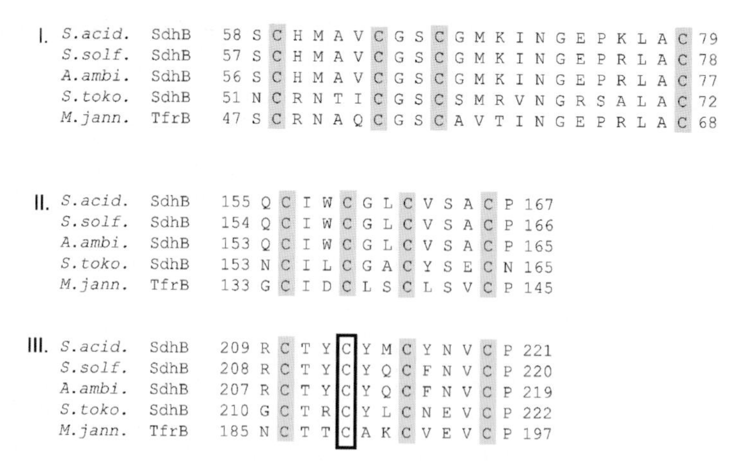

Fig. 4. Sequence alignments of the three iron binding sites I, II, and III in subunit B from archaeal SQR/QFR complexes. The boxed residues indicate the additional cysteine of the second [4Fe-4S] cluster. Species abbreviations: S.acid., *S. acidocaldarius*; S.solf., *S. solfataricus*; A.ambi., *A. ambivalens*; S.toko., *S. tokodaii*; M.jann., *Methanococcus jannaschii*. The latter sequence refers to subunit B of the thiol-fumarate-reductase of *M. jannaschii*. Sequences are taken from gene bank.

different and functionally unrelated oxidoreductases included in the alignment of Fig. 5. This motif occurs as a tandem with the general signature –GC-X_{32}-CC(G/P)-X_{35}-CxxC-(repeat I)-X_{71}-(G/P)C-X_{39}-CCG-X_{39}-CXXC-(repeat II); the latter example is taken from *S. acidocaldarius*. The second repeat within this tandem contains a strictly conserved glycin preceding the CCG motif by 12 residues.

EPR spectroscopy of isolated *S. acidocaldarius* complex-II gave no hint to the existence of additional iron clusters involving these cysteines. Therefore, a possible involvement in disulfide bonds, or in redox reactions with participation of thiol radicals has been discussed (Janssen et al., 1997). Actually, it should be emphasized that isolated SDH from *S. acidocaldarius* which belongs to this novel SQR type is strongly inhibited by the stable radical forming 2,3,5,6-tetra-chloro-benzoquinone (Moll, 1991) suggesting direct interaction with the inhibitor of a free radical involved in the electron transport reaction.

A recent study with subunit C from *S. tokodaii* produced recombinantly in *E. coli* revealed, however, that it contains a new $[2Fe2S]^{2+}$ cluster completely liganded by cysteines as well as an isolated zinc site (Iwasaki et al., 2002). These new metal centers were characterized by EPR and EXAF spectroscopy. The [2Fe2S] center is thought to represent as a novel electron entry site from SDH to the respiratory chain of the organism. In this context it is of interest that the chemically determined Fe content of isolated complex-II from *S. acidocaldarius* significantly exceeded the amount determined by EPR quantification (Moll, 1991). Thus, in these preparations the novel center C might have lost its integrity; that could also explain the obvious inactivity of these preparations with the natural quinone as electron acceptor.

A phylogenetic analysis has suggested a coevolution of subunits A and B of SQRs (Schäfer et al., 2002). Interestingly, in an unrooted tree of the iron-sulfur hosting subunit B (Fig. 6) the archaeal domain originates from a common branch point splitting into 2 subdomains (shaded area). These subdomains clearly reflect the division into 'classical' complexes bearing the heme-liganding histidines in their small subunits, and the 'novel' type sharing the unusual cysteine motif as well as the [4Fe-4S]-cluster instead of a regular S3.

VI. Does a bc_1 Complex Exist in Archaea?

Generally quinol:cytochrome-*c*-oxidoreductases (complex III of respiratory chains) are functioning as efficient redox driven proton pumps based on a sophisticated mechanism, termed as Q-cycle (Mitchell, 1976; Trumpower, 1990; Talfournier et al., 1998). The essential components are the Rieske [2Fe-2S] iron-sulfur protein, *b*-type cytochromes, and an intermediate electron acceptor (Cyt c_1) between this core complex and cytochrome *c*. Thus, answering the above question has to rely on the proof of whether or not the respective constituents are present in Archaea. Actually, the answer can only be both, 'yes' and 'no', for several reasons. 'yes' because for a number of

```
S.acid.   SdhC  1       MAYAYYPGCTAHGLSKDIDIATKKVFETLGLKLDEVKDWNCCGG-GFYDEYDEVGH
S.solf.   SdhC  1    MIGMKIAYYPGCATHGLSKDVDIATKKVAEVLGVELVEVPDWNCCGG-GFYDEYDEVGH
A.ambi.   SdhC  1       MKVAYYPGCATHGLSKDVDIATKKVADVLGLELVEVEDWNCCGG-GFLDEYNEKAH
M.jann.   HdrB  1       MKYAFFLGCIMPHRYPGVEKATKIVMEELGVELEYMPGASCCPAPGVFGSFDQKTW
M.ther.   HdrB  1       MEIAYFLGCIMNNRYPGIEKATRVLFDKLGIELKDMEGAFCCPAPGVFGSFDKTTW
S.toko.   SdhC  1   MITALEYAYFPGCVAQGACGELHLATTALSKALGIKLLELKKASCCGS-GTFKEDSQLLE
A.fulg.   HdrB  1     MFMKYALFPGCKIAFERPDLELAMREVLTALDVPFEYLSDFSCCPTWASVPSFDIEAW
M.ther.   TfrB  259     SRIGFFTGCLVDYRMPDVGMALLRVLREHGFEVDVPDGQVCCGS-PMIRTGQLDIV
E.coli.   G3PDH 170     DQVAFFHGCFVNYNHPQLGKDLIKVLNAMGTGVQLLSKEKCCGVPLIANGFTDKAR
E.cli.    GOX   168     RRVLMLEGCAQPTLSPNTNAATARVLDRLGISVMPANEAGCCGAVDYHLNAQEKGL
                               **               :              **

S.acid.   SdhC       VALNLRNLSIVEKMGYQK--MVTPCSVCLQSHRLATHKYKENKD----LKKEVDDRIKG-
S.solf.   SdhC       VALNLRNLSQVEKMGLTK--MVTPCSVCLHSHRLATYKYKEDKD----IKRKTDKRLEG-
A.ambi.   SdhC       VALNLRNLSTVERMGMDK--MVTPCSVCLQSHRLAAYKYNENKD----LRKEVDKKLKE-
M.jann.   HdrB       LTLAARNLCIAEEMGLD---IVTVCNGCYGSLFEAAHILHENKE----ALDFVNEKLDK-
M.ther.   HdrB       AAIAARNITIAEEMGSD---VMTECNGCFGSLFEANHLLKEDEE----MRAKINEILKE-
S.toko.   SdhC       DSVNARNIALAEQLNLP---LLTHCSTCQGVIAHVDERLKKAQKDDPAYVEQINGYLKKE
A.fulg.   HdrB       LAISARNISLAEEKGLD---IVVGCGDCYSVLNHARDMLKREE-----WRERVNRILAK-
M.ther.   TfrB       EDLVERNRRALE--GYDT--IITVCAGCGATLKKDYPRYGVE------------LNV--
E.coli.   G3PDH      KQAITNVESIREAVGVKGIPVIATSSTCTFALRDEYPEVLNVD----------NKGLR--
E.coli.   GOX        ARARNNIDAWWPAIEAGAEAILQTASGCGAFVKEYGQMLKND--------ALYADKAR--
                                               ::    *
```

Fig. 5. Partial sequence alignment of polypeptides bearing the unusual 10-cysteine motif. Only the first repeat is shown. Sequences are from succinate dehydrogenases subunit C (SdhC), heterodisulfide reductases subunit B (HdrB), thiol-fumarate reductase subunit B(TfrB), Glycerol-3-phosphate dehydrogenase, and glucose oxidase; the latter two sequences are from *E. coli* for comparison. The archaeal sequences are from: S.acid., *S. acidocaldarius*; S.solf., *S. solfataricus*; A.ambi., *A. ambivalens*; M.ther., *M. thermoautotrophicum*; M.jann., *M. jannaschii*; S.toko., *S. tokodaii*; A.fulg., *A.fulgidus*. The sequences were extracted from gene bank. Numbers in column 3 indicate the starting position within the respective sequence.

archaea the existence of one or even two respiratory Rieske FeS-proteins has been demonstrated, and 'no' because the association of these Rieske proteins with cytochromes deviates significantly from orthodox bc_1-complexes. Though in Crenarchraeota like the thermoacidophilic *Sulfolobales* Rieske proteins were detected spectroscopically (Anemüller et al., 1993; Lübben et al., 1994a; Gomes et al., 1998), the complete absence of *c*-type cytochromes is incompatible with the existence of a regular *bc*-complex. Further, the usually associated di-heme cytochrome is hosting heme A_S. Also the terminal electron acceptor c_1 is apparently replaced for example by a blue copper protein or another cytochrome. And finally, the sensitivity against specific inhibitors like antimycin-A or myxothiazol, commonly considered as discriminating inhibitors for Rieske/*bc*-complexes, is absent. The situation appears less divergent with euryarchaeota for example the *Halobacteriales*, since both, *b*- (Hallberg Gradin and Baltscheffsky, 1981; Hallberg Gradin and Colmsjö, 1989; Kuhn and Ward, 1998; Tanaka et al., 2002) and *c*-type cytochromes have been found, as well as EPR spectroscopic evidence for a Rieske protein has been demonstrated (Sreeramulu et al., 1998). Figure 7 shows spectra of a partially enriched membrane fraction from *H. salinarum* (strain JW5) strongly suggesting the presence of a bc_1-like complex; additional support comes from the reported sensitivity to antimycin. From none of the archaeal organisms a functional complex could be isolated so far. However, a functional equivalent to a respiratory complex III is obviously included in the supercomplex SoxM as to be discussed in a subsequent paragraph. In the following, single constituents of possible archaeal equivalents to complex III are discussed as well as their functional significance.

A. *Archaeal Rieske Proteins*

The unique electronic structure of the typical FeS cluster in respiratory Rieske proteins is generated by the mixed ligandation of two Fe ions by two histidines on one side, and two cysteines on the other, with two sulfur ions bridging the Fe ions. This coordination gives rise to a characteristic EPR spectrum which is strongly different from those of other iron-sulfur clusters. The highly resolved X-ray structure of mitochondrial bc_1 complexes could be determined (Xia et al., 1997; Berry et al., 1999; Hunte, 2001), revealing two or even three (Montoya et al., 1999) quinol/quinone binding sites and the surprising fact, that the primary electron acceptor for quinol

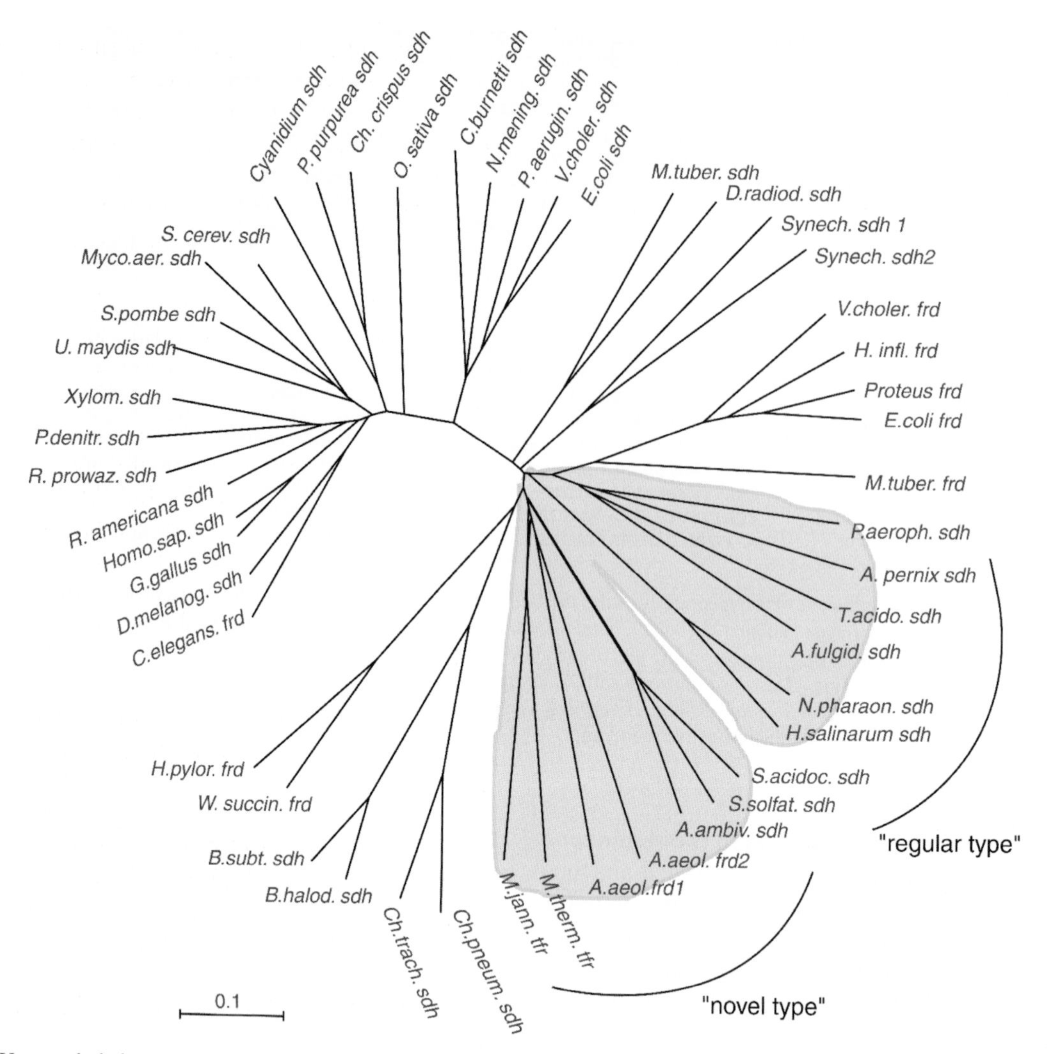

Fig. 6. Unrooted phylogenetic tree of the flavin hosting polypeptides A of SQRs and QFRs. The Archaea are clustering in two major branches (shaded) one comprising those organisms with regular ligands for the S3 [3Fe-4S] cluster, the other comprising the novel type with a second [4Fe-4S] cluster. All other (non-archaeal) sequences contain also the 'regular' type. All sequence data extracted from gene bank and the archaeal genome projects.

oxidation, the Rieske protein, acts as a reorienting redox domain between the site of electron input and the acceptor cytochrome c_1 (Snyder et al., 1999). The second electron of a quinol is transferred to the *b*-type cytochrome (for further mechanistic details see Xia et al., 1998; Yu et al., 1999; Darrouzet et al., 2001; Gutierrez-Cirlos and Trumpower, 2002; see also Chapter 3 by Cooley et al., Vol. 1).

The first archaeal respiratory Rieske protein was purified from *S. acidocaldarius* (Schmidt et al., 1995). Surprisingly, a second Rieske protein was detected in the same organism as a constituent of the SoxM respiratory complex (Castresana et al., 1995; Schmidt et al., 1996). These, the SoxL and SoxF Rieske proteins were named according to their respective genes. Both could be produced heterologously and their redox potentials, redox linked pK_a values, and detailed spectroscopic properties were determined (Anemüller et al., 1994b; Boekema et al., 1997; Shima et al., 1999). From the cloned gene also the Rieske protein from the neutrophilic and hyperthermophilic crenarchaeon *Pyrobaculum aerophilum* could be produced (Henninger et al., 1999). The only other archaeal Rieske proteins identified in native material are that from *Sulfolobus* sp. strain 7 (now renamed *S. tokodaii)* (Iwasaki et al., 1995a) and that from *Halobacterium salinarum* (DSM3754,strain JW5) in a partially enriched form (Sreeramulu et al., 1998).

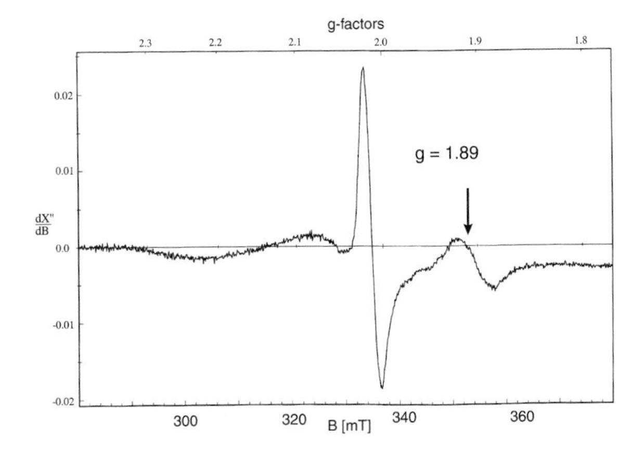

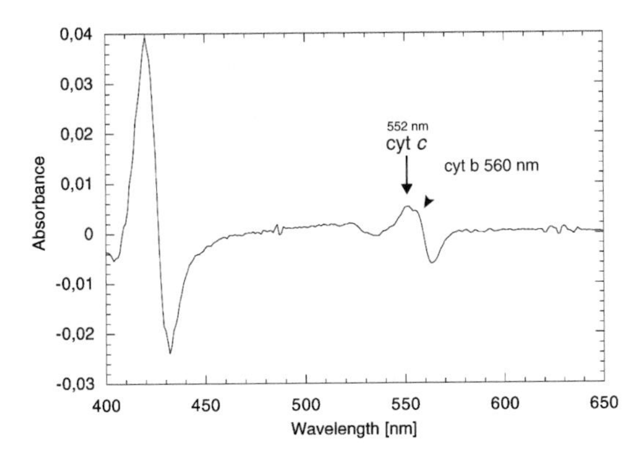

Fig. 7. Spectroscopic evidence for a bc_1 complex in *H. salinarum*. The spectra were produced with a partially purified fraction of solubilized plasma membranes. Top: EPR spectrum of the ascorbate reduced preparation clearly indicating the presence of a Rieske-like FeS-center. Bottom: First derivative optical spectrum of the same ascorbate reduced fraction indicating the presence of both, *b*- and *c*-type cytochromes. Details are given in the text and (Sreeramulu et al., 1998).

Also in membrane preparations from *S. metallicus* clear evidence for a Rieske-type protein emerged from EPR spectra which in addition indicate a novel FeS center with high reduction potential (+350 mV) reminiscent of HIPIP iron-sulfur proteins (Gomes et al., 1998).

Table 2 gives a survey of experimental data for archaeal Rieske proteins. A number of further Archaeal Rieske proteins is also suggested from the recently completed and ongoing genome projects, including *H. salinarum*, *H .marismortui*, *Ae. pernix*, and *A. ambivalens*. A comparison of archaeal Rieske proteins with their bacterial and eukaryal counterparts reveals that Archaeal Rieske proteins are usually much larger and are bearing an extended insertion between the iron-cluster binding sites. For illustration a typical partial alignment is shown in Fig. 8. A comprehensive review on sequence features and the phylogenetic relation of Rieske proteins, Rieske-*type* proteins and Rieske-like ferredoxins is given in (Schmidt and Shaw, 2001).

The only 3D structure with atomic resolution of an Archaeal Rieske protein has recently been determined for SoxF from *S. acidocaldarius*; the N-terminal membrane anchor has been truncated (Bönisch et al., 2000; Bönisch et al., 2002). This 1.1Å structure gives a hint on the structural differences imposed by the large sequence insertion (see Fig. 8) as demonstrated in Fig. 9. An additional β-sheet (βC) is inserted between (normally present) β5 and β6 as well as an additional disulfide bridge between Cys76 and Cys247. This enables a firm fixation of the C-terminus together with the H-bond interaction between β5-β11 and may contribute to the enormous heat stability. Importantly, the N-terminus and the adjacent hinge region, necessary for a large-scale domain movement (Iwata et al., 1999) are not fixed in the same way. Further, a 3/10-helix not found in other known Rieske proteins (Iwata et al., 1996; Xia et al., 1997; Hunte, 2001) is present in this Archaeal protein. For a detailed structural alignment the reader is referred to (Darrouzet et al., 2001). In comparison to the Archaeal Rieske protein those from spinach b_6f or bovine bc_1 complexes display a much lower packing density of their 3D structure.

B. Associated b-Cytochromes and a Novel Complex

For a decisive answer to the above question, the archaeal Rieske proteins have to be seen in a more general context along with their protein chemical and genetic environment. Again, the best investigated example is the thermoacidophilic archaeon *Sulfolobus acidocaldarius,* where genetic, transcriptional, protein chemical, and functional studies are supporting each other.

Earlier a mono-heme *b*-type cytochrome from *S. acidocaldarius* [Cyt $b_{558/562}$] has been described together with its genes *cbsA* and *cbsB* (Hettmann et al., 1998; Zähringer et al., 2000). The function of this highly glycosylated protein was not known. This cytochrome is an integral membrane protein with its heme bearing domain facing the outer surface of the plasma membrane; it is abundant in *Sulfolobus* membranes growing under limited oxygen supply. The discovery that it is part of a gene cluster also

Table 2. Comparative list of authentic and recombinant archaeal Rieske proteins including a few cases of Rieske proteins identified only in situ by EPR spectroscopy. The data were extracted from the literature quoted in the text.

Organism	Protein auth./recomb.	amino acids	M_w (kDa)	$E_m^{7-7.5}$ (mV)	pK_a	EPR characteristics g_z	g_y	g_x
S. acidocaldarius	SoxF auth.	250	28.6	+320	6.2/8.5	2.036	1.889	1.781
S. acidocaldarius	SoxL mature precursor	238 314	25.0 33.8	+320	6.2/8.5	2.035	1.895	1.768
S. acidocaldarius	SoxF recomb.	250	26.8	–	–	2.042	1.890	1.785
S. acidocaldarius	SoxF recomb. Δ1–46	204	21.7	+374	–	2.042	1.895	1.785
S. acidocaldarius	SoxL recomb.	238	25.0	+347	–	2.032	1.892	1.764
S. metallicus	in membrane	–	–	+320	–	2.028	1.90	1.74
P. aerophilum	ParR auth.	186	19.7	–	–	n.d.		
P. aerophilum	ParR recomb. Δ1-23	163	17.2	+229	8.1/9.8	2.03	1.888	1.795
P. aerophilum	PaR recomb. Δ1–42F137W	145	15.5	+170	–	–	–	–
H. salinarum	in membrane	–	–	–	–	2.03	1.89	n.d.
S. tokodaii	in membrane	–	–	n.d.	n.d.	–	1.89	–
S. tokodaii	sol. fragmt.	114	12.9	+125	6.2/8.6	2.02	1.89	1.79

```
                                  SGQLTASEPDQLTAAALLAARQANV
Sulf.acid.SoxL   AICQHLGCTPPYIHFYPPNYVN         PALIHCDCH-GSTYDPYHGASVLTGPTVR
Acid.ambi.       AICQHLGCQPPYIHFYPPNHVN-X25-PAIFHCDCH-GSTYDPYHGAAVLTGPTVR
Sulf.acid.SoxF   DVCVHLGCQLPAQVIVSSESDPGLYAKGADLHCPCH-GSIYALKDGGVVVSGPAPR
Pyrb.aero.       AICTHFGCPVN--------------------NCPCHGSIFAIX14 LEMYVSGGPAP
Halo.salin.      NKCTHFCCAPQGFRTSNYEGAED------KIYCQCHQSIYDPX7KKSFVAFARPEN
Halo.maris.      KVCTHAGCMVSDREDLV--------------VCPCHFGKFNVLEGAA-VSGGPPGR
Ferr.acid.       GICSHLKCILNVSEDK-------------HVICPCHNAKFELDTGK--MVEPPFIA

Bovin mito.      GVCTHLGC-VPIANA-GD---------FGGYYCPCH-GSHYDA--SGRIRKGPAPL
Spinach clpl.    AVCTHLGCVVPFNAA------------ENKFICPCH-GSQYNN--QGRVVRGPAPL
Tobacco clpl.    AVCTHLGCVVPFNAA------------ENKFICPCH-GSQYNN--QGRVVRGPAPL
Chlor.lim.       AVCTHLGCLVNWVDA------------DNQYFCPCH-GAKYKL--TGII-SGPQPL
                 Cluster motif 1                Cluster motif 2          P-loop
```

Fig. 8. Comparative partial alignment of the cluster binding domains from archaeal Rieske proteins. Several exhibit a long insertion as compared to the mitochondrial or chloroplast Rieske proteins. The sequences from *P. aerophilum* and *H. salinarum* have an additional long insertion (X_{14} and X_7) between the second cluster binding motif and the proline-loop. SoxL and SoxF are both from *S. acidocaldarius* (see text). Unusual is the presence of an aspartate in cluster binding motif 2 of the SoxL protein. Interestingly, the sequence from *A. acidianus* reveals 70% over-all identity to SoxL from *Sulfolobus*.

encoding the Rieske protein SoxL together with a not yet purified *b*-type cytochrome, SoxN, and a small soluble protein strongly suggests a functional coordination of the respective gene products into a novel complex, tentatively termed the SoxLN complex. Figure 10 illustrates the gene order together with hydrophobicity profiles. Cytochrome SoxN clearly reveals 2 heme binding sites between the putative transmembrane helices II and IV, thus perfectly resembling cytochrome-*b* of known bc_1-complexes. It has a C-terminal extension of about 250 amino acid residues providing two additional transmembrane helices. Very recent studies (Hiller, 2002; Hiller et al., 2003) have shown that all genes are transcribed,

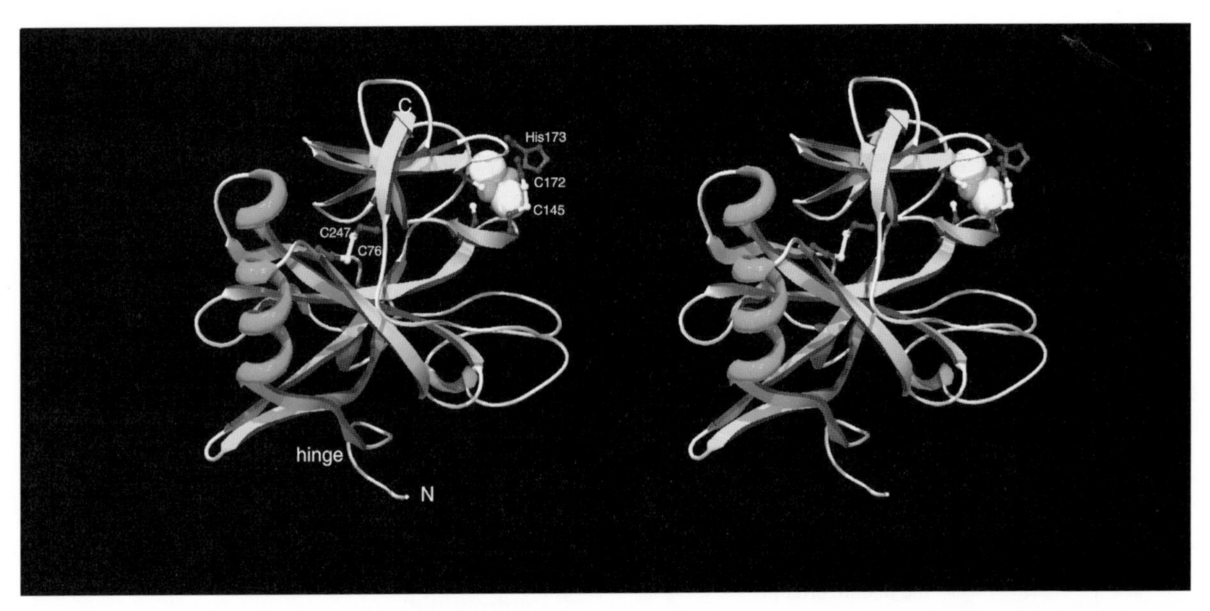

Fig. 9. Stereo representation of the SoxF Rieske protein from *S. acidocaldarius* structure (view with crossed eyes, see also Color Plate 1). The structure was resolved with 1.1 Å. The additional disulfide bridge (C247-76) and the fixation of the C-terminus onto an additional β-sheet are presumably important for the extreme thermostability. For details see text.

partly individually, and partly as multicystronic transcripts.

It is justified therefore to propose the existence of a novel *b*/FeS complex equivalent to classical bc_1-complexes in these thermoacidophilic archaea. In particular it is suggested that cytochrome-$b_{558/562}$ (product of the *cbsA* gene) replaces the function of cytochrome c_1 and possibly even that of cytochrome *c*. The positive redox potential of cytochrome-$b_{558/562}$ is well in line with this assumption. More important, it has been shown by EPR spectroscopy of native *Sulfolobus* membranes that its heme bearing head group can fluctuate between two limit orientations (Schoepp-Cothenet et al., 2001) theoretically enabling the shuffling of electrons between distant donor- and acceptor sites, respectively.

The occurrence of similar complexes in *S. solfataricus and S. tokodaii* is very likely due to the obvious presence of homologous genes (Kawarabayashi, 2001; She et al., 2001). Nevertheless, the nature of the oxidizing electron acceptor for the new complex is unclear at present. Possible candidates are the SoxABCD complex or an additional as yet undetected terminal oxidase. However, no indications for the latter can be extracted from the completed *S. solfataricus* genome for example.

Similarly to *S. acidocaldarius*, a gene cluster has been found in *H. salinarum* (strain R1) forming a transcriptional unit of three genes (Pfeiffer and Oesterhelt, 2001); one encoding for an iron sulfur protein homologous to Rieske proteins and another which encodes a *b*-type cytochrome. Thus, the core elements for a complex analogous to SoxLN are also present in *H. salinarum*. It should be emphasized, however, that no cytochrome *c* homologue has been found in the genome. This does not exclude the presence of heme-C. Actually, membranes from *H. salinarum* (strain JW5) display an absorption maximum at 552 nm in reduced-oxidized difference spectra, indicative of heme-C (Moll et al., 1999). In the same study the occurrence of cytochrome *d* was also suggested which agrees with recent genomic data (Pfeiffer and Oesterhelt, 2001).

Acidianus ambivalens is a special case. Though no respective spectroscopic or protein chemical data from laboratory grown cultures are available, the ongoing genome project revealed both, a gene coding for a Rieske FeS protein with 70% identity to SoxL, as well as a *b*-type cytochrome homologous to SoxC or SoxN, respectively (A. Kletzin, personal communication). Therefore, the functional formation of a SoxLN-like complex appears possible under conditions when these genes are expressed.

In summary, the existence of ancient complex-III precursors consisting of a cytochrome *b*/[2Fe2S]-protein assembly is evident and also indicates that

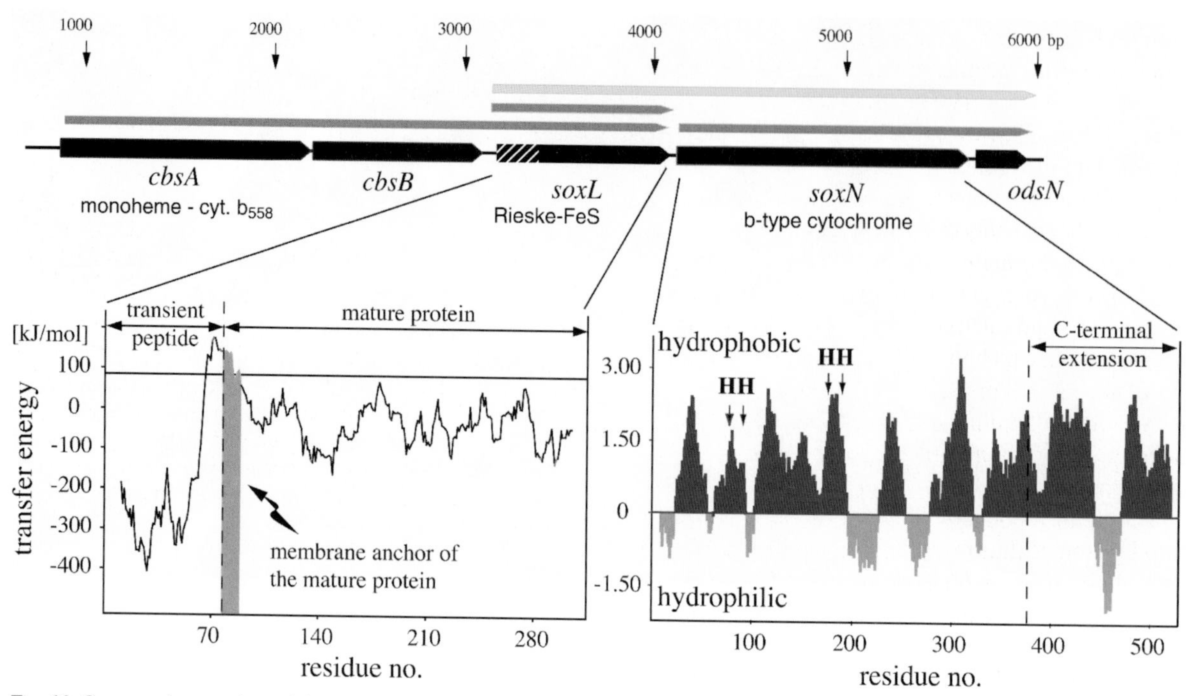

Fig. 10. Genes and transcripts of the SoxLN complex from *S. acidocaldarius*. The narrow arrows indicate transcript lengths; the light grey transcript occurs only in low abundance. The genes are indicated by the bold arrows below. *cbsA* and *cbsB* are encoding cytochrome $b_{558/562}$ with a mono-heme binding site in *cbsA*. *SoxL* encodes the Rieske protein exhibiting a targeting peptide and a membrane anchor, illustrated in the hydropathy profile. *SoxN* encodes an analog to diheme *b*-type cytochromes with a long C-terminal extension as indicated in the hydropathy plot. For functional attribution see text.

archaeal complexes are much less sophisticated than those from higher organelles like mitochondria or chloroplasts.

VII. Respiratory Chains and Supercomplexes

The organization of electron transport components from aerobic Archaea into structured energy transducing redox chains has been resolved for only very few species. In most cases a number of possible constituents has been studied which—based on their redox potentials—can be put into a likely order as for *N. pharaonis* (Scharf et al., 1997). Other constituents like terminal oxidases from extreme halophiles could be isolated in catalytically inactive form (Fujiwara et al., 1989;Denda et al., 1991; Scharf et al., 1997). A recent study characterizes an aa_3-type oxidase from *H. volcanii* (Tanaka et al., 2002) with cytochrome *c* oxidizing activity. Taking into account that an equivalent to respiratory complex-I has not been verified in Archaea it becomes evident that the oxygen reducing redox chains are limited to quinol reoxidizing devices and thus also to a maximum of two energy coupling sites, provided a complex-III equivalent and an ion translocating terminal oxidase are functioning. The use of protons as coupling ions has been shown (Moll and Schäfer, 1988) and is likely to be the general case. It is not known whether also sodium translocating respiratory complexes exist in Archaea as they have been shown for *E. coli* (Avetisyan et al., 1992) or *V. alginolyticus* (Steuber et al., 1997) and *K. pneumoniae* (Krebs et al., 1999) (for reviews see Dimroth, 1997; Skulachev, 1994), for example, though it might be useful tools for alkaliphilic organisms. So far archaeal sodium based bioenergetics are known only in methanogens (reviewed in Schäfer et al., 1999).

Interestingly, the two outmost extremes of respiratory chains have been reported for Archaea; a minimum respiratory system on the one hand, and electron transporting supercomplexes on the other as to be illustrated below.

A. The Minimal System of Acidianus ambivalens

In previous reviews we have already postulated a

minimal respiratory chain for the extremely thermoacidophilic archaeon *Acidianus ambivalens* (Anemüller et al., 1994a; Schäfer, 1996; Schäfer et al., 1996a), however with a number of question marks. The idea relies on spectroscopic data which in plasma membranes of aerobically grown *A. ambivalens* showed only one type of heme-A-containing cytochromes; other cytochromes were not detected. In addition, the chromophore bearing complex, an aa_3-type quinol oxidase, could be isolated and characterized biochemically and genetically (Giuffre et al., 1997; Purschke et al., 1997). The enzyme uses caldariellaquinone as substrate. On the other hand, a caldariellaquinone reducing type-II NADH dehydrogenase has been purified from this organism and could be reconstituted with the terminal oxidase into liposomes thus resembling the minimal respiratory chain of *A. ambivalens* in vitro (Gomes et al., 2001b). Together with the succinate dehydrogenase (Lemos et al., 2001) a respiratory electron flow can be postulated as illustrated in the scheme of Fig. 11. The unique feature of this minimal system is the presence of only one energy coupling site, i.e. the terminal oxidase which was shown to act as a proton pump (Gilderson et al., 2001). It appears conceivable, that additionally 'chemical' (vectorial) protons are contributing to the large proton gradient across the membrane of this extreme acidophile, because the release of protons from reduced quinol might occur on the outside of the plasma membrane during QH_2 oxidation. However, an exact analysis of the proton stoichiometry in whole cells is still an open issue. *A. ambivalens* is an obligate chemolithoautotroph and sulfur oxidizer. Though the oxidation of elemental sulfur to sulfuric acid ($2S^0 + 2H_2O + 3O_2 \rightarrow 2H_2SO_4$) proceeds with a large change of free energy ($\Delta G = -1014$ kJ/mol) the efficiency of the respiratory system providing only one energy conservation site is surprisingly low.

Whereas the flux of reducing equivalents from NADH or succinate to oxygen is the usual case in most organisms, the recent detection of a membrane bound thiosulfate:quinone-reductase in *A. ambivalens* is an important novelty (Kletzin et al., 2003). It provides new insight how oxidative sulfur metabolism is directly linked to oxygen respiration. Of significant interest is the co-purification of the enzyme with subunit-II (SoxA) from the terminal aa_3-type oxidase suggesting the existence of a functional metabolic unit.

The terminal oxidase—which by itself is the full

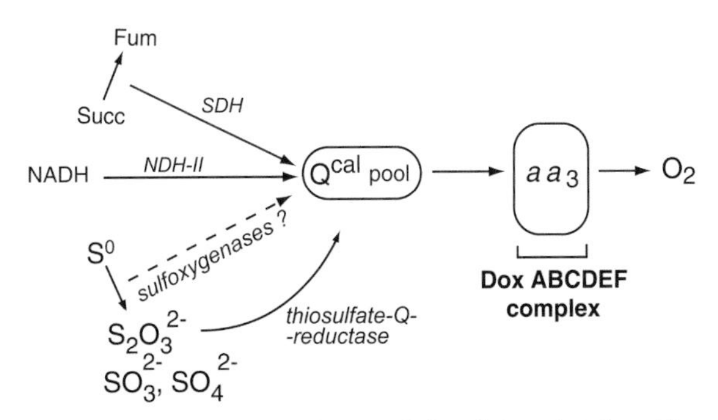

Fig. 11. The minimal respiratory chain of *A. ambivalens*. The only quinol oxidizing complex is DoxABCDEF. It is named after *Desulfurolobus ambivalens* the previous name of *Acidianus ambivalens*. The quinone pool is reduced by NADH and succinate; how electrons from oxidation of elemental sulfur are transferred to the quinone pool is still unknown, except for the thiosulfate: quinone reductase (Kletzin et al., 2003).

respiratory 'chain'—is composed of at least five different polypeptides of 64.9, 38, 20.4, 18.8, and 7.2 kDa. These are encoded within two operons (DoxA and DoxB) (Purschke et al., 1997) which are both present in duplicate on distant loci of the *A. ambivalens* genome. Only the 64 kDa polypeptide (product of *doxB* gene) exhibits clear homology to heme/Cu-oxidases; it represents the heme bearing subunit-I with two heme binding motifs. Polypeptide doxA was classified as pseudosubunit-II with a possible quinone binding site, and polypeptide doxC as a pseudosubunit-III by weak analogies.

The enzyme shows highest activity with caldariellaquinol (2.4 U/nmol heme *a*) and is effectively inhibited by quinolone analogs (~95% at 1 μM) as reported also for *E. coli* quinol oxidase (Meunier et al., 1995). Interestingly, also stigmatellin is an inhibitor but with much weaker activity (~78% at 65 μM). The finding of an essentially non-competitive type of inhibition (Komorowski, 1996) was interpreted as indicative of two different quinol binding sites. It was proposed that a tightly bound quinone may replace the function of Cu_A of regular cytochrome *c* oxidases, serving as the primary electron acceptor and capable of mediating the stepwise transfer of two electrons into the 1-electron accepting heme-*a* site of the enzyme. Independent spectroscopic and kinetic studies came to the same conclusion (Iwasaki and Oshima, 1997) that a tight quinone may represent the fourth redox center of typical terminal oxidases. The same studies report on the midpoint potentials of the heme centers at pH 5.4, and on the affinity versus

CO (K_a=5.7 × 10^4 M^{-1}) which is significantly lower than with mammalian enzymes. The redox potentials of the heme centers are 215 ± 20 mV (heme *a*), and 415 ± 20 mV (heme a_3), in very good agreement with other terminal heme/Cu oxidases.

In contrast to the sulfur oxidizing pathway neither the genetically predicted Rieske FeS protein nor the putative *b*-type cytochrome have been included in the scheme of Fig. 11 because neither one could be detected in laboratory grown cultures so far. This does not imply that under as yet unknown growth conditions *A. ambivalens* might express a more complicated respiratory system.

B. The Supercomplexes of Sulfolobus

Unusual and large respiratory complexes have been first detected in *Sulfolobales* comprising the Sox-ABCD complex and the SoxM complex from *S. acidocaldarius*, as well as the supercomplex described from *S. tokodaii* (formerly *Sulfolobus* sp. strain 7). Whether a previously described terminal oxidase from *S. solfataricus* with cytochrome *c* oxidizing activity (Wakagi et al., 1989) is a fragment also of a supercomplex is not clear; however, by comparing the genome data of *S. solfataricus* (She et al., 2001) this conclusion is not far from being likely.

1. The SoxABCD Complex

Dithionite reduced *S. acidocaldarius* cells exhibit a rather unusual reduced-oxidized difference spectrum in the visible region with the most prominent absorption band at 586–587 nm. This band has first been considered to indicate an a_1-type cytochrome (Anemüller et al., 1985) which could never be isolated individually, however. Instead, it turned out to be difficult to get rid of this heme compound when attempting to purify the putative aa_3-type terminal oxidase displaying a typical absorption at 604 nm. Initially a 'single subunit' aa_3-type oxidase was described (Anemüller and Schäfer, 1989; Anemüller and Schäfer, 1990; Anemüller et al., 1992) capable to oxidize caldariella quinone in vitro with high activity and, as a pseudo-substrate, TMPD. Later, after the respective gene locus had been sequenced (Lübben et al., 1992; Lübben et al., 1994b) this preparation turned out to be a kind of 'over purification,' namely: a fragment of a thus far unseen terminal oxidase supercomplex containing five redox centers, 4 hemes A_S and 1 Cu_B. This complex seems to combine features of respiratory complex-III and complex IV (Lübben et al., 1992). According to the respective genes the complex was assigned as SoxABCD; it is composed of four polypeptides and has very high in vitro activity as a caldariella quinol oxidase. Figure 12 gives a sketch of its composition. Subunit SoxB represents the aa_3-type heme/Cu-oxidase with 12 putative membrane spanning helices; subunit SoxA is a typical homolog to subunit-II of cytochrome *c* oxidases but lacking a Cu_A binding motif. Polypeptide SoxC is by sequence analogy a homolog to *b*-type cytochromes with two heme binding sites in this case hosting 2 hemes A_S. It is responsible for the prominent absorption band at 587 nm. The four heme A_S centers could be also characterized by unique resonance-Raman spectra which differ significantly from those of known aa_3-type cytochrome c oxidases (Gerscher et al., 1996). An unusual feature is the occurrence of a penta-co-ordinated high-spin signature for the a_3 heme.

A compilation of spectral and kinetic data comparing the single entity fragment (SoxB) and the entire SoxABCD complex is given in Table 3. The kinetic data on CO binding and reduction/reoxidation kinetics were verified with intact membranes of *Sulfolobus acidocaldarius* (Giuffrè et al., 1994). As an important observation it has been detected that at least two different pools of cytochromes absorbing at 586–597 nm are present in membranes, as well as a second CO binding site with properties unusual for terminal aa_3-type oxidases.

Both, the single entity fragment (SoxB, or SoxBA) and the intact SoxABCD complex could be reconstituted into liposomes and were shown to generate a proton motive force upon reductant pulses (Driessen, 1994; Gleissner et al., 1997). The mechanism of proton translocation was initially discussed to possibly involve a Q-cycle because a H^+/e^- ratio slightly above one was measured. Recent results on terminal oxidases of the SoxB-type (Gomes et al., 2001a) suggest, however, that also or probably only the SoxB moiety of this complex represents a proton pump.

As to the role of cytochrome *a*587 (SoxC) it appears conceivable to function as a 2-electron acceptor or an intermediate electron storage site according to the scheme:

$$\underset{(QH_2?)}{\text{electron donor}} \xrightarrow[2e]{} \underset{[\,SoxC\,]}{\text{Cyt } a_{587}} \xrightarrow[2x\,1]{} \underset{[\,SoxA/SoxB\,]}{\text{Cyt } a \rightarrow \text{Cyt } a_3/Cu_B} \rightarrow O_2$$

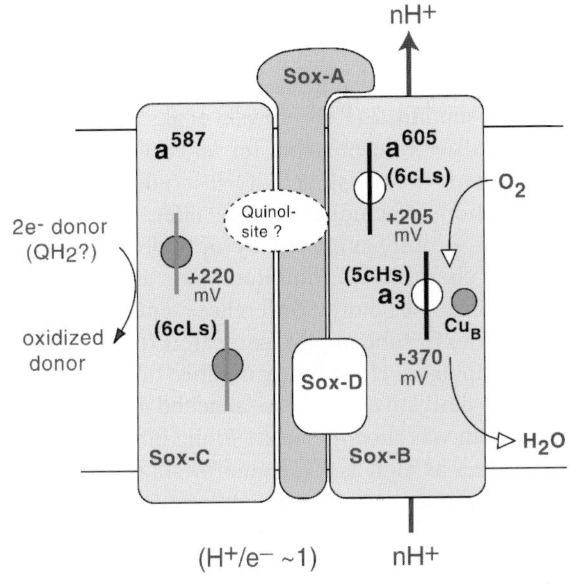

Fig. 12. Schematic representation of the SoxABCD supercomplex of *S. acidocaldarius*. The membrane residing polypeptides are not drawn to scale. For functional details see text. This complex was the first example of a proton pumping heme/Cu-oxidase lacking the D-channel and the Glu[278]. The presence of a native quinol oxidizing site is questionable; for details see text.

Whether or not the SoxABCD complex is really a quinol oxidase in vivo is an open topic to be addressed further below.

2. The SoxM Complex

This alternative terminal oxidase to the SoxABCD complex was first described for *S. acidocaldarius* (Lübben et al., 1994a) and with respect to its constituents has similarities to a respiratory supercomplex described from *S. tokodaii* (formerly *Sulfolobus* sp. strain 7)(Iwasaki et al., 1995a). The initially isolated complex, named SoxM had very little catalytic activity, obviously due to the loss of essential constituents. With the aid of the genomic information (Castresana et al., 1995) and the study of heterologous expressed single constituents (Schäfer et al., 1996b; Komorowski et al., 2001) it was recently possible to isolate and characterize the entire supercomplex in a catalytically fully competent form (Komorowski, 2001; Komorowski et al., 2002). The complex contains a complete respiratory chain between quinol and oxygen. A sketch of its composition and function is illustrated in Fig. 13.

The supercomplex consists of two functional subcomplexes, an Archaeal homologue to bc_1 complexes and a terminal oxidase subcomplex. Both are linked by the blue copper protein Sulfocyanin (see also section III.C). According to a 1:1 stoichiometry of its constituents the complex contains eight metal redox centers with a total of 6 mol Fe and 4 mol Cu per mol of complex. In contrast to the SoxABCD complex SoxM is hosting 2 hemes-*b* and thus is a bb_3-oxidase;

Table 3. Comparison of molecular properties between the isolated aa_3-type terminal oxidase moiety SoxB, and the intact quinol oxidase complex SoxABCD from *Sulfolobus acidocaldarius*. Data from the references quoted in the text.

Parameter	SoxB / single entity oxidase	SoxABCD complex
mol. mass DNA derived (kDa)	57.9	18.8; 57.9; 62.8; 4.5
polypeptide subunits	1 (soxB)	soxA; soxB; soxC; soxD
app. mass on denat. gels	38–40	25 38–40 39 ?
subunit stoichiometry	–	1 : 1 : 1 : 1(?)
cofactors / mol	1 heme a, 1 heme a_3, 1 Cu	3 heme a, 1 heme a_3, 1 Cu,
vis. absorption bamds (red-oxid.) nm	441, 604	440, 587, 605
redox potentials (optical titr.) mV	+208 (a); +365 (a_3)	+220 (a_{587}); + 205 (a); +370 (a_3)
redox potentials (EPR titr.) mV	+200; +370	n.d.
g values LS heme	g_z=3.02; g_y=2.23; g_x=1.45	g_z=2.93; g_y=2.28; g_x=1.55 (Cyt *a*) g_z=2.80; g_y=2.37; g_x=1.72 (Cyt a_{587})
g values HS heme	g_y=6.03; g_x=5.97; g_z=2.00	gy=6.03; - -
electron donor(s)	caldariella quinol, TMPD,	caldariella quinol; (SoxL FeS)
activation energy (kJ/mol/	31.6	n.d.
turnover s^{-1} (70°C)	540	1300
inhibitors	CO, CN^{-1}, N_3^{-1}, S^{2-}	CO, CN^{-1}, N_3^{-1}, S^{2-}
CO binding	$k=2.5 \cdot 10^4\ M^{-1}s^{-1}$	$k=5.5 \cdot 10^4\ M^{-1}s^{-1}$; $K_a=5 \cdot 10^4$

in addition it has a Cu_A center in its subunit-II (SoxH) like typical cytochrome *c* oxidases. Because *c*-type cytochromes are absent from *S. acidocaldarius* an alternate natural reductant has to be used; this is sulfocyanin which is anchored in the membrane by a long hydrophobic N-terminal segment. The exposed peripheral domain bearing the blue copper center is thought to exert pivoting fluctuations between the two subcomplex moieties. Thus, the blue copper protein simultaneously serves as the electron acceptor for the Rieske-FeS/SoxG subcomplex. SoxG is a structural analog to *b*-type cytochromes of bc_1-complexes but in this case is bearing two hemes A_S which are responsible for the prominent absorption at 587 nm in reduced-oxidized difference spectra. The physicochemical properties of all redox centers could be determined either with the intact complex, or in part with its heterologous expressed constituents, and are summarized in Table 4.

Though highest catalytic activities are reported with its natural reductant caldariella quinol the isolated complex can use cytochrome *c* as a pseudosubstrate. This is presumably due to the ability to reduce either sulfocyanin as well as Cu_A in the isolated form of subunit-II (Komorowski et al., 2001, 2002). It may also be responsible for an earlier observed cytochrome *c* oxidase activity in crude membrane preparations(Anemüller and Schäfer, 1990). The complex is fully inhibited by cyanide but not by typical complex-III inhibitors like antimycin-A. Thus, clear structural differences emerge of the Rieske/SoxG subcomplex compared to the known bc_1 complexes. However, a number of aurachin-C and -D derivatives with an attached hydrophobic side chain was shown to exert strong inhibition with pI_{50} values as high as >7 (Komorowski et al., 2002). These inhibitors are quinone analogs which presumably have high affinity to the quinol binding sites of the subcomplex.

Based on the sequence analyses of the constituents and its overall composition the SoxM supercomplex should include two putative proton pumping sites;

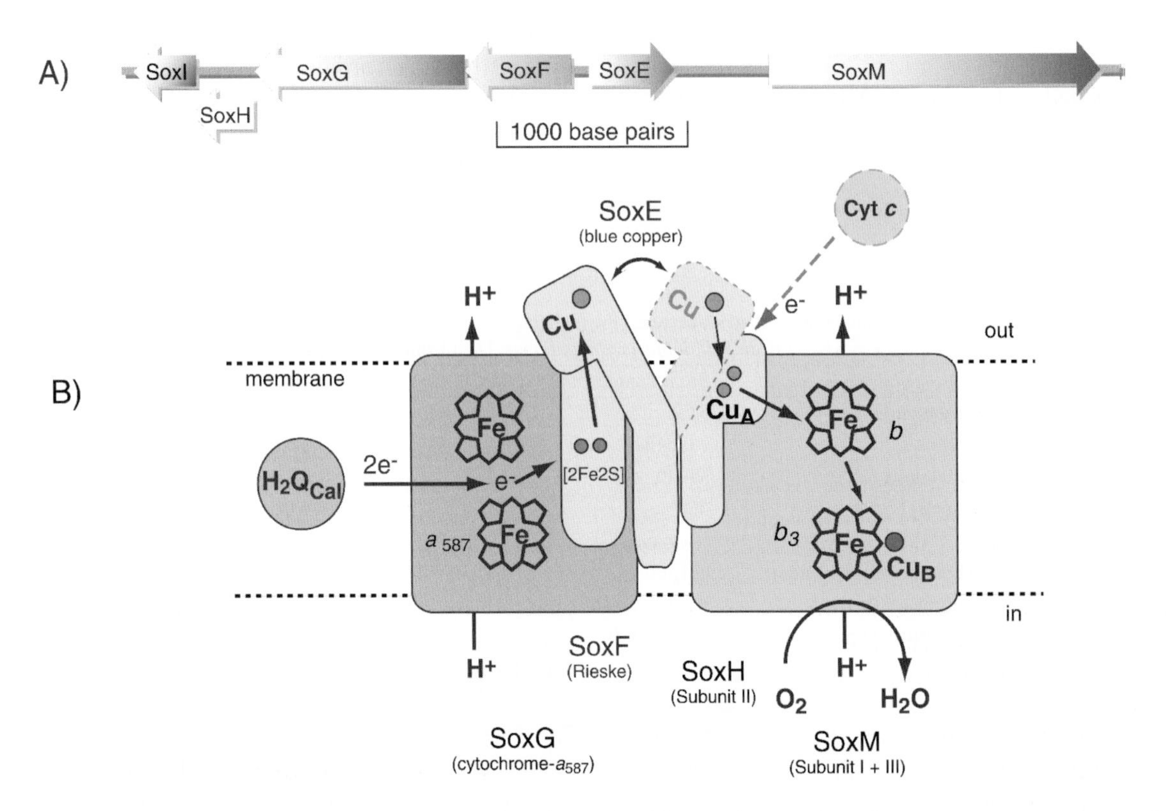

Fig. 13. Schematic representation of the SoxM supercomplex of *S. acidocaldarius*. A) gene order and arrangement on opposite strands of the DNA. B) Functional arrangement of the constituent polypeptides (not drawn to scale) and the putative proton translocating sites. The SoxGFE moiety functions as a bc_1 analogous subcomplex. SoxE as a reorienting intermediate electron carrier is thought to reduce the terminal oxidase subcomplex SoxHM. Cytochrome c can serve in vitro as a pseudosubstrate. For further details see text.

Table 4. Synopsis of physical and biochemical properties of the respiratory supercomplex SoxM from *Sulfolobus acidocaldarius*. Data from Komorowski et al. (2002).

Mol.mass (DNA derived) [kD]:		224.7 (without cofactors)
polypeptide subunits		6 (M, G, E, F, H, I)
subunit mol.mass (DNA) [kD]:		87, 56.7, 21, 26.8, 16.3, 16.8
subunit mol.mass appar. (SDS) [kD]:		45, 39, 30, 27, 19, 14
subunit stoichiometry:		1 : 1 : 1 : 1 : 1 : 1(?)
		M G E F H I
spectral properties:		$\lambda_{max}\alpha = \mathit{561}; \underline{587}; \lambda_{max}\beta = \mathit{517}; \underline{531}$ [nm]
(red.-oxid.)		$\lambda_{max}\gamma = \mathit{436}; \underline{441}$ (sh) [nm]
		(*italic* = Cyt *b*; underlined = Cyt *a*)
CO-diff.-spectra:		$\lambda_{max} = 418; \lambda_{min} = 440; \lambda_{min} = 562$ [nm]
differential absorption		heme *a* ε^{587}_=, 19900 $M^{-1}cm^{-1}$
-coefficients (red.-oxid.)		heme *b* ε^{561} = 15400 $M^{-1}cm^{-1}$
EPR signatures:	hemes:	high spin g = 6.03 5Hs Fe
		low spin g = 3.03 6Ls Fe
	blue-Cu (Sox E):	$g_{\|} = 2.022$; 2.06 A = 116 $10^{-4}cm^{-1}$
	Cu_A (Sox H):	$g_{\|} = 2.25$; 2.03
	FeS (Sox F) :	$g_y = 1.89$ (Rieske 2 Fe2S) (20K)
	Cu_B (SoxM):	EPR-silent
Cofactors:		2 Heme A_s
		2 Heme B (*b* and b_3)
		4 Cu $^{1+;2+}$ (1 Cu^I; 2 Cu_A; 1 Cu_B)
		6 Fe (4 *heme*-Fe; 1 [2Fe-2S])
$E^{6.5}$ [mV]:		heme a^{587} +30/+100; heme bb_3 +200/+350
		FeS +310; Cu^I +300; Cu_A +237
substrate (native):	caldariella quinol:	$K_M = 36\ \mu M$
		V_{max} = 42 (nmol substr.) s^{-1} (nmol ox.)$^{-1}$
pseudo-substrate:	cytochrome *c*:	$K_M = 57\ \mu M$
		V_{max} = 32 (nmol substr.) s^{-1} (nmol ox.)$^{-1}$
pH-optimum:		pH 5
inhibitors:		CN^-; aurachines (pI_{50} = 4.9–7.3);
		HPO_4^{2-} (at pH 7.5 90%); EDTA (cI_{50} = 21mM).

one in the Rieske/SoxG subcomplex, and one in the terminal oxidase subcomplex. Proton pumping in a reconstituted vesicle system remains to be verified, however.

The deficiency of catalytic activity as previously reported can be explained by the unexpected property of subunit-II to dissociate easily from the complex. Two reasons may be involved. One is the relatively short hydrophobic N-terminal anchor of this polypeptide and the presence of a number of polar amino acid residues in this domain; another is an obvious proteolytic cleavage when membranes were prepared at near neutral pH; its loss can be totally avoided by a novel isolation protocol (Komorowski, 2001).

3. Is SoxABCD a Quinol Oxidase in vivo?

Several reasons support the proposal that the quinol oxidizing function of detergent isolated SoxABCD might be an artifactual in vitro activity. Instead, an alternate proposal for the electron transport chain of *S. acidocalarius* is depicted in Fig. 14. According to this novel scenario SoxABCD is the oxidant of the reduced SoxLN-complex (cf. Section VI.B*)* in vivo with cytochrome $b_{558/562}$ serving as a link replacing the function of cytochrome c_1/c of 'classical' respiratory chains, or of sulfocyanin of the SoxM supercomplex. Cytochrome $b_{558/562}$ appears as an ideal candidate due to its positive redox potential and its high stability especially at acidic pH; also its high degree of glycosylation suggests a topological orientation to the outer surface of the membrane like cytochrome *c*.

The reasons supporting this hypothesis are the following: i. Detergent solubilized SoxABCD can accept electrons from totally unusual donors as for example the purified Rieske protein SoxL; ii. at low

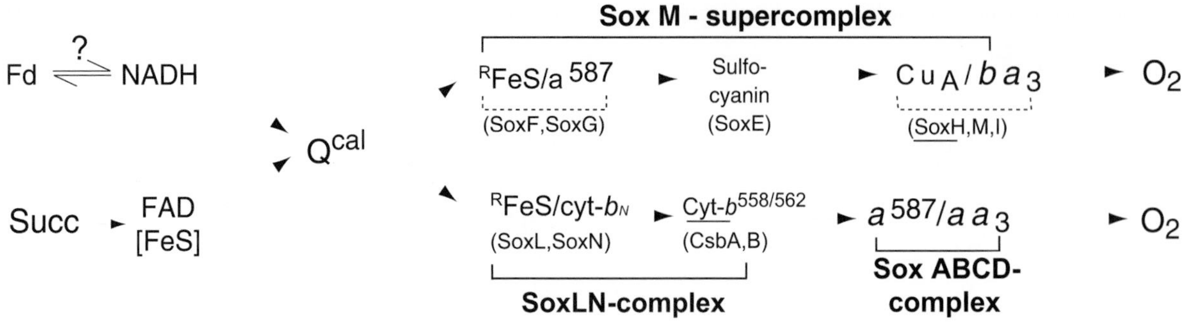

Sox M - supercomplex

Fd ⇌? NADH

^{R}FeS/a 587 ► Sulfo-cyanin ► Cu_A/ba_3 ► O_2

(SoxF,SoxG) (SoxE) (SoxH,M,I)

Q^{cal}

Succ ► FAD [FeS]

^{R}FeS/cyt-b_N ► Cyt-$b^{558/562}$ ► a^{587}/aa_3 ► O_2

(SoxL,SoxN) (CsbA,B)

SoxLN-complex

Sox ABCD-complex

Fig. 14. Revised scheme of the branched respiratory chains of *S. acidocaldarius*. The upper branch is depicting the hypothesis on the organization of electron flow from caldariella quinone through the novel SoxLN complex/$b_{558/562}$ and SoxABCD to oxygen. The lower branch is depicting the function of the SoxM supercomplex as discussed in the text. Underlined components are discussed to exert regulatory functions in electron flow.

pH the quinol oxidizing activity is only marginal (unpublished); iii. under the usual test conditions at near neutral pH caldariella quinol is rather labile (autoxidable) and may unspecifically reduce cytochrome a_{587} of the complex; iv. SoxABCD has a prominent TMPD oxidase activity which is unusual for quinol oxidases; v. SoxABCD is practically insensitive to Q-analogs acting as inhibitors of quinol oxidases; and finally, vi. the fully sequenced *Sulfolobus* genomes do not encode any additional terminal oxidase, besides SoxABCD and the SoxM complex, proposed to serve as an alternate reoxidant for the SoxLN complex. This latter argument is certainly the most convincing one.

VIII. Regulatory Adaptations

The question whether or even how archaeal species with a branched respiratory system respond to environmental changes has not been answered in any of the known cases. Our own attempts to influence the relative abundance of either SoxABCD or the SoxM complex in *S. acidocaldarius* by variations of oxygen supply, failed to yield significant effects. The only obvious difference was found for the amount of Cyt b_{558} which was expressed in highest amounts at limiting oxygen concentrations but also was influenced by the type of carbon source (Hettmann et al., 1998). If that cytochrome operates as a link between the SoxLN complex and its putative oxidizing complex SoxABCD, the adaptation of its abundance might influence the flux of electrons through the respective branch of the respiratory chain.

An interesting novel case is the proposed function of subunit-II (SoxH) of the SoxM supercomplex as a pH sensor and regulator of electron flow to oxygen through this pathway. With the recombinantly produced subunit SoxH a hitherto unusual pH dependence of the UV/Vis- and the EPR-spectra was observed (Komorowski et al., 2001) (shown in the spectra given in Fig. 15). In a pH range from acidic to alkaline, a reversible change of spectral properties occurs; the reversibility of the process (between pH 7.0 to 3.0) is of fundamental importance. Conversely, while at intermediate pHs (4.6–5.8) a visible absorption spectrum is associated with Cu_A-centers, more acidic or alkaline pH conditions generate aberrant spectra indicating variations in the ligand field of the binuclear Cu_A-center. The EPR spectra reflect these changes equally well. The molecular rearrangements producing the spectrum seen at alkaline pH are still unknown. However, at acidic pH the spectrum shows properties of mono-nuclear Cu-centers similar to blue copper proteins as well as hyperfine splitting indicating free copper. Such dramatic rearrangements will certainly result in alterations of the redox properties.

A titration of the pH effects suggests the participation of one of the cluster liganding histidines with a pK_a of 6.4. The pH optimum was found at ~4.8. Interestingly this coincides perfectly with the pH optimum of the activity profile of the SoxM complex (Table 4). It appears conceivable therefore that subunit-II (SoxH) can act as both, a pH sensor and a regulator of activity of the SoxM supercomplex in a physiological pH range between pH 3.0–5.0

The compelling aspect of this hypothesis is that under normal acidic conditions a basic respiratory system, for example SoxABCD, is sufficient to sus-

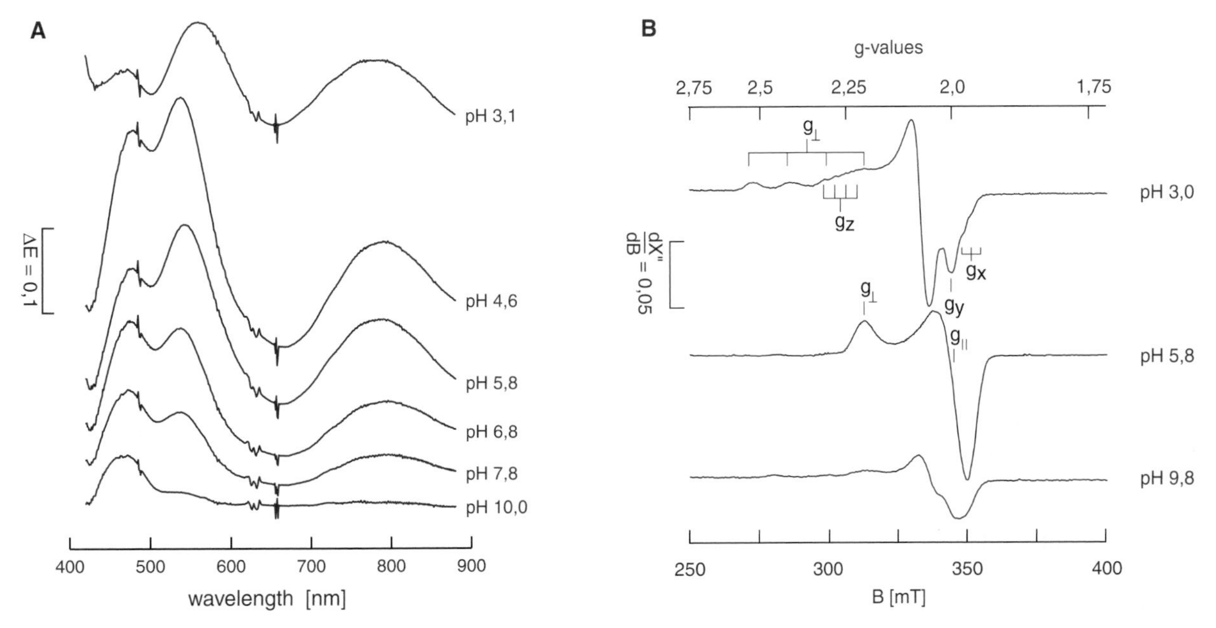

Fig. 15. pH dependence of the binuclear Cu-center in subunit II (SoxH) of the SoxM supercomplex. A) visible spectra, and B) EPR spectra of the recombinant protein taken at a series of different pH conditions. For details see Komorowski et al. (2001) and text.

tain the energy metabolism. If the environmental pH increases above 3.0 the SoxM supercomplex is activated and provides two additional proton pumps supporting a rapid restoration of the optimum ambient pH for *S. acidocaldarius*. At pH above 6.5 none of the respiratory systems is active any longer and, in line with experimental experience, the organism can not survive.

IX. Are All Archaeal Heme/Cu-Oxidases Proton Pumps?

When the first archaeal terminal oxidase gene had been sequenced from *S. acidocaldarius* an obvious contradiction to the established paradigm became evident. That was the absence of functionally essential amino acid residues of subunit-I which bears the heme/Cu center. (Lübben et al., 1992). In all known archaeal sequences this subunit has 12 or more putative transmembrane helices as well as all residues to bind hemes and Cu located in helices II and X (low spin heme site) and VI/VII (heme/Cu center), respectively. From site directed amino acid exchanges in the terminal oxidases from *Rh. sphaeroides* (aa_3 oxidase) and *E. coli* (bo_3-oxidase) the identification of two channels for proton delivery to the binuclear oxygen reduction site and for active proton translocation was concluded; the D- and the K-channel (Holländer et al., 1977; Garcia-Horsman et al., 1995; Hellwig et al., 1998; Pfitzner et al., 1998). The D-channel provides a proton entry site containing a typical N-X_{10}-$\underline{D}$-X_6-N motif connected via tyrosin and serins to an essential glutamate residue E278 (numbering of *P. denitrificans*) close to the binuclear reaction center. The K-channel (starting with K354) leads directly to the binuclear site. The likely involvement of these channels in proton delivery and transduction was strongly supported by the first high resolution structures of cytochrome *c* oxidases from *P. denitrificans* and from bovine heart mitochondria (Fetter et al., 1995; Iwata et al., 1995). Whereas in the novel archaeal oxidases the residues to form the K-channel were conserved at identical or very close positions, the essential glutamate E278 as well as aspartate D124 (*P. denitr.* numbering) were missing. Consequently it was proposed that the respective terminal oxidases do not pump protons but may generate a proton motive potential simply by chemical charge separation (Gleissner et al., 1994; Schäfer et al., 1996a, 1999; Purschke et al., 1997); that occurs by release of substrate protons (from QH_2) at the membrane outer surface while taking up protons for oxygen reduction from the membrane inside. Also alternate mechanisms have been discussed, based on the principles of a Q-cycle catalyzed by the terminal

Q-oxidase complex (Lübben et al., 1992; Schäfer et al., 1996a) because it was evident, that Archaea can acidify the medium by proton extrusion very rapidly upon oxygen pulses (Moll and Schäfer, 1988).

The construction of similarity trees from multiple sequence alignments suggested indeed the evolution of two types of terminal oxidases, those with and the other without the D-channel signatures. Both originated presumably from the duplication of one ancestral gene. A critical aspect emerged, however, from the fact that these branches did not reflect the evolutionary trees based on 16s-rRNA sequences, but rather disclosed a scattered occurrence of both, Archaea and Bacteria, in either group, with or without a D-channel. Moreover, some Archaeal and bacterial organisms were found to possess both types of terminal oxidases. This is exemplified by the partial sequence alignment of helix VI of subunit-I from Archaeal and two bacterial species (Fig. 16) depicting the locus of the 'essential' Glu278.

In addition, it has recently been demonstrated that also those terminal oxidases which lack the essential Glu278 as well as the D-channel [Asp124] can actively translocate protons like that from *Th. thermophilus* ba_3 (Kannt et al., 1998), or from *A. ambivalens* aa_3 with a H^+/e^- stoichiometry of close to 1 (Gomes et al., 2000, 2001a). Therefore the previous paradigm had to be abandoned and one has to assume that all heme/Cu oxidases can actively translocate protons. Obviously a variety of possible proton pathways exists which can be verified only by further high resolution X-ray structures and mutagenesis studies with the respective enzymes (Aagaard et al., 2000).

Taking these results into account also the previous interpretation of a possible Q-cycle (Gleissner et al., 1997) has to be revised and most likely only subunit-I (SoxB) of the SoxABCD oxidase from *S. acidocaldarius* is responsible for proton pumping. Moreover, the determined proton stoichiometry of 1.2 H^+/e^- is closer to that of a typical heme/Cu proton pump ($1H^+/e^-$) than to a Q-cycle ($2H^+/e^-$). This latter scenario supports the above conclusion that in the SoxABCD oxidase SoxC serves as a two-electron acceptor or electron buffer.

A detailed analysis of the functional subunits I and II of heme/Cu terminal oxidases including 26 amino acid sequences has been recently compiled (Pereira et al., 2001). Interestingly, 12 out of 26 sequences reveal a lack of the D-channel, and only 8 sequences contain the very Glu278 in a –XGHP*E*V– motif previously assigned as indispensable

```
                                transmembrane helix VI
A.ambiv.    aa3       ANQIAFWIFGHAVVYMAWLPAV
N.phara.    ba3       LTRTLFWYFGHAVVYFWLMPAY
S.acido.    aa3       LWAILFWFYGHPVVYYVPFPLF
T.therm.    ba3       VARTLFWWTGHPIVYFWLLPAY
Ae.pern.     ox2      LARTLFWWFGHPLVYFWLLPAV

S.acido.    ba3       LWQQLFWFFGHPEVYILILPAM
P.aeroph.   aa3       LFQHIFWFFGHPEVIILVLPAM
H.salin.    aa3       FWQHLFWFFGHPEVYVLVLPPM
T.therm.   caa3       LFQQFFWFYSHPEVYVMLLPYL
Ae.pern.    ox1       LWDHLFWFFGHPEVYILLFPAL
```

Fig. 16. Partial alignment of the essential helix VI motif of terminal oxidases. The top group combines various Archaea and *T.therm.* which are lacking the essential glutamate in the D-channel. The bottom group shows alternate oxidases of Archaea and *T. thermophilum* which have this residue as it is also conserved in cytochrome oxidases of mitochondria or purple bacteria ox1 and ox2 refer to the two terminal oxidases of *Aerophilum pernix* (Kawarabayashi, 1999). The other organisms are *Acidianus ambivalens, Natronomonas pharaonis, Sulfolobus acidocaldarius, Halobacterium salinarum*, and *Pyrobaculum aerophilum*. For detailed discussion see text.

for proton pumping. For 7 sequences an alternate –YSHPXV– motif was suggested to functionally replace the former. Finally, 11 remaining sequences contain neither one of these motifs. A protein based unrooted phylogenetic tree supports the proposal of three evolutionary groups: group A having both, the D- and K-channels; group B having an alternate K-channel but no D-channel; a novel group C with no D-channel and only a rudimentary K-channel and also the Tyr280 (*P. denitrificans* numbering) missing which in all other subunits-I is covalently linked to the Cu liganding histidine. Actually, Archaea and Bacteria appear again scattered between the groups A and B. In summary this analysis underlines the discrepancy between RNA based phylogeny and phylogenetic trees based on amino acid sequences. One should remember that the latter do not represent real evolutionary pathways but simply reflect distance matrices of sequence similarities.

X. Conclusions and Perspectives

In recent years, it became evident that the core structures of archaeal respiratory electron transport systems are found even in the lowest branches of the phyolgenetic tree. An immediate conclusion would suggest that in essence these as well as the chemiosmotic principle of energy conservation were present already in progenotic precursors prior to the split into the three kingdoms of organisms and prior to the origin of oxygenic photosynthesis (see also Chapter 1 by Castresana, Vol. 1). This issue is still matter of debate and other hypotheses assume an acquisition of heme/Cu oxidases by lateral gene transfer from Bacteria to Archaea (Pereira et al., 2001). But aerobiosis may also be seen as an adaptation simply by changes of specificities with respect to electron donors and –acceptors. Strong evidence comes from the relation between hydrogenases and elements of NADH dehydrogenases (Albracht et al., 1997; Friedrich, 2001; Kerscher et al., 2001), or of the family of heme/Cu oxidoreductases including both, NO reductases and oxygen reductases (Hendriks et al., 1998a; Hendriks et al., 1998b; Van der Oost et al., 1994). Similarly the succinate dehydrogenases and fumarate reductases are members of a family with common origin as are the components of Rieske-FeS/cytochrome complexes acting as quinol oxidoreductases and redox driven proton pumps (Schmidt and Shaw, 2001). A comparison of phylogenetic trees even suggests a coevolution of respiratory *b*-cytochromes and Rieske FeS-proteins.

The investigation of Archaea brought also into light that presumably all terminal heme/Cu oxidases are acting as proton pumps, thus raising the question for other proton conducting pathways than those suggested from previous paradigms. Another surprising result was the discovery of integrated supercomplexes in respiratory systems, which might be an adaptation to high temperatures or to the specific membrane architecture of hyperthermophilic Archaea. However, a large number of problems regarding structure and function of the archaeal electron transport modules remains to be solved in future studies.

One of these is the function of genes whose products have never been chemically proved. Are these genes encoding putative respiratory electron transport modules, actually transcribed? If so, under which conditions? Furthermore, the occurrence of multiple copies of genes like those of the terminal oxidase from *A. ambivalens* (Purschke et al., 1997) leads to the question of their regulation. Also unknown of *A. ambivalens* and other sulfur oxidizers is the pathway how the six electrons are delivered to the respiratory chain during oxidation of S^0 to H_2SO_4.

An ambitious goal will be the structural resolution of the supercomplexes. To come up with this, a massive production of homogeneous preparations is required; this point is however hampered by a number of difficulties, namely: a) dissociation of the complexes or loss of single constituents upon solubilization of plasma membranes; b) the equilibration of ionic strength and pH upon cell disruption; conversely, under native conditions these complexes, i.e. from thermoacidophiles, are exposed to dramatically asymmetric environments on opposite membrane surfaces ($\Delta pH \geq 4$); c) the glycosylation of surface exposed components may cause a non-homogeneity and also exclude the heterologous expression in bacterial systems along with a correct assembly of the supercomplexes. Finally, also the specific archaeal lipid environment is a critical parameter influencing structure and function of archaeal respiratory complexes.

Functional and regulatory properties of single electron transport pathways from organisms with a branched respiratory system could be best investigated in specially designed deletion mutants or overproducers. Therefore the development of an efficient genetic manipulation systems is a prerequisite to further functional and structural studies. Indeed, whereas the genetic manipulation of *Haloferax* has been achieved, all attempts failed so far to establish a stable genetic system for thermoacidophilic *Sulfolobales*. Neither, useful shuttle vectors replicating in both, *E. coli* and *Sulfolobus*, nor reliable selection markers are presently available. Successful genetic manipulation of extreme thermoacidophilic Archaea would be a meritorious contribution for the elucidation of the basic molecular mechanisms of respiration in prokaryotes.

Acknowledgments

I am grateful to Dr. S. Anemüller and Dr. C.L. Schmidt for critical advice, fruitful discussions and careful inspection of the manuscript, as well as for providing some of their data prior to publication. I am indebted also to Dr. F. Pfeiffer and Dr. D. Oesterhelt (Max Planck Institute, Martinsried) for giving access to unpublished genomic data from *H. salinarum*; I am grateful to Dr. A. Kletzin (Institute of Microbiology, TU Damstadt) for providing unpublished data on sulfur metabolism of *A. ambivalens* and also for insight into the unpublished *A. acidianus* genome project. I thank Dr. P. Palm (Max Planck Institute, Martinsried) for providing the rhodopsin and carotenoid deficient *H. salinarum* strain JW5 for our experiments.

References

Aagaard A, Gilderson G, Mills DA, Ferguson-Miller S and Brzezinski P (2000) Redesign of the proton-pumping machinery of cytochrome *c* oxidase: Proton pumping does not require Glu(I-286). Biochemistry 39: 15847–15850

Albracht SPJ and Hedderich R (2000) Learning from hydrogenases: Location of a proton pump and of a second FMN in bovine NADH-ubiquinone oxidoreductase (Complex I). FEBS Lett 485: 1–6

Albracht SPJ, Mariette A and De Jong P (1997) Bovine-heart NADH:ubiquinone oxidoreductase is a monomer with 8 Fe-S clusters and 2 FMN groups. Biochim Biophys Acta1318: 92–106

Anemüller S and Schäfer G (1989) Cytochrome aa_3 from the thermoacidophilic archaebacterium *Sulfolobus acidocaldarius*. FEBS Lett 244: 451–455

Anemüller S and Schäfer G (1990) Cytochrome aa3 from *Sulfolobus acidocaldarius*. Eur J Biochem, 191: 297–3–05

Anemüller S, Lübben M and Schäfer G (1985) The respiratory system of *Sulfolobus acidocaldarius*, a thermoacidophilic archaebacterium. FEBS Lett, 193:83–87

Anemüller S, Bill E, Schäfer G, Trautwein AX and Teixeira M (1992) EPR studies of cytochrome aa_3 from *Sulfolobus acidocaldarius*—Evidence for a binuclear center in archaebacterial terminal oxidase. Eur J Biochem 210: 133–138

Anemüller S, Schmidt CL, Schäfer G, and Teixeira M (1993) Evidence for a Rieske-type FeS center in the thermoacidophilic archaebacterium *Sulfolobus acidocaldarius*. FEBS Lett 318: 61–64

Anemüller S, Schmidt CL, Pacheco I, Schäfer G and Teixeira M (1994a) A cytochrome *aa*3-type quinol oxidase from *Desulfurolobus ambivalens*, the most acidophilic archaeon. FEMS Microbiol Lett 117: 275–280

Anemüller S, Schmidt CL, Schäfer G, Bill E, Trautwein AX and Teixeira M (1994b) Evidence for a two proton dependent redox equilibrium in an Archaeal Rieske iron-sulfur cluster. Biochem Biophys Res Commun 202: 252–257

Anemüller S, Hettmann Th, Moll R, Teixeira M and Schäfer G (1995) EPR characterization of an Archaeal succinate dehydrogenase in the membrane bound state. Eur J Biochem 232:563–568

Aono S, Bryant FO and Adams MWW (1989) A novel and remarkably thermostable ferredoxin from the hyperthermophilic archaebacterium *Pyrococcus furiosus*. J Bacteriol 171: 3433–3439

Avetisyan AV, Bogachev AV, Murtasina RA and Skulachev VP (1992) Involvement of a *d*-type oxidase in the Na^+-motive respiratory chain of *Escherichia coli* growing under low $\Delta\mu_{H^+}$ conditions. FEBS Lett 306:, 199–202

Barns SM, Fundyga RE, Jeffries MW and Pace NR (1994) Remarkable Archaeal diversity detected in a Yellowstone National Park hot spring environment. Proc Natl Acad Sci USA 91: 609–1613

Barns SM, Delwiche CF, Palmer JD and Pace NR (1996) Perspectives on Archaeal diversity, thermophily and monophyly from environmental rRNA sequences. Proc Natl Acad Sci USA 93: 9188–9193

Berry EA, Huang LS, Zhang ZL and Kim SH (1999) Structure of the avian mitochondrial cytochrome bc_1 complex. J Bioenerg Biomembr 31: 177–190

Blöchl E, Burggraf S, Fiala G, Lauerer G, Huber G, Huber R, Rachel R, Segerer A, Stetter KO and Völkl P (1996) Isolation, taxonomy and phylogeny of hyperthermophilic microorganisms. World J Microbiol Biotech 11: 9–16

Bock A-K, Kunow J, Glasemacher J and Schönheit P (1996) Catalytic properties, molecular composition and sequence alignments of pyruvate:ferredoxin oxidoreductase from the methanogenic archaeon *Methanosarcina barkeri* (strain Fusaro). Eur J Biochem 237: 35–44

Boekema EJ, Ubbink-Kok T, Lolkema JS, Brisson A and Konings WN (1997) Visualization of a peripheral stalk in V-type ATPase: Evidence for the stator structure essential to rotational catalysis. Proc Natl Acad Sci USA 94: 14291–14293

Bönisch H, Schmidt CL, Schäfer G and Ladenstein R (2000) Crystallization and preliminary crystallographic analysis of Rieske iron-sulfur protein II (SoxF) from *Sulfolobus acidocaldarius*. Acta Cryst D-56: 643–644

Bönisch H, Schmidt CL, Schäfer G and Ladenstein R (2002) The structure of the soluble domain of the Archaeal Rieske protein SoxF from *Sulfolobus acidocaldarius* at 1.11 Å resolution. J Mol Biol 319: 791–785

Boone DR, Whitman WB and Rouviere P (1993) Diversity and taxonomy of methanogens. In: Ferry JG (ed) Methanogenesis, pp 33–80. Chapman and Hall, London

Breton JL, Duff JLC, Butt JN, Armstrong FA, George SJ, Pétillot Y, Forest E, Schäfer G and Thomson AJ (1995) Identification of the iron-sulfur clusters in a ferredoxin from the archaeon *Sulfolobus acidocaldarius*—Evidence for a reduced [3Fe-4S] cluster with pH-dependent electronic properties. Eur J Biochem 233: 937–946

Brischwein M, Scharf B, Engelhard M and Mäntele W (1993) Analysis of the redox reaction of an archaebacterial copper protein, halocyanin, by electrochemistry and FTIR difference spectroscopy. Biochemistry 32: 13710–13717

Castresana J and Saraste M (1995) Evolution of energetic metabolism: The respiration-early hypothesis. Trends Biochem Sci 20: 443–448

Castresana J, Lübben M and Saraste M (1995) New archaebacterial genes coding for redox proteins: Implications for the evolution of aerobic metabolism. J Mol Biol 250: 202–210

Collins MD (1985) Analysis of isoprenoid quinones. In: Gottschalk G (ed) Methods in Microbiology, pp 329–366. Academy Press, New York

Collins MD and Langworthy TA (1983) Respiratory quinone composition of some acidophilic bacteria. Syst Appl Microbiol 4: 295–304

Danson MJ (1993) Central metabolism of the Archaea. In: Kates M, Kushner DJ, Matheson AT (eds) The Biochemistry of Archaea, pp 1–24. Elsevier, Amsterdam

Danson MJ and Hough DW (1998) Structure, function and stability of enzymes from the Archaea. Trends Microbiol 6: 307–314

Darrouzet E, Moser CC, Dutton PL and Daldal F (2001) Large scale domain movement in cytochrome bc_1: A new device for electron transfer in proteins. TIBS 26: 445–451

De Rosa M, De Rosa S, Gambacorta A, Minale L, Thomson RH and Worthington RD (1977) Caldariella quinone, a unique benzo-b-thiophen-4,7-quinone from *Caldariella acidophila*, an extremely thermophilic and acidophilic bacterium. J Chem Soc Perkin Trans I: 653–657

Denda K, Fujiwara T, Seki M, Yoshida M, Fukumori Y and Yamanaka T (1991) Molecular cloning of the cytochrome aa_3 gene from the archaeon (Archaebacterium) *Halobacterium halobium*. Biochem Biophys Res Commun 181: 316–322

Dimroth P (1997) Primary sodium ion translocating enzymes. Biochim Biophys Acta 1318: 11–51

Doolittle WF (1996) At the core of the Archaea. Proc Natl Acad Sci USA 93: 8797–8799

Driessen AJM (1994) How proteins cross the bacterial cytoplasmic membrane. J Membr Biol 142: 145–159

Fetter JR, Qian J, Shapleigh J, Thomas JW, Garcia-Horsman A, Schmidt E, Hosler JP, Babcock GT, Gennis RB and Ferguson-Miller S (1995) Possible proton relay pathways in cytochrome *c* oxidase. Proc Natl Acad Sci USA 92: 1604–1608

Fitz-Gibbon S, Choi AJ, Miller JH, Stetter KO, Simon MI, Swanson R and Kim U (1997) A fosmid-based genomic map and identification of 474 genes of the hyperthermophilic archaeon *Pyrobaculum aerophilum*. Extremophiles 1: 36–51

Fitz-Gibbon S, Ladner H, Kim U, Stetter KO, Simon MI and Miller JH (2002) Genome sequence of the hyperthermophilic crenarchaeon *Pyrobaculum aerophilum*. Proc Natl Acad Sci USA 99: 984–989

Friedrich T (1998) The NADH:ubiquinone oxidoreductase (complex I) from *Escherichia coli*. Biochim Biophys Acta 1364: 134–146

Friedrich T (2001) Complex I: A chimaera of a redox and conformation-driven proton pump? J Bioener Biomembr 33: 169–177

Fujii T, Hata Y, Wakagi T, Tanaka N and Oshima T (1996) Novel zinc-binding centre in thermoacidophilic Archaeal ferredoxins. Nature Struct Biol 3: 834–837

Fujiwara T, Fukumori Y and Yamanaka T (1989) Purification and properties of *Halobacterium halobium* 'cytochrome aa_3' which lacks Cu_A and Cu_B. Journal of Biochemistry 105: 287–292

Garcia-Horsman JA, Puustinen A, Gennis RB and Wikström M (1995) Proton transfer in cytochrome bo_3 ubiquinol oxidase of *Escherichia coli*: Second-site mutations in subunit I that restore proton pumping in the mutant Asp135 → Asn. Biochemistry 34: 4428–4433

George GN, Prince RC, Mukund S and Adams MWW (1992) Aldehyde Ferredoxin Oxidoreductase from the hyperthermophilic archaebacterium *Pyrococcus furiosus* contains a tungsten oxothiolate center. J Am Chem Soc 114: 3521–3523

Gerscher S, Döpner S, Hildebrandt P, Gleissner M and Schäfer G (1996) Resonance Raman spectroscopy of the integral quinol oxidase complex of *Sulfolobus acidocaldarius*. Biochemistry 35: 12796–12803

Gilderson G, Aagaard A, Gomes CM, Ädelroth P, Teixeira M and Brzezinski P (2001) Kinetics of electron and proton transfer during O_2 reduction in cytochrome aa_3 from *A-ambivalens*: An enzyme lacking Glu(I-286). Biochim Biophys Acta 1503: 261–270

Giuffrè A, Antonini G, Brunori M, D'Itri E, Malatesta F, Nicoletti F, Anemüller S, Gleissner M and Schäfer G (1994) *Sulfolobus acidocaldarius* terminal oxidase. A kinetic investigation and its structural interpretation. J Biol Chem 269: 31006–31011

Giuffrè A, Gomes C, Antonini G, D'Itri E, Teixeira M and Brunori M (1997) Functional properties of the quinol oxidase from *Acidianus ambivalens* and the possible role of its electron donor: Studies in the membrane-integrated and purified enzyme. Eur J Biochem 250: 383–388

Gleissner M, Elferink MGL, Driessen AJM, Konings WN, Anemüller S and Schäfer G (1994) Generation of proton-motive force by an Archaeal terminal quinol oxidase from *Sulfolobus acidocaldarius*. Eur J Biochem 224: 983–990

Gleissner M, Kaiser U, Antonopuolos E and Schäfer G (1997) The Archaeal SoxABCD-complex is a proton pump in *Sulfolobus acidocaldarius*. J Biol Chem 272: 8417–8426

Gomes CM and Teixeira M (1998) The NADH oxidase from the thermoacidophilic Archaea *Acidianus ambivalens*: Isolation and physicochemical characterisation. Biochem Biophys Res Commun 243: 412–415

Gomes CM, Huber H, Stetter KO and Teixeira M (1998) Evidence for a novel type of iron cluster in the respiratory chain of the archaeon *Sulfolobus metallicus*. FEBS Lett 432: 99–102

Gomes CM, Lemos RS, Teixeira M, Kletzin A, Huber H, Stetter KO, Schäfer G and Anemüller S (1999) The unusual iron sulfur composition of the *Acidianus ambivalens* succinate dehydrogenase complex. Biochim Biophys Acta 1411: 134–141

Gomes CM, Verkhovskaya ML, Teixeira M, Wikström M and Verkhovsky MI (2000) Canonical proton pumping by an enzyme that lacks both D- and K-channels. EBEC Short Reports 11: 247

Gomes CM, Backgren C, Teixeira M, Puustinen A, Vervkhovskaya ML, Wikström M and Verkhovsky MI (2001a) Heme-copper oxidases with modified D- and K-pathways are yet efficient proton pumps. FEBS Lett 497: 159–164

Gomes CM, Bandeiras TM and Teixeira M (2001b) A new type-II NADH dehydrogenase from the archaeon *Acidianus ambivalens*: Characterization and in vitro reconstitution of the respiratory chain. J Bioenerg Biomembr 33: 1–8

Gomes C, Faria A, Carita JC, Mendes J, Regalla M, Chicau P, Huber R, Stetter KO and Teixeira M (2002) Dicluster seven-iron ferredoxins from hyperthermophilic *Sulfolobales*. J Bio-Inorg Chem 3: 499–507

Gutierrez-Cirlos EB and Trumpower BL (2002) Inhibitory analogs of ubiquinol act anti-cooperatively on the yeast cytochrome bc_1 complex—Evidence for an alternating, half-of-the-sites mechanism of ubiquinol oxidation. J Biol Chem 277: 1195–1202

Hallberg Gradin C and Baltscheffsky H (1981) Solubilization and separation of two *b*-type cytochromes from a carotenoid mutant of *Halobacterium halobium*. FEBS Lett 125: 201–204

Hallberg Gradin C and Colmsjö A (1989) Four different b-type cytochromes in the halophilic archaebacterium, *Halobacterium halobium*. Biochim Biophys Acta 272: 130–136

Hallberg Gradin C and Hederstedt L (1981) Succinate dehydrogenase activity and the succinate reducible cytochrome in *Halobacterium halobiun*. Acta Chemica Scandin B 35: 6601–6605

Hellwig P, Behr J, Ostermeier C, Richter OMH, Pfitzner U, Odenwald A, Ludwig B, Michel H and Mäntele W (1998) Involvement of glutamic acid 278 in the redox reaction of the cytochrome *c* oxidase from *Paracoccus denitrificans* investigated by FTIR spectroscopy. Biochemistry 37: 7390–7399

Hendriks J, Gohlke U and Saraste M (1998a) From NO to OO: Nitric oxide and dioxygen in bacterial respiration. J Bioenerg Biomembr 30: 15–24

Hendriks J, Warne A, Gohlke U, Haltia T, Ludovici C, Lübben M and Saraste M (1998b) The active site of the bacterial nitric oxide reductase is a dinuclear iron center. Biochemistry 37: 13102–13109

Henninger T, Anemüller S, Fitz-Gibbon S, Miller JH, Schäfer G and Schmidt CL (1999) A novel Rieske iron-sulfur protein from the hyperthermophilic crenarchaeon *Pyrobaculum aerophilum*: Sequencing of the gene, expression in *E. coli* and characterization of the protein. J Bioenerg Biomembr 31: 119–128

Hershberger KL, Barns SM, Reysenbach A-L, Dawson SC and Pace NR (1996) Wide diversity of crenarchaeota. Nature 384: 420

Hettmann Th, Schmidt CL, Anemüller S, Zähringer U, Moll H, Petersen A and Schäfer G (1998) Cytochrome *b*-558/566 from the archaeon *Sulfolobus acidocaldarius*: A novel, highly glycosylated, membrane bound B-type hemoprotein. J Biol Chem 273: 12032–12040

Hildebrandt P, Matysik J, Schrader B, Scharf B and Engelhard M (1994) Raman spectroscopic study of the blue copper protein halocyanin from *Natronobacterium pharaonis*. Biochemistry 33: 11426–11431

Hiller A (2002) The bc1-equivalent SoxLN complex of *Sulfolobus acidocaldarius*. Thesis/Dissertation, University of Luebeck, Germany

Hiller A, Henninger T, Schafer G, Schmidt CL (2003) New genes encoding subunits of a cytochrome bc_1-analogous complex in the respiratory chain of the hyperthermoacidophilic crenarchaeon *Sulfolobus acidocaldarius*. J Bioeng Biomembr 35: 121–131

Holländer R, Wolf G and Mannheim W (1977) Lipoquinones of some bacteria and mycoplasmas, with considerations on their functional significance. Ant v Leeuwenhoek 43: 177–185

Hunte C (2001) Insights from the structure of the yeast cytochrome bc_1 complex: Crystallization of membrane proteins with antibody fragments. FEBS Lett 504: 126–132

Iwasaki T and Oshima T (1996) Role of cytochrome b563 in the Archaeal aerobic respiratory chain of *Sulfolobus* sp. strain7. FEMS Microbiol Lett 144: 259–266

Iwasaki T and Oshima T (1997) A stable intermediate product of the Archaeal zinc-containing 7Fe ferredoxin from Sulfolobus sp. strain 7 by artificial oxidative conversion. FEBS Lett 417: 223–226

Iwasaki T and Oshima T (2001) Ferredoxin and related enzymes from *Sulfolobus*. Methods in Enzymology 334: 3–22

Iwasaki T, Wakagi T and Oshima T (1993) The ferredoxin dependent redox system of the thermoacidophilic archaeon *Sulfolobus* sp. strain7. International Workshop on Molecular Biology and Biotechnology of Archaebacteria, pp 69–70. RIKEN, Wakao, Japan

Iwasaki T, Matsuura K and Oshima T (1995a) Resolution of the aerobic respiratory system of the thermoacidophilic archaeon, *Sulfolobus* sp strain 7 .1. The Archaeal, terminal oxidase supercomplex is a functional fusion of respiratory complexes III and IV with no c-type cytochromes. J Biol Chem 270: 30881–30892

Iwasaki T, Wakagi T and Oshima T (1995b) Resolution of the aerobic respiratory system of the thermoacidophilic archaeon, *Sulfolobus* sp strain 7 .3. The Archaeal novel respiratory complex II (succinate:caldariellaquinone oxidoreductase complex) inherently lacks heme group. J Biol Chem 270: 30902–30908

Iwasaki T, Suzuki T, Kon T, Imai T, Urushiyama A, Ohmori D and Oshima T (1997) Novel zinc-containing ferredoxin family in thermoacidophilic Archaea. J Biol Chem 272: 3453–3458

Iwasaki T, Kuonosu A, Aoshima M, Ohmori D, Imai T, Urushiyama A, Cosper NJ and Scott RA (2002) Novel [2Fe2S]-type redox center C in SdhC of archaeal respiratory complex II from *Sulfolobus tokodaii* strain 7. J Biol Chem 277: 39642–39648

Iwata S, Ostermeier C, Ludwig B and Michel H (1995) Structure at 2.8 Å resolution of cytochrome *c* oxidase from *Paracoccus denitrificans*. Nature 376: 660–669

Iwata S, Saynovits M, Link TA and Michel H (1996) Structure of a water soluble fragment of the Rieske iron-sulfur protein of bovine heart mitochondrial cytochrome bc_1 complex determined by MAD phasing at 1,5 Ångstrom resolution. Structure 4: 567–579

Iwata M, Björkman J and Iwata S (1999) Conformational change of the Rieske [2Fe-2S] protein in cytochrome bc_1 complex. J Bioenerg Biomembr 31: 169–175

Janssen S, Schäfer G, Anemüller S and Moll R (1997) A succinate dehydrogenase with novel structure and properties from the hyperthermophilic archaeon *Sulfolobus acidocaldarius*: Genetic and biophysical characterization. J Bacteriol 179: 5560–5569

Janssen S, Trincao J, Teixeira M, Schäfer G and Anemüller S (2001) Ferredoxins from the archaeon *Acidianus ambivalens*: Overexpression and characterization of the non-zinc-containing ferredoxin FdB. Biol Chemistry 382: 1501–1507

Kannt A, Soulimane T, Buse G, Becker A, Bamberg E and Michel H (1998) Electrical current generation and proton pumping catalyzed by the ba_3-type cytochrome *c* oxidase from *Thermus thermophilus*. FEBS Lett 434: 17–22

Kawarabayashi Y, Hino Y, Horikawa H, Yamazaki S, Haikawa Y, Jin-no K, Takahashi M, Sekine M, Baba S-I, Ankai A, Kosugi H, Hosoyama A, Fukui S, Nagai Y, Nishijima K, Nakazawa H, Takamiya M, Masuda S, Funahashi T, Tanaka T, Kudoh Y, Yamazaki J, Kushida N, Oguchi A, Aoki K-I, Kubota K, Nakamura Y, Nomura N, Sako Y and Kikuchi H (1999) Complete genome sequence of an aerobic hyperthermophilic crenarchaeon, *Aeropyrum pernix* K1. DNA Res 6: 83–101, 145–152

Kawarabayashi Y, Hino Y, Horikawa H, Jin-no K, Takahashi M, Sekine M, Baba S-I, Ankai A, Kosugi H, Hosoyama A, Fukui S, Nagai Y, Nishijima K, Otsuka R, Nakazawa H, Takamiya M, Kato Y, Yoshizawa T, Tanaka T, Kudoh Y, Yamazaki Y, Kushida N, Oguchi A, Aoki K, Masuda S, Yanagii M, Nishimura M, Yamagishi A, Oshima T and Kikuchi H (2001) Complete genome sequence of an aerobic thermophilic crenarchaeon *Sulfolobus tokodaii,* strain 7. DNA Res 8: 123–140

Kerscher L, Nowitzki S and Oesterhelt D (1982) Thermoacidophilic Archaebacteria contain bacterial-type ferredoxins acting as electron acceptors of 2-oxoacid:ferredoxin oxidoreductases. Eur J Biochem 128: 223–230

Kerscher S, Kashani-Poor N, Zwicker K, Zickermann V and Brandt U (2001) Exploring the catalytic core of complex I by *Yarrowia lipolytica* yeast genetics. J Bioener Biomembr 33: 187–196

Kletzin A, Urich T, Müller F, Bandeiras T and Gomes CM (2004) Dissimilatory oxidation and reduction of elemental sulfur in thermophilic archaea. J Bioenerg Biomembr 36: 77–91

Kojoh K, Matsuzawa H and Wakagi T (1999) Zinc and an N-terminal extra stretch of the ferredoxin from a thermoacidophilic archaeon stabilize the molecule at high temperature. Eur J Biochem 264: 85–91

Komorowski L (1996) Combative characterization of the quinone-binding sites of cytochrome *aa3* from *A. ambivalens* and the terminal oxidase complex SoxABCD from *S. acidocaldarius*. Thesis, University of Luebeck, Germany

Komorowski L (2001) The SoxM-complex from *S. acidocaldarius*.

Thesis, University of Luebeck, Germany
Komorowski L and Schäfer G (2001) Sulfocyanin and subunit II, two copper proteins with novel features, provide new insight into the Archaeal SoxM oxidase supercomplex. FEBS Lett 487: 351–355
Komorowski L, Anemüller S and Schäfer G (2001) First expression and characterization of a recombinant Cu_A- containing subunit II from an Archaeal terminal oxidase complex. J Bioenerg Biomembr 33: 27–34
Komorowski L, Verheyen W and Schäfer G (2002) The archaeal respiratory supercomplex SoxM from *S. acidocaldarius* combines features of quinole- and cytochrome *c*-oxidases. Biol Chemistry 383: 1791–1799
Krebs W, Steuber J, Gemperli AC and Dimroth P (1999) Na^+ translocation by the NADH:ubiquinone oxidoreductase (complex I) from *Klebsiella pneumoniae*. Mol Microbiol 33: 590–598
Kuhn NJ and Ward S (1998) Purification, properties, and multiple forms of a manganese- activated inorganic pyrophosphatase from *Bacillus subtilis*. Arch Biochem Biophys 354: 47–56
Lancaster CRD, Kröger A, Auer M and Michel H (1999) Structure of fumarate reductase from *Wolinella succinogenes* at 2.2 Å resolution. Nature 402: 377–385
Lanyi JK (1969) Studies of the electron transport chain of extremely halophilic bacteria. II. Salt dependence of reduced diphosphopyridine nucleotide oxidase. J Biol Chem 244: 2864–2869
Lanyi JK (1972) Studies of the electron transport chain of extremely halophilic bacteria. VII. Solubilization properties of menadione reductase. J Biol Chem 247: 3001–3007
Lemos RS, Gomes CM and Teixeira M (2001) *Acidianus ambivalens* complex II typifies a novel family of succinate dehydrogenase. Biochem Biophys Res Commun 281: 141–150
Lübben M (1995) Cytochromes of Archaeal electron transfer chains. Biochim Biophys Acta 1229: 1–22
Lübben M and Schäfer G (1989) Chemiosmotic energy conversion of the thermoacidophile *Sulfolobus acidocaldarius*: Oxidative phosphorylation and the presence of an Fo-related DCCD-binding proteolipid. J Bacteriol 171: 6106–6116
Lübben M, and Morand K (1994) Novel prenylated hemes as cofactors of cytochrome oxidases. J Biol Chem 269: 21473–21479
Lübben M, Kolmerer B and Saraste M (1992) An archaebacterial terminal oxidase combines core structures of two mitochondrial respiratory complexes. EMBO J 11: 805–812
Lübben M, Arnaud S, Castresana J, Warne A, Albracht SPJ and Saraste M (1994a) A second terminal oxidase in *Sulfolobus acidocaldarius*. Eur J Biochem 224: 151–159
Lübben M, Castresana J and Warne A (1994b) Terminal oxidases of *Sulfolobus*: Genes and proteins. System Appl Microbiol 16: 556–559
Mattar S, Scharf B, Kent SBH, Rodewald K, Oesterhelt D and Engelhard M (1994) The primary structure of halocyanin, an Archaeal blue copper protein, predicts a lipid anchor for membrane fixation. J Biol Chem 269: 14939–14945
Meunier B, Madgwick SA, Reil E, Oettmeier W and Rich PR (1995) New inhibitors of the quinol oxidation sites of bacterial cytochromes *bo* and *bd*. Biochemistry 34: 1076–1083
Minami Y, Wakabayashi S, Wada K, Matsubara H, Kerscher L and Oesterhelt D (1985) Amino Acid Sequence of a Ferredoxin from Thermoacidophilic Archaebacterium, *Sulfolobus acidocaldarius*. Presence of an N6- Monomethyllysine and Phyletic Consideration of Archaebacteria. J Biochem 97: 745–753
Mitchell P (1976) Possible molecular mechanisms of the proton motive function of cytochrome systems. J Theor Biol 62: 327–367
Moll R (1991) Isolierung und Charakterisierung des Succinat Dehydrogenase Komplexes aus *Sulfolobus acidocaldarius*. Thesis, University of Lübeck, Germany
Moll R and Schäfer G (1988) Chemiosmotic H^+ cycling across the plasma membrane of the thermoacidophilic archaebacterium *Sulfolobus acidocaldarius*. FEBS Lett 232: 359–363
Moll R, Schmidtke S and Schäfer G (1999) Domain structure, GTP-hydrolyzing activity and 7S RNA binding of *Acidianus ambivalens* Ffh-homologous protein suggest an SRP-like complex in Archaea. Eur J Biochem 259: 441–448
Montoya G, Te Kaat K, Rodgers S, Nitschke W and Sinning I (1999) The cytochrome bc_1 complex from *Rhodovulum sulfidophilum* is a dimer with six quinones per monomer and an additional 6-kDa component. Eur J Biochem 259: 709–718
Mukund S and Adams MWW (1993) Characterization of a novel tungsten-containing formaldehyde ferredoxin oxidoreductase from the hyperthermophilic archaeon, *Thermococcus litoralis*. A role for tungsten in peptide catabolism. J Biol Chem 268: 13592–13600
Ng WV, Kennedy SP, Mahairas GG, Berquist B, Pan M, Shukla HD, Lasky SR, Baliga NS, Thorsson V, Sbrogna J, Swartzell S, Weir D, Hall J, DAhl TA, Welti R, Goo JA, Leithauser B, Keller K, Cruz R, Danson MJ, Hough DW, Maddocks DG, Jablonski PE, Krebs MP, Angevine CM, Dale H, Isenbarger TA, Peck RF, Pohlschroder M, Spudich JL, Jung K-H, Alam M, Freitas T, Hou S, Daniels CJ, Dennis PP, Omer AD, Ebhardt H, Lowe TM, Liang P, Riley M, Hood L and DasSarma S (2000) Genome sequence of *Halobacterium* species NRC-1. Proc Natl Acad Sci USA 97: 12176–12181
Nicolaus B, Trincone A, Lama L, Palmieri G and Gambacorta A (1992) Quinone composition in *Sulfolobus acidocaldarius* grown under different conditions. System Appl Microbiol 15: 18–20
Pereira MM, Carita JN and Teixeira M (1999) Membrane-bound electron transfer chain of the thermohalophilic bacterium *Rhodothermus marinus*: Characterization of the iron-sulfur centers from the dehydrogenases and investigation of the high-potential iron-sulfur protein function by in vitro reconstitution of the respiratory chain. Biochemistry 38: 1276–1283
Pereira MM, Santana M and Teixeira M (2001) A novel scenario for the evolution of haem-copper oxygen reductases. Biochim Biophys Acta1505: 185–208
Pfeiffer F and Oesterhelt D (2001) HaloLex Databank of Genome from *Halobacterium salinrum*. RefType: [http://www.halolex.mpg.de. (2002)]
Pfeiffer F, Griffig J and Oesterhelt D (1993) The fdx gene encoding the (2Fe-2S)ferredoxin of *Halobacterium salinarium* (*H. halobium*). Mol Gen Genet 239: 66–71
Pfitzner U, Odenwald A, Ostermann T, Weingard L, Ludwig B and Richter OMH (1998) Cytochrome *c* oxidase (Heme aa_3) from *Paracoccus denitrificans*: Analysis of mutations in putative proton channels of subunit I. J Bioenerg Biomembr 30: 89–97
Purschke W, Schmidt CL, Petersen A and Schäfer G (1997) The terminal quinol oxidase of the hyperthermophilic archaeon *Desulfurolobus ambivalens* exhibits unusual subunit structure and gene organization. J Bacteriol 179: 1344–1353

Scharf B and Engelhard M (1993) Halocyanin, an archaebacterial blue copper protein (type I) from *Natronobacterium pharaonis*. Biochemistry 32: 12894–12900

Scharf B, Wittenberg R and Engelhard M (1997) Electron transfer proteins from a haloalkaliphilic archaebacterium: Main components of the respiratory chain *of Natronobacterium pharaonis* are cytochrome bc and cytochrome *ba*3. Biochemistry 36: 4471–4479

Schäfer G (1996) Bioenergetics of the archaebacterium *Sulfolobus*. Biochim Biophys Acta 1277: 163–200

Schäfer G, Anemüller S, Moll R, Gleissner M and Schmidt CL (1994) Has *Sulfolobus* an archaic respiratory system? Structure, function and genes of its components. Syst Appl Microbiol 16: 544–555

Schäfer G, Purschke WG, Gleissner M and Schmidt CL (1996a) Respiratory chains of Archaea and extremophiles. Biochim Biophys Acta 1275: 16–20

Schäfer T, Bönisch H, Kardinahl S, Schmidt CL and Schäfer G (1996b) Three extremely thermostable proteins from *Sulfolobus* and a reappraisal of the traffic rules. Biol Chem 377: 505–512

Schäfer G, Engelhard M and Müller V (1999) Bioenergetics of the Archaea. Microbiol Molbiol Rev 63: 570–620

Schäfer G, Anemüller S and Moll R (2002) Archaeal complex II: Classical and non-classical succinate:quinone reductases with unusual features. Biochim Biophys Act a1553: 57–73

Schmidt CL and Shaw L (2001) A comprehensive phylogenetic analysis of Rieske and Rieske-type iron-sulfur proteins. J Bioenerg Biomembr 33: 9–26

Schmidt CL, Anemüller S, Teixeira M and Schäfer G (1995) Purification and characterization of the Rieske iron-sulfur protein from the thermoacidophilic crenarchaeon *Sulfolobus acidocaldarius*. FEBS Lett 359: 239–243

Schmidt CL, Anemüller S and Schäfer G (1996) Two different respiratory Rieske proteins are expressed in the extreme thermoacidophilic crenarchaeon *Sulfolobus acidocaldarius*: Cloning and sequencing of their genes. FEBS Lett 388: 43–46

Schmidt CL, Hatzfeld OM, Petersen A, Link TA and Schäfer G (1997) Expression of the *Sulfolobus acidocaldarius* Rieske-iron sulfur protein II (SoxF) with the correctly inserted [2Fe2S] cluster in *Escherichia coli*. Biochem Biophys Res Commun 234: 283–287

Schoepp-Cothenet B, Schütz M, Baymann F, Brugna M, Nitschke W, Myllykallio H and Schmidt C (2001) The membrane-extrinsic domain of cytochrome $b_{558/566}$ from the Archaeon *Sulfolobus acidocaldarius* performs pivoting movements with respect to the membrane surface. FEBS Lett 487: 372–376

Schönheit P and Schäfer T (1995) Metabolism of hyperthermophiles. World J Microbiol 11: 26–57

She Q, Singh RK, Confalonieri F, Zivanovic Y, Allard G, Awayez MJ, Chan-Weiher CCY, Clausen IG, Curtis BA, De Moors A, Erauso G, Fletcher C, Gordon PMK, Heikamp-de Jong I, Jeffries AC, Kozera CJ, Medina N, Peng X, Thi-Ngoc HP, Redder P, Schenk ME, Theriault C, Tolstrup N and Charlebois RL (2001) The complete genome of the crenarchaeon *Sulfolobus solfataricus* P2. Proc Natl Acad Sci USA 98: 7835–7840

Shima S, Netrusov A, Sordel M, Wicke M, Hartmann GC and Thauer RK (1999) Purification, characterization, and primary structure of a monofunctional catalase from *Methanosarcina barkeri*. Arch Microbiol 171: 317–323

Shimada H, Shida Y, Nemoto N, Oshima T and Yamagishi A (2001) Quinone profiles of *Thermoplasma acidophilum* HO-62. J Bacteriol 183: 1462–1465

Skulachev VP (1994) The latest news from the sodium world. Biochim Biophys Acta 1187: 216–221

Smith DC (2001) Marine biology—Expansion of the marine Archaea. Science 293: 56–57

Snyder CH, Merbitz-Zahradnik T, Link TA and Trumpower BL (1999) Role of the Rieske iron-sulfur protein midpoint potential in the protonmotive Q-cycle mechanism of the cytochrome bc_1 complex. J Bioenerg Biomemb 31: 235–242

Sreeramulu K, Schmidt CL, Schäfer G and Anemüller S (1998) Studies of the electron transport chain of the euryarcheon *Halobacterium salinarum*: Indications for a type II NADH dehydrogenase and a complex III analog. J Bioenerg Biomemb 30: 443–453

Stetter KO (1996) Hyperthermophilic procaryotes. FEMS Microb Rev 18: 149–158

Stetter KO (1999) Extremophiles and their adaptation to hot environments. FEBS Lett 452: 22–25

Steuber J, Krebs W and Dimroth P (1997) The Na^+-translocating NADH:ubiquinone oxidoreductase from *Vibrio alginolyticus*—Redox states of the FAD prosthetic group and mechanism of Ag^+ inhibition. Eur J Biochem 249: 770–776

Sykes A (1990) Plastocyanin and the blue copper proteins. Structure and Bonding 75: 177–224

Takai K and Sako Y (1999) A molecular view of Archaeal diversity in marine and terrestrial hot water environments. FEMS Microbiol Ecol 28: 177–188

Talfournier F, Colloc'h N, Mornon JP and Branlant G (1998) Comparative study of the catalytic domain of phosphorylating glyceraldehyde-3-phosphate dehydrogenases from bacteria and Archaea via essential cysteine probes and site-directed mutagenesis. Eur J Biochem 252: 447–457

Tanaka M, Ogawa N, Ihara K, Sugiyama Y and Mukohata Y (2002) Cytochrome aa_3 in *Haloferax volcanii*. J Bacteriol 184: 840–845

Teixeira M, Batista R, Campos AP, Gomes C, Mendes J, Pacheco I, Anemüller S and Hagen WR (1995) A seven-iron ferredoxin from the thermoacidophilic archaeon *Desulfurolobus ambivalens*. Eur J Biochem 227: 322–327

Trincone A, Nicolaus B, Palmieri G, De Rosa M, Huber R, Stetter KO and Gambacorta A (1992) Distribution of complex core lipids within new hyperthermophilic members of the Archaea domain. System Appl Microbiol 15: 11–17

Trumpower BL (1990) The protonmotive Q-cycle. J Biol Chem 265: 11409–11412

Van de Vossenberg JCLM, Driessen AJM, Zillig W and Konings WN (1998) Bioenergetics and cytoplasmic membrane stability of the extreme acidophilic thermophilic archaeon *Picrophilus oshimae*. Extremophiles 2: 67–74

Van der Oost J, De Boer APN, De Gier J-WL, Zumft WG, Stouthamer AH and Van Spanning RJM (1994) The heme-copper oxidase family consists of three distinct types of terminal oxidases and is related to nitric oxide reductase. FEMS Microbiol Lett 121: 1–10

Wakagi T and Oshima T (1987) Energy metabolism of a thermoacidophilic archaebacterium *Sulfolobus acidocaldarius*. Orig Live 17: 391–399

Wakagi T, Yamauchi T, Oshima T, Mueller M, Azzi A and Sone N (1989) A novel *a*-type terminal oxidase from *Sulfolobus acidocaldarius* with cytochrome *c* oxidase activity. Biochem

Biophys Res Commun 165: 1110–1114
Wakao H, Wakagi T and Oshima T (1987) Purification and properties of nadh dehydrogenase from a thermoacidophilic archaebacterium, *Sulfolobus acidocaldarius*. J Biochem (Jap) 102: 255–262
Ward DE, Donnelly CJ, Mullendore ME, Van der Oost J, De Vos WM and Crane EJ (2001) The NADH oxidase from *Pyrococcus furiosus*—Implications for the protection of anaerobic hyperthermophiles against oxidative stress. Eur J Biochem 268: 5816–5823
Weiss H, Friedrich T, Hofhaus G and Preis D (1991) The respiratory-chain NADH dehydrogenase (complex I) of mitochondria. Eur J Biochem, 197: 563–576
Wheelis ML, Kandler O and Woese CR (1992) On the nature of global classification. Proc Natl Acad Sci USA 89: 2930–2034
Wittung-Stafshede P, Gomes C and Teixeira M (2000) Stability and folding of the ferredoxin of the hyperthermophilic archaeon *Acidianus ambivalens*. J Bioinorg Chem 78: 35–41
Woese CR (1998) The universal ancestor. Proc Natl Acad Sci USA 95: 6854–6859
Woese CR and Fox GE (1977) Phylogenetic structure of the procaryotic domain: The primary kingdoms. Proc Natl Acad Sci USA 74: 5088–5090
Woese CR, Kandler O and Wheelis ML (1990) Towards a natural system of organisms: Proposal for the domains Archaea, bacteria, and eucarya. Proc Natl Acad Sci USA 87: 4576–4579
Xia D, Yu CA, Kim H, Xian JZ, Kachurin AM, Zhang L, Yu L and Deisenhofer J (1997) Crystal structure of the cytochrome bc_1 complex from bovine heart mitochondria. Science 277: 60–66
Xia D, Kim H, Yu CA, Yu L, Kachurin A, Zhang L and Deisenhofer J (1998) A novel electron transfer mechanism suggested by crystallographic studies of mitochondrial cytochrome *bc1* complex. Biochem Cell Biol 76: 673–679
Yu CA, Tian H, Zhang L, Deng KP, Shenoy SK, Yu L, Xia D, Kim H and Deisenhofer J (1999) Structural basis of multifunctional bovine mitochondrial cytochrome bc_1 complex. J Bioenerg Biomembr 31:, 191–199
Zähringer U, Moll H, Hettmann T, Knirel VA and Schäfer G (2000) Cytochrome $b_{558/566}$ from the archaeon *Sulfolobus acidocaldarius* has a unique Asn-linked highly branched hexasaccharide chain containing 6-sulfoquinovose. Eur J Biochem 267: 4144–4149
Zhang Q, Iwasaki T, Wakagi T and Oshima T (1996) 2-oxoacid:ferredoxin oxidoreductase from the thermoacidophilic archaeon, *Sulfolobus* sp strain 7. J Biochem (Tokyo) 120: 587–599

Chapter 2

Aerobic Respiration in the Gram-Positive Bacteria

Nobuhito Sone[1,3]*, Cecilia Hägerhäll[2] and Junshi Sakamoto[1]
[1]Dept. of Biochemical Engineering and Science, Kyushu Institute of Technology, Iizuka, Japan; [2]Dept. of Biochemistry, Center for Chemistry and Chemical Engineering, Lund University, Lund, Sweden; [3]ATP System Project, Japan Science and Technology Corporation, Nagatsuta (5-800-2), Yokohama (226-0026), Japan

*Author for correspondence, email: nsone-ra@res.titech.as.jp

Davide Zannoni (ed): Respiration in Archaea and Bacteria. Vol 2: Diversity of Prokaryotic Respiratory Systems, pp. 35–62.

Summary

The group of Gram-positive bacteria is a major phylum of prokaryotes, including several typical saprophytic aerobes. Their respiratory chains are apparently similar to those of eukaryotic mitochondria, but in several points are different from them. The respiratory chain of Gram-positives, like many bacteria, contains branched electron transfer pathways, usually 1-3 heme-Cu oxidases and 1-2 cytochrome *bd*-type quinol oxidases. Most heme-Cu oxidases are SoxM-type as in the mitochondrial cytochrome *c* oxidases, but SoxB-type cytochrome *c* oxidases (cytochrome $b(a/o)_3$) are working under air-limited conditions in thermophilic and alkaliphilic *Bacillus* species. On the other hand, SoxM-type quinol oxidase (cytochrome aa_3) is mainly working in mesophilic *Bacillus subtilis*. Quinol-cytochrome *c* reductase of low G+C content Gram-positives (Firmicutes) is composed of a Rieske Fe-S subunit, a split cytochrome *b* (b_6-type) and a novel cytochrome c_1 subunit formed from 'subunit IV' and a kind of class-I cytochrome *c*, while that of high G+C content Gram-positives (Actinobacteria) has a cytochrome *b*, a Fe-S protein with three transmembrane helices and a novel diheme cytochrome *cc*, without any small cytochromes *c* in the cell. Some of these characteristics are due to the lack of outer membrane of Gram-positives, in which any *c*-type cytochrome should be anchored to the cell membrane, and others are probably due to the evolutionary history of bacteria.

I. Overview

A. Two Distinct Groups of Gram-Positive Bacteria

The group of Gram-positive bacteria is a major phylum of the eubacteria, which includes many species of scientific, medical or industrial importance. Molecular studies of the bacterial respiratory chain has relatively been focused on proteobacteria, such as *Paracoccus denitrificans*, *Rhodobacter sphaeroides* and *Escherichia coli*, since their main respiratory enzymes are similar to those of mitochondria and that these organisms are susceptible to genetic manipulation. However, Gram-positive bacteria is one of the major group of prokaryotes, comparable to or even larger than proteobacteria in terms of the number of species included. Thus, it should be fruitful to study biochemistry and molecular biology of their respiratory chains, both in order to reveal the variety of energy metabolism of these important organisms and to find the universal cores of energetic transducing mechanisms.

The Gram-positive bacteria are subdivided into two groups based on the G+C content of the chromosomal DNA. The low G+C group (firmicutes) includes endospore-forming rods (such as *Bacillus* species), nonspore-forming cocci (such as *Staphylococcus aureus* and *Enterococcus faecalis*), photosynthetic bacteria (such as *Heliobacillus mobilis*), and cell wall-less bacteria (mycoplasmas), while the high G+C Gram-positive bacteria (actinobacteria) includes coryne-form bacteria (*Corynebacterium*), rod-shaped mycolic acid-containing organisms (*Mycobacterium*), and filamentous actinobacteria (or Actinomyces). The molecular composition of the cells, including the enzyme components of the respiratory chain, is substantially different in these two groups (Table 1).

Acetogenetic low G+C Gram-positive bacteria, such as *Acetobacterium woodii* and *Clostridium aceticum*, can grow either chemoorganotrophically by fermentation of sugar or chemolithotrophically

Abbreviations: $C_{12}E_8$ – octo-polyoxyethylene lauryl ether; Complex I – NADH dehydrogenase type I and II; Complex II – SQR, succinate dehydrogenase; Complex III – QcR, quinol:cytochrome *c* reductase; Complex IV – CcO, cytochrome *c* oxidase; Cyt – cytochrome; MALDI-TOF – matrix-assisted laser desorption ionization-time of flight; MK – menaquinone; NDH-I – NADH dehydrogenase type I; NDH-II – NADH dehydrogenase type II; PMF – proton motive force; SQR – succinate:menaquinone oxidoreductase; TMPD – *N,N,N′,N′*-tetramethyl-*p*-phenylene diamine

Table 1. Respiratory enzymes encoded in the whole genomes of Gram-positive bacteria

Enzyme	NADH deh		SQR	QcR	Cyt *c*	Terminal oxidase			
Subtype	NAD-1	NDH-2				SoxM-type; SoxB-type Cyt *bd*			
(other name	(complex I)		(SDH)	(complex III)		(complex IV)			
Substrate	NADH		Succinate	QH_2		Cyt *c*	QH_2	Cyt *c*	QH_2
Low G+C species									
B. subtilis	X	0	0	0	0	0^c	0	X	$0^{\#}$
B. halodurans	X	0	0	0	0	0^c	0	0	$0^{\#}$
S. aureus	X	0	0	X	X	X	0	X	0
High G+C species									
C. glutamicum	X	0	0	0^c	X	0	X	X	0
M. tubercolosi	0	0	0	0^c	X	0	X	X	0
M. leprae	X	0	0	0^c	X	0	X	X	X
Str. Coelicolor	0	0	0	0^c	X	0	X	X	X

Genus abbreviations: *B., Bacillus; S., Staphylococcus; C., Corynebacterium; M., Mycobacterium; Str., Streptomyces.* Non standard abbreviations: deh, dehydrogenase; QH_2, quinol. Symbols: 0^c, subunit with fused Cyt *c* domain; $0^{\#}$, two genes for Cyt

through the reduction of CO_2 to acetate with H_2 as electron donor. The latter process results in the generation of ion gradient across the membrane, either of H^+ or Na^+, which drives ATP synthesis. This is an autotrophic metabolism called anaerobic respiration. Other Gram-positive bacteria are facultative aerobes, and can use alternative terminal electron acceptors, such as nitrate, under anaerobic conditions. Furthermore, various Gram-positive bacteria can grow by fermentation relying solely on substrate-level phosphorylation to meet their energy needs. However, in this chapter we will focus solely on aerobic respiration in Gram-positive bacteria.

B. Main Architectures of the Respiratory Chain

Detailed molecular studies of the respiratory chain have been carried out in several *Bacillus* species of the low G+C group and in *Corynebacterium glutamicum* of the high G+C group. These studies will be highlighted in the following respective sections. In addition, whole genome sequences have become available for various aerobic or facultatively anaerobic Gram-positive organisms from both groups, revealing which respiratory enzymes are encoded in their genomes (summarized in Table 1). The low G+C firmicutes investigated to date contain one or more NADH dehydrogenases of type-II (NDH-II), a single subunit membrane-associated enzyme which does not conserve redox energy, but lack a proton pumping, multi-subunit NADH dehydrogenase type-I (NDH I) or Complex I. In the high G+C actinobacteria, complex I is present in *Mycobacterium tuberculosis* and *Streptomyces coelicolor*, but absent in *Mycobacterium leprae* and *C. glutamicum*. Genes encoding succinate dehydrogenase, SQR or Complex II are found in the genomes of all the sequenced species, both in the low- and high-G+C groups. The quinone used by the Gram-positive bacteria is typically menaquinone (MK), with the isoprenoid side chain length varying between seven to nine units among the species. MKs in high G+C Gram positives are partially reduced. When comparing the respiratory chain of Gram-positive bacteria to that of other organisms using ubiquinone, it is important to remember that the redox potential of menaquinone is almost 200 mV lower than that of ubiquinone.

Like many bacteria, and in contrast to mitochondria, the Gram-positive bacteria often contain branched respiratory chains. One is the Cyt-*c* reductase-oxidase branch composed of two consecutive enzymes, Complex III (Cyt bc_1-type quinol-Cyt *c* oxidoreductase or QcR) and Complex IV (Cyt *c* oxidase or CcO). The main CcO is a Cyt caa_3-type heme-copper oxidase (SoxM-type), but the extremophilic species, such as the thermophilic *Bacillus thermodenitrificans* and the alkalophilic *B. halodurans* contain a second CcO, a SoxB-type heme-copper oxidase. The second branch bypasses QcR and consists of quinol oxidases of either heme-copper oxidase or Cyt *bd* type. The latter type of quinol oxidases form a distinct family not related to the heme-copper oxidases. They lack copper and do not pump protons, thus less energy is conserved when a Cyt *bd* is used. *B. subtilis* and *B. halodurans* possess two putative gene sets for Cyt *bd*. In the facultative aerobe *Staphylococcus aureus*, the QcR-CcO branch is absent and only quinol oxidases, a SoxM-type enzyme and a Cyt *bd*, are present. Interestingly, *Lactococcus lactis*, which obtains energy only through anaerobic lactate fermentation, still contains a gene set for a Cyt *bd*-type quinol oxidase. This may explain aerotolerance, which is a remarkable feature of this otherwise anaerobic organism.

C. Relationship of the Membrane Structure and the Respiratory Chain

It is also important to contemplate that the Gram-positive bacteria, in contrast to mitochondria and Gram-negative bacteria, contain only one cell membrane, corresponding to the inner membrane in the former organelles/organisms, and lack an outer membrane and thus a periplasmic space. As a result, it is common that the small Cyts *c* in Gram-positives are membrane anchored, either via a transmembrane helix or via a lipid moiety. Another remarkable difference may be the requirement of an energized membrane for the respiratory chain enzyme to function. A positive respiratory control ratio is a common feature of energy coupled enzymatic reactions, where in a coupled membrane system the rate of respiration increases several folds in the presence of uncouplers or ionophores. In the Gram-positive *B. subtilis*, exactly the opposite was observed, and respiration was instead slowed down (Barsky et al., 1989; Lemma et al., 1990; Samuilov and Khakimov, 1991; Schirawski and Unden, 1998). In the presence of both a protonophore and a K^+ ionophore, respiration was shown to shut down almost completely. This respiratory chain requirement for an energized state was not a specific feature of succinate oxidation, and occurred for several different substrates (Azarkina and Konstantinov, 2002). It remains to be investigated to what extent this phenomenon is present in other Gram-positive bacteria, and the underlying molecular mechanisms for this respiratory chain behavior.

A common feature of the respiratory chains in actinobacteria, examined so far, is that they completely lack a small cytochrome *c*, and have a new type of QcR, whose *c*-type cytochrome subunit contains two hemes *c* and thus two *c*-type Cyt domains (Table I). One of the two domains might be equivalent to the peripheral small Cyt *c* of the mitochondrial type respiratory chain, which has been genetically fused to the reductase at some stage of the molecular evolution. This difference is reflected also in the CcO, where the gene set for SoxM-type CcO is present and homologous to the regular type of Cyt aa_3 oxidase, but contains a particular inserted segment in the Cyt *c*-binding domain. In addition to reductase-oxidase branch, many of these organisms have a Cyt *bd*-type oxidase constituting an alternative electron-transfer pathway. One exception is *Myc. leprae*, which lack a Cyt *bd*.

II. Thermophilic *Bacillus*

A. The Overall Respiratory Chain

Several *Bacillus* species are moderately thermophilic and grow at temperatures between 50 °C and 75 °C. *B. caldotenax* and *B. caldolyticus* were isolated from a hot spring in Yellowstone National Park and can grow at temperatures up to 75 °C. *Bacillus* sp. PS3 was isolated from a hot spring in Japan, and grow well at 65–70 °C. Many thermophilic *Bacillus* strains found in soil are classified as *B. stearothermophilus*. They usually grow at 55–65 °C. *B. thermodenitrificans* strain K1041 (Studholme et al., 1999), that was isolated from soil in Japan as a transformable strain, was at first classified as a strain of *B. stearothermophilus* (Narumi et al., 1992), and can grow at 45–60 °C. These bacteria are aerobic, and in rich media their doubling time is as short as 10–15 min. They are rich in Cyts and colored brown; the Cyt content in the membrane can be higher than 1 nmol /mg membrane protein, especially when the bacteria are grown at high temperatures. The thermostability of the cells also facilitates purification of the intact enzymes from the respiratory chain.

There is no whole genome sequence available for any thermophilic *Bacillus*, but the branched respiratory chain of a thermophilic *Bacillus,* as we know it, is outlined in Fig. 1. The quinone pool, consisting of menaquinone-7 (MK-7), is reduced by several dehydrogenases such as NADH, succinate, malate and glycerolphosphate dehydrogenases. The NADH dehydrogenase is of type-II, a single polypeptide with FAD as the prosthetic group (Xu et al., 1989, 1991), and does not translocate H^+. Energy conservation occurs upon menaquinol oxidation catalyzed by the b_6c_1 complex (QcR) and Cyt caa_3 (CcO), which is

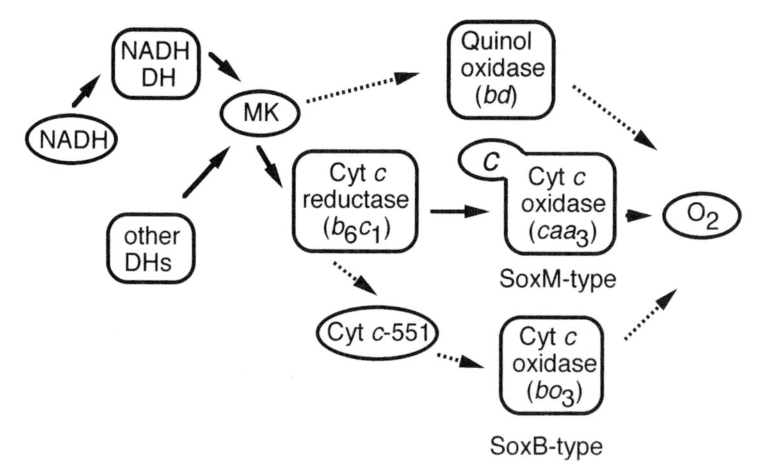

Fig. 1. Putative respiratory chain of thermophilic *Bacillus* K1041. Solid arrows indicate the main electron-transfer branch operating under highly aerated conditions, and broken arrows the alternative branches under air-limited conditions.

illustrated by the solid line in the figure. Aerobically grown thermophilic *Bacillus* cells showed Cyt *a*, *b* and *c*-type spectrum similar to that of mammalian mitochondria (Sone et al., 1983), which is mostly attributable to the presence of QcR and CcO. Notable differences to the mitochondrial respiratory chain is the absence of NADH dehydrogenase type I, the presence of a quinol oxidase and a SoxB type CcO, and that a Cyt *c* is fused to subunit II of the SoxM type CcO (Sone and Yanagita, 1982; Ishizuka et al., 1990). Especially during growth of thermophiles, the oxygen concentration of the culture medium often becomes very low in spite of shaking. At high temperatures, the concentration of dissolved oxygen is low in combination with that the oxygen consumption of the bacteria is very rapid, giving rise to an air-limited culture condition. After several hours of growth under these conditions, the Cyt pattern of the cells changes. There are two molecular events involved in the change; one is caused by heme O accumulation due to O_2 limitation, which results in the conversion of the caa_3 CcO to cytochrome cao_3 (see below), and another is the appearance of quinol oxidases, the alternative pathways to use O_2, illustrated by the broken lines in Fig. 1.

B. Cytochrome caa_3-type CcO

The structural genes encoding the caa_3-type CcO were cloned and sequenced from *Bacillus* PS3 (Sone et al., 1988; Ishizuka et al., 1990) and from *B. thermodenitrificans* 1041 (Kusano et al., 1996). The genes (*ctaCDEF*) are clustered with *ctaB* and *ctaA*, just like those of *B. subtilis* (Saraste et al., 1991). CtaB is a heme O synthase, or farnesyl transferase (Saiki et al., 1994) comparable to *E. coli* CoxE (Saiki et al., 1993). On the other hand, CtaA, which is encoded from the other strand and separately expressed, turned out to be a heme A synthase, or heme O oxidase/oxygenase, which converts the 8-methyl group of porphyrin into a formyl group (Sakamoto et al., 1999). *CtaCDEF* are structural genes encoding subunits II, I, III and IV, respectively of the CcO (Fig. 2a). Subunit I contains 14 transmembrane helices and bears heme *a* and the heme a_3-Cu_B binuclear center where dioxygen is reduced. Subunit II consists of N-terminal hydrophobic domain composed of two transmembrane segments, and a C-terminal hydrophilic domain containing the Cu_A-center and a fused cytochrome c domain (Ishizuka et al., 1990). The subunit III was rather short and contained 5 transmembrane helices. The role of subunit III and IV (Sone et al., 1990) have not been elucidated, but since some bacterial oxidase preparations lack subunit(s) IV and/or III (Haltia et al., 1991), they are proposed to be necessary only for assembly of the catalytic portion composed of subunits I and II,

TMPD, $Ru(NH_3)_6$, and phenazine methosulfate function as artificial electron donors for the CcO (Sone and Yanagita, 1982) and Cyts *c* from several different species such as PS3 Cyt *c*-551, *T. thermophilus* Cyt *c*-552, *Candida krusei* Cyt *c* and *Sachharomyces cerevisiae* Cyt *c* are oxidized well, but not equine Cyt *c*. However, these substrates show respective K_m's of several μM, indicating that the high affinity site for Cyt *c* is occupied by the intrinsic Cyt *c* domain

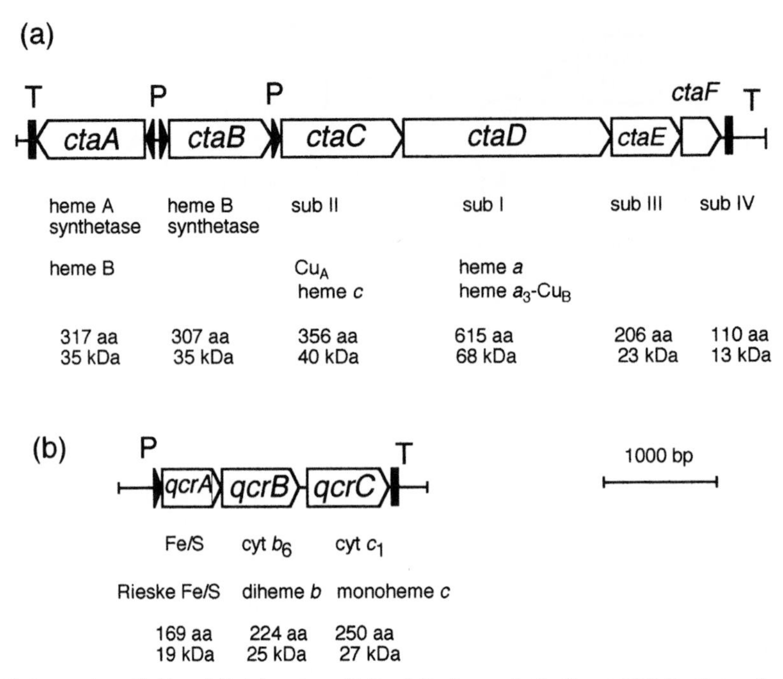

Fig. 2. Gene map for Cyt caa_3-type CcO and Cyt b_6c_1-type QcR of *B. thermodenitrificans* K1041. Gene clusters for Cyt caa_3-type CcO (a) and for QcR (b). Symbols used: empty arrows (⇨), genes. P (▶), putative promoters; T (⊢), putative terminators with a palindrome sequence.

(Nicholls and Sone, 1984). It is thus likely that the electrons from the reductant reach the binuclear site through the intrinsic Cyt *c* and are then passed on to $Cu_A \rightarrow$ Cyt $a \rightarrow$ Cyt a_3-Cu_B as in the mitochondrial enzyme. However, radiation inactivation experiments indicated that a minor portion of the donated electrons may reduce the binuclear site without mediation of Cyt *a* (Sone and Kosako, 1986).

The CcO enzyme from PS3 not only pumps H^+ (Sone and Hinkle, 1982; Sone and Yanagita, 1984), but also shows many properties similar to the mitochondrial enzyme; it performs resting to pulsed form conversion (Sone et al., 1984a) and ligand-binding with cyanide and azide (Sone and Nicholls, 1984). However, K_i for cyanide in the PS3 enzyme is 1.6 μM, which is about 16-fold less sensitive than for the beef heart enzyme, mainly due to a faster rate of dissociation of cyanide. The PS3 enzyme formed intermediates A (oxygenated) and B (peroxy) upon flash photolysis of the CO-reduced form in the presence of O_2 at low temperatures (–100 °C and –80 °C, respectively) as the bovine enzyme (Sone et al., 1984b). Flash photolysis experiments at the room temperature showed the oxygenated form with a time constant of 10 μs as the bovine enzyme, but neither the second (peroxy) nor the third (ferryl) was detected (Hirota et al., 1996). The reason for this result was due to the electron flow from the intrinsic Cyt *c* harboring the fifth electron in the fully reduced enzyme. Flash-induced CO dissociation from mixed-valence enzyme showed that more electrons were back-transferred from heme a_3 to heme *a* than in the case of the bovine enzyme, suggesting a higher E_m' of the PS3 Cyt *a*.

Resonance Raman spectroscopy shows that the structure of the PS3 enzyme in the vicinity of the chromophore is very similar to that of the bovine enzyme, including the 214 cm^{-1} line due to Fe-histidine stretching mode. This Raman line was found too susceptible to selective-heat inactivation which destroys H^+-pump activity, without severe loss of the oxidase activity (Ogura et al.,1984; Sone et al., 1986). Alkaline treatment, which also induce loss of H^+-pumping activity in the bovine enzyme, did not destroy the 214 cm^{-1} line and had no effect on the Raman line or H^+-pumping in the PS3 enzyme (Sone et al., 1986). It should be noted that alkalinization induced monomerization of the dimeric bovine enzyme, whereas the monomeric bacterial enzyme (Sone and Takagi, 1990) may pump H^+.

The Cyt cao_3-type PS3 enzyme with heme O at the high-spin heme site pumped H^+ as the caa_3-type, and showed almost the same Fe-histidine ($\nu_{Fe\text{-}His}$) and Fe-CO stretching frequencies of resonance Raman, but the formyl stretching ($\nu_{CH=O}$) band of the high spin heme was absent (Sone et al., 1994). The K_i value for cyanide of Cyt *cao*$_3$ was four times smaller than that of Cyt *caa*$_3$, which is mainly due to the slower 'off' constant (k_{off}) of the former.

C. Cytochrome b_6c_1-type QcR

The *qcrABC* operon (Fig 2b) was cloned from *B. thermodenitrificans*, and sequenced (Sone et al., 1995,1996). The b_6c_1 complex, equivalent to QcR, is catalyzing menaquinol-dependent Cyt *c* reductase activity, and is composed of three subunits, although a forth band was usually found in the final preparation (Kutoh and Sone, 1988). QcrA contains cysteine and histidine ligands for a Rieske-type FeS cluster. The very hydrophobic 27 kDa (21 kDa in SDS-PAGE) QcrB subunit is homologous to Cyt b_6 of the plastid b_6f complex and corresponds the N-terminal half of mitochondrial Cyt *b*. Notably, heme b_H of the *B. thermodenitrificans* Cyt b_6 appears to be covalently bound to protein (Kutoh and Sone, 1988). The 29-kDa QcrC subunit contains a class-I Cyt *c* (Sone and Toh, 1994), but its N-terminal sequence is homologous to subunit IV of the plastid b_6f complex (Widger et al., 1984).

The Cyt b_6c_1 complex from PS3 has two quinone binding sites. In preparations of detergent-solubilized and purified QcR, almost all the endogenous MK-7 had been lost, and was rather difficult to reconstitute due to low solubility of MK-7. However, preincubating QcR with MK-3 is more feasible. When MK-3 was used, K_m's for the low and high affinity sites were 0.1 and 0.02 μM, respectively. Using the MK-3 preincubated enzyme, the K_m of the low affinity site for duroquinol and UQ-1 was 10.0 and 3.7 μM, respectively. The enzyme donates electrons to Cyt *c* via its intrinsic Cyt c_1 with rather low substrate specificity; equine Cyt *c* and yeast Cyt *c* can be reduced as well as Cyt *c*-551 from *Bacillus* PS3. QcR activity was inhibited by myxothiazol, 1-heptyl-4-hydroxyquinoline *N*-oxide and abietic acid, but not by antimycin A, as in the case of plastid Cyt b_6f complexes.

D. Super-complex of QcR and CcO

The two major respiratory complexes, Cyt b_6c_1 and Cyt *caa*$_3$, seem to form a super-complex having

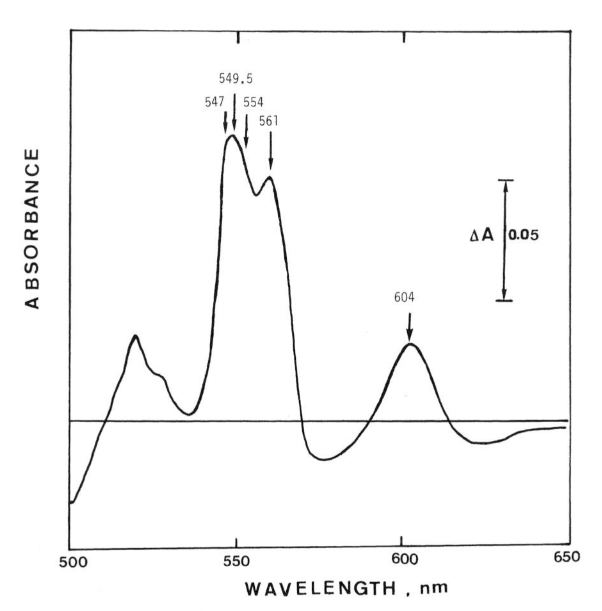

Fig. 3. Dithionite-reduced minus air-oxidized spectrum of Cyt b_6c-*caa*$_3$-type quinol oxidase supercomplex from *Bacillus* PS3..

menaquinol oxidase activity (Sone et al., 1987). The enzymes (see Fig. 3) comigrated during gel filtration after solubilization by a mild detergent such as $C_{12}E_8$. QcR coprecipitated with CcO when a PS3 membrane fraction was solubilized with heptylthioglucoside, and immuno-precipitated with antiserum against CcO. Both the chromatographically purified sample and the immunoprecipitate showed TMPD oxidase, quinol-dependent Cyt *c* reductase and quinol oxidase activities. However, the quinol oxidase activity of either super-complex preparations was not accelerated by addition of Cyt *c*-551 from the same bacterium (Sone et al., 1989). It is thus likely that the presence of the super-complex is physiologically important. This super-complex seems to be formed in most strains of thermophilic *Bacillus*. Inter-complex cross-linking between QcR and CcO, especially between subunits I and II of CcO and Cyt b_6 of QcR was confirmed by bifunctional crosslinkers such as 3,3′-dithiobis(succinimidyl-propionate (DSP), which was not observed in the presence of 0.5% Triton X-100 (Tanaka et al., 1996). A prolonged incubation in the presence of a high concentration of Triton X-100 is necessary to obtain pure Cyt b_6c_1 complex and *caa*$_3$-type CcO separately (Sone and Yanagita, 1982; Kutoh and Sone, 1988).

Table 2. Effects of aeration on cytochrome content of the thermophilic *Bacillus* PS3. Data are taken from Sone et al. (1983).

Cytochrome	Wavelength pair (nm)	Content (nmol/ mg membrane protein) Vigorously aerated	Oxygen-limited
b	562 – 575	1.09	2.15
c	551 – 540	1.12	1.88
a	604 – 630	0.33	0.11
o_3	569 - 582*	0.0	0.08
a_3	592 - 611*	0.22	0.00

*CO-difference spectrum

E. Effect of Air-Limitation on the Cytochrome Expression Pattern

The Cyt pattern of thermophilic *Bacillus* cells cultured under air-limited conditions is very different from that of well-aerated cells. The reduced *minus* oxidized, and CO-difference spectra of cells cultured under both conditions are compared and the amount of cytochromes is summarized in Table 2. Cells grown with low aeration synthesized higher amounts of Cyt *c* and Cyt *b*, and lower amounts of Cyt *a*. The CO difference spectra indicate that the terminal oxidase contains Cyt *o* under air-limited conditions and a Cyt caa_3 in well aerated cells (Sone et al., 1983; Sone and Fujiwara, 1991b).

Heme O, 2-hydroxyfarnesylated protoheme IX, is the precursor of heme A, having a methyl-group at the 8-position, not formyl. Heme O functions as the prosthetic group of Cyt bo_3 of *E. coli* (Puustinen and Wikström, 1991; Wu et al., 1992). Under air-limited conditions, the conversion of heme O to heme A, which is catalyzed by an oxygenase/oxidase, may be hampered by the low concentration of dissolved O_2 in the culture medium (Sone and Fujiwara, 1991a). Biochemical analyses of the CcO indicated that the protein part of the oxidase was the same, even if the absolute amount become less under air-limited conditions. Thus, the Cyt caa_3-type oxidase was changed to a cao_3 containing the same protein parts (Sone and Fujiwara, 1991a). This change also affected the turnover number of the enzyme, several kinetic parameters of the reaction with cyanide and the formyl stretching mode in the Resonance Raman spectrum (Sone et al., 1994).

The cytochrome content changes observed by down-shift of aeration (Table 2) is due to increased formation of several Cyts of *b*- and *c*-type, The concentration of $ca(a/o)_3$-type oxidase became lower, as mentioned, but the concentration of Cyt *c*-551 increased from about 0.15 nmol/mg membrane protein in well aerated cells to 0.5–1.5 nmol/mg membrane protein in air-limited cells (Sone et al., 1989). Cyt *c*-551 is a membrane-bound *c*-type Cyt of 10.5 kDa (Fig. 1). The amount of *b*-type Cyts was also increased (Sone et al., 1983), which was later shown to be due to appearance of Cyt *bd* quinol oxidase similar to that of *E. coli*, and a Cyt $b(a/o)_3$ SoxB-type Cyt *c*-551 oxidase (Fig. 1)

F. Cytochrome bd Type Quinol Oxidase

Several strains of *B. thermodenitrificans* lacking Cyt $ca(a/o)_3$ were selected after random mutagenesis (Sakamoto et al., 1996). Two alternative terminal oxidases were purified from membranes of one such mutant (K17) by ion-exchange and hydroxylapatite chromatographies.

CydAB, encoding the two subunit Cyt *bd*, was also cloned and sequenced (Fig. 4a). Both subunits I and II are very hydrophobic proteins and have eight and nine transmembrane helices, respectively (Sakamoto et al., 1999a). Cyt *bd* from *B. thermodenitrificans* contained 14.5 nmol heme B and 8.5 nmol heme D per mg protein, and showed an absorbance peak at 618 nm of reduced form (Fig. 5a) and a peak of the CO difference spectrum at 625 nm, which is about 10 nm shorter wavelengths than what was found in the *E. coli* and *A. vinelandii* Cyt *bd*. The difference seems to be due to structural differences of the surrounding polypeptide, since the pyridine-ferrohemochrome spectrum of heme D is normal. The purified Cyt *bd* functioned as a quinol oxidase after pre-incubation for 2 h with UQ-1 or MK-2; duroquinol oxidase activity of about 10-30 s^{-1} was detected at 30 °C (Sakamoto et al., 1999a). The activity was effectively inhibited by *p*-benzoquinone, 2,6-dimethyl benzoquinone and $ZnCl_2$ (I_{50} of 0.12, 0.065 and 0.20, mM, respectively), but only slightly inhibited by NaCN (I_{50} of 0.5 mM).

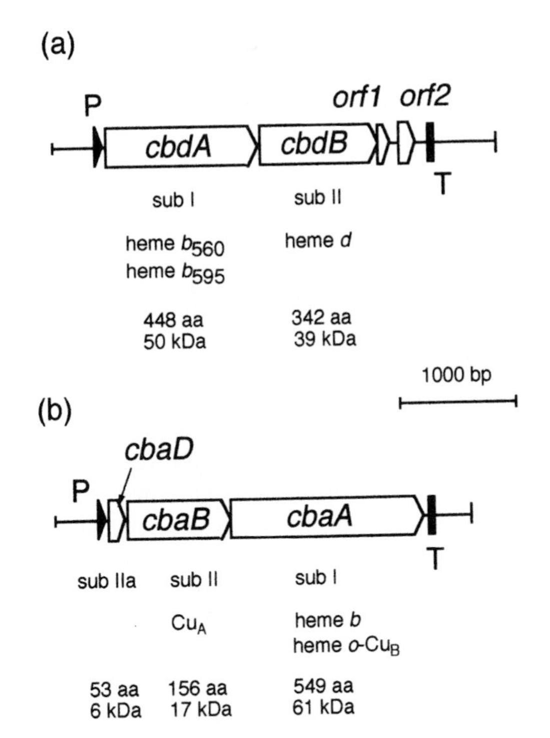

Fig. 4. Gene map for Cyt *bd*-type quinol oxidase and Cyt bo_3-type CcO of *B. thermodenitrificans* K1041. Gene clusters for Cyt *bd*-type quinol oxidase (a) and for Cyt bo_3-type CcO (b). Symbols are as for Fig. 2.

G. Sox B-type CcO

The gene and subunit structure of Cyt $b(o/a)_3$ from *B. thermodenitrificans* is summarized in Fig. 4b. *CbaB* encodes a short subunit II the Cu_A-motif, while *cbaA* encodes subunit I which contains the six His residues that are ligands for the low spin heme, and O_2-reducing binuclear center (Nikaido et al., 1998). Cyt $b(o/a)_3$ from *B. thermodenitrificans* contained heme B as the low spin heme, heme O or heme A as the high spin heme, and Cu_A and Cu_B, since heme B; heme O:heme A:Cu was 1.0:0.7:0.2:3.0, when the enzyme was prepared from the *B. thermodenitrificans* K17 strain which lacks Cyt $ca(a/o)_3$ (Fig. 5b). The heme B content of about 10 nmol/mg membrane protein is slightly lower than that expectable from the Mr of 75,000. The enzyme showed a high TMPD oxidase activity (190 s^{-1} at 30°C), when Cyt *c*-551 was supplied and the reaction medium is rich in salt (Fig. 6; Sakamoto et al., 1997). Analysis of substrate specificity indicated that there is a high affinity site which is Cyt *c*-551-specific, and a low affinity site where TMPD and mitochondrial Cyt *c* can react, reminiscent of in mitochondrial aa_3-type CcO. However, in *B. thermodenitrificans* high concentration of salt activates the reaction, instead of inhibiting

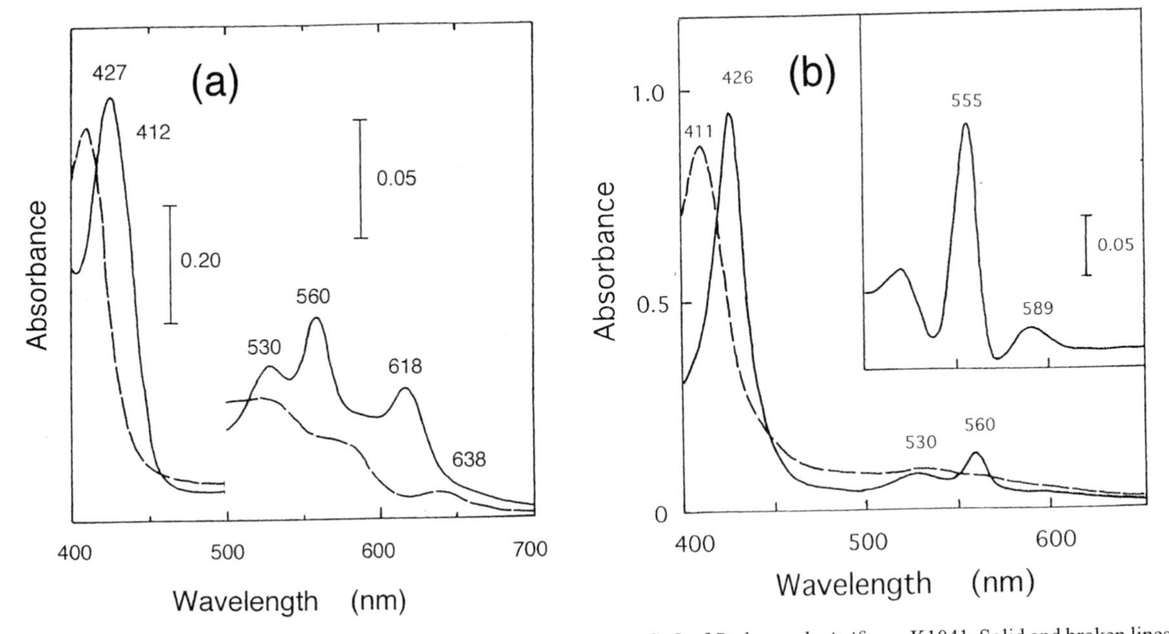

Fig. 5. Absorption spectra of Cyt *bd*-type quinol oxidase and Cyt bo_3-type CcO of *B. thermodenitrificans* K1041. Solid and broken lines are for reduced and oxidized forms, respectively, of Cyt *bd*-type quinol oxidase (*a*, Sakamoto et al., 1999a) and Cyt bo_3-type CcO (*b*, Sakamoto et al., 1997). Inset: redox difference spectra of pyridine hemochrome of the CcO.

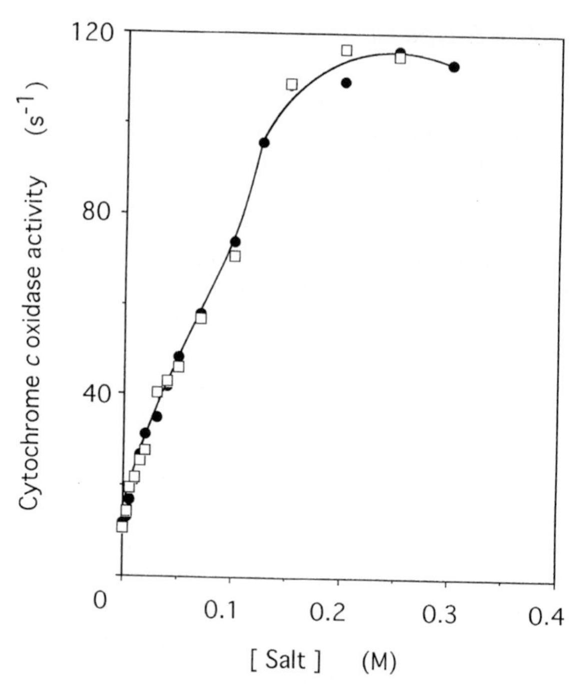

Fig. 6. Salt dependency of Cyt bo_3 enzyme activity. CcO activity was measured in the presence of the indicated concentrations of NaCl (□) or KCl (●) (Sakamoto et al., 1997).

it. The enzyme was susceptible to cyanide (K_i of 19 μM) and azide (K_i of 0.5 mM). The amounts of Cyt $b(o/a)_3$ found in strain K17 was variable from culture to culture, and was often too low to attempt purification. A mutant, K17q8, having very low quinol oxidase activity, was created by random mutagenesis and selection in the presence of 0.02% *p*-benzoquinone (Sone et al., 2000). An expression vector for Cyt $b(o/a)_3$ was also constructed, using the natural promoter and the structural genes *cbaAB* (Nikaido et al., 2000). The *B. thermodenitrificans* cells containing the vector produced as 2 nmol Cyt bo_3 per mg membrane protein under air-restricted growth conditions. The enzyme over-expressed under these conditions was named Cyt bo_3, since the heme at the high-spin site is heme O (Nikaido et al., 2000), just as in the case of Cyt cao_3. The subunit structure and amino acid sequences indicate that this CcO is a SoxB-type heme-Cu oxidase. SoxB-type oxidases are often found in *Archaea* and are quinol oxidases (cf. Chapter 5 by Sakamoto and Sone, Vol. 1), although a few SoxB-type CcO have been found in eubacteria. The active site environment of *B. thermodenitrificans* Cyt bo_3 was examined by EPR and resonance Raman spectroscopies (Uchida et al., 2000), and found to be very similar to the SoxM-type bovine enzyme. The most pronounced difference was found in the Fe-CO stretching mode ($\nu_{Fe\text{-}CO}$), which appeared at 510 cm^{-1}, suggesting that the effect of Cu_B to the Fe-CO binding was much lower. The EPR data of the CO-reduced enzyme form also suggested a lower interaction between high spin heme and Cu_B.

H. The Small Cytochrome c-551

Bacillus PS3 Cyt *c*-551 is a small membrane protein of about 10 kDa (Sone et al., 1989). The DNA fragment encoding Cyt *c*-551 apoprotein, composed of 111 amino acids, was cloned (Fujiwara et al., 1993). The hydropathy plot indicated that the protein should not be membrane-bound, but lipase-treatment greatly diminished the hydrophobicity of the protein. The mass number of 10037.7 after deacylation suggested that the N-terminal signal peptide of 18 amino acid residues is cleaved, probably by the type-II signal peptidase, and subsequently the new N-terminal Cys is diacylglycerated via thioether and the amino-group is blocked by acetylation, as previously shown for the Cyt *c* subunit of photosynthetic reaction center from *Rhodopseudomonas viridis* (Weyer et al., 1987). A Cyt *c*-551 expression vector was then constructed. The transformed cells of *B. thermodenitrificans* expressed as much as 5 nmol PS3 Cyt *c*-551 per mg of membrane protein. The molecular mass determination and treatment with *Rhizopus* lipase indicated that the same processes, cleavage of signal peptidase, blocking of the N-terminal group and diacyl glycerylation of thiol, took place in the over-expression systems (Noguchi et al., 1994). As described, the thermophilic bacilli contain two alternative branches in the respiratory chain (Fig. 1). Cyt *c*-551 seems to be the natural physiological substrate of the $b(o/a)_3$-type oxidase, since the K_m of Cyt bo_3 for this Cyt *c* was very low (below 0.1 μM), and the diacylglycerol moiety of Cyt *c*-551 specifically activates the oxidase (Sakamoto et al., 1997).

Using the over-expressed Cyt bo_3 SoxB-type oxidase, interaction with Cyt *c*-551 was analyzed in more detail (Nikaido et al., 2000), and subsequently almost all Lys residues in Cyt c-551 were changed to Ala or Ser (Kagekawa et al., 2002). Table 3 summarizes these results. Most of changes did not affect the activity much, but the mutant Cyt c such as K69A K73A, R87A and K96A, had lower K_{m1} than the wild-type, indicating that removal of positive charge close to the

Table 3. Kinetic constants of wild-type and mutant cyt *c*-551 as a substrate for cyt bo_3 in the presence of TMPD (Kagakawa et al., 2002).

Substrate	K_m (nM)	V_{max} (s^{-1})
Wild-type	74	270
K69A	40	190
K73A	45	230
K80A	97	220
R87A	56	220
K96A	45	220
K102S	140	140
W106F	82	220
K111*	270	110
C19A	n.d.[a]	5>

[a]Not detectable

oxidase-binding domain resulted in increment of the affinity. The C19A mutant, that lacks the hydrophobic diacylglycerol group, also lost its substrate activity. The importance of the diacyl group is also shown by the fact that di-delipidated *c*-551 also lost activity, while mono-delipidated *c*-551 retained most of the activity (Kagekawa et al., 2002). Recently, soluble Cyt *c*-553 from *Bacillus pasteurii* was purified and crystallized (Benini et al., 2000). The sequence of *B. pasteurii* Cyt *c* is homologous to Cyt c-551 of PS3, but is cleaved by unphysiological proteolysis, and thus lost the lipid anchor. However, the crystal structure of Cyt *c*-553 showed that the front face of the heme crevice is mostly covered with hydrophobic amino acid residues, suggesting hydrophobic interaction between the cytochrome *c* and subunit II of the Cyt bo_3-type oxidase.

I. Individual Roles of the Three Terminal Oxidases

The most important points of this issue are the K_m for O_2 and the H^+-translocating efficiency of the enzymes. Table 4 summarizes several data on these points. The K_m values for O_2 can be determined by tracing the time courses of O_2 consumption with an oxygen electrode, if the K_m is above 0.2 μM. For lower values, like the K_m of the *bd*-type oxidase from *E. coli* that was around 10 nM, rheghemoglobin can be used as oxygen sensor (D'mello et al., 1996). In the case of the *Bacillus* PS3 or *B. thermodenitrificans* enzymes, the K_m of Cyt *caa*$_3$ is higher than that of Cyt bo_3, and that of Cyt *bd* is very low. The second key value, $H^+/2e^-$ (or H^+/O) ratio is usually measured using proteoliposomes reconstituted from purified enzyme and phospholipids. There is, however, a possibility that the enzyme may loose H^+-pumping activity while keeping its oxidase activity intact (Sone and Nicholls, 1984; Sone et al., 1986), or that the reconstitution was not appropriate. Another possible method is to measure the H^+/O ratio of resting cells, although using this approach there is some ambiguity as to which respiratory complex(es) participate(s) and what substrate is used. It is usually assumed that NADH and/or NADPH are used to reduce O_2 during the rapid oxygen pulse given to the anaerobic cell suspension (Sone, 1991). In *Bacillus* species, the former problem is simplified, since proton pumping NDH-I is absent and QcR translocate H^+ using the quinone cycle as in the mitochondrial QcR (Trumpower, 1990). The H^+/O ratios of resting cells of *B. thermodenitrificans* K1041 (mainly using branch 1), K17 (mainly using branch 2 without having Cyt *caa*$_3$) and K17q8 (using Cyt, *b(o/a)*$_3$ without having Cyt *caa*$_3$ or Cyt *bd*) were about 5.0 (Sone et al., 1999, 2000). Thus, the expected $H^+/2e^-$ ratios for three terminal oxidases were almost in accordance

Table 4. Comparison of three terminal oxidases in *B. thermodenitrificans*

Feature	Cyt *caa*$_3$	Cyt bo_3	Cyt *bd*
Electron donor	QcR (Cyt b_6c_1)	Cyt *c*-551	MK-7
K_m for O_2 (μM)	0.6	≅0.2	0.1 >
H^+/O of whole cells[a]	6-7	4-5	2-3
H^+/O (reconstituted)	2-3[b]	1[b]	1[c]
Cyanide sensitivity (I_{50}, μM)[d]	30	4	600

[a]Cells contain mainly the indicated terminal oxidase but the numbers listed represent the H^+/O stoichiometry of the whole respiratory chain. [b]Data were taken from Sone and Yanagita (1984) for Cyt caa$_3$ and from Nikaido et al. (2000) for Cyt bo$_3$. [c]Not determined. The value shown is the one expected from chemical proton. [d]Obtained from a titration curve of cell respiration (Sone et al., 2000).

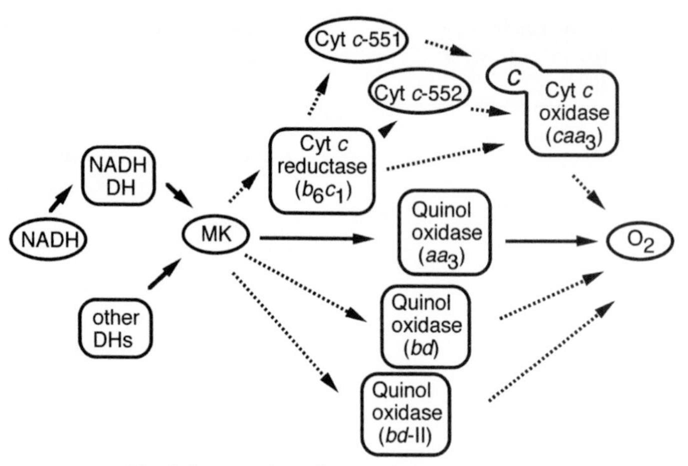

Fig. 7. Suggested respiratory chain of *B. subtilis*.

with the values obtained using the respective purified enzyme inlayed in liposomes. It is thus likely that three terminal oxidases have different roles. When the O_2 concentration is sufficient, the most efficient energy-conserving oxidase, Cyt caa_3 is expressed, but when O_2 concentration is very low, the more wasteful quinol oxidase, Cyt *bd* is working, whereas the SoxB-type Cyt bo_3 is intermediate. It seems likely that one of the main roles of Cyt *bd* is to remove oxygen which may cause injury to cells.

III. *Bacillus subtilis*

Among the mesophilic Gram-positive bacteria, the respiratory chain of *B. subtilis* is the by far most extensively studied. The complete genome of *B. subtilis* strain 168 has been sequenced (Kunst et al., 1997) and many of the components that constitute the *B. subtilis* respiratory chain (Table 1) has also been studied biochemically in great detail. Like the thermophile *Bacillus*, *B. subtilis* contains a branched aerobic respiratory chain with a Cyt *c* reductase-oxidase branch and quinol oxidase branches (Fig. 7).

A. NADH Dehydrogenase

NADH dehydrogenase type II was isolated from *B. subtilis* strain W23 and characterized by Bergsma et al., 1982. The enzyme is a single, membrane-associated polypeptide containing FAD as the prosthetic group. The genome of *B. subtilis* 168 contains three putative NDH-II encoding genes, *yjlD*, *yutJ* and *yumB*. It is presently unknown which one that may correspond to the enzyme isolated from strain W23. Expression of YjlD was repressed under anaerobic growth conditions (Marino et al., 2000). No genes encoding a Complex I/NDH-I are present in the *B. subtilis* genome. A gene that is denoted *ndh5* in the databases, due to primary sequence similarity with the Complex I ND5/NuoL subunit, encodes an antiporter-like polypeptide of unknown function (Mathiesen and Hägerhäll, 2002).

B. Succinate Dehydrogenase

The succinate:menaquinone oxidoreductase (SQR) in *B. subtilis* consists of three polypeptides, encoded by the *sdhCAB* operon. Those are a flavoprotein (584 residues) encoded by *sdhA* containing covalently bound FAD (FP), an iron-sulfur protein (251 residues) encoded by *sdhB* containing three iron-sulfur clusters (IP) and a diheme Cyt *b* membrane anchor encoded by *sdhC* (201 residues). Structurally, the enzyme is very similar to that of the *Wolinella succinogenes* fumarate reductase, that also possesses a one polypeptide diheme membrane anchor, for which the three-dimensional structure has been solved (Lancaster et al., 1999). It should be emphasized though, that other members of this enzyme family that have two polypeptide anchors, containing one heme or lacking heme, also have a conserved structure (Hägerhäll and Hederstedt, 1996; Iverson et al., 1999, Lancaster et al., 1999). The two hemes in *B. subtilis* SQR have different optical and EPR spectral properties, and a high (+16 mV) and low (–132 mV) redox midpoint potential (Hägerhäll et al., 1992). The two hemes have bis-histidine axial ligands. The high potential heme (heme b_P) is located towards the negative side of the membrane, near iron-sulfur cluster S3, and is ligated by H70 and H155, and the low potential heme (heme b_D) is located towards the positive side of the membrane, ligated by H28 and H113 (Hägerhäll et al., 1995). Heme is required for proper assembly of the SQR enzyme in *B. subtilis* (Hederstedt and Rutberg, 1980). The enzyme is a succinate dehydrogenase of Class 3 (Hägerhäll, 1997), catalyzing oxidation of succinate (E_m' = +30 mV) and reduction of MK (E_m' = –75 mV), a seemingly energetically unfavorable reaction. It was observed that succinate oxidation activities were severely reduced when the bacteria were disrupted or treated with uncoupler (Lemma et al, 1990; Schirawski and Unden, 1998). It has thus been postulated that the Class 3 enzymes catalyzing MK reduction by succinate are energy coupled, and rely on the transmembrane electrochemical potential for function. All Class 3 SQRs investigated so far contain two hemes in the membrane anchor, whereas enzymes of Class 1 (ubiquinone reducing SQRs) or Class 2 (fumarate reductases) can have different compositions (Hägerhäll, 1997). Other features that identify a Class 3 SQR is an E_m' of the binuclear iron-sulfur cluster S1 higher than that of the succinate/fumarate couple (+80 mV in *B. subtilis*) and a lower E_m' of the trinuclear iron-sulfur cluster S3 (–25 mV in *B. subtilis*). The MK-like inhibitor HQNO was shown to bind at the distal quinone binding site (Q_D) near heme b_D, in *B. subtilis* SQR (Smirnova et al., 1995). The H28Y and H113Y mutants used in identification of the heme b_D axial ligands do not assemble a SQR enzyme with FP and IP, but subsequent mutations at the same positions, H28L and H113M, gave rise to assembled SQR. The latter mutants contained heme b_D with altered properties and/or exhibited no or very poor quinone reductase activity (Mattsson et al., 2000). This strongly suggests that MK reduction occurs at Q_D. In *E. coli* fumarate reductase, a Class 2 enzyme that carry out the reverse, but energetically favorable reaction compared to the *B. subtilis* enzyme, HQNO binds at the proximal quinone binding site (Q_P) close to iron-sulfur cluster FR3 (Hägerhäll et al., 1999). Taken together, the data fits a model where in B. *subtilis* SQR, electrons are transferred from succinate via FAD → S1 → S2 → S3 → heme b_P → heme b_D to MK at the distal quinone binding site, using the PMF to drive the MK reductase reaction (Hederstedt 2002). If this holds true, we expect that fumarate reduction by a Class 3 SQR should create a membrane potential. Schnorpfeil et al. (2001) indeed demonstrated that fumarate reduction by NADH was coupled to membrane potential generation in *B. subtilis*. In contrast to succinate respiration, this activity was not inhibited by uncouplers. In the structurally very similar fumarate reductase from *W. succinogenes* menaquinol oxidation by fumarate appears to be an electroneutral process (Kröger et al., 2002; Lancaster, 2002). Finally, it should be emphasized that the requirement for membrane energization for respiratory activity in *B. subtilis* is not a specific feature for succinate oxidation using the low potential menaquinone, but was also observed in a mutant strain lacking SQR and using other substrates than succinate (Azarkina and Konstantinov, 2002).

C. QcR

The *qcrABC* operon of *B. subtilis* is very similar to that of *B. thermodenitrificans* already described, and encodes a QcR closely related to but distinct from bc_1 and b_6f complexes. QcrA is similar to Rieske-type iron-sulfur proteins, QcrB resembles the cytochrome *b* of b_6f complexes and QcrC appears to be a fusion of a subunit IV (of b_6f complexes) and a Cyt *c* domain (Yu et al, 1995). The QcrB and QcrC subunits give rise to bands at 22 and 29 kDa respectively on SDS-PAGE, (Yu and LeBrun, 1998). Interestingly, not only the 29-kDa band, but also the 22-kDa band retained heme after denaturation. It was suggested that heme b_H is covalently linked to the polypeptide *via* Cys 43, which is conserved in several *Bacillus* species and cyanobacteria/plastids (Kutoh and Sone, 1988, Yu et al., 2000). In *B. subtilis*, QcR expression is induced at the end of the exponential growth phase and reduced as the bacteria move from the transition state into sporulation (Yu et al., 1995)

D. Small Cyt c

B. subtilis contains two small *c*-type cytochromes with homologous heme domains but different mode of membrane anchoring. CccA, or cytochrome *c*-550, contains an N-terminal transmembrane polypeptide segment (von Wachenfeldt and Hederstedt, 1990) whereas CccB, or cytochrome *c*-551 is a lipoprotein (Bengtsson et al., 1999b) similar to the previously described *Bacillus* PS3 Cyt *c*-551. The *cccB* gene is cotranscribed with the *yvjA* gene, but YvjA is not required for Cyt c-551 synthesis and its function remains unknown. The small Cyts *c* should be part of the QcR-CcO branch of the respiratory chain, but individual specific functions have not been identified. Mutants deleted for *cccA* or *cccB* show no apparent phenotypic differences (von Wachenfeldt and Hederstedt, 1992, Bengtsson et al., 1999a).

E. Terminal Oxidases

The genome of *B. subtilis* contains genes for four potential oxidases (Table 1, Fig. 7), two heme-copper oxidases (van der Oost et al., 1991) of which caa_3 is a CcO and aa_3 is a quinol oxidase (Lauraeus et al., 1991,) and one, possibly two Cyt *bd*-type oxidases. The caa_3-type CcO is encoded by *ctaCDEF* (Saraste et al., 1991). CtaC, that encodes subunit II of the CcO, was shown to be a lipoprotein but the lipid modification was seemingly not required for formation of a functional enzyme (Bengtsson et al., 1999b). The quinol oxidase aa_3 is encoded by *qox-ABCD* (Santana et al., 1992). Two additional genes, *ctaA* and *ctaB* are required for production of the two heme-copper oxidases since their gene products are involved in heme A biosynthesis (Svensson et al., 1993; Svensson and Hederstedt, 1994; Sturr et al., 1996). Cyt *bd* is encoded by the *cydABCD* operon, where *cydA* and *B* constitute the structural genes encoding the oxidase and CydC and D resemble a membrane-bound ABC-transporter and are required for expression of a functional Cyt *bd* (Winstedt et al., 1998). The *cyd* operon is mainly expressed under low oxygen growth conditions. The other putative cytochrome *bd* is predicted from DNA sequence analyses. The genes *ythA* and *ythB* are closely related to *cbdA* and *cbdB* from *B. thermodenitrificans* encoding a Cyt *bd* (Sakamoto et al., 1999a), but there is no direct experimental evidence for the presence of this oxidase in *B. subtilis*. Each of the oxidases can be deleted individually without any noticeable effect on growth. However, it has been demonstrated that one quinol oxidase, either aa_3 or Cyt *bd* is required for aerobic growth, and that one heme-copper oxidase, either caa_3 or aa_3 is needed for efficient sporulation (Winstedt and von Wachenfeldt, 2000). In addition, there is spectroscopic data indicating the presence of a *bb′*-type oxidase in *B. subtilis*, but the genes encoding such an oxidase have not been identified (Azarkina et al., 1999).

The QcR-CcO branch (Cyt b_6c_1 complex and Cyt caa_3) is only expressed in very low amount in vegetative *B. subtilis* cells, since such cells lacked absorbance features around 550 nm (von Wachenfeldt and Hederstedt, 1992) indicating a very low *c*-type Cyt content. The H^+/O ratio of *B. subtilis* has been determined to be around 4.0 (Jones, 1975), which is in agreement with the value expected for the Cyt aa_3-type quinol oxidase branch. It is thus likely that the main terminal oxidase in *B. subtilis* is Cyt aa_3, although the a combination of a Cyt b_6c_1-caa_3 branch (H^+/O of 6.0) and a Cyt *bd* quinol oxidase (H^+/O of 2.0) could account for the observed H^+/O ratio. Both Cyt b_6c_1 and Cyt caa_3, with a clear absorbance shoulder at around 550 nm was observed in membranes from *B. subtilis* grown in succinate medium, and purified Cyt caa_3 was prepared from such cells (Lauraeus et al., 1991; Henning et al., 1995). The enzyme contains a mixed-valence, binuclear Cu_A copper center (von Wachenfeldt and Hederstedt, 1993). In addition, the presence of heme-Cu terminal oxidases has been shown in several other mesophilic *Bacillus* species. Almost equal amounts of Cyt caa_3 and Cyt aa_3 were found to be present in *B. cereus* (Garcia-Horsman et al., 1991) and *B. brevis* (Yaginuma et al., 1997). Interestingly, the H^+/O ratio of these bacteria is clearly higher than 4.0.

IV. Alkaliphilic *Bacillus*

Several *Bacillus* species are known to grow at pH as high as 10-11. Some of them are obligate alkaliphiles like *Bacillus* YN-1, *B. alcalophilus* and *B. firmus* RAB, while several species are facultative, and also able to grow at neutral pH. Examples of the latter are *Bacillus* YN-2000, *B. halodurans* C-125 and *B. firmus* OF4.

A. Energy Coupling in the Alkaliphiles

It has been demonstrated that a Na^+ cycle plays a central role for pH homeostasis in the alkaliphiles (Krulwich et al., 1998), that often maintains the internal pH up to two pH units more acidic than the exterior pH. This reverse-direction ΔpH poses a fundamental problem for a traditional Mitchellian energy coupling. Yet alkaliphilic *Bacillus* thrive on non-fermentable substrates, conserving redox energy in the form of PMF, and the ATP synthase is H^+-dependent, although most of secondary translocators and the flagellar motor are Na^+-driven. It might thus be expected that the respiratory chain components of these bacteria are specially adapted for function under these conditions (Hicks and Krulwich, 1995). A systematic investigation of the components of the

Table 5. E_m' of *c*-type cytochromes of alkaliphiles in comparison with neutrophilic *Bacillus* species. The data were taken from the following references: Cyt c_1 of the thermophilic *Bacillus* PS3 (Kutoh and Sone, 1988); Cyt *c*-552 from *B. firmus* RAB (Davidson et al., 1988); Cyt *c*-553 from *Bacillus* YN-2000 (Yumoto et al., 1991); Cyt *c*-550 from *B. subtilis* (von Wachenfeldt and Hederstedt, 1993); Cyt *c* in cyt. caa_3 from *Bacillus* YN-2000 (Yumoto et al., 1993); Cyt *c* in Cyt caa_3 of the thermophilic *Bacillus* PS3 (Poole et al., 1983).

	Mid point potential (E_m', mV)	
	Alkaliphile	*Neutrophile*
QcR (cyt c_1)	85[a] (OF4)	200 (PS3)
Small Cyt *c*	66 (RAB *c*-552)	225 (PS3 *c*-551)
	87 (YN *c*-553)	178 (Bsu *c*-550)
CcO (Cyt *c* of caa_3)	95 (YN)	229 (PS3)

[a] E_m' of the Rieske Fe-S center is shown (Riedle et al., 1993).

PMF has been carried out by Krulwich's group, when the pH of the culture medium of *B. firmus* OF4 was changed (summarized in Table 5, for reviews see Hicks and Krulwich, 1995: Krulwich et al., 1997). For example, PMF across *B. firmus* membrane are –60 mV at pH 9.5 and –45 mV at pH 10.5, since pH_{in} are 7.5 and 8.2, and $\Delta\Psi$ are –182 and –183 mV, respectively. On the contrary, the cells to cultured at pH 7.5 form $\Delta\Psi$ of –140/150 mV without ΔpH. The measured PMF was far smaller than 150 mV, which is commonly found in cells that synthesize one ATP per three H^+. It is thus possible either that the H^+/ATP ratio may be different or changeable in alkaliphiles or that there may be some space, which is not in equilibrium with the bulk phase through which H^+ is translocated between the redox complex and H^+-ATP synthase. They tend to postulate super-complex formation between CcO and H^+-ATP synthase (Krulwich et al., 1998).

B. Respiratory Chain Components in the Alkaliphiles

Recently, the whole genome sequence of the alkaliphilic *B. halodurans* was reported (Takami et al., 2000). Unfortunately, very little, if any, biochemical work has been carried out on the respiratory chain enzymes of *B. halodurans*. In *B. firmus* the main respiratory chain components were reported to be NADH dehydrogenase → MK-7 → QcR → Cyt caa_3 → O_2, with an alternative quinol oxidase branch using a *bd*-type quinol oxidase (Gilmour and Krulwich, 1996). The latter oxidase is mainly expressed in stationary growth phase. In *B. firmus* neither the presence of a SoxB-type CcO nor an aa_3-type quinol oxidase have been reported. In the genome of *B. halodurans* there are five gene clusters encoding putative different terminal oxidases; a Cyt caa_3–type CcO, a quinol oxidase Cyt aa_3, a SoxB-type CcO and two Cyt *bd*. It remains to be confirmed whether *B. firmus* OF4 lacks the genes for a Cyt aa_3-type quinol oxidase and a SoxB-type CcO or if it only did not express them in detectable amounts.

1. NADH Dehydrogenase

Hicks and Krulwich (1995) postulate that *B. firmus* contain an NADH dehydrogenase type I, since deamino-NADH was oxidized by the membrane fraction of *B. firmus*. The genome of *B. halodurans* (Table 1) did not contain NDH-I encoding genes. However, since the substrate specificity of NDH-II depend on the organism, and that NDH-II from *T. thermophilus* was shown to be able to oxidize deamino-NADH (Yagi et al., 1988), it remains to be established if *B. firmus* contains a proton translocating NDH-I. Xu et al. (1989, 1991) have isolated and characterized NDH-II and its gene from *Bacillus* YN-1. The enzyme, containing one FAD, has molecular mass of 56 kDa and its sequence has high similarity to thioredoxin reductase from *E. coli*.

2. SQR

B. halodurans genome also tells that its SQR should be very similar to that of *B. subtilis*. The enzyme was purified from *B. firmus* OF4, which was composed of three subunits (Gilmour and Krulwich, 1996).

3. QcR

A QcR has not been isolated or directly demonstrated in any alkaliphile *Bacillus* species. However, a Rieske-type Fe-S center was detected in *B. firmus* OF4 membranes (Riedle et al., 1993), and a band at 28 kDa was detected when a SDS-PAGE of the same membranes was stained for heme (Hicks and Krulwich, 1995). A *qcrABC* operon is present in the *B. halodurans* genome, and its deduced amino acid sequence is very similar to that of *B. thermodenitrificans*.

C. Small Cytochrome c

Two membrane-bound *c*-type Cyts have been isolated from *Bacillus* YN-2000 (Yumoto et al., 1991). One, the Cyt *c*-552 is a multi-subunit protein and its function has not been elucidated. The other, Cyt *c*-553, is a small 10.5-kDa polypeptide with p*I* 3.4, which was shown to be an effective electron donor to the cao_3 oxidase in the presence of poly-L-lysine. The *B. halodurans* genome contains one gene for a small Cyt *c*, potentially encoding a 122 amino-acid polypeptide with one heme C-motif. The primary sequence is more similar to *B. subtilis* CccA than to CccB, and the essential Cys residue in the N-terminal region is absent. Thus this Cyt *c* is most likely not a lipoprotein but may be membrane-bound by an N-terminal hydrophobic signal sequence as in *B. subtilis* Cyt *c*-550 (von Wachenfeldt and Hederstedt, 1990). The *c*-type Cyts of alkaliphiles are very acidic, probably because they are located on the outer surface of the cytoplasmic membrane. Another peculiarity of the *c*-type Cyts in alkaliphiles is that their redox midpoint potentials are much lower than those of neutrophile and thermophile *Bacillus* (Table 5). A possible reason for this could be that the low E_m' of the *c*-type Cyts may facilitate electron transfer to the active centers of the oxidases, located inside of the membrane. Otherwise, the large negative-inside membrane potential found in alkaliphilic *Bacillus* (Table 5) might retard the reaction (Yumoto, et al., 1991).

D. Terminal Oxidases

Only two oxidases from alkaliphilic *Bacillus* have been characterized biochemically, namely:

1. Cyt caa_3-type CcO

A Cyt caa_3 CcO purified from *B. firmus* OF4 was composed of three subunits (44, 37.5 and 22.5 kDa) and in addition a possible candidate for the small subunit IV was found. The enzyme showed very similar properties to the CcO from the thermophilic *Bacillus* PS3 (Quirk et al., 1993). The gene organization is also very similar; with *ctaA* and *ctaB* preceding the four structural genes, *ctaCDEF*. The deduced amino acid sequence is also very similar to that of *Bacillus* PS3, except for a characteristic feature of membrane spanning enzymes of alkaliphilic bacteria. A number of otherwise conserved basic residues predicted to be facing the outside of the membrane are in alkaliphiles substituted by neutral or acidic residues. For example, the Cyt *c*-binding domain of subunit II from *B. firmus* has a predicted p*I* of 4.0, which is much lower than the 8.0 or above found for subunit II in *B. subtilis* and *Bacillus* PS3 (Hicks and Krulwich, 1995).

A Cyt cao_3-type CcO was purified from *Bacillus* YN-2000 grown under air-limited conditions at pH 10 (Qureshi, 1990). The Cyt consisted of three polypeptides (50, 41 and 22 kDa and the heme composition of the enzyme was 1:1:1 for heme A, heme C and heme O. The enzyme contained about 2 Cu, 2.5 Fe and 1.8 Mg per mole of heme A. The presence of Cu_A was shown by EPR and Resonance Raman spectra indicated that the environment around heme O was similar to that of heme a_3 of the mitochondrial and *Bacillus* PS3 enzymes (Yumoto et al., 1993). These molecular features indicate that this is a typical *Bacillus* CcO of Cyt caa_3-type, except for that the heme at the high spin site was substituted by heme O as also seen in the *Bacillus* PS3 enzyme (Sone and Fujiwara, 1991a).

2. Cyt bd-type Quinol Oxidase

A Cyt *bd* quinol oxidase is found in *B. firmus* OF4 grown at high pH when reaching the stationary phase (Hicks et al., 1991). This oxidase was purified from a strain of OF4 with a disrupted *cta* gene cluster, in which the amount of Cyt *bd* increased several times, although the mutant cells can grow only with fermentable substrate, or at neutral pH with a non-fermentable C-source (Gilmour and Krulwich, 1997). The isolated enzyme consisted of two polypeptide subunits of 35.5 and 23.5 kDa, and showed an absorption peak at 619 nm upon reduction. These properties are very similar to those of the *B. thermodenitrificans* Cyt *bd* (Sakamoto et al., 1996,1999a).

V. *Corynebacterium glutamicum* and Other High G+C Gram-Positive Bacteria

C. glutamicum is an aerobic Gram-positive high-G+C bacterium which is of industrial importance in producing amino acids used as nutritious additives to food and feed. The organisms previously named *Brevibacterium lactofermentum* and *Br. flavum* are found to be closely related to *C. glutamicum*, and are thus now classified as subspecies. In early studies, Cyts *a*, *b*, and *c* have been identified spectroscopically (Kawahara et al., 1988). The *Corynebacterium* cells contain di- or tetrahydrogenated menaquinone-8 (MK-8(2H) or MK-8(4H)), as reported for *C. diphtheriae* (Scholes and King, 1965) and *Br. lipolyticum* (Yamada et al., 1976). A remarkable biochemical feature was that the

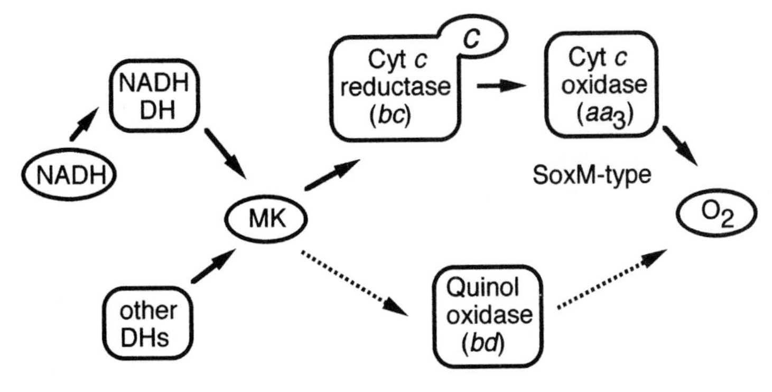

Fig. 8. Suggested respiratory chain of *C. glutamicum*.

organism has only one polypeptide with covalently bound heme C whose apparent molecular weight is about 24 kDa, as judged by SDS-PAGE of the total proteins and heme-staining of the gel. In a typical Cyt *bc-c-aa*$_3$ respiratory pathway, there are two *c*-type Cyts, a peripheral, small Cyt *c* and the Cyt c_1 subunit of the QcR, while in a quinol oxidase branch, there are no *c*-type Cyts. Therefore, the presence of a sole *c*-type Cyt had been mysterious (Fig. 8).

A. The Cytochrome bcc-type QcR

This single *c*-type Cyt was solubilized with detergents and purified to homogeneity from *C. glutamicum*, and found to contain two moles of heme C per mole polypeptide (Sone et al, 2001). The N-terminal amino acid residue was not detected with Edman degradation, suggesting that its amino group was modified, while two internal peptide sequences could be obtained. In parallel, the structural genes for the QcR, *qcrCAB,* were cloned, and the amino acid sequence deduced from the *qcrC* gene matched the partial peptide sequences from the purified *c*-type Cyt (Fig. 9a). Indeed, QcrC contains two heme-C binding motifs, CXXCHX$_n$M (Sone et al, 2001; Niebisch and Bott, 2001). In addition, QcrC has a hydrophobic segment at the C-terminus, which could form a transmembrane α-helix. Taken together, these findings clearly indicate that this Cyt is a hydrophobic diheme protein, composed of two Cyt *c* domains fused in tandem, and thus, the protein was named Cyt *cc*.

The molecular weight of Cyt *cc* (26,214) measured with matrix-assisted laser desorption ionization-time of flight (MALDI-TOF) mass spectrometry was smaller than that of the total weight of two heme C molecules *plus* a QcrC polypeptide (31,074) deduced from the nucleotide sequence, suggesting a posttranslational modification (Sakamoto et al., unpublished). The primary sequence of Cyt *cc* is not homologous to the Cyt c_1 subunit of regular Cyt bc_1 complexes. Instead, the N-terminal domain is most similar to *Bacillus* small Cyt *c* such as Cyt *c*-551 from thermophilic *Bacillus* PS3 and Cyt *c*-550 from *B. subtilis* (CccA). The C-terminal domain resembles to Cyt c_6 from cyanobacteria (Sone et al., 2001).

The QcrA, comprising the Rieske FeS protein, and QcrB, forming the cytochrome *b*, in *C. glutamicum* are homologous to the corresponding subunits of regular Cyt bc_1 complex in aerobic organisms. However, these two proteins from *C. glutamicum*, together with the counterparts from other actinobacteria, are most similar to those from radio-resistant Gram negative bacteria, *Thermus thermophilus* and *Deinococcus radiodurans*, and compose a new group. It is also noteworthy that the Cyt *b* and Rieske FeS protein compose relatively conservative core of QcR, whereas a different Cyt *c* is adopted as the subunit containing heme C in an ancestral organism of each phylogenetic group, or sometimes two such as in the case of *C. glutamicum*,

The QcR subunits easily dissociate in the presence of various detergents and thus it is difficult to purify the intact protein complex (Sone et al., 2001). On the other hand, a heterogeneous sample of QcR, partially purified in a mild detergent, also contains Cyt *aa*$_3$ (see also the following section). This observation suggests that the two enzymes compose a loosely associated super-complex. One of the polypeptides in this latter preparation was identified as QcrA, the Rieske Fe/S protein, by determining the N-terminal amino acid sequence, which lacks the initial Met but starts with the second residue (NNNDKQY…; Sakamoto et al., unpublished). Very recently, Niebisch and Bott (2003) showed that the two enzymes indeed form a super-complex using affinity chromatography of strep-tagged Cyt *b* (QcrB) and strep-tagged subunit I (CtaD). Either of the two subunits expressed in the cells were always copurified with all the subunits of Cyt *bcc* complex and Cyt aa_3, including the fourth subunit (CtaF) of the latter enzyme. A unique feature of the Rieske Fe/S protein in the high G+C Gram-positives is that it contains three N-terminal transmembrane α-helices, whereas in all other organisms only one transmembrane segment is present.

Since no small Cyts *c* is present and the only *c*-type Cyt is the diheme-C Cyt subunit of the QcR, it is likely that one of the two *c*-type Cyt domains of QcR replaces the small Cyt *c* in mediating electron transfer. In other words, the small Cyt *c* has been genetically fused to the ancestral mono-heme C subunit of the QcR resulting in a Cyt *bcc* complex (Fig. 9a). Consequently, it is likely that the Cyt *bcc* complex must donate electrons directly to the terminal oxidase (Fig. 10), and thus the presence of a super-complex is not surprising.

B. Cytochrome aa_3-type CcO

Cyt aa_3 was purified from the *C. glutamicum* membranes as a three-subunit Cyt *c* oxidase (Sakamoto et al, 2001). In addition, all the three structural genes have been cloned (Fig. 9a,b). The molecular weights of subunits I and III determined with MALDI-TOF mass spectrometry were identical within the experimental error to those deduced from the nucleotide sequence, whereas that measured for subunit II was lower than that deduced from the primary sequence. Edman degradation combined with data from mass spectrometry indicated that the N-terminal 28 residues of subunit II are cleaved off and Cys29 is diacylglycerated (Sakamoto et al., 2001). The A-type heme in this enzyme is heme As, which has a geranylgeranyl

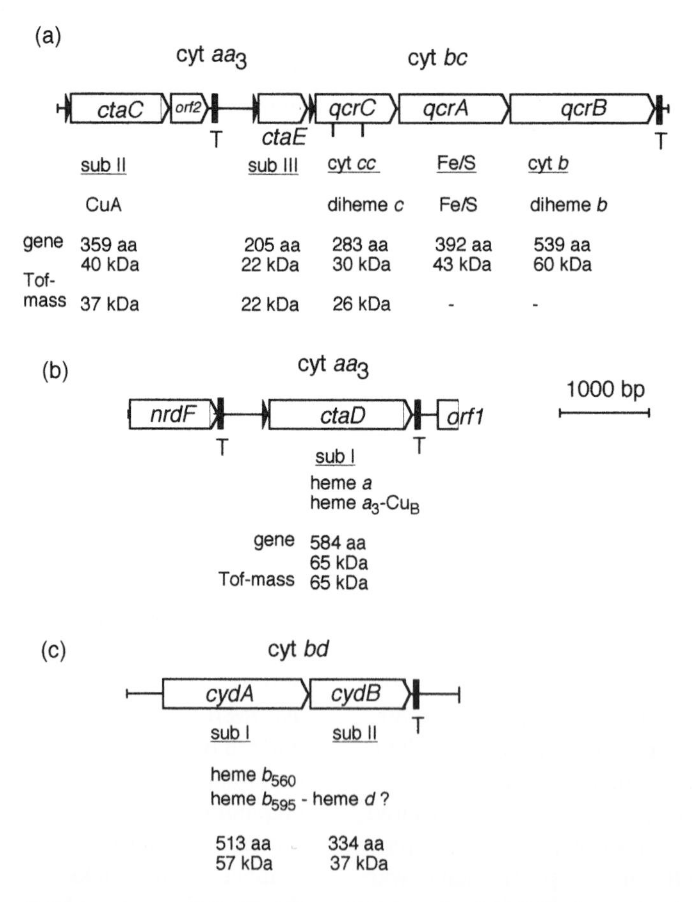

Fig. 9. Gene maps for the respiratory enzymes of *C. glutamicum*. Gene clusters for all the three subunits of Cyt b_6c_1-type QcR and subunits II, III and IV (orf2) of Cyt caa_3-type CcO (a), for subunit I of the CcO (b) and for both of the two subunits of Cyt *bd*-type quinol oxidase.

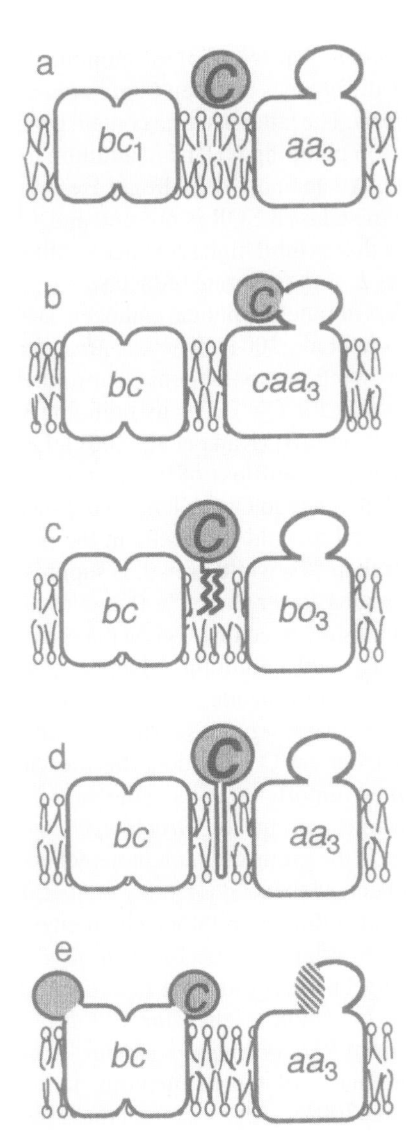

Fig. 10. Five examples of Cyt *c*-membrane interaction. (a) Soluble Cyt *c* as in the mitochondrial respiratory chain; (b) Cyt *c* fused to CcO; (c) Cyt *c* having a lipid anchor as in the *Bacillus* species; (d) Cyt *c* having an N-terminal peptide anchor as in *B. subtilis*; and, (e) Cyt *c* fused to QcR as in Gram-positive bacteria with high G+C.

group side chain instead of a farnesyl. Heme As was first identified in Archaea (Lübben and Morand; 1994). The enzyme is similar to other SoxM-type Cyt *c* oxidases, however, an extra charged amino-acid cluster is inserted between the β2 and β4 strands in the substrate-binding domain of subunit II. The β2-β4 loop of this oxidase is about 30 residues longer than that of the major SoxM-type CcOs, and rich in both acidic and basic residues. This extra charged cluster may play a role in the specific interaction of the oxidase with the Cyt *c* subunit of the new type of *bcc* complex. As described above, no *c*-type Cyts or intrinsic substrates for the Cyt aa_3 oxidase have been found in the cells besides the Cyt *bcc* complex. Oxidase activities measured using Cyts *c* from yeast and horse hearts, Cyt *c*-551 from thermophilic *Bacillus* or other Gram-positive bacteria, and the artificial substrate *N,N,N',N'*-tetramethyl-*p*-phenylene diamine (TMPD) are all very low (Sakamoto et al., 2001). The extra charged inserted in subunit II might contribute to the substrate specificity of the oxidase. The genes encoding subunits II (*ctaC*) and III (*ctaE*) of the oxidase are located directly upstream of the *qcrCAB* operon, while that for subunit I (*ctaD*) is located separately. The close location of the genes for Cyt *bcc* complex and subunit II of Cyt aa_3 is a further indication of the close relationship in the electron transfer between the Cyt *c* reductase and the oxidase.

C. Cytochrome bd-type Quinol Oxidase

A second terminal oxidase, the quinol oxidase Cyt *bd* has been purified from *C. glutamicum* cells and was shown to have high menaquinol oxidase activity (Kusumoto et al., 2000). The genes (*cydAB*) encoding subunits I and II have also been cloned (Fig. 9c). The Cyts *bd* can be divided into two subfamilies (Sakamoto et al., 1999a). Those from many proteobacteria including *E. coli* have a long hydrophilic loop, called 'Q loop', which is suggested to be important for quinol binding and oxidation, between 6th and 7th transmembrane segment, while those from Gram positives, cyanobacteria, archaea and some proteobacteria have a shorter Q loop. Although the spectral properties and the size of subunit I of *C. glutamicum* Cyt *bd* are similar to those of *E. coli* and *A. vinelandii* Cyt *bd*, the enzymatic properties and deduced amino acid sequence clearly indicate that the Cyt is functionally and structurally closer to *B. thermodenitrificans* Cyt *bd* and belongs to the latter subfamily.

D. Respiratory Chain Components and Proton Pumping

The whole genome sequence of *C. glutamicum* became available in the public DNA database of DDBJ/EMBL/GenBank in 2002. The sequence confirms that the respiratory chain contains only two electron-transfer branches, the *bcc*-aa_3 route and the Cyt *bd*-type quinol oxidase route downstream of the quinone pool, as far as the conventional homology searches can detect. A cyanide insensitive oxidase activity has been reported (Matsushita et al. 1998; Kusumoto et al., 2000), but its molecular entity has not been identified. Upstream of the quinone pool, a proton pumping NDH-I/complex I is absent in *C. glutamicum*, while NDH-II is present. Succinate dehydrogenase is present, albeit with an unusual putative SdhC membrane anchor polypeptide, with very little primary sequence similarity to known SdhC and/or SdhD polypeptides from organelles, bacteria or Archea. Malate-menaquinone oxidoreductase or membrane-bound malate dehydrogenase has been reported to play an important role in the citric acid cycle, although there is also a conventional soluble succinate dehydrogenase, which transfers electrons to NAD^+ (Molenaar et al., 1998, 2000). However, there seem to be no proton-pumping enzymes upstream of the quinone pool.

There are two electron-transfer branches, a *bcc-aa*$_3$ route (or more generally a *bc-c-aa*$_3$ route) and *bd* route down stream of the quinone pool, which must exhibit different of energy conservation efficiency. Cyt *bc* and Cyt *aa*$_3$ have been shown to pump protons in several organisms, while Cyt *bd* does not pump but merely translocate chemical protons (Sone, 1990). Altogether, the enzymes in the *bc-c-aa*$_3$ route are expected to translocate about three protons per electron, while the Cyt *bd* route translocates only one proton per electron. The efficiency of energy metabolism must thus depend on which respiratory pathway is mainly operating. Thus, genetic manipulations of respiratory chain components may enhance the cell growth. Very recently, it was found that destruction of the *cydAB* genes increased the cell yield in stationary growth phase, whereas the growth rate was decreased during the exponential growth phase (Sakamoto et al., unpublished). This observation is in agreement with the above considerations about cellular energy metabolism.

E. The Respiratory Chain of Other High G+C Gram-Positive Bacteria

The whole genome sequence of *Mycobacterium tuberculosis,* a major human pathogen, (Cole et al., 1998) indicates that it contains two electron-transfer pathways, *bcc-aa*$_3$ route and *bd* route, similar to what was described for *C. glutamicum*. The structural genes for Cyt *bcc* (*qcrCAB*), *aa*$_3$ (*ctaCDE*) and *bd* (*cydAB*) are highly homologous to the counterparts in *C. glutamicum*. Upstream of the quinone pool, there are many differences between *C. glutamicum* and *M. tuberculosis*. The latter bacteria contain genes encoding a proton pumping NDH-I in addition to NDH-II (Table 1). A succinate dehydrogenase predicted to be a diheme Class 3 SQR is present, and a fumarate reductase that exhibit high sequence similarity to the heme-less *E. coli* fumarate reductase.

Myc. leprae, another human pathogen, lacks *cydAB* genes (Cole et al., 2001), whereas *Myc. smegmatis*, a fast-growing nonpathogenic saprophyte, have two gene sets for Cyt *bd* or its homologue, *cydAB* and *cbdAB* (*ythAB*) (Kana et al., 2001). Deletion of *cydAB* genes caused loss of the absorption peak at 631 nm that is due to Cyt *d*. This *cydAB*-less mutant grows as well as wild type cells in the presence of O_2 higher than 5%, while growth is suppressed at O_2 concentrations lower than 1% (Kana et al., 2001). The authors suggested that Cyt *bd* plays a role under air-limited growth conditions where an oxidase with high affinity to O_2 is needed, as is the case with *E. coli* Cyt *bd*. Since *Myc. leprae* is an intracellular parasite that lives under low O_2 conditions, the presence of Cyt *bd* might be important for its pathogenesis.

The whole genome of *Streptomyces coelicolor,* belonging to the group of filamentous actinomycetes, has also been sequenced (Bentley et al., 2002). The gene sets encoding respiratory chain enzyme complexes are similar to those found in *Myc. tuberculosis* (Table 1), except that the *E. coli*-like fumarate reductase is absent. Genes for Cyt *bcc* was also isolated from *Rhodococcus rhodochrous*, belonging to Nocardiaceae of the filamentous actinomycetes (Sone et al., 2003), suggesting that the diheme *c*-Cyt subunit is a common feature of the whole high G+C actinobacteria.

VI. Molecular Evolution of the Respiration of the Gram-Positives

A. Enzymes Upstream of the Quinone Pool

The Gram-positive bacteria commonly contain a NDH-II type NADH dehydrogenase that does not pump protons, whereas a proton pumping NDH-I is found in some, but not all members of the high G+C group (Table 1).

All Gram-positive bacteria investigated so far seem to have a Class 3 SQR, containing a diheme membrane anchor, possibly with the exception of *C. glutamicum*

(Table 1, see also Section III). As mentioned, two hemes seem to be required to carry out succinate oxidation using a low potential quinone in all species. From the low G+C group, *B. subtilis, B. halodurans, Paenibacillus macerans* and *S. aureus* have only one Class 3 SQR, and a diheme membrane anchor very similar to that of *B. subtilis* (Hederstedt, 2002). In the High G+C group, the primary sequence similarity to other SQRs is lower, and a second enzyme, a fumarate reductase, can be present.

B. Multiple Terminal Oxidases

Aerobic *Bacillus* species, belonging to the low G+C group, typically have more than three oxidases. They commonly contain Cyt caa_3-type (SoxM-type) Cyt *c* oxidase and Cyt *bd*-type quinol oxidase. In addition to these two oxidases, the thermophilic *B. thermodenitrificans* (Sakamoto et al., 1997; Nikaido et al., 1998) and the alkaliphilic *B. halodurans* (Takami et al., 2000) have SoxB-type Cyt *c* oxidase, while the mesophilic *B. subtilis* does not (Kunst et al., 1997). SoxB-type oxidases and its genes have mainly, if not exclusively, been identified in archaea and eubacteria living in more or less extreme environments, including Gram-negative bacteria such as *Thermus thermophilus*. Distribution of SoxB-type oxidase is thus not parallel to the phylogeny of the living organisms. This is 'epidemiological' evidence for the hypothesis that this type of oxidase may be somehow suitable to extremophiles and the genes have been laterally transferred during the evolution. The combination of the Cyt bc-caa_3 oxidase branch and the alternative SoxB-type CcO branch, found in the thermophilic and alkalophilic *Bacillus* species, is the same as that found in *T. thermophilus*, while no similar combination is found in the high G+C Gram-positives.

SoxM-type (Cyt aa_3-type) quinol oxidase is the main terminal oxidase of *B. subtilis* and a homologous gene is also found in the *B. halodurans* genome (Table 1). The genomes of these two *Bacillus* species also contain two gene sets for Cyt *bd*. Altogether, *B. halodurans* has five terminal oxidases, while *B. subtilis* seems to have four. *S. aureus*, contains two quinol oxidases, a SoxM-type heme-copper oxidase and Cyt *bd* (Kuroda et al., 2001; Baba et al., 2002). The different oxidases have various affinities to dioxygen, and the multiple electron-transferring branches show different efficiencies of chemiosmotic energy transduction or various ratios of proton transported per electron transfer, as discussed in the previous sections. The multiple branches are not just redundant but selectively expressed and operating depending on the environmental conditions, especially dioxygen concentration.

On the other hand, high G+C actinobacteria seem to have a Cyt bcc-caa_3 branch and Cyt *bd* branch as the fundamental combination of enzymes in the respiratory chain, except for *Myc. leprae* that lack genes for Cyt *bd* in the genome (Table 1).

C. Divergence of Cytochrome bc Complex

Cyt *bc*-type QcR is not only present in the respiratory chain of aerobic organisms (Cyt bc_1) but also in the electron-transfer chain of photosynthetic organisms (Cyt b_6f). Sequence analysis of Cyt b/b_6 in a wide range of organisms indicates that they can be divided into five groups (Fig. 11; Sone et al., 2001). First, Cyts *b* from proteobacteria and mitochondria compose a distinct group. Secondly, Cyts b_6 of Cyt b_6f complex from cyanobacteria and chloroplast compose another discrete group. Sequence comparison show that Cyt b_6 lacks a portion corresponding to the C-terminal part of Cyt *b* containing three transmembrane helices. Instead, Cyt b_6f complex have an additional subunit, subunit IV, encoded just downstream of the gene for Cyt b_6, and its sequence is similar to the C-terminal part of Cyt *b*. In other words, Cyt *b* is split into Cyt b_6 and subunit IV in this group. Thirdly, Cyts *b* from the low G+C firmicutes composes another group. Most of these Cyt *bc* complexes operate in the respiratory chain, however, the Cyt *b* sequence is more similar to that of Cyt b_6 than to mitochondrial or proteobacterial Cyt *b*. The major difference of Cyt *bc* in this group from Cyt b_6f is that subunit IV is not present in an isolated polypeptide but genetically fused to the N-terminus of the Cyt c_1 subunit. *Heliobacillus mobilis* is a photosynthetic low G+C firmicute. Its Cyt *bc* complex, encoded from a large gene cluster for photosynthetic apparatus, is more similar to a Cyt b_6f complex and has subunit IV encoded by a separate gene (Xiong et al., 1998). The forth group, Cyts *b* from the high G+C actinobacteria is similar to those from radiation-resistant bacteria, *T. thermophilus* and *Deinococcus radiodurans*. This group is distinct, but relatively similar to Cyt *b* of the first group. Archaeal Cyts *b* compose a fifth distinct group.

The phylogenetic tree drawn for the Rieske Fe/S protein is quite similar to that for Cyt *b* (Fig. 3; Sone

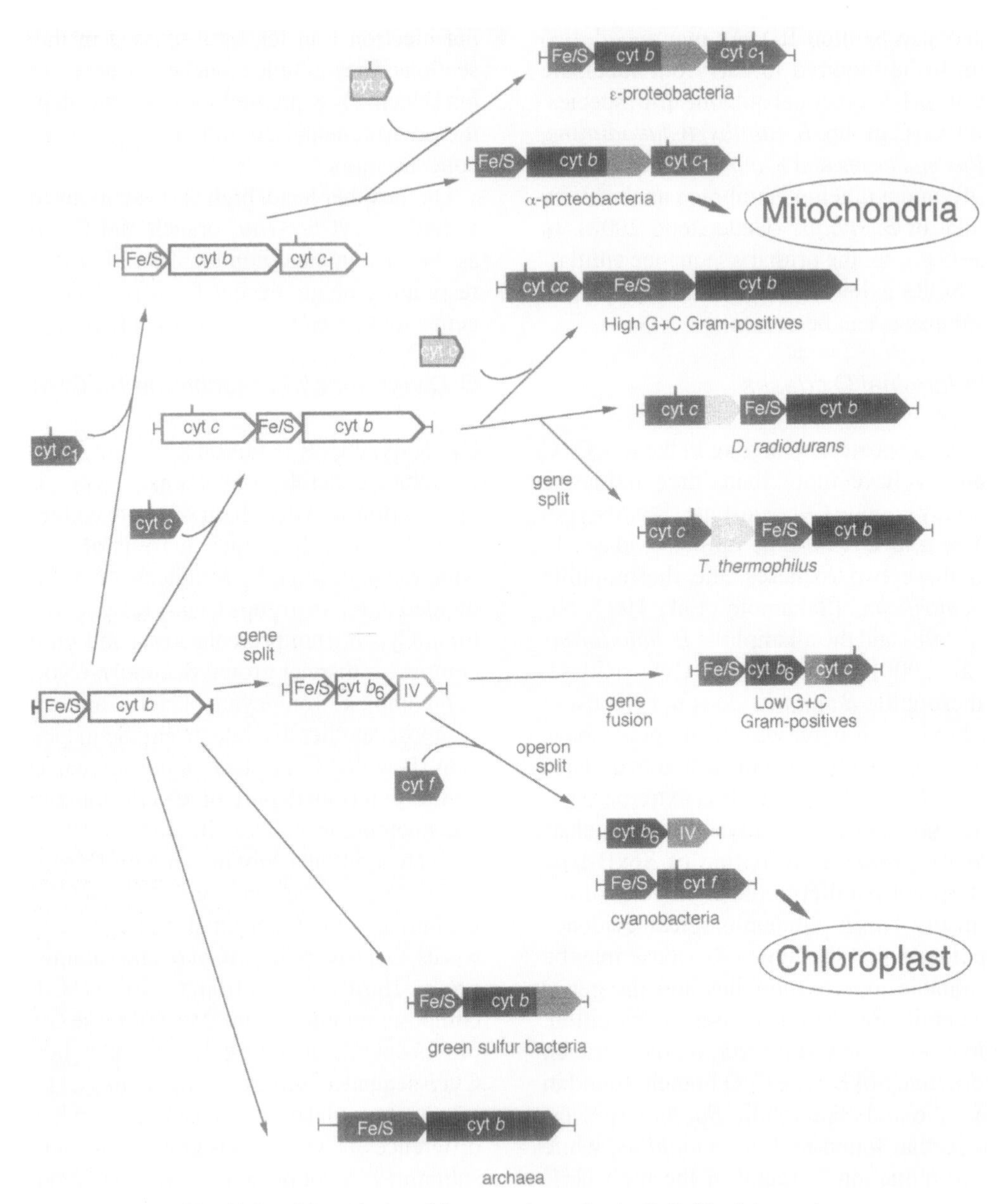

Fig. 11. Possible evolution of the gene cluster for the QcR (Cyt *bc*) complex.

et al., 2003), suggesting that genes for these two proteins have descended and diverged together in the genome during the molecular evolution of Cyt *bc* complex. A remarkable feature of the Fe/S protein from the high G+C actinobacteria is that it has three transmembrane helices at the N-terminal, in contrast to those of other origins, which have only one. Taken together, the total size of Fe/S protein and that of Cyt *b*, of the high G+C group is much larger than in the counterparts from other eubacteria.

Although there are divergent subgroups of Cyt *b* and Fe/S, all the members of the each protein group are so similar that they can be related in a single phylogenetic tree. In contrast, the *c*-type Cyt subunit of *bc* complex is highly divergent. It is likely that different peripheral small-sized Cyt *c* were adopted as the '*c* subunit' of QcR independently in each phylogenetic group of organisms. In the high G+C Gram-positives, two Cyt *c* were adopted to QcR. Genes for such a Cyt *cc* QcR subunit are found in the genomes of ε-proteobacteria; *Helicobacter pylori* (Tomb et al., 1997) and *Campylobacter jejuni* (Parkhill et al.,

2000). A Cyt *cc* gene is also found in a large gene cluster for photosynthesis in the low G+C firmicute *Heliobacillus mobilis* (Xiong et al., 1998), but none has been detected in the respiratory chain of the low-G+C group. It should also be noticed that no sequence similarity is seen between the three groups of Cyts *cc*; the aerobic high G+C actinobacteria, the ε-proteobacteria and the photosynthetic low G+C firmicute, suggesting that these diheme-*c* proteins have emerged independently during molecular evolution of QcR.

'Molecular adoption' of Cyt *c* has also taken place in terminal oxidases, resulting in the Cyt caa_3-type oxidase in the low G+C firmicutes, the radiation-resistant Gram-negatives, in Flavobacteria-Bacteroides species, and in some proteobacteria. Sequence comparison of the Cyt *c* domains suggests that the genetic fusion of Cyt *c* has occurred independently in these groups.

VII. Concluding Remarks

The molecular studies on the respiratory enzyme complexes of Gram-positive bacteria have focused on a handful of model species, as described in this chapter. In addition, the recent whole genome-sequencing projects have provided comprehensive information on the composition of the respiratory chain in particular Gram-positive bacteria. The biochemical studies and primary sequence comparisons have demonstrated a high divergence of the enzyme complexes and how they are combined into respiratory chain of Gram-positive prokaryotes. These results clearly indicate that each of the two groups, the low- and the high-G+C containing Gram-positives, constitutes a distinct class, both in terms of similarities in amino-acid sequences and the combinations of the respiratory enzyme complexes being present. There are few ties enabling us to combine the two groups into an ur-group at a higher level. Further studies on the respiration of these large diverged groups will provide insights into the molecular mechanism of the energy transduction and possibility to manipulate the energy metabolism of such useful organisms.

References

Azarkina N and Konstantinov AA (2002) Stimulation of menaquinone-dependent electron transfer in the respiratory chain of *Bacillus subtilis* by membrane energization. J Bacteriol 184: 5339–5347

Azarkina N, Siletsky S, Borisov V, von Wachenfeldt C, Hederstedt L and Konstantinov AA (1999) A cytochrome *bb′*-type quinol oxidase in *Bacillus subtilis* strain 168. J Biol Chem 274: 32810–32817

Baba T, Takeuchi F, Kuroda M, Yuzawa H, Aoki K, Oguchi A, Nagai Y, Iwama N, Asano K, Naimi T, Kuroda H, Cui L, Yamamoto K, Hiramatsu K (2002) Genome and virulence determinants of high virulence community-acquired MRSA. Lancet 359: 1819–1827

Barsky EL, Nazqarenko AV, Samuilov VD and Khakimov SA (1989) Inhibition of respiration in *Bacillus subtilis* cells by acidification of the cytoplasm. Biol. Membr. (Moscow) 6:720–724

Bengtsson J, Rivolta C, Hederstedt L and Karamata D (1999a) *Bacillus subtilis* contains two small *c*-type cytochromes with homologous heme domains but different types of membrane anchors. J Biol Chem 274: 26179–26184

Bengtsson J, Tjalsma H, Rivolta C and Hederstedt L (1999b) Subunit II of *Bacillus subtilis* cytochrome *c* oxidase is a lipoprotein. J Bacteriol 181: 685–688

Benini S, Gonzalez A, Rypniewski WR Wilson K, Van Beeumen J and Ciurli S (2000) Crystal structure of oxidized *Bacillus pasteurii* cytochrome c-$_{553}$ at 0.97 Å resolution. Biochemistry 39: 13115–13125

Bentley SD, Chater KF, Cerdeno-Tarraga AM, Challis GL, Thomson NR, James KD, Harris DE, Quail MA, Kieser H, Harper D, Bateman A, Brown S, Chandra G, Chen CW, Collins M, Cronin A, Fraser A, Goble A, Hidalgo J, Hornsby T, Howarth S, Huang CH, Kieser T, Larke L, Murphy L, Oliver K, O'Neil S, Rabbinowitsch E, Rajandream MA, Rutherford K, Rutter S, Seeger K, Saunders D, Sharp S, Squares R, Squares S, Taylor K, Warren T, Wietzorrek A, Woodward J, Barrell BG, Parkhill J and Hopwood DA (2002) Complete genome sequence of the model actinomycete *Streptomyces coelicolor* A3(2). Nature 417: 141–147

Bergsma J, van Dongen MBM and Konings WN (1982) Purification and characterization of NADH dehydrogenase from *Bacillus subtilis*. Eur J Biochem 28 (1): 151–157

Cole ST, Brosch R, Parkhill J, Garnier T, Churcher C, Harris D, Gordon SV, Eiglmeier K, Gas S, Barry CE 3rd, Tekaia F, Badcock K, Basham D, Brown D, Chillingworth T, Connor R, Davies R, Devlin K, Feltwell T, Gentles S, Hamlin N, Holroyd S, Hornsby T, Jagels K, Krogh A, McLean S, Moule S, Murphy L, Oliver K, Osborne J, Quail MA, Rajandream M-A, Rogers J, Rutter S, Seeger K, Skelton J, Squares R, Squares S, Sulston JE, Taylor K, Whitehead S and Barrell BG (1998) Deciphering the biology of *Mycobacterium tuberculosis* from the complete genome sequence. Nature 393: 537–544

Cole ST, Eiglmeier K, Parkhill J, James KD, Thomson NR, Wheeler PR, Honore N, Garnier T, Churcher C, Harris D, Mungall K, Basham D, Brown D, Chillingworth T, Connor R, Davies RM, Devlin K, Duthoy S, Feltwell T, Fraser A, Hamlin N, Holroyd S, Hornsby T, Jagels K, Lacroix C, Maclean J, Moule S, Murphy L, Oliver K, Quail MA, Rajandream MA, Rutherford KM, Rutter S, Seeger K, Simon S, Simmonds M, Skelton J, Squares R, Squares S, Stevens K, Taylor K, Whitehead S, Woodward JR and Barrell BG (2001) Massive gene decay in the leprosy bacillus. Nature 409: 1007–1011

Davidson MW, Gray KA, Knaff DB and Krulwich TA (1988)

Purification and characterization of two soluble cytochromes from the alkalophile *Bacillus firmus* RAB. Biochim Biophys Acta 933: 470–477

D'mello R, Hill S and Poole RK (1996) The cytochrome *bd* quinol oxidase in *Escherichia coli* has an extremely high oxygen affinity and two oxygen-binding haems: Implications for regulation of activity in vitro by oxygen inhibition. Microbiology 142: 755–763

Fujiwara Y, Oka M, Hamamoto T and Sone N (1993) Cytochrome *c*-551 of the thermophilic bacterium PS3, DNA sequence and analysis of the mature cytochrome. Biochim Biophys Acta 1144: 213–219

Garcia-Horsman JA, Barquera B and Escamilla JE (1991) Two different aa_3-type cytochromes can be purified from the bacterium *Bacillus cereus*. Eur J Biochem 199: 761–768

Gilmour R and Krulwich TA (1996) Purification and characterization of the succinate dehydrogenases complex and CO-reactive *b*-type cytochromes from the facultative alkaliphile *Bacillus firmus* OF4. Biochim Biophys Acta 1276: 57–63

Gilmour R and Krulwich TA(1997) Construction and characterization of a mutant of alkaliphilic *Bacillus firmus* OF4 with a disrupted *cta* operon and purification of a novel cytochrome *bd*. Bacteriol 179: 863–870

Hägerhäll C (1997) Succinate:quinone oxidoreductases, variations on a conserved theme. Biochim Biophys Acta 1320: 107–141

Hägerhäll C and Hederstedt L (1996) A structural model for the membrane-integral domain of succinate:quinone oxidoreductases. FEBS Lett 389: 25–31.

Hägerhäll C, Aasa R, von Wachenfeldt C and Hederstedt L (1992) Two hemes in *Bacillus subtilis* Succinate: Menaquinone oxidase. Biochemistry 31: 7411–7421

Hägerhäll C, Friden H, Aasa R and Hederstedt L (1995) Transmembrane topology and axial ligands to hemes in cytochrome *b*558 subunit of *Bacillus subtilis* succinate:menaquinone reductase. Biochemistry 34: 11080–11089

Hägerhäll C, Magnitsky S Sled, V, Schröder, I, Gunsalus, R.P., Cecchini, G. and Ohnishi, T. (1999) An *Escherichia coli* mutant quinol:fumarate reductase contains an EPR detectable semiquinone stabilized at the proximal quinone binding site. J Biol Chem 274 : 26157–26164

Haltia T, Saraste M and Wikström M (1991) Subunit III of cytochrome *c* oxidase is not involved in proton translocation: A site-directed mutagenesis study. EMBO J 10: 2015–2021

Hederstedt L (2002) Succinate:quinone oxidoreductase in the bacteria *Paracoccus denitrificans* and *Bacillus subtilis*. Biochim Biophys Acta 1553: 73–83

Hederstedt L and Rutberg L (1980) Biosynthesis and membrane binding of succinate dehydrogenase in *Bacillus subtilis*. J Bacteriol 144: 941–951

Henning W, Vo L, Albanese J and Hill BC (1995) High-yield purification of cytochrome aa_3 and cytochrome caa_3 oxidases from *Bacillus subtilis* plasma membranes. Biochem J 309: 279–283

Hicks DB and Krulwich TA (1995) The respiratory chain of alkaliphilic bacilli. Biochim Biophys Acta 1229: 303–314

Hicks DB, Plass RJ and Quirk PG (1991) Evidence for multiple terminal oxidases, including cytochrome d, in facultatively alkaliphilic *Bacillus firmus* OF4. J Bacteriol 173: 5010–5016

Hirota S, Svensson B, Adelroth P, Sone N, Nilsson T, Malmström BG and Brezezinski P (1996) A flash-photolysis study of the reactions of a *caa*3-type cytochrome *c* oxidase with dioxygen and carbon monoxide. J Bioenerg Biomembr 28: 495–501

Ishizuka M, Machida K, Shimada S, Mogi A, Tsuchiya T, Ohmori T, Souma Y Gonda M and Sone N (1990) Nucleotide sequence of the gene coding for four subunits of cytochrome *c* oxidase from the thermophilic bacterium PS3. J Biochem 108: 866–873

Iverson TM, Luna-Chavez C, Cecchini G and Rees DC (1999) Structure of the *Escherichia coli* fumarate reductase respiratory complex. Science 284: 1961–1966

Jones CW, Brice JM, Downs AJ and Drozd JW (1975) Bacterial respiration-linked proton translocation and its relationship to respiratory chain composition. Eur J Biochem 52: 265–271

Kagekawa S, Mizukami M, Noguchi S, Sakamoto J and Sone N (2002) Importance of hydrophobic interaction between a SoxB-type cytochrome *c* oxidase with its natural substrate cytochrome *c*-551 and its mutants. J Biochem 132: 189–195

Kana BD, Weinstein EA, Avarbock D, Dawes SS, Rubin H, Mizrahi V. (2001) Characterization of the *cydAB*-encoded cytochrome *bd* oxidase from *Mycobacterium smegmatis*. J Bacteriol 183: 7076–7086

Kawahara Y, Tanaka T, Ikeda S and Sone N (1988) Coupling site of the respiratory chain of *Brevibacterium lactofermentum*. Agric Biol Chem 52(8): 197–1983

Koyanagi S, Nagata K, Tamura T, Tsukita S and Sone N. (2000) Purification and characterization of cytochrome *c*-553 from *Helicobacter pylori*. J Biochem 128: 371–375

Kröger A, Biel S, Simon J, Gross R, Unden G and Lancaster CRD (2002) Fumarate respiration of *Wolinella succinogenes*: Enzymology, energetics and coupling mechanism. Biochim Biophys Acta 1553: 23–38

Krulwich TA, Ito M, Gilmour R, Hicks DB and Guffanti AA(1998) Energetics of alkaliphilic *Bacillus* species: Physiology and molecules. Adv Microbial Physiol 40:40–438

Kunst F, Ogasawara N, Moszer I, Albertini AM, Alloni G, Azevedo V, Bertero MG, Bessieres P, Bolotin A, Borchert S, Borriss R, Boursier L, Brans A, Braun M, Brignell SC, Bron S, Brouillet S, Bruschi CV, Caldwell B, Capuano V, Carter NM, Choi SK, Codani JJ, Connerton IF, Cummings NJ, Daniel RA, Denizot F, Devine KM, Durterhoft A, Ehrlich SD, Emmerson PT, Entian KD, Errington J, Fabret C, Ferrari E, Foulger D, Fritz C, Fujita M, Fujita Y, Fuma S, Galizzi A, Galleron N, Ghim S-Y, Glaser P, Goggeau A, Golightly EJ, Grandi G, Guiseppi G, Guy BJ, Haga K, Haiech J, Harwood CR, Henaut A, Hilbert H, Holsappel S, Hosono S, Hullo M-F, Itaya M, Jones L, Joris B, Karamata D, Kasahara Y, Klabber-Blanchard M, Klein C, Kobayashi Y, Koetter P, Koningstein G, Krough S, Kumano M, Kurita K, Lapidus A, Lardinois S, Lauber J, Lazarevic V, Lee S-M, Levine A, Liu H, Masuda S, Mauel C, Medigue C, Medina N, Mellado RP, Mizuno M, Moestl D, Nakai S, Noback M, Noone D, O'Reilly M, Ogawa K, Ogiwara A, Oudega B, Park S-H, Parrot V, Pohl TM, Portetelle D, Porwollik S, Prescott AM, Presecan E, Pujic P, Purnelle B, Rapoport G, Rey M, Reynolds S, Rieger M, Rivolta C, Rocha E, Roche B, Rose M, Sadaie Y, Sato T, Scanlan E, Schleich S, Schroeter R, Scoffone F, Sekiguchi J, Sekowska A, Seror SJ, Shin B.S, Soldo B, Sorokin A, Tacconi E, Takagi T, Takahashi H, Takemaru K, Takeuchi M, Tamakoshi A, Tanaka T, Terpstra P, Tognoni A, Tosato V, Uchiyama S, Vandenbol M, Vannier F, Vassarotti A, Viari A, Wanbutt R, Wedler E, Wedler H, Weitzenegger T, Winters P, Wipat A, Yamamoto H, Yamane K, Yasumoto K, Yata K, Yoshida K, Yoshikawa H-F, Zumstein E, Yoshikawa H and Danchin A, (1997) The complete genome sequence of the Gram-positive bacterium *Bacillus subtilis*. Nature 390: 249–256

Kuroda M, Ohta T, Uchiyama I, Baba T, Yuzawa H, Kobayashi I, Cui L, Oguchi A, Aoki K, Nagai Y, Lian J, Ito T, Kanamori M, Matsumaru H, Maruyama A, Murakami H, Hosoyama A, Mizutani-Ui Y, Takahashi NK, Sawano T, Inoue R, Kaito C, Sekimizu K, Hirakawa H, Kuhara S, Goto S, Yabuzaki J, Kanehisa M, Yamashita A, Oshima K, Furuya K, Yoshino C, Shiba T, Hattori M, Ogasawara N, Hayashi H and Hiramatsu K (2001) Whole genome sequencing of meticillin-resistant *Staphylococcus aureus*. Lancet 357: 1225–1240

Kusano T, Kuge S, Sakamoto J, Noguchi S and Sone N (1996) Nucleotide and amino acid sequence for cytochrome *caa*3-type oxidase of *Bacillus stearothermophilus* K1041. Biochim Biophys Acta, 1273: 129–138

Kusumoto K, Sakiyama M, Sakamoto J, Noguchi S and Sone N (2000) Menaquinol oxidase activity and primary structure of cytochrome *bd* from the amino-acid fermenting bacterium *Corynebacterium glutamicum*. Arch Microbiol 173: 390–397

Kutoh E and Sone N (1988) Quinol-cytochrome *c* oxidoreductase from the thermophilic bacterium PS3. J Biol Chem 263: 9020–9026

Lancaster CR (2002) *Wolinella succinogenes* quinol:fumarate reductase-2.2-Å resolution crystal structure and the E-pathway hypothesis of coupled transmembrane proton and electron transfer. Biochim Biophys Acta 1565: 215–231

Lancaster CR, Kröger A, Auer M and Michel H(1999) Structure of fumarate reductase from *Wolinella succinogenes* at 2.2 Å resolution. Nature 402: 377–385

Lauraeus M, Haltia T, Saraste M and Wikström M (1991) *Bacillus subtilis* expresses two kinds of haem-A-containing terminal oxidases. Eur J Biochem 197: 699–705

Lemma E, Unden G and Kröger A (1990) Menaquinone is an obligatory component of the chain catalyzing succinate respiration in *Bacillus subtilis*. Arch Mikrobiol 155: 62–67

Lübben M and Morand K (1994) Novel prenylated hemes as cofactors of cytochrome oxidases. Archaea have modified hemes A and O. J Biol Chem 269: 21473–21479

Marino M, Hoffmann T, Schmid R, Möbitz H and Jahn D (2000) Changes in protein synthesis during the adaptation of *Bacillus subtilis* to anaerobic growth conditions. Microbiology 146: 97–105

Mathiessen C and Hägerhäll C (2002) Transmembrane topology of the NuoL, M and N subunits of NADH:quinone oxidoreductase and their homologues among membrane-bound hydrogenases and *bona fide* antiporters. Biochim Biophys Acta 1556: 121–132

Matsson M, Tolstoy D, Aasa R and Hederstedt L (2000) The distal heme center in *Bacillus subtilis* succinate:quinone reductase is crucial for electron transfer to menaquinone. Biochemistry 39: 8617–8624

Matsushita K, Yamamoto T, Toyama H and Adachi O (1998) NADPH oxidase system as a superoxide-generating cyanide-resistant pathway in the respiratory chain of *Corynebacterium glutamicum*. Biosci Biotechnol Biochem 62: 1968–1977

Molenaar D, van der Rest ME, Petrovic S (1998) Biochemical and genetic characterization of the membrane-associated malate dehydrogenase (acceptor) from *Corynebacterium glutamicum*. Eur J Biochem 254: 395–403

Molenaar D, van der Rest ME, Drysch A, Yucel R (2000) Functions of the membrane-associated and cytoplasmic malate dehydrogenases in the citric acid cycle of *Corynebacterium glutamicum*. J Bacteriol. 182: 6884–6891

Narumi I, Sawakami K, Nakamoto S, Nakayama N, Yanagisawa T, Takahashi N and Kihara H (1992) A newly isolated *Bacillus stearothermophilus* K1041 and its tranformation by electroporation. Biotechnol Biotechniques 6: 83–86

Nicholls P and Sone N (1984) Kinetics of cytochrome *c* and TMPD oxidation by cytochrome *c* oxidase from the thermophilic bacterium PS3. Biochim Biophys Acta 767: 240–247

Niebisch A and Bott M (2001) Molecular analysis of the cytochrome bc_1-aa_3 branch of the Corynebacterium glutamicum respiratory chain containing an unusual diheme cytochrome c_1. Arch Microbiol. 175: 282–94

Niebisch A and Bott M (2003) Purification of a cytochrome bc-aa_3 supercomplex with quinol oxidase activity from *Corynebacterium glutamicum*. Identitfication of a fourth subunit of cytochrome aa_3 oxidase and mutational analysis of diheme cytochrome c_1. J Biol Chem 278: 4339–4346

Nikaido K, Sakamoto J, Handa Y and Sone N (1998) The cbaAB genes for bo_3-type cytochrome *c* oxidase in *Bacillus stearothermophilus*. Biochim Biophys Acta 1397: 262–267

Nikaido K, Sakamoto J, Noguchi S and Sone N (2000) Over-expression of *cbaAB* gene of *Bacillus stearothermophilus* produces a two-subunit SoxB-type cytochrome *c* oxidase with proton pumping activity. Biochim Biophys Acta 1456: 35–44

Noguchi S, Yamazaki T, Yaginuma A, Sakamoto J and Sone N (1994) Over-expression of membrane-bound cytochrome c-$_{551}$ from thermophilic *Bacillus* PS3 in *Bacillus stearothermophilus* K1041. Biochim Biophys Acta 1188: 302–310

Ogura T, Sone N, Tagawa K, Kitagawa T (1984) Resonance Raman study of the *aa*3-type cytochrome oxidase of thermophilic bacterium PS3. Biochemistry 23:2826–2831

Parkhill J, Wren BW, Mungall K, Ketley JM, Churcher C, Basham D, Chillingworth T, Davies RM, Feltwell T, Holroyd S, Jagels K, Karlyshev AV, Moule S, Pallen MJ, Penn CW. Quail MA, Rajandream MA, Rutherford KM, van Vliet AH, Whitehead S, Barrell BG (2000) The genome sequence of the food-borne pathogen *Campylobacter jejuni* reveals hypervariable sequences. Nature 403: 665–668

Poole RK, von Wielink JE, Baines BS, Rejinders WNM, Salmon I and Oltmann LF (1983) The membrane-bound cytochromes of an aerobically grown, extremely thermophilic bacterium, PS3: Characterization by spectral deconvolution coupled with potentiometric analysis. J General Microbiol 129: 2163–2173

Puustinen A and Wikström M (1991) The heme groups cytochrome *o* from *Escherichia coli*. Biochemistry 88: 6122–6126

Quirk PG, Hicks DB and Krulwich TA (1993) Cloning of the operon from alkaliphilic *Bacillus firmus* OF4 and characterization of the pH-regulated cytochrome caa_3 oxidase it encodes. J Biol Chem 268: 678–685

Qureshi MH, Yumoto I, Fujiwara T Fukumori Y and Yamanaka T (1990) A novel *aco*-type cytochrome-*c* oxidase from a facultative alkaliphilic *Bacillus*: Purification and some molecular and enzymatic features. J Biochem 107: 480–485

Riedel A, Kellner E, Grodsitski D, Liebl Y Hauska G, Muller A, Rutherford AW and Nitschke W (1993) The [2Fe-2S] centre of the cytochrome *bc* complex in *Bacillus firmus* OF4 in EPR: An example of a menaquinol-oxidizing Rieske centre. Biochim Biophys Acta 1183: 26–268

Saiki K, Mogi T, Ogura K and Anraku Y (1993) In vitro heme O synthesis by the *cyoE* gene product from *Escherichia coli*.

J Biol Chem 268: 26041–26045
Saiki K, Mogi T, Ishizuka M, Anraku Y (1994) An *Escherichia coli cyoE* gene homologue in thermophilic *Bacillus* PS3 encodes a thermotolerant heme O synthase. FEBS Lett. 351: 385–388
Sakamoto J, Matsumoto A, Oobuchi K and Sone N (1996) Cytochrome *bd*-type quinol oxidase in a mutant of *Bacillus stearothermophilus* deficient in caa$_3$-type cytochrome *c* oxidase. FEMS Microbiol Lett 143: 151–158
Sakamoto J, Handa Y and Sone N (1997) A novel Cytochrome *b(o/a)*$_3$-type oxidase from *Bacillus stearothermophilus* catalyzes cytochrome *c*-551 oxidation. J Biochem 122: 764–771
Sakamoto J, Koga T, Mizuta T, Sato C, Noguchi S and Sone N (1999a) Gene structure and quinol oxidase activity of a cytochrome *bd*-type oxidase from *Bacillus stearothermophilus*. Biochim Biophys Acta 1411: 147–158
Sakamoto J, Hayakawa A, Uehara T, Noguchi S and Sone N (1999b) Cloning of *Bacillus stearothermophilus* ctaA and heme synthesis with the ctaA protein produced in *Escherichia coli*. Biosci Biotechnol Biochem 63: 96–103
Sakamoto J, Shibata T, Mine T, Miyahara R, Torigoe T, Noguchi S, Matsushita K and Sone N (2001) Cytochrome *c* oxidase contains an extra charged amino acid cluster in a new type of respiratory chain in the amino acid-producing gram-positive bacterium *Corynebacterium glutamicum*. Microbiology 147: 2865–2871
Samuilov VD and Khakimov SA (1991) Dependence of respiration of *Bacillus subtilis* cells on monovalent cations. Biokhimia 56: 1209–1214
Santana M, Kunst F, Hullo MF, Rapoport G, Danchin A and Glaser P (1992) Molecular cloning, sequencing, and physiological characterization of the *qox* operon from *Bacillus subtilis* encoding the *aa*$_3$-600 quinol oxidase. J Biol Chem 267: 10225–10231
Saraste M, Metso T, Nakari T, Jalli T, Lauraeus M and Van der Oost J (1991) The *Bacillus subtilis* cytochrome-*c* oxidase. Variations on a conserved protein theme. Eur J Biochem 195: 517–525
Schirawski J and Unden G (1998) Menaquinone-dependent succinate dehydrogenase of bacteria catalyzes reversed electron transfer driven by the proton potential. Eur J Biochem 257: 210–215
Scholes PB, and King HK (1965) Isolation of a naphthoquinone with partly hydrogenated side chain from *Corynebacterium diphtheriae*. Biochem J 97: 766–768
Schnorpfeil M, Janausch IG, Biel S, Kroger A and Unden G (2001) Generation of a proton potential by succinate dehydrogenase of *Bacillus subtilis* functioning as a fumarate reductase. Eur J Biochem 268: 3069–3074
Smirnova I, Hägerhäll C, Konstantinov A and Hederstedt L (1995) HOQNO interaction with cytochrome *b* in succinate: menaquinone reductase from *Bacillus subtilis*. FEBS Lett 359: 23–26
Sone N (1990) Respiration-driven proton pumps. In: Krulwich TA (ed) The Bacteria, Vol 12, pp 1–32. Academic Press, New York
Sone N and Fujiwara Y (1991a) Effects of aeration during growth of *Bacillus stearothermophilus* on proton pumping activity and change of terminal oxidases. J Biochem 110: 1016–1021
Sone N and Fujiwara Y (1991b) Haem O can replace haem A in the active site of cytochrome *c* oxidase from thermophilic bacterium PS3. FEBS Lett 288: 154–158
Sone N and Hinkle PC (1982) Proton transport by cytochrome *c* oxidase from the thermophilic bacterium PS3 reconstituted in liposome. J Biol Chem 257: 1579–1582
Sone N and Kosako T (1986) Evidence for dimer structure of proton-pumping cytochrome *c* oxidase, an analysis by radiation inactivation. EMBO J 5: 1515–1519
Sone N and Nicholls P (1984) Effect of heat treatment on oxidase activity and proton-pumping capability of proteoliposome-incorporated beef heart cytochrome *aa*$_3$. Biochemistry 23: 6550-6554
Sone N and Takagi T (1990) Monomer-dimer structure of cytochrome-*c* oxidase and cytochrome *bc*$_1$ complex from the thermophilic bacterium PS3. Biochim Biophys Acta 1020: 207–212
Sone N and Toh H (1994) Membrane-bound *Bacillus* cytochromes *c* and their phylogenetic position among bacterial Class I cytochromes *c*. FEMS Microbiol Lett 122: 203–210
Sone N and Yanagita Y (1982) A cytochrome *aa*3-type terminal oxidase of a thermophilic bacterium. Purification, properties and proton pumping. Biochim Biophys Acta 682: 216–226
Sone N and Yanagita Y (1984) High vectorial proton stoichiometry by cytochrome *c* oxidase from the thermophilic bacterium PS3 reconstituted in liposomes. J Biol Chem 259: 1405–1408
Sone N, Kagawa Y and Orii Y (1983) Carbon monooxide-binding cytochromes in the respiratory chain of thermophilic bacterium PS3 grown with sufficient or limited aeration. J Biochem 93: 1329–1336
Sone N, Naqui A, Kumar C, Chance B (1984a) Pulsed cytochrome *c* oxidase from the thermophilic bacterium PS3. Biochem J 223: 809–813.
Sone N, Naqui A, Kumar C and Chance B (1984b) Reaction of *caa*$_3$-type terminal cytochrome oxidase from the thermophilic bacterium PS3 with oxygen and carbon monoxide at low temperatures. Biochem J 221: 529–533
Sone N, Ogura T and Kitagawa T (1986) Iron-histidine stretching Raman line and enzymatic activities of bovine and bacterial cytochrome *c* oxidase. Biochim Biophys Acta 850: 139–145
Sone N, Sekimachi M and Kutoh E (1987) Identification and properties of a quinol oxidase super-complex composed of a *bc*$_1$ complex and cytochrome oxidase in the thermophilic bacterium PS3. J Biol Chem 262: 15386–15391
Sone N, Yokoi J, Fu T, Ohta S, MetsoT, Raitio M and Saraste M (1988) Nucleotide sequence of the gene coding for cytochrome oxidase subunit I from the thermophilic bacterium PS3. J Biochem 103: 606–610
Sone N, Kutoh E and Yanagita Y (1989) Cytochrome *c*-$_{551}$ from the thermophilic bacterium PS3. Biochim Biophys Acta 1977: 329–334
Sone N, Shimada S, Ohmori T, Souma Y, Gonda M and Ishizuka M (1990) A fourth subunit is present in cytochrome *c* oxidase from the thermophilic bacterium PS3. FEBS Lett 262: 249–252
Sone N, Ogura T, Noguchi S and Kitagawa T (1994) Proton pumping activity and visible absorption and resonance Raman spectra of a *cao*-type cytochrome *c* oxidase isolated from thermophilic bacterium *Bacillus* PS3. Biochemistry 33: 849–85
Sone N, Sawa G, Sone T and Noguchi S (1995) Thermophilic Bacilli have split cytochrome *b* genes for cytochrome *b6* and subunit IV. J Biol Chem. 270: 10612–10617
Sone N, Tsuchiya N, Inoue M and Noguchi, S (1996) *Bacillus stearothermophilus qcr* operon encoding Rieske FeS protein, cytochrome b_6 and a novel-type cytochrome c_1 of quuinol- cy-

tochrome *c* reductase. J Biol Chem, 271: 12457–12462
Sone N, Tsukita S and Sakamoto J (1999) Direct correlationship between proton translocation and growth yield: An analysis of the respiratory chain of *Bacillus stearothermophilus*. J Biosci and Bioengineer 87: 495–499
Sone N, Koyanagi S, and Sakamoto J (2000) Energy-yielding properties of SoxB-type cytochrome bo_3 oxidase: Analyses using *Bacillus stearothermophilus* K1041 and its mutant strain. J Biochem 127: 551–513
Sone N, Nagata K, Kojima H, Tajima J, Kodera Y Kanamaru T, Noguchi S and Sakamoto J (2001) A novel hydrophobic diheme *c*-type cytochrome. Purification from *Corynebacterium glutamicum* and analysis of the qcrCAB operon encoding three subunit proteins of a putative cytochrome reductase complex. Biochim Biophys Acta 1503: 279–290
Sone N, Fukuda M, Katayama S, Jyoudai A, Syugyou M, Noguchi S and Sakamoto J (2003) QcrCAB operon of a nocardia-form actinomycete *Rhodococcus rhodochrous* encodes cytochrome reductase complex with diheme cytocrome *cc* subunit. Biochim Biophys Acta 1557: 125–131
Studholme DJ, Jackson RA and Leak DJ (1999) Phylogenetic analysis of transformable strains of thermophilic *Bacillus* species. FEMS Microbiol Lett 172: 85-90
Sturr MG, Krulwich TA and Hicks DB (1996) Purification of a cytochrome *bd* terminal oxidase encoded by the *Escherichia coli app* locus from a *cyo cyd* strain complemented by genes from *Bacillus firmus* OF4. J Bacteriol 176: 174–1749
Svensson B and Hederstedt (1994) *Bacillus subtilis* CtaA is a heme-containing membrane protein involved in heme A biosynthesis. J Bacteriol 176: 6663–6671
Svensson B, Lubben M and Hederstedt L (1993) *Bacillus subtilis* CtaA and CtaB function in haem A biosynthesis. Molec Microbiol 10: 193–201
Takami H, Nakasone K, Takaki Y, Maeno G, Sasaki R, Masui N, Fuji F, Hirama C, Nakamura Y, Ogasawara N, Kuhara S, Horikoshi K (2000) Complete genome sequence of the alkaliphilic bacterium *Bacillus halodurans* and genomic sequence comparison with *Bacillus subtilis*. Nucleic Acids Res 28: 4317–4331
Tanaka T, Inoue M, Sakamoto J and Sone N (1996) Intra- and Inter-complex cross-linking of subunits in the quinol oxidase super-complex from thermophilic *Bacillus* PS3. J Biochem 119: 482–486
Tomb JF, White O, Kerlavage AR, Clayton RA, Sutton GG, Fleischmann RD, Ketchum KA, Klenk HP, Gill S, Dougherty BA, Nelson K, Quackenbush J, Zhou L, Kirkness EF, PetersonS, Loftus B, Richardson D, Dodson R, Khalak HG, Glodek A, McKenney K, Fitzegerald LM, Lee N, Adams MD, Hickey E, Berg DE, Gocayne JD, Utterback TR, Peterson JD, Kelley JM, Cotton MD, Eidman JM, Fujii C, Bowman C, Watthey L, Wallin E, Hayes WS, Borodovsky M, Karp PD, Smith HO, Fraser CM and Venter JC (1997) The complete genome sequence of the gastric pathogen *Helicobacter pylori*. Nature 388: 539–547
Trumpower BL (1990) The protonmotive Q cycle. Energy transduction by coupling of proton translocation to electron transfer by the cytochrome bc_1 complex. J Biol Chem 265: 11409–11412
Uchida T, Tsubaki M, Kurokawa T, Hori H, Sakamoto J, Kitagawa T and Sone N (2000) Active site structure of SoxB-type cytochrome bo_3 oxidase from thermophilic *Bacillus*. J Inorganic Biochem 82: 65–72
van der Oost J, von Wachenfeld C, Hederstedt L, Saraste M (1991) *Bacillus subtilis* cytochrome oxidase mutants: Biochemical analysis and genetic evidence for two aa_3-type oxidases. Mol Microbiol 5: 2063–2072
von Wachenfeldt C and Hederstedt L (1990) *Bacillus subtilis* 13-kilodalton cytochrome *c*-550 encoded by *cccA* consists of the membrane-anchor and a heme domain. J Biol Chem 265: 13939–13948
von Wachenfeldt C and Hederstedt L (1992) Molecular biology of *Bacillus subtilis* cytochromes. FEMS Lett 100: 91–100
von Wachenfeldt C and Hederstedt L (1993) Physico-chemical characterisation of membrane bound and water-soluble forms of *Bacillus subtilis* cytochrome c_{-550}. Eur J Biochem 212(2): 499–509
Weyer KA, Schafer M, Lottspeich F and Michel H (1987) The cytochrome subunit of the photosynthetic reaction center from *Rhodopseudomonas viridis* is a lipoprotein. Biochemistry 26: 2909–2914
Widger WR, Cramer WA, Herrmann RG and Trebst A (1984) Sequence homology and structural similarity between cytochrome *b* of mitochondrial complex III and the chloroplast b_6-f complex: Position of the cytochrome *b* hemes in the membrane. Proc Natl Acad Sci USA 81: 674–678
Winstedt L and von Wachenfeldt C (2000) Terminal oxidases of *Bacillus subtilis* strain 168: One quinol oxidase, cytochrome aa_3 or cytochrome *bd*, is required for aerobic growth. J Bacteriol 182: 6557–6564
Winstedt L, Yoshida K, Fujita Y, von Wachenfeldt C (1998) Cytochrome *bd* biosynthesis in *Bacillus subtilis*: characterization of the *cyd*ABCD operon. J Bacteriol 180(24): 6571–6580
Xiong J, Inoue K, and Bauer CE (1998) Tracking molecular evolution of photosynthesis by characterization of a major photosynthesis gene cluster from *Heliobacillus mobilis*. Proc Natl Acad Sci USA 95: 14851–14856
Xu XM, Kanaya S, Koyama N, Sekiguchi T, Nosoh Y, Ohashi S and Tsuda K. (1989) Tryptic digestion of NADH dehydrogenase from alkalophilic *Bacillus*. J Biochem (Tokyo) 105: 626–632.
Xu XM, Koyama N, Cui M, Yamagishi A, Nosoh Y, Oshima T(1991) Nucleotide sequence of the gene encoding NADH dehydrogenase from an alkalophile, *Bacillus sp*. strain YN-1. J Biochem 109: 678–683
Yagi T, Honnami K, Ohnishi T (1988) Purification and characterization of 2 types of NADH-quinone reductase from *Thermus thermophilus* HB-8. Biochem 27: 2008–2013
Yaginuma A, Tsukita S, Sakamoto J and Sone N (1997) Characterization of two terminal oxidase in *Bacillus brevis* and efficiency of energy conservation of the respiratory chain. J Biochem 122: 967–976
Yamada Y, Inouye G, Tahara Y, Kondo K (1976) On the chemical structure of menaquinones with the tetrahydrogenated isoprenoid side chain. Biochim Biophys Acta 486: 195–203
Yu J and Le Brun NE (1998) Studies of the cytochrome subunits of menaquinone: Cytochrome *c* reductase (*bc* complex) of *Bacillus subtilis*. J Biol Chem 273(15): 8860–8866
Yu J, Hederstedt L and Piggot PJ (1995) The Cyt *bc* complex (menaquinone: Cyt *c* oxidoreductase) in *Bacillus subtilis* has a non traditional subunit organization. J Bacteriol 177: 6751—6760
Yu J, Vassiliev IR, Jung YS, Golbeck JH and McIntosh L (1997) Strains of *Synechocystis* sp. PCCC 6803 with altered PsaC.

I. Mutations incorporated in the cysteine ligands of the two [Fe-4S] clusters FA and FB of photosystem I. J Biol Chem 272: 8032–8039
Yumoto I, Fukumori Y and Yamanaka T (1991) Purification and characterization of two membrane-bound *c*-type cytochromes from a facultatively alkalophilic *Bacillus*. J Biochem 110: 26–273
Yumoto I, Takahashi S, Kitagawa T, Fukumori Y and Yamanaka T (1993) The molecular features and catalytic activity of Cu_A-containing aco_3-type cytochrome *c* oxidase from a facultative alkalophilic *Bacillus*. J Biochem 114: 88–95

Chapter 3

Respiratory Electron Transport in *Helicobacter* and *Campylobacter*

Jonathan D. Myers and David J. Kelly*
Department of Molecular Biology and Biotechnology, University of Sheffield, Firth Court, Western Bank, Sheffield S10 2TN, U.K.

*Author for correspondence, email: d.kelly@sheffield.ac.uk

Davide Zannoni (ed): Respiration in Archaea and Bacteria. Vol 2: Diversity of Prokaryotic Respiratory Systems, pp. 63–80.

Summary

The microaerophilic, human gastro-intestinal pathogens, *Campylobacter jejuni* and *Helicobacter pylori* are closely related phylogentically, yet distinct in some major aspects of their physiology, especially with regard to electron transport. *C. jejuni* is a more versatile and metabolically active pathogen, with a complete citric-acid cycle, and a complex and highly branched respiratory chain allowing the use of a variety of electron donors such as formate, hydrogen, D-lactate, succinate, malate, NAD(P)H and alternative electron acceptors to oxygen, including fumarate, nitrate, nitrite, N- or S-oxides and hydrogen peroxide. *H. pylori* is a more specialized pathogen, largely restricted to the human stomach, with an incomplete citric-acid cycle and a simpler respiratory chain. Alternative electron acceptors to oxygen include fumarate, hydrogen peroxide and possibly N- or S-oxides. Both organisms contain a novel type of complex I, which lacks the two subunits NQO1 (NuoE) and NQO2 (NuoF), responsible for NADH binding and initial electron transfer reactions in other bacteria. The nature of the electron donor to complex I has yet to be identified. In both bacteria, reducing equivalents are transferred to the sole quinone, menaquinone-6. Menaquinol reduces the cytochrome bc_1 complex, which in turn reduces periplasmic cytochrome *c*. Unusually for a bacterium, *H. pylori* contains only a single terminal oxidase (a *cb*-type cytochrome *c* oxidase). However, *C. jejuni* also possesses a *bd*-like quinol oxidase in addition to the *cb*-type oxidase, allowing an additional but less coupled pathway of electron transfer to oxygen.

I. Introduction

Many species in the related genera *Helicobacter* and *Campylobacter* are extremely important gastrointestinal pathogens which have come to prominence in recent years as the causative agents of major animal and human diseases. They are Gram-negative, spiral-shaped bacteria which are fastidious, microaerophilic and which generally also require elevated levels of carbon dioxide for growth. Discovered in 1982, studies on the human helicobacter species, *H. pylori*, have expanded rapidly, and there are now many aspects of the biology of this bacterium that are understood in greater detail than in *Campylobacter* species, even though members of the latter genus have been known for many decades. This is particularly true since the genome sequence of *H. pylori* 26695 was published (Tomb et al., 1997). The sequence of another strain of *H. pylori* (J99) has also been determined (Alm et al., 1999), making direct comparisons of genome structure and the identification of strain specific genes possible. However, studies on the biology of campylobacters have been given new impetus with the publication of the genome sequence of *C. jejuni* (Parkhill et al., 2000).

Despite the importance of these bacteria in human and animal disease, studies of their metabolism and physiology have lagged behind work on properties related to virulence. Nevertheless, several reviews have been published which have collated the available biochemical information from the literature and predicted physiological properties, including electron transport pathways, from genome sequence information (Kelly, 1998, 2001; Doig et al., 1999; Marais et al., 1999; Kelly et al., 2001). In this chapter, we describe the current state of knowledge about the composition and function of the respiratory electron transport pathways in these bacteria.

Abbreviations: CCP – cytochrome *c* peroxidase; DMSO – dimethylsulphoxide; MK – menaquinone; NDH – NADH dehydrogenase; NQO – NADH quinol:oxidase; RNR – ribonucleotide reductase; TMAO – trimethylamine-N-oxide

II. Characteristics and Pathogenicity of *Campylobacter jejuni* and *Helicobacter pylori*

Campylobacter jejuni is the leading cause of acute bacterial gastroenteritis in both the western world and in developing countries (Friedman et al., 2000), where it is predominantly acquired by ingesting contaminated food, milk or water. *C. jejuni* is a commensal inhabitant of the gastrointestinal tract of poultry and other birds (Friedman et al., 2000) and as a consequence poultry serves as the primary source of contamination. Acute symptoms of *C. jejuni* infection in humans include diarrhea, fever and abdominal pain but the complications can include reactive arthritis and neurological disturbances such as the Miller-Fisher and the Guillaine-Barré syndromes (Skirrow and Blaser, 2000). The pathogenic mechanisms of *C. jejuni* are mediated by a number of virulence factors (Leach, 1997), including motility, adhesion and the ability to invade host cells, as well as the production of several toxins (at present not well characterized; Leach, 1997) which are likely to be responsible for many of the acute manifestations of infection.

H. pylori is closely related to *C. jejuni*. However, it is not a food-borne pathogen, and there is little evidence that it can survive or grow in the environment outside of a suitable host. It colonizes the gastric or duodenal mucosa, rather than the intestinal tract (although other helicobacters do colonize the lower GI tract of animals) and is thought to be transmitted directly from person to person via the oral-oral or fecal-oral route (Mitchell, 2001). *H. pylori* is one of the most common infections in Man and is now recognized as the major etiological factor in chronic active type B gastritis, gastric and duodenal ulceration and as a risk factor for gastric cancer (Dixon, 2001). Major virulence factors include motility, the production of a potent urease which is involved in acid resistance, a vacuolating cytotoxin (VacA) and the products of the *cag* pathogenicity island (Censini et al., 1996).

The microaerophilic nature of *C. jejuni* and *H. pylori* means that the majority of strains have to be cultured in an atmosphere containing 5–10% (v/v) oxygen and 5–10% (v/v) carbon dioxide. They are nutritionally fastidious bacteria, routinely cultured in complex media with additional growth supplements. Blood derivatives are often added, but the growth of *H. pylori* can be improved with charcoal, starch, bovine serum albumin (BSA), catalase, or β-cyclodextrin (Hazell et al., 1989; Olivieri et al., 1993). The function of these supplements may be to reduce oxidative damage or to adsorb potentially toxic long-chain fatty acids. The role of media supplements in preventing oxidative damage is well established for *Campylobacter* (Bolton et al., 1984). Although defined media have been developed for both *C. jejuni* (Leach et al., 1997) and *H. pylori* (Nedenskov, 1994; Reynolds and Penn, 1994), which contain several essential amino-acids and vitamins, some kind of supplement is often still required.

After several days of growth in vitro, both *C. jejuni* and *H. pylori* batch cultures undergo a morphological change from spiral-bacillary to coccoid forms. This stationary-phase morphological conversion is characterized by the loss of culturability. There is controversy over whether these coccoid forms are 'alive,' and whether they could be a means of transmission (Mitchell, 2001). There is evidence that the formation of coccoid cells of *C. jejuni* is not an active process and represents a degenerate form resulting from oxidative damage (Harvey and Leach, 1998). For *H. pylori* Kusters et al. (1997) showed that inhibition of protein or RNA synthesis did not affect conversion to the coccoid form and, moreover, that coccoid cells did not exhibit a measurable cytoplasmic membrane potential. These data strongly indicate that the coccoid cells are dead, and are the end result of a passive conversion from the spiral form.

III. Microaerophily

Physiological explanations for the microaerophilic nature of *C. jejuni* and *H. pylori* have long been sought and it is likely that are several contributors to this phenotype. The possession of oxygen sensitive enzymes is one possibility. Both bacteria possess pyruvate and 2-oxoglutarate oxidoreductases, which are normally associated with obligately anaerobic bacteria (Kelly and Hughes, 2001). Hughes et al. (1995, 1998) found that the *H. pylori* enzymes were very oxygen-sensitive, and the presence of oxygen and the omission of dithiothreitol from purification buffers resulted in their rapid inactivation. It seems likely that this property could be a major contributor to the microaerophilic growth phenotype, as both enzymes are essential for viability (Hughes et al., 1998), but there may be some form of protection to prevent inactivation by oxygen in vivo.

The stepwise one-electron reduction of O_2 results in the formation of the superoxide radical (O_2^-) and hydrogen peroxide (H_2O_2). *Campylobacter jejuni* and *H. pylori* possess well known major defense mechanisms against such oxidative stress, including proteins such as superoxide dismutase, catalase and ferritin, yet both organisms are still oxygen sensitive. The early work of Hoffman et al. (1979a,b) suggested that campylobacters may be more sensitive to *exogenous* superoxide and peroxides (which can be easily generated spontaneously in growth media by e.g. exposure to light) than are aerotolerant bacteria, despite their possession of these protective enzymes. This is consistent with observations that estimates of oxygen tolerance can vary widely depending upon the media used (Hodge and Krieg, 1994). The importance of alkylhydroperoxide reductase (AhpC) in aerotolerance and oxidative stress resistance has been demonstrated in *C. jejuni* (Baillon et al., 1999) by the analysis of an *ahpC* mutant, which was considerably more sensitive to oxygen and organic hydroperoxides (but not hydrogen peroxide) than its isogenic parent. The *ahpC* and catalase (*katA*) genes in *C. jejuni* are regulated by iron and also by PerR, a Fur-like protein (van Vliet et al., 1999). *H. pylori* also possesses AhpC, which has been shown to be an essential peroxiredoxin and which plays a similar role (Baker et al., 2001). There are some additional proteins predicted by the *C. jejuni* genome sequence (Parkhill et al., 2000), which are potentially involved in oxygen metabolism, but which are not shared with *H. pylori*. These include a bacterial globin homologue, and the proteins ruberythrin and haemerythrin, which could be oxygen-binding proteins. The roles of these proteins are currently unknown.

IV. Electron Transport Chains

A. Overview

The most important characteristics of bacterial respiratory chains are (i) branching at both 'dehydrogenase' and 'reductase' ends, (ii) the use of oxygen as well as alternative electron acceptors, (iii) the presence of numerous types of cytochromes and, often, more than one type of quinone, (iv) 'cross-talk' between pathways optimizing the possibility of each reductant being paired with a wide choice of oxidants, and (v) concomitant proton translocation and energy transduction. Simple, linear pathways involving a small number of dehydrogenases, a quinone and a single terminal oxidase or reductase are uncommon in bacteria. Nevertheless, genome and biochemical analyses (Marcelli et al., 1996; Kelly, 1998; Chen et al., 1999; Doig et al., 1999) do suggest a surprisingly simple organization (Fig. 1a) for the respiratory apparatus of *H. pylori*, in which dehydrogenases pass reducing equivalents to menaquinone only, then through a cytochrome bc_1 complex and a soluble cytochrome *c* to a single terminal oxidase.

In contrast, even early work on the respiratory physiology of campylobacters led to the identification of a rich complement of cytochromes (Lascelles and Calder, 1985), and the work of Hoffman and Goodman (1982) and Carlone and Lascelles (1982) indicated that the respiratory chain of *C. jejuni* was complex. Their work can now be interpreted in terms of the available genome sequence of strain 11168 (Parkhill et al., 2000) Electron input to the quinone pool is via a complex I-type membrane bound quinone reductase, a hydrogenase and a variety of other primary substrate dehydrogenases (Fig. 1b). In terms of oxygen dependent electron transport, a *cb*-type cytochrome *c*-oxidase is present (Parkhill et al., 2000), but *C. jejuni* also has an alternative quinol oxidase (CydAB homologue) and two separate genes encoding cytochrome *c*-peroxidases which may be important in detoxifying hydrogen peroxide in the periplasm. Moreover, the genome sequence indicates the presence of several alternative electron transport pathways to electron acceptors such as fumarate, nitrate, nitrite, TMAO and DMSO (Fig. 1b). The respiratory chain in *C. jejuni* appears to be highly branched and is significantly more complex than that in *H. pylori*.

B. Electron donors

1. Hydrogen

Uptake hydrogenase mediates the transfer of electrons from hydrogen to the menaquinone pool. The enzyme is membrane bound and consists of three polypeptides, HydA, HydB, and HydC. Both *H. pylori* and *C. jejuni* have been shown to possess hydrogenase activity in membrane fractions (Carlone and Lascelles, 1982; Hoffman and Goodman, 1982; Maier et al., 1996) Using antisera raised against the *Bradyrhizobium japonicum* uptake hydrogenase, Maier et al. (1996) identified *H. pylori* hydrogenase polypeptides of 65 and 26 kDa. These are the products of the *hydB* (HP0632/JHP575) and *hydA* (HP0631/JHP574) genes respectively, which are part of a *hydABCD* operon also encoding a cytochrome subunit (HydC) and a maturation protein (HydD). Gilbert et al. (1995) cloned a *hypB* homologue, which in *Bradyrhizobium japonicum* encodes a nickel binding-protein necessary for hydrogenase biosynthesis. The *hypB* gene (HP0900/JHP8837) is co-transcribed with two further biogenesis proteins encoded by the *hypDC* genes. Additional proteins needed for hydrogenase assembly are encoded by *hypA* (HP0869/JHP803) and *hypEF* (HP0047, HP0048/JHP40, JHP41) homologues. Olson et al. (2001) have shown that HypA and HypB, are required for the full activity of both the hydrogenase and the urease of *H. pylori*. However urease activity was normal in *hypD* or *hypF* mutants. These results indicate some sharing of the accessory proteins necessary for the correct incorporation of nickel into both of these enzymes. Hydrogenase has been shown to have an important role in sustaining growth of *H. pylori* in vivo, as hydrogenase deficient mutants

(a) Electron transport chains in *Helicobacter pylori*

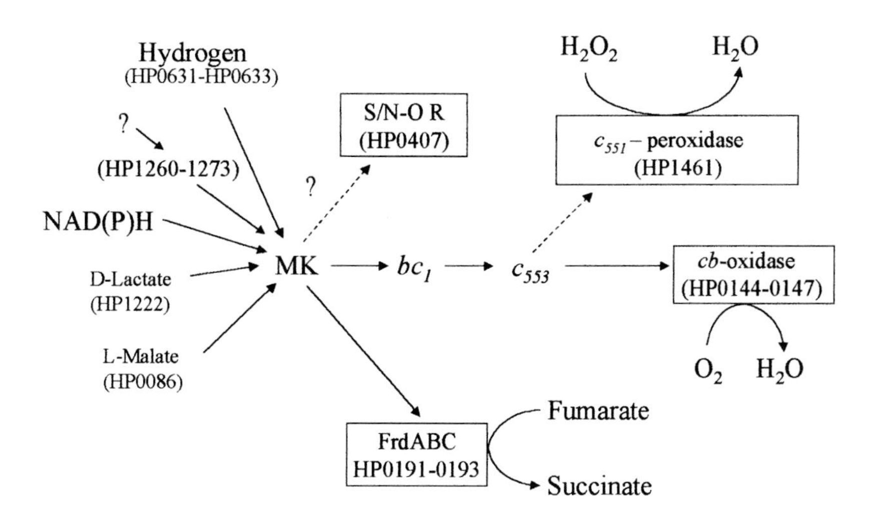

(b) Electron Transport Chains in *Campylobacter jejuni*

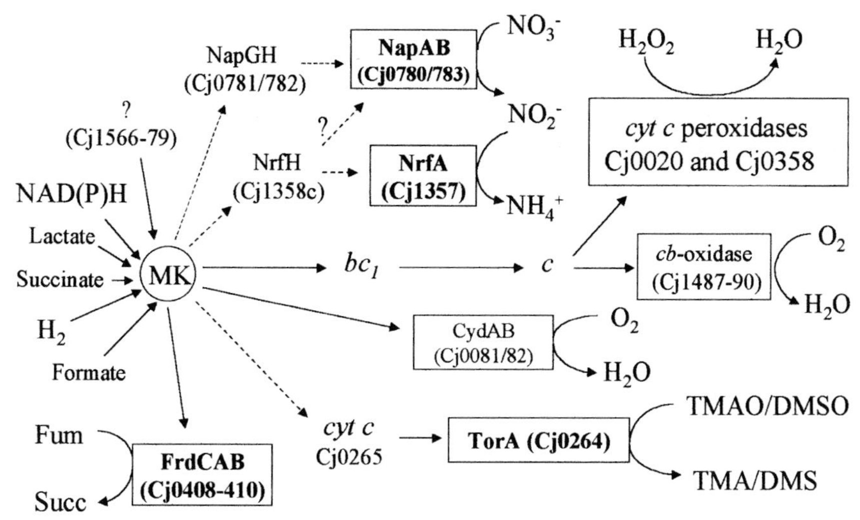

Fig. 1. Comparison of predicted electron transport chains in (a) *H. pylori* and (b) *C. jejuni*. In (a) integral membrane oxidoreductases include an NDH-1 like complex (HP1260-1273), the electron donor to which is unknown, and hydrogenase. Peripherally associated oxidoreductases include (among several others) malate:quinone oxidoreductase and a D-lactate dehydrogenase. Reducing equivalents are transferred to the sole quinone, menaquinone-6 (MK) in the lipid bilayer of the inner membrane. Menaquinol reduces the trimeric cytochrome bc_1 complex, which in turn reduces periplasmic cytochrome c_{553}. Cytochrome c is reoxidized by the sole terminal oxidase, a cb-type cytochrome c oxidase. Cytochrome c may also be reoxidized by hydrogen peroxide in the periplasm through the activity of a single cytochrome c peroxidase. Fumarate reductase (FrdABC) catalyses electron transfer from menaquinol to fumarate as terminal acceptor. HP0407 may encode an S- or N-oxide oxidoreductase. In *C. jejuni* (b), a similar range of primary dehydrogenases is present, with the notable addition of formate and succinate dehydrogenases. A cb-type cytochrome c oxidase is present in addition to a CydAB-like quinol oxidase (not present in *H. pylori*). There are also two separate peroxidases in *C. jejuni*. Several alternative reductases are predicted in *C. jejuni*. These include nitrate, nitrite and a TMAO/DMSO reductase, TorA. In both (a) and (b), solid lines indicate experimentally established or highly likely routes of electron transport, while dotted lines indicate uncertainty as to the exact route, possibly with the participation of unidentified additional redox proteins.

were unable to colonize the stomachs of mice (Olson and Maier 2002), and evidence was obtained that the concentration of hydrogen (derived from colon fermentation reactions) in stomach tissues was well above the K_M value of the enzyme (Olson and Maier 2002). *C. jejuni* contains a similar set of structural genes for a NiFe-type hydrogenase, as well as accessory genes for nickel incorporation, but no functional studies have yet been done on these genes.

2. Formate

Formate dependent oxygen respiration in *C. jejuni* has been demonstrated and characterized by Hoffman and Goodman (1982) and Carlone and Lascelles (1982). The respiratory activities determined with membrane vesicles were 50 to100 times greater with formate or hydrogen as substrates when compared to the rates achieved with succinate, lactate, malate or NADH, suggesting that the former are excellent electron donors (Hoffman and Goodman, 1982). In *E. coli*, formate dehydrogenase-N (formate: Quinone Oxidoreductase-N) consists of three subunits, a large selenomolybdoprotein containing the catalytic site (FdnG), a smaller iron-sulfur protein (FdnH) and a *b*-type cytochrome (FdnI) (Gennis and Stewart, 1996). The three subunits α, β, and γ, are 110, 32, and 20 kDa respectively (Enoch and Lester, 1975; Berg et al., 1991). Formate dehydrogenase-N catalyses trans-membrane proton translocation from the cytoplasm to the periplasm (Jones, 1980). The genome sequence of *C. jejuni* reveals an operon encoding putative formate dehydrogenase subunits (*fdhA-D*/Cj1511c-Cj1508c). *C. jejuni fdhA* (Cj1511c) encodes a large 104 kDa selenocysteine containing molybdo-protein equivalent to the *E. coli* 110 kDa FdnG (α) subunit. *fdhB* (Cj1510c) encodes a 24 kDa iron-sulfur subunit equivalent to the *E. coli* 32 kDa FdnH (β) subunit. *fdhC* (Cj1509c) encodes a 35 kDa cytochrome *b* subunit equivalent to the *E. coli* 20 kDa FdnI (γ) subunit. In addition to these three subunits, both *E. coli* and *C. jejuni* encode an FdhD protein (29 kDa) required for activity of the formate dehydrogenase enzyme complex (Berg et al., 1991). In contrast to *C. jejuni*, *H. pylori* does not contain any orthologous genes encoding formate dehydrogenase subunits (Tomb et al., 1997; Alm et al., 1999) and formate respiration has not been demonstrated. This difference may well be related to the different niches that the bacteria occupy in vivo; formate may be produced by the rich intestinal flora and thus be available to *C. jejuni*, but not to *H. pylori* living in the stomach.

3. Lactate

Hoffman and Goodman (1982) demonstrated the activity of a lactate dehydrogenase in oxygen-linked respiration of *C. jejuni.* A H^+/O ratio of 2.12 was found for lactate in whole cell proton-pulse assays (Hoffman and Goodman, 1982). Analysis of the genome sequence has yet to produce a candidate locus for a membrane associated lactate dehydrogenase enzyme in *C. jejuni*, although an annotated L-lactate dehydrogenase (Cj1167) is present which is probably a fermentative enzyme. *Helicobacter pylori* also shows lactate dehydrogenase activity, and a putative flavoprotein D-lactate dehydrogenase (Dld; HP1222/JHP1143) could account for the D-lactate dependent respiration observed in this bacterium (Chang et al., 1995).

4. Malate

Malate can reduce *c*-type cytochromes in *H. pylori* and a membrane bound NADH independent (dye-linked) activity suggested the presence of a flavoprotein type malate oxidoreductase (Kelly, 1998). Kather et al. (2000) characterized this activity fully and showed that it was due to a malate:quinone oxidoreductase (Mqo), a 51 kDa FAD-dependant membrane-associated enzyme that donates electrons to quinone and which could also participate in the citric acid cycle (Kather et al., 2000). The enzyme was unequivocally shown to be encoded by HP0086 in *H. pylori* 26695, but the deduced protein has fairly low sequence similarity with other Mqo enzymes (Kather et al., 2000). Analysis of the *C. jejuni* genome reveals a putative oxidoreductase (Cj0393c) with 24.6% identity to *E. coli* malate:quinone oxidoreductase and 49.3% identity to the closely related Mqo of *H. pylori,* indicating that both bacteria use a similar enzyme for malate oxidation. This is consistent with the observations of Hoffman and Goodman (1982), who demonstrated that malate stimulated respiration in *C. jejuni*, producing a H^+/O ratio of 2.03.

5. NAD(P)H

It is widely held that the function of NADH in bioenergetics is to act as the major electron donor for oxidative phosphorylation, via interaction with a proton-translocating quinone oxidoreductase (Complex I or NDH-1), while the role of NADPH is as a source of electrons for biosynthetic reactions. NDH-1 is a large enzyme with a multitude of redox centers and is most commonly made up of 14 different subunits (Friedrich, 1998; Yagi et al., 1998). Electrons are passed from NADH, via 1 FMN and 9 Fe-S centers in NDH-1, to ubiquinone/menaquinone in the respiratory chain. Four protons are translocated across the membrane for every two electrons transferred by NDH-1 according to the overall equation:

$$NADH + Q + 5H^{+} \rightarrow NAD^{+} + QH_2 + 4H^{+} \qquad (1)$$

Several studies have shown that membrane preparations of *H. pylori* exhibit low or insignificant rates of NADH oxidation, but that the rates with NADPH as electron donor are much higher (Chang et al., 1995; Hughes et al., 1998; Chen et al., 1999). This unusual situation suggests that NADPH is the physiological electron donor to the respiratory chain in *H. pylori*, rather than NADH as in the majority of bacteria. The genome sequence of *H. pylori* and *C. jejuni* shows that these bacteria contain a cluster of genes encoding a potential NADH-quinone oxidoreductase of the NDH-1 type. However, an examination of the deduced proteins encoded by this gene cluster led Finel (1998) to conclude that the complex may not actually oxidize NADH, because of the lack of the NQO1 (NuoE) and NQO2 (NuoF) subunits. These subunits are thought to be essential components for the function of the NDH-1 complex. NuoF binds NADH, and also possesses a bound FMN and an Fe-S center. NuoE, in conjuction with NuoG, B and I, have cysteine residues that probably coordinate Fe-S clusters (Smith et al., 2000). In the place of the expected *nuoE* and *nuoF* genes, *C. jejuni* and *H. pylori* have ORFs of unknown function. Cj1574c from *C. jejuni* encodes a protein 230 aa in length (26 kDa) which has 26.5% identity to the *H. pylori* equivalent (HP1265) which encodes a protein 328 aa in length (37 kDa). In addition, Cj1575c from *C. jejuni* encodes a protein 75 aa in length (8.7 kDa) with 52.0% identity to the *H. pylori* equivalent (HP1264) which encodes a protein 76 aa in length (8.9 kDa) (Tomb et al., 1997; Parkhill et al., 2000). Significantly, these proteins do not contain any obvious NADPH-binding motif, and they are apparently unique to these two bacteria.

An obvious possibility is that electrons from NADPH are transferred to Complex I via an intermediate that interacts with the NQO1 and NQO2 replacements (Finel, 1998; Smith et al., 2000). Alternatively, coupling of NADPH with the respiratory chain may not occur via the NDH-1 homologue at all, but through an alternative quinone reductase which may not be proton-translocating. It should be noted that there is no obvious homologue of an NDH-2 type protein which may fulfill this role in either *H. pylori* or *C. jejuni*. Finel (1998) suggested that HP1264/1265 could act as a docking site for a protein that delivers electrons directly to the FeS cluster of the NQO3 homologue. Although the identity of such a protein is unknown, one possibility is flavodoxin or ferredoxin, reduced by the activities of the pyruvate and 2-oxoglutarate oxidoreductases which are present in place of the usual NADH-producing multi-enzyme dehydrogenase complexes found in conventional aerobes (Hughes et al., 1995, 1998). Further evidence that the NDH-1 homologue is unusual and is not a conventional NAD(P)H quinone oxidoreductase, has come from the observation that NADH oxidation in *H. pylori* is insensitive to the classical Complex I inhibitor, rotenone (Chen et al., 1999). Whether this type of enzyme is proton translocating is an important unresolved question for understanding the bioenergetics of these bacteria.

6. Succinate

In *E. coli*, the interconversion of fumarate and succinate can be carried out by two related enzymes; succinate dehydrogenase, which is expressed under aerobic conditions, or fumarate reductase, which is induced under anaerobiosis (Spiro and Guest, 1991). Succinate dehydrogenase (succinate:quinone oxidoreductase) catalyses electron transfer from succinate to ubiquinone, where succinate is oxidized to fumarate as part of the tricarboxylic acid cycle. *H. pylori* has an incomplete citric acid cycle and lacks succinyl CoA synthetase and succinate dehydrogenase (Tomb et al., 1997; Alm et al., 1999; Pitson et al., 1999), while the presence of a fumarate reductase has been confirmed by biochemical analysis (Mendz and Hazell, 1993; Birkholz et al., 1994; Pitson et al., 1999) and sequence data (Ge et al., 1997; Tomb et al., 1997; Alm et al., 1999). However, *H. pylori* can respire succinate; succinate respiration results in the production of acetate, suggesting reversal of fumarate reductase, and participation of enzymes to convert fumarate to pyruvate and then acetate (Kelly, 1998).

C. jejuni appears to have a complete citric-acid cycle with genes encoding clearly identifiable homologues of all of the conventional enzymes (Parkhill et al., 2000), including separate *sdh* and *frd* operons. In *E. coli*, the four subunits of succinate dehydrogenase are encoded by the *sdhCDAB* operon. SdhA (64 kDa) contains the active site and covalently bound FAD. SdhB (26 kDa) contains three iron-sulfur clusters and is responsible for the transfer of electrons from SdhCD to SdhA. Subunits SdhC and D (14 and 12 kDa respectively) combine to share a Heme b_{556} and are responsible for the electron transfer from quinone to the SdhB membrane anchor (Gennis and Stewart, 1996). The *C. jejuni* genome reveals an *sdhABC* operon (Parkhill et al., 2000). *sdhA* (Cj0437) encodes a 66 kDa protein with 37.3% identity to *E. coli* SdhA and contains an FAD binding domain. *sdhB* (Cj0438) encodes a 35.8 KDa putative iron-sulfur protein. *sdhC* (Cj0439) encodes a 31.4 kDa putative succinate dehydrogenase subunit C which is a probable functional equivalent to *E. coli* succinate dehydrogenase subunits C and D.

7. Other Organic Compounds as Electron Donors

Chang et al. (1995) were unable to detect oxygen uptake when acetate, glycerol, L-lactate, oxaloacetate, 2-oxobutyrate and several amino acids including aspartate and glutamate were added to *H. pylori* cells. Homologues of proline dehydrogenase (PutA; HP0056/JHP48), glycolate oxidase (GlcD; HP0509/JHP459) and a D-amino-acid dehydrogenase (DadA; HP0943/JHP8878) provide additional possibilities for substrate-derived electrons to be donated to the membrane-bound electron transport chain, and Nagata et al., (2003) have demonstrated that L-serine, D-alanine and L-proline can indeed act as electron donors for whole cell oxygen-linked respiration in *H. pylori*.

C. The Quinone Pool

E. coli utilizes the higher potential ubiquinone (E_m = +110 mV) during aerobic respiration, but under anaerobic conditions it switches to using the lower midpoint potential menaquinone (E_m = –75 mV; Ingeldew and Poole, 1984). *Helicobacter pylori* and *C. jejuni* do not contain ubiquinone; menaquinone is the sole isoprenoid quinone present (Carlone and Anet, 1983; Collins et al., 1984; Marcelli et al., 1996). Marcelli et al. (1996) demonstrated that MK-6 is the dominant form of menaquinone with traces (~10%) of MK-4 in several strains of *H. pylori* tested. No alteration in quinone type was detected during growth with various O_2 levels between 2–15% (v/v) and levels of menaquinone were also found to be highest at optimum growth concentrations of oxygen of 5 to 10% (v/v). In *C. jejuni* the menaquinone also contains six isoprene units. In addition to MK-6 *Campylobacter* ssp. have been found to contain a novel methyl substituted MK-6 that has not been reported in *H. pylori* (Carlone and Anet, 1983; Collins et al., 1984).

D. The Cytochrome bc_1 Complex

The cytochrome bc_1 complex is an oligomeric membrane protein complex containing three subunits and four redox centers which couples electron transfer from quinone to proton translocation across the membrane, leading to the formation of an electrochemical proton gradient (Trumpower, 1990). The bc_1 complex is found in many Gram-negative and Gram-positive bacteria, aerobic, anaerobic and photosynthetic bacteria, and in mitochondria of lower and higher eukaryotes although it is not present in *E. coli* (Ingeldew and Poole, 1984). It is usually found in respiratory chains that contain ubiquinone; an important consequence of the use of menaquinone in *H. pylori* and *C. jejuni* is that the bc_1 complex must be able to oxidize the lower potential menaquinol rather than ubiquinol.

Evidence for the operation of the cytochrome bc_1 complex in *H. pylori* has come from respiratory inhibition studies and genome sequence information. Lactate respiration in intact cells is inhibited by the specific cytochrome bc_1 complex inhibitors antimycin A and myxothiazol (Alderson et al., 1997). In addition, succinate: cytochrome *c* reductase activity was inhibited by these reagents in sonicates (Chen et al., 1999). In *H. pylori,* genes encoding the three subunits of the complex are present in an operon. Cytochromes *b* and c_1 are present as separate polypeptide subunits (encoded by HP1538/JHP1461 and HP1539/JHP1460 respectively), along with an Fe-S protein (the Rieske protein) containing a 2Fe-2S cluster (encoded by HP1540/JHP1459).

No studies to date have been published on the properties or operation of the *C. jejuni* cytochrome bc_1 complex, although the cytochrome *b* and cytochrome *c* redox centers of this complex were detected in the early work of Carlone and Lascelles (1982). Genes encoding the three subunits of the bc_1 complex can be found in *C. jejuni* (Cj1184c-1186c).

E. c-type Cytochromes in H. pylori *and* C. jejuni

Cytochromes *c* are proteins that contain covalently bound heme and function in a range of redox reactions in bacteria. *c*-type cytochromes occur either as soluble periplasmic proteins or anchored to the cytoplasmic membrane via N-terminal hydrophobic extensions (Thöny-Meyer, 1997). Biogenesis of *c*-type cytochromes is complex; the biosynthesis of heme, export of the apoprotein to the periplasm, and covalent linkage of the heme moiety to the apo-protein are all essential steps in this process (Thöny-Meyer, 1997). There are interesting differences in the biogenesis of cytochromes *c* between *H. pylori* and *C. jejuni*, which are apparent from an analysis of the respective genome sequences. *C. jejuni* contains homologues of the classical *ccm* genes that are present in many Gram-negative bacteria and which constitute the complex 'System I' pathway. *H. pylori*, however, employs the much simpler 'System II' pathway, which is characteristic of Gram-positive bacteria (Kranz et al., 1998).

Sonicates of *H. pylori* were shown to have ascorbic acid-oxidizing activity (Ødum and Anderson, 1995), which was proposed to be responsible for the destruction of gastric vitamin C seen in *H. pylori*-infected patients. A water-soluble component of the *H. pylori* sonicate was shown to be responsible for the ascorbate oxidizing activity and this was tentatively assigned to a low molecular mass (<14 kDa) cytochrome *c* (Ødum and Anderson, 1995). Spectroscopic studies on *H. pylori* (Marcelli et al., 1996; Nagata et al., 1996) confirmed the presence of *c*-type cytochromes in the organism. The purification and N-terminal sequence of a periplasmic, soluble *c*-type cytochrome has been reported (Evans and Evans, 1997; Koyanagi et al., 2000), and characterized spectroscopically as cytochrome c_{553}. Cytochrome c_{553} has been identified as a potential electron donor to the *cb*-type cytochrome *c* oxidase (Tsukita et al., 1999; Koyanagi et al., 2000). The redox potential of the protein was reported to be +170 mV (Koyanagi et al., 2000). The genome sequences of *H. pylori* 26695 and J99 contain only one annotated *c*-type cytochrome gene (*cycA*; HP1227/JHP1148), which encodes the protein identified in these biochemical studies. Interestingly, the *cycA* gene is transcribed divergently from a homologue of the *hemN* gene, encoding an important enzyme in the anaerobic biosynthesis of heme.

Searching the whole genome sequence for genes that encode proteins containing the CxxCH motif has identified a second, putative low molecular mass cytochrome *c* (Kelly et al., 2001). The gene encoding this putative cytochrome (HP0236, designated *cycB*) is found as the distal gene of an eight gene operon. Significantly, however, in addition to the putative cytochrome gene, there are two genes within this operon that are involved in heme biosynthesis, *hemA* and *hemC*. Both CycA and CycB contain a single CxxCH motif indicating that they are monoheme *c* type cytochromes. The role of CycB in electron transport is unknown.

Analysis of the *C. jejuni* genome reveals a putative periplasmic mono-heme cytochrome *c* (Cj1153) as the most likely candidate to be an electron donor to the cytochrome *c*-oxidase, with 39% identity with the *H. pylori* cytochrome c_{553} and 34.6% identity to the cytochrome c_{553} of *Desulfovibrio vulgaris* (Koyanagi et al., 2000; Parkhill et al., 2000). Unlike *H. pylori, C. jejuni* contains many proteins with a potential heme C binding motif (CXXCH), only some of which have been annotated as cytochromes in the genome database (Parkhill et al., 2000). Many of these are associated with terminal reductases for alternative electron acceptors and are discussed in Section H, but the others have as yet unknown functions.

F. Terminal Oxidases for Oxygen Dependent Respiration

Most bacteria possess at least two terminal oxidases, often a quinol oxidase and a cytochrome *c* oxidase (Poole and Cook, 2000), with different catalytic properties and transcriptional controls. Genome analysis of *C. jejuni* and *H. pylori* indicates the absence of any homologues for an aa_3-type cytochrome *c* oxidase. Only one terminal oxidase, a *cb* (cbb_3)-type cytochrome *c*-oxidase has been found to terminate the respiratory chain of *Helicobacter pylori* (Marcelli et al., 1996; Nagata et al., 1996). It is encoded by the *ccoNOQP* operon, and the subunit structure is essentially the same as in all other organisms from which this enzyme has been identified, except that the CcoN subunit from *H. pylori* is truncated at the N-terminus when compared to CcoN subunits from other bacteria. The amino acid sequences of the *ccoO* and *ccoP* genes reveal conserved motifs for the binding of heme C (CXXCH). CcoO is capable of binding a single heme C and CcoP is capable of binding two hemes C. The *H. pylori* oxidase has been reported to have a K_M for oxygen of 0.4 μM (Nagata et al., 1996) or 0.04 μM (Tsukita et al., 1999), but the values were determined using a relatively insensitive membrane-covered O_2 electrode, which probably underestimates the true affinity. The oxidase has been purified (Tsukita et al., 1999) and contains three hemes C, and two protohemes (one high-spin, one low-spin) as predicted. However, surprisingly, CcoN and CcOO appear to form a protein complex even in the presence of SDS. There is evidence that the enzyme can pump protons, although the H^+ pumping activity by reconstituted proteoliposomes was low (Tsukita et al., 1999).

Campylobacter jejuni has two terminal oxidases, as clearly revealed by genome analysis (Parkhill et al., 2000). This confirms the earlier spectroscopic work of Carlone and Lascelles (1982) and Hoffman and Goodman (1982). One oxidase is a *cb*-type cytochrome *c*-oxidase, very similar to that described above in *H. pylori*. Subunit I (*ccoN*/Cj1490c) has a molecular weight of 56 kDa and contains the heme-copper oxidase catalytic center and a copper B binding region signature. Subunit II (*ccoO*/Cj1489c) has a molecular weight of 25 kDa and contains a cytochrome *c* family heme-binding site signature. Subunit III (*ccoP*/Cj1487c) has a molecular weight of 31 kDa and contains two heme C binding motifs. Subunit IV (*ccoQ*/Cj1488c) has a molecular weight of 10.4 kDa with 35.1% identity to the corresponding *H. pylori* subunit. No functional studies have yet been reported on this oxidase.

Cytochrome *bd* encoded by the *cydAB* operon is well characterized in *E. coli* (Green et al., 1988), where it is expressed under microaerobic conditions (Fu et al., 1991) as a heterodimer located in the cytoplasmic membrane (Miller et al., 1988). The enzyme oxidizes ubiquinol in the lipid bilayer and reduces oxygen to water, contributing to the generation of a proton motive force (H^+/O ratio = 2) (Miller and Gennis, 1985). Lorence et al.(1986) identified the three heme prosthetic groups in the enzyme; heme b_{558}, heme b_{595} and heme *d*. Quinol oxidation is likely to be carried out by heme b_{558} and oxygen binding at heme *d* (Hata-Tanaka et al., 1987). Carlone and Lascelles (1982) demonstrated an alternative terminal oxidase in *C. jejuni* and analysis of the genome sequence reveals the presence of a *cydAB*-like operon encoding subunits I and II. Subunit I (58 kDa) is encoded by *cydA* (Cj0081), and has 45.7% identity to *E. coli* cytochrome *bd* oxidase subunit I. Subunit II (42 kDa) is encoded by *cydB* (Cj0082) and is similar to the *E. coli* subunit II with 27.7% identity (Parkhill et al., 2000). The physiological roles of the two oxidases in *C. jejuni* are unknown.

G. Hydrogen peroxide as an Electron Acceptor

Hydrogen peroxide is produced as a byproduct in the reduction of molecular oxygen to water. It is a partially reduced species that is toxic to the cell. Hydrogen peroxide can be degraded to H_2O and O_2 by the cytoplasmic enzyme catalase, but in the periplasm, H_2O_2 can be broken down to water alone by a periplasmic cytochrome *c* peroxidase, with the requirement of reduced cytochrome *c* as an electron donor:

$$H_2O_2 + 2e^- c_{red} + 2H^+ \rightarrow 2H_2O + 2c_{ox} \quad (2)$$

Cytochrome *c* peroxidase (CCP) has been isolated from a range of bacteria including *Nitrosomonas europaea* (Arciero and Hooper, 1994), *Rhodobacter capsulatus* (Hu et al., 1998), *Paracoccus denitrificans* (Gilmour et al., 1993) and *Pseudomonas aeruginosa* (Rönnberg et al., 1989). Studies of CCP in these bacteria have revealed that the enzyme has a molecular weight of 34–37 kDa and contains two hemes C, one high potential and one low potential. The high potential heme is the source of the second electron for H_2O_2 reduction and the low potential heme acts as a peroxidatic center. Analysis of the *C. jejuni* genome reveals two separate genes encoding putative cytochrome *c* peroxidase homologues (Parkhill et al., 2000). Cj0020c and Cj0358 are 34 kDa and 37 kDa respectively with both proteins containing two cytochrome *c* heme-binding site signatures. Cj0020c is similar to the cytochrome *c* peroxidase of *Nitrosomonas europaea* and *Pseudomonas aeruginosa* with 46% and 42% identity respectively (Parkhill et al., 2000). Cj0358 is similar to the *H. pylori* putative cytochrome *c* peroxidase homologue (HP1461) with 58.5% identity, and to the cytochrome *c* peroxidase of *Pseudomonas aeruginosa* with 48.9% identity (Parkhill et al., 2000). The physiological roles and regulation of expression of the peroxidases have yet to be studied. However, peroxidase has been studied in the related microaerophile *Campylobacter mucosalis*, where the oxidation of formate leads to the generation of periplasmic H_2O_2. This is reduced by CCP, using electrons from cytochrome c_{553} that has been reduced by the bc_1 complex. Thus, removal of H_2O_2 from the periplasm leads to net proton extrusion and energy conservation (Goodhew et al., 1988).

H. Respiration with Alternative Terminal Electron Acceptors to Oxygen

1. Fumarate as Electron Acceptor

In the absence of oxygen, fumarate can be used as an alternative terminal electron acceptor for the proton-translocating electron transport chain, and is important in ATP generation in many anaerobic bacteria. *H. pylori* fumarate reductase is very similar to that of the related anaerobe *Wolinella succinogenes* and is encoded by three genes *frdCAB* (HP0193-0191/JHP179-177) encoding polypeptides of 27, 81 and 31 kDa respectively (Ge at al., 1997; Tomb et al., 1997). FrdA and FrdB display the amino acid motifs for FAD and Fe-S binding respectively, in common with other fumarate reductases and also succinate dehydrogenase enzymes. The *frdC* gene encodes a hydrophobic, di-heme cytochrome *b*, which may serve as a membrane anchor for FrdA and B. Like the fumarate reductase of *W. succinogenes*, activity was localized in the membrane fraction (Birkholz et al., 1994; Ge et al., 1997). Mutants in *frdA* have been generated, indicating this enzyme is not essential for *H. pylori*. However, the mutants demonstrated a prolonged lag-phase on standard growth medium under microaerobic conditions (Ge et al., 1997). Expression of fumarate reductase activity appears to be constitutive in *H. pylori*, as unlike *E. coli,* levels of activity did not markedly change in cells grown under varying O_2 concentrations (Davidson et al., 1993).

Although the presence of fumarate reductase provides evidence of anaerobic-type respiration, *H. pylori* has not been successfully cultured under strictly anaerobic conditions in the presence of fumarate as a terminal electron acceptor, and Veron et al. (1981) reported anaerobic growth with fumarate to be absent in *C. jejuni.* Yet cytochrome *b*-linked fumarate reductase activity was demonstrated in the membranes of both *C. fetus* and *C. jejuni* (Carlone and Lascelles, 1982; Harvey and Lascelles, 1980). Analysis of the *C. jejuni* genome reveals a *frdABC* operon, encoding subunits similar to the thoroughly investigated fumarate reductase of *W. succinogenes* and to *H. pylori*. Why then are *H. pylori* and *C. jejuni* apparently unable to grow by anaerobic respiration of fumarate? Sellars et al. (2002) found that although growth with fumarate (and other electron acceptors such as nitrate, nitrite and TMAO) was insignificant under strictly anaerobic conditions, electron-acceptor dependent growth was possible under severely oxygen-limited conditions. Their results indicated that some oxygen requiring metabolic reaction(s) prevented anaerobic growth. As *C. jejuni* only contains genes for an oxygen-requiring class I type of ribonucleotide reductase, it was suggested that the inability to synthesize DNA anaerobically is the most likely explanation. Consistent with this, cells incubated anaerobically with electron acceptors did not divide properly but formed filaments analogous to those seen after treatment of aerobic cells with the RNR inhibitor hydroxyurea (Sellars et al., 2002). Thus, *C. jejuni* can use alternative electron acceptors in energy conserving reactions, but only if some oxygen is present to satisfy the requirement for deoxyribonucleotide production. A similar explanation for *H. pylori* is likely as it too only contains the class I RNR genes. A model for fumarate respiration coupled to energy conservation in *C. jejuni* is shown in Fig 2a. The active site of fumarate reduction is on the cytoplasmic side of the membrane, but only if quinol oxidation releases protons into the periplasm will fumarate reductase be an electrogenic enzyme and contribute to the generation of a proton-motive force. This has yet to be proven. However, this is definitely not the case for the other reductases discussed below, which are all periplasmic in location (see Fig. 2b–d).

2. Nitrate as Electron Acceptor

There are three types of bacterial nitrate reductases (i) soluble, cytoplasmic, assimilatory nitrate reductases (NAS); (ii) membrane-associated 'respiratory' nitrate reductases (NAR); (iii) soluble, periplasmic, 'dissimilatory' nitrate reductases (NAP) (Potter et al., 2001). The NAP class of nitrate reductase is a two-subunit complex, located in the periplasm, which is coupled to quinol oxidation via a membrane anchored tetraheme cytochrome. Although *H. pylori* does not have genes for any of the known types of nitrate reductase, *C. jejuni* is predicted to possess a periplasmic NAP enzyme, encoded by a *napAGHBD* operon (Parkhill et al., 2000; Kelly 2001; Sellars et al., 2002) and nitrate reductase activity has been demonstrated in intact cells by nitrite accumulation assays (Sellars et al., 2002). As with fumarate, nitrate-dependent growth in *C. jejuni* can only be demonstrated under oxygen-limited, not strictly anaerobic conditions (Sellars et al., 2002).

NapA is a ~90 kDa catalytic subunit which binds a bis-MGD co-factor and a [4Fe4S] cluster. NapB is

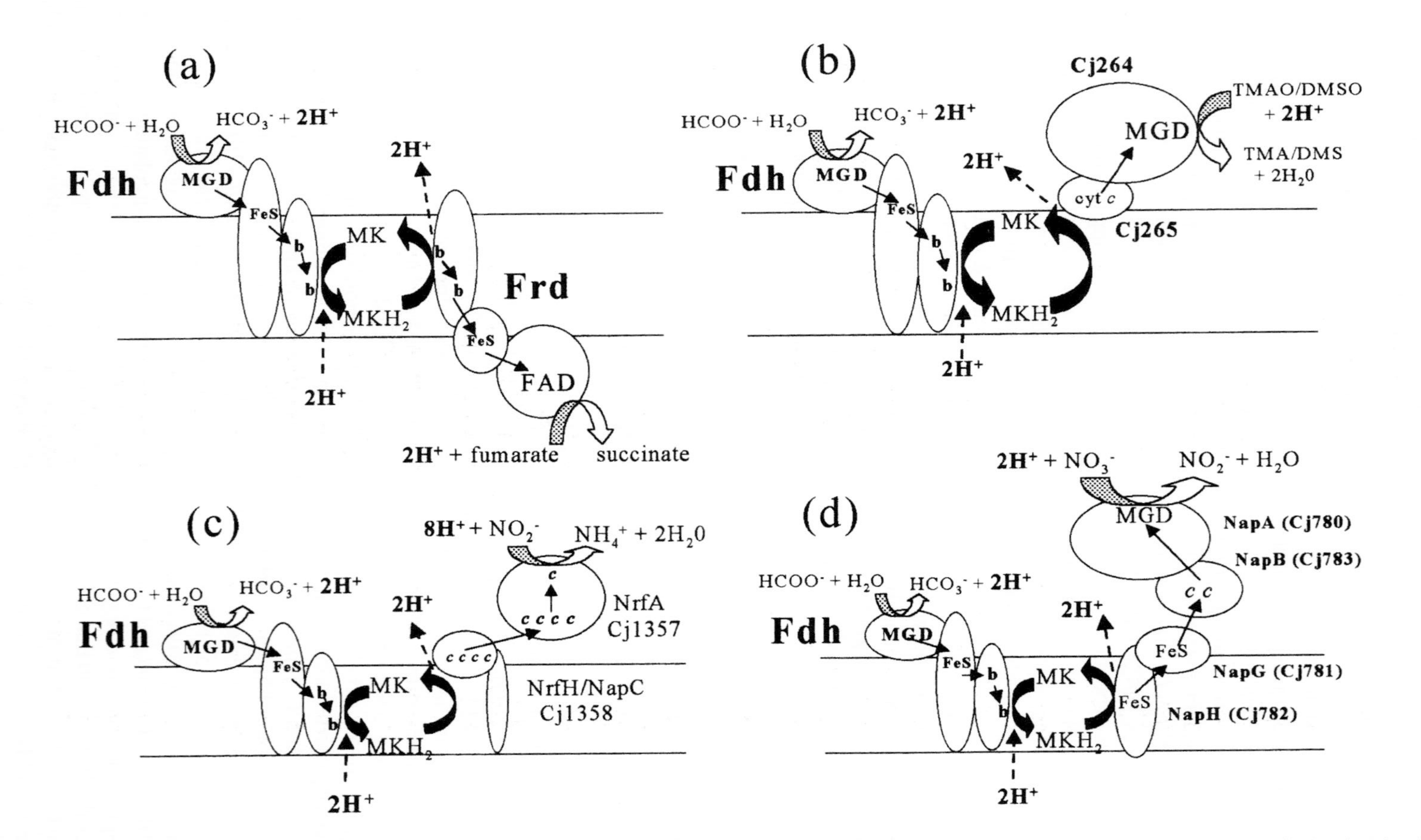

Fig. 2. Predicted topological organization and consequences for energy conservation of alternative electron transport chains in *C. jejuni*. In each case, formate is depicted as the electron donor to the quinone pool, through the action of formate dehydrogenase (Fdh). (a) fumarate respiration may result in the generation of a proton-motive force by fumarate reductase (Frd) if the sites of quinol oxidation and fumarate reduction are on opposite sides of the membrane. With TMAO/DMSO (b), nitrite (c) and nitrate (d), the terminal reductases are all predicted to be periplasmic, so that reduction of the electron acceptor and quinol oxidation occur on the same (P-phase) side of the membrane. In these cases a proton-motive force can only be generated at the level of the primary dehydrogenase.

a ~16 kDa electron transfer sub-unit which in other bacteria binds two *c*-type hemes (Richardson et al., 2001). NapD (13 kDa) is proposed to be involved in maturation of NapA prior to export to the periplasm (Berks et al., 1995; Potter and Cole, 1999). The *nap* operon of *C. jejuni* does not contain *napC*, encoding the NapC subunit. NapC acts as an electron donor to the NapAB complex, a role which was shown to be essential in the function of nitrate reductases in many bacteria, including *E. coli* and *Paracoccus pantotrophus* (Berks et al., 1995; Potter and Cole, 1999). Analysis of the *C. jejuni* genome shows that it does encode a NapC homologue (Cj1358c), but it is upstream of the *nrfA* nitrite reductase gene. The product of Cj1358c is a NapC/NirT-type cytochrome *c* having 34.5% identity to the *E. coli* K-12 NapC subunit (Parkhill et al., 2000). Despite this, the operonal location of Cj1358c implies that it is part of the nitrite reductase system, and thus may not be involved in electron transfer to the nitrate reductase. A potential role of Cj1358c in nitrite reduction is suggested by studies on the *nrf* operon in *Wolinella succinogenes,* which encodes a NapC-like subunit (NrfH) similar to the product of Cj1358c (Simon et al., 2000). NrfH is the mediator between the quinone pool and the cytochrome *c* nitrite reductase (NrfA) of *W. succinogenes* (Simon et al., 2000). This leaves the question of how the NapAB complex is coupled to menaquinol oxidation. One possibility is that the NapG and NapH subunits could function in this role (Sellars et al., 2002). NapG (27 kDa) is predicted to bind up to four Fe-S clusters. NapH (30 kDa) is predicted to be an integral membrane protein with four transmembrane helices with both the N- and C-terminus on the cytoplasmic side of the membrane. A model for nitrate respiration in *C. jejuni* is shown in Fig. 2d. Because of the periplasmic location of Nap, a proton-motive force can only be generated at the level of the primary dehydrogenase and not from quinol oxidation.

There is increasing evidence that nitrate respiration may play a significant role in the growth of human and animal pathogens in vivo. In *Mycobacterium bovis*, membrane bound nitrate reductase (Nar) activity has been shown to contribute to virulence (Weber et al., 2000), and *nar* mutants were unable to colonize a mouse animal model. In a wide range of Gram-negative bacterial pathogens, including *C. jejuni*, periplasmic nitrate reductases are more commonly present than Nar-type enzymes (Potter et al., 2001). Nitrate concentrations in human body fluids are in the range 10–50 μM (Potter et al., 2001). Significantly, it has been shown that in *E. coli* (which has both Nar and Nap) the Nap enzyme has a much higher affinity for nitrate compared to the membrane bound Nar (Potter et al., 1999), making Nap ideally suited to a role in scavenging the low nitrate concentrations encountered in vivo.

3. Nitrite as Electron Acceptor

Three distinct enzymes in enterobacteria catalyze the reduction of nitrite to ammonia: (i) the anaerobic nitrite reductase (NirBD), which is present in the majority of enterobacteria, is responsible for the regeneration of NAD^+ and the detoxification of nitrite; (ii) assimilatory nitrite reductase (NasB) is found in enterobaceria such as *Klebsiella* spp. but it is not found in *E. coli* K-12 or *S. typhimurium*; (iii) respiratory nitrite reductase (NrfAB), found in *E. coli* K-12 (Gennis and Stewart, 1996).

The nitrite-reducing enzyme of interest in *C. jejuni* is cytochrome *c* nitrite reductase (NrfA), which is the terminal enzyme in the six-electron dissimilatory reduction of nitrate to ammonia:

$$NO_2^- + 8H^+ + 6e^- \rightarrow NH_4^+ + 2H_2O \quad (3)$$

This is a periplasmic enzyme which was detected in intact cells of *C. jejuni* using methyl viologen as an electron donor (Sellars et al., 2002).

Einsle et al. (2000) proposed that three basic elements are required to form a functional periplasmic nitrite reductase complex, (i) the NrfA enzyme; (ii) a system to oxidize membranous menaquinol and transport electrons to NrfA; (iii) a modified heme lyase needed for the covalent attachment of the active site heme group. The cytochrome *c* nitrite reductases (NrfA) are pentaheme enzymes with a molecular mass of 55–65 kDa. NrfA acts as a homodimer, with each monomer presumed to be functional and to act independently. The active site is located at heme 1, with a Ca^{2+} ion in close proximity (Einsle et al., 2000). There are two distinct systems for the transfer of electrons resulting from menaquinol oxidation to NrfA enzyme, the NrfH tetraheme protein described by Simon et al. (2000) for *Wolinella succinogenes*, and the NrfBCD system of e.g. *E. coli* K-12, in which the NrfA is connected to the quinol oxidase NrfCD via the soluble penta-heme cytochrome *c* NrfB (Hussain et al., 1994).

The putative nitrite-reducing enzyme in *C. jejuni*

(Cj1357c) and the upstream gene encoding a putative NapC/NirT-type cytochrome *c* (Cj1358c), indicates similarity to the *Wolinella succinogenes* nitrite-reducing system in which NrfA accepts electrons from a membrane anchored tetra-heme cytochrome, NrfH (Simon et al., 2000). NrfH is a tetraheme cytochrome *c* acting as a quinol oxidase to receive electrons from the quinone pool and interacts with the positively charged environment around heme 2 of the NrfA enzyme (Einsle et al., 2000). NrfA contains an unusual CXXCK motif for the heme 1 binding site, instead of the CXXCH motif found in many heme containing enzymes. This substitution in NrfA may explain the requirement for a modified heme lyase for covalent attachment of the active site heme The *C. jejuni* NrfA homologue, Cj1357c, is slightly larger at 69 kDa when compared to the *W. succinogenes* NrfA (58 kDa), and contains a novel ligation at heme 1, with a CXXCH motif instead of CXXCK. This would explain the absence in *C. jejuni* of genes encoding heme lyases and assembly proteins that are found in other bacteria (Einsle et al., 2000). Although the functional consequences of this substitution are unknown at present, it is possible it could have important implications for the catalytic activity of the protein. The possible mechanism of nitrite reduction in *C. jejuni* is shown in Fig. 2c.

4. S- or N-oxides as Electron Acceptors

The most comprehensive studies of bacterial DMSO/TMAO reductases are those from *Rhodobacter* spp. and *Escherichia coli.* The reduction of TMAO and DMSO may be carried out by two different reductase enzymes as in *E. coli* (Gennis and Stewart, 1996), or reduction of both TMAO and DMSO may be carried out by a single reductase as in the Dor system of *Rhodobacter* (Kelly et al., 1988; Knablein et al. 1996). In either case the reduction is a two electron transfer process:

$$(CH_3)_2SO + 2H^+ + 2e^- \rightarrow (CH_3)_2S + H_2O \quad (4)$$

$$(CH_3)_3NO + 2H^+ + 2e^- \rightarrow (CH_3)_3N + H_2O \quad (5)$$

The *torCAD* operon in *E. coli* encodes the TMAO reductase complex, consisting of a large periplasmic molybdoprotein (TorA, 94 kDa), a membrane anchored periplasmic cytochrome *c* (TorC, 43 kDa) and a presumed intrinsic membrane subunit (TorD, 22kDa) (Gennis and Stewart, 1996). The DMSO reductase complex contains three subunits, consisting of a large extrinsic membrane bound molybdoprotein, facing the cytoplasm (DmsA, 82.6 kDa), a smaller iron-sulfur subunit (DmsB, 23.6 kDa), and an intrinsic membrane subunit (DmsC, 22.6 kDa) (Bilous and Weiner, 1988; Weiner et al., 1988; Gennis and Stewart, 1996).

Both TMAO and DMSO reductase activities can be measured in intact cells of *C. jejuni* using methyl viologen linked assays and TMAO and DMSO dependent growth under oxygen limited (but not anaerobic) conditions has been demonstrated (Sellars et al., 2002). Analysis of the *C. jejuni* genome reveals a gene encoding a 93 KDa molybdoprotein DorA/TorA homologue (Cj264c) containing conserved residues for the binding of a molybdenum guanosine dinucleoside (MGD) co-factor and an N-terminal motif (RRKFLK), similar to the 'twin-arginine' translocase recognition motif (S/TRRXFLK), characteristic of extracytoplasmic proteins containing complex redox co-factors (Berks, 1996). The Cj264c gene was mutated, and the mutant was found to deficient in both TMAO and DMSO reductase activity, indicating that a single enzyme is responsible (Sellars et al., 2002). In addition there is an upstream gene encoding a 22 kDa mono-heme *c* type cytochrome (Cj265c) with similarity to the C-terminus of the membrane-anchored pentaheme *c* type cytochrome TorC of *E. coli* (Gon et al., 2001; Sellars et al., 2002). TorC contains a C-terminal extension (TorC(C)) with an additional heme-binding site, responsible for the electron transfer to TorA. The tetraheme N-terminus (TorC(N)) is involved in the binding of TorC to TorA (Gon et al., 2001). The mono-heme *c* type cytochrome (Cj265c) of *C. jejuni,* may be involved in electron transport to the DorA/TorA homologue (Cj264c), as with the C-terminal part of the TorC subunit in *E. coli*. The possible topological organization of the electron transport chain to TMAO/DMSO is shown in Fig. 2b, although additional redox proteins are likely to be involved in electron transfer from quinol to Cj265. *H. pylori* may also be capable of TMAO/DMSO respiration, as it contains a putative molybdoprotein homologue of Cj265 (HP0407) which has a twin-arginine motif at the N-terminus indicating a periplasmic location (see Fig. 1a).

V. Conclusions

Although closely related phylogenetically and sharing many similarities, *C. jejuni* and *H. pylori* are distinct in some major aspects of their physiology, especially with regard to electron transport. *C. jejuni* is emerging as a more versatile and metabolically active pathogen, with a complete citric-acid cycle, and a complex and highly branched respiratory chain allowing the use of a variety of alternative electron acceptors. These properties enable it to colonize and survive in a number of environments in addition to the mammalian or avian gut. However, it should be noted that there are many un-annotated redox proteins in the genome sequence which have not been reviewed here, the function of which will undoubtedly reveal more about the emerging metabolic versatility of *C. jejuni*. *H. pylori* is a more specialized pathogen, largely restricted to the human stomach, with an incomplete citric-acid cycle, a simple respiratory chain and few regulatory systems. Both bacteria share the property of being microaerophiles, but there are differences in how they respond to oxygen and oxidative stress, and while there is likely to be no single explanation for their oxygen sensitivity, the possession of several oxygen-sensitive enzymes may be one important factor. However, there are also some interesting parallels between these bacteria in terms of electron transport. For example, both have an apparently unique type of complex I which appears to use an as yet unidentified electron donor. This novel system deserves further study.

Acknowledgments

Work on *Helicobacter* and *Campylobacter* in the laboratory of DJK is supported by the UK Biotechnology and Biological Sciences Research Council, including a studentship to JDM.

References

Alderson J, Clayton CL and Kelly DJ (1997) Investigations into the aerobic respiratory chain of *Helicobacter pylori*. Gut 41(S1): A7

Alm RA, Ling LS, Moir DT, King BL, Brown ED, Doig PC, Smith DR, Noonan B, Guild BC, deJonge BL, Carmel G, Tummino PJ, Caruso A, Uria-Nickelsen M, Mills DM, Ives C, Gibson R, Merberg D, Mills SD, Jiang Q, Taylor DE, Vovis GF and Trust TJ (1999) Genomic-sequence comparison of two unrelated isolates of the human gastric pathogen *Helicobacter pylori*. Nature 397: 176–180

Arciero DM and Hooper AB (1994) A di-heme cytochrome *c* peroxidase from *Nitrosomonas europaea* catalytically active in both the oxidized and half-reduced states. J Biol Chem 269: 11878–11886

Baillon ML, van Vliet AH, Ketley JM, Constaninidou C and Penn CW (1999) An iron-regulated alkyl hydroperoxide reductase (AhpC) confers aerotolerance and oxidative stress resistance to the microaerophilic pathogen *Campylobacter jejuni*. J Bacteriol 181: 4798–4804

Baker LMS, Raudonikiene A, Hoffman PS and Poole LB (2001) Essential thioredoxin-dependent peroxiredoxin system from *Helicobacter pylori*: Genetic and kinetic characterization. J Bacteriol 183: 1961–1973

Berg BL, Li J, Heider J and Stewart V (1991) Nitrate-inducible formate dehydrogenase in *Escherichia coli* K–12. I. Nucletide sequence of the *fdnGHI* operon and evidence that opal (UGA) encodes selenocysteine. J Biol Chem 266: 22380–22385

Berks BC, Richardson DJ, Reilly A, Willis AC and Ferguson SJ (1995) The *napEDABC* gene cluster encoding the periplasmic nitrate reductase system of *Thiosphaera pantotropha*. Biochem J 309: 983–992

Berks BC (1996) A common export pathway for proteins binding complex redox cofactors? Mol Microbiol 22: 393–404

Bilous PT and Weiner JH (1988) Molecular cloning and expression of the *Escherichia coli* DMSO reductase operon. J Bacteriol 170: 1511–1518

Birkholz S, Knipp U, Lemma E, Kroger A and Opferkuch W (1994) Fumarate reductase of *Helicobacter pylori*—an immunogenic protein. J Med Microbiol 42: 56–62

Bolton FJ, Coates D and Hutchinson DN (1984) The ability of *Campylobacter* media supplements to neutralise photochemically induced toxicity and hydrogen peroxide. J Appl Bacteriol 56: 151–157

Carlone GM and Anet FAL (1983) Detection of menaquinone-6 and a novel methyl-substituted menaquinone-6 in *Campylobacter jejuni* and *Campylobacter fetus* subsp. *fetus*. J Gen Microbiol 129: 3385–3393

Carlone GM and Lascelles J (1982) Aerobic and anaerobic respiratory systems in *campylobacter fetus* subsp. *jejuni* grown in atmospheres containing hydrogen. J Bacteriol 152: 306–314

Censini S, Lange C, Xiang ZY, Crabtree JE, Ghirra P, Borodovsky M, Rappuoli R and Covacci A (1996) Cag, a pathogenicity island of *Helicobacter pylori*, encodes type I-specific and disease associated virulence factors. Proc Natl Acad Sci USA 93: 14648–14653.

Chang HT, Marcelli SW, Davison AA, Chalk PA, Poole RK and Miles R J (1995). Kinetics of substrate oxidation by whole cells and cell membranes of *Helicobacter pylori*. FEMS Microbiol Lett 129: 33–38

Chen M, Andersen LP, Zhai L and Kharazmi A (1999) Characterization of the respiratory chain of *Helicobacter pylori*. FEMS Immunol Med Microbiol 24: 169–174

Collins MD, Costas M and Owen JR (1984) Isoprenoid quinone composition of representatives of the genus *Campylobacter*. Arch Microbiol 137: 168–170

Davidson AA, Kelly DJ, White PJ and Chalk PA (1993) Citric-acid cycle enzymes and respiratory metabolism in *H. pylori*. Acta Gastro-Enterol Belg 56S: 96

Dixon MF (2001) Pathology of gastritis and peptic ulceration. In: Mobley HLT, Mendz GL, and Hazell SL (ed) *Helicobacter pylori*: Physiology and Genetics, pp 459–469. ASM Press, Washington, D. C.

Doig P, de Jong BL, Alm RA, Brown ED, Uria-Nickelson M,

Noonan B et al. (1999) Helicobacter pylori physiology predicted from genomic comparisons of two strains. Microbiol Mol Biol Rev 63: 675–707

Einsle O, Stach P, Messerschmidt A, Simon J, Kröger A, Huber R and Kroneck MH (2000) Cytochrome *c* nitrite reductase from *Wolinella succinogenes.* J Biol Chem 275: 39608–39616

Enoch HG and Lester RL (1975) The purification and properties of formate dehydrogenase and nitrate reductase from *Escherichia coli.* J Biol Chem 250: 6693–6705

Evans DJ Jr. and Evans DG (1997) Identification of a formate dehydrogenase associated cytochrome c_{553} in *Helicobacter pylori.* Gut 41(S): A6

Finel M (1998) Does NADH play a central role in energy metabolism in *Helicobacter pylori*? Trends Biochem Sci 23: 412–414

Friedman CR, Neiman J, Wegener HC and Tauxe RV (2000) Epidemiology of *Campylobacter jejuni* in the United States and other industrialized nations. In: Nachamkin I and Blaser MJ (ed) *Campylobacter,* 2nd Ed, pp 121–138. ASM Press, Washington D.C.

Friedrich T (1998) The NADH:ubiquinone oxidoreductase (complex I) from *Escherichia coli.* Biochim Biophys Acta 1364: 134–146

Fu HA, Iuchi S and Lin ECC (1991) The requirement of ArcA and Fnr for peak expression of the *cyd* operon in *Escherichia coli* under microaerobic conditions. Mol Gen Genet 226: 209–213

Ge Z, Jiang Q, Kalisiak MS and Taylor DE (1997) Cloning and functional characterization of *Helicobacter pylori* fumarate reductase operon comprising three structural genes coding for subunits C, A and B. Gene 204: 227–234

Gennis RB and Stewart V (1996) Respiration. In: Neidhardt FC (ed) *Escherichia coli* and *Salmonella typhimurium*, 2nd Ed, Vol 1, pp 217–261, ASM Press, Washington D.C.

Gilbert JV, Ramakrishna J, Sunderman Jr. FW, Wright A and Plaut AG (1995) Protein Hpn: Cloning and characterisation of a histidine-rich metal binding polypeptide in *Helicobacter pylori* and *Helicobacter mustelae*. Infect Immun 62: 2682–2688

Gilmour R, Goodhew CF, Pettigrew GW, Prazeres S, Moura I and Moura JJ (1993) Spectroscopic characterization of cytochrome *c* peroxidase from *Paracoccus denitrificans*. Biochem J 294: 745–752

Gon S, Guidici-Orticoni MT, Mejean V and Iobbi-Nivol C (2001) Electron transfer and binding of the *c*-type cytochrome TorC to the trimethylamine N-oxide reductase in *Escherichia coli.* J Biol Chem 276: 11545–11551

Goodhew CF, ElKurdi AB and Pettigrew GW (1988) The microaerophilic respiration of *Campylobacter mucosalis.* Biochim Biophys Acta 933: 114–123

Green GN, Fang H, Lin R-J, Newton G, Mather M, Georgiou CD and Gennis RB (1988) The nucleotide sequence of the *cyd* locus encoding the two subunits of the cytochrome *d* terminal oxidase complex of *Escherichia coli.* J Biol Chem 263: 13138–13143

Harvey P and Leach S (1998) Analysis of coccal cell formation by *Campylobacter jejuni* using continuous culture techniques, and the importance of oxidative stress. J Appl Microbiol 85: 398–404

Harvey S and Lascelles J (1980) Respiration systems and cytochromes in *Campylobacter fetus* subsp. *intestinalis.* J Bacteriol 144: 917–922

Hata-Tanaka A, Matsuura K, Itoh S and Anraku Y (1987) Electron flow and heme-heme interaction between cytochromes *b-558*, *b-595* and *d* in a terminal oxidase of *Escherichia coli.* Biochim Biophys Acta 893: 289–295

Hazell SL, Markesich DC, Evans DJ, Evans DG and Graham DY (1989) Influence of media supplements on growth and survival of *Campylobacter pylori*. Eur J Clin Microbiol Infect Dis 8: 597–602

Hodge JP and Krieg NR (1994) Oxygen tolerance estimates in *Campylobacter* species depend upon the testing medium. J Appl Bacteriol 77: 6666–6763

Hoffman PS and Goodman TG (1982) Respiratory physiology and energy conservation efficiency of *Campylobacter jejuni.* J Bacteriol 150: 319–326

Hoffman PS, George HA, Krieg NR and Smibert RM (1979b) Studies on the microarophilic nature of *Campylobacter fetus* subsp. *jejuni*. II. Role of exogenous superoxide anions and hydrogen peroxide. Can J Microbiol 25: 8–16

Hoffman PS, Krieg NR and Smibert RM (1979a) Studies on the micraerophilic nature of *Campylobacter fetus* subsp. *jejuni.* I. Physiological aspects of enhanced aerotolerance. Can J Microbiol 25: 1–7

Hu W, de Smet L, van Driessche G, Bartsch RG, Meyer TE, Cusanovich MA and van Beeumen J (1998) Characterization of cytochrome *c-556* from the purple phototrophic bacterium *Rhodobacter capsulatus* as a cytochrome-*c* peroxidase. Eur J Biochem 258: 29–36

Hughes NJ, Chalk PA, Clayton CL and Kelly DJ (1995) Identification of carboxylation enzymes and characterization of a novel four-subunit pyruvate: flavodoxin oxidoreductase from *Helicobacter pylori.* J Bacteriol 177: 3953–3959

Hughes NJ, Clayton CL, Chalk PA and Kelly DJ (1998) *Helicobacter pylori porCDAB* and *oorDABC* genes encode distinct pyruvate:flavodoxin and 2-oxoglutarate:acceptor oxidoreductases which mediate electron transport to NADP. J Bacteriol 180: 1119–1128

Hussain H, Grove J, Griffiths L, Busby S and Cole J (1994) A 7-gene operon essential for formate-dependant nitrite reduction to ammonia by enteric bacteria. Mol Microbiol 12: 153–163

Ingeldew WJ and Poole RK (1984) The respiratory chains of *Escherichia coli.* Microbiol Rev 48: 222–271

Jones RW (1980) Proton translocation by the membrane-bound formate dehydrogenase of *Escherichia coli.* FEMS Microbiol Lett 8: 167–172

Kather B, Stingl K, van der Rest ME, Altendorf K and Molenaar D (2000) Another unusual type of citric acid cycle enzyme in *Helicobacter pylori:* Malate:quinone oxidoreductase. J Bacteriol 182: 3204–3209

Kelly DJ (1998) The physiology and metabolism of the human gastric pathogen *Helicobacter pylori.* Adv Microb Physiol 40: 137–189

Kelly DJ (2001) The physiology and metabolism of *Campylobacter jejuni* and *Helicobacter pylori.* J Appl Microbiol 90: 16S–24S

Kelly DJ and Hughes NJ (2001) The citric acid cycle and fatty acid biosynthesis. In: Mobley HLT, Mendz GL and Hazell SL (ed) *Helicobacter pylori*: Physiology and Genetics, pp 135–146. ASM Press, Washington, D. C.

Kelly DJ, Richardson DJ, Ferguson SJ and Jackson JB (1988) Isolation of transposon Tn5 insertion mutants of *Rhodobacter capsulatus* unable to reduce trimethylamine-N-oxide and dimethylsulphoxide. Arch Microbiol 150: 138–144

Kelly DJ, Hughes NJ and Poole RK (2001) Microaerobic physiology: Aerobic respiration, Anaerobic respiration, and carbon dioxide metabolism. In: Mobley HLT, Mendz GL and Hazell SL (ed) *Helicobacter pylori*: Physiology and Genetics, pp 113–124. ASM Press, Washington, D. C.

Knablein J, Mann K, Ehlert S, Fonstein M, Huber R and Schneider F (1996) Isolation, cloning, sequence analysis and localization of the operon encoding dimethyl sulfoxide/trimethylamine N-oxide reductase from *Rhodobacter capsulatus*. J Mol Biol 263: 40–52

Koyanagi S, Nagata, K, Tamura T, Tsukita S and Sone N (2000) Purification and characterisation of cytochrome *c-553* from *Helicobacter pylori* J. Biochem 128: 371–375

Kranz RG, Lill R, Goldman B, Bonnard G and Merchant S. (1998). Molecular mechanisms of cytochrome *c* biogenesis: Three distinct systems. Mol Microbiol 29: 383–396.

Kusters JG, Gerrits MM, Van Strijp JAG and Vandenbroucke-Grauls CMJE (1997) Coccoid forms of *Helicobacter pylori* are the morphologic manifestation of cell death. Infect Immun 65: 3672–3679

Lascelles J and Calder K (1985) Participation of cytochromes in some oxidation-reduction systems in *Campylobacter fetus*. J Bacteriol 164: 401–409

Leach SA (1997) Growth, survival and pathogenicity of enteric campylobacters. Rev Med Microbiol 8: 113–124

Leach S, Harvey P and Wait R (1997) Changes with growth rate in the membrane lipid composition of and amino-acid utilisation by continuous cultures of *Campylobacter jejuni*. J Appl Microbiol 82: 631–640

Lorence RM, Koland JG and Gennis RB (1986) Coulometric and spectroscopic analysis of purified cytochrome *d* complex of *Escherichia coli*: Evidence for the identification of 'cytochrome a_1' as cytochrome b_{595}. Biochemistry 25: 2314–2321.

Maier RJ, Fu C, Gilbert J, Moshiri F, Olson J and Plaut AG (1996). Hydrogen uptake hydrogenase in *Helicobacter pylori.* FEMS Microbiol Lett 141: 71–76

Marais A, Mendz GL, Hazell SL and Megraud F (1999) Metabolism and genetics of Helicobacter pylori: The genome era. Microbiol Mol Biol Rev 63: 642–674

Marcelli SW, Chang H-T, Chapman T, Chalk PA, Miles RJ and Poole RK (1996) The respiratory chain of *Helicobacter pylori*: Identification of cytochromes and the effects of oxygen on cytochrome and menaquinone levels. FEMS Microbiol Lett 138: 59–64

Mendz GL and Hazell SL (1993) Fumarate catabolism in *Helicobacter pylori*. Biochem Mol Biol Int 31: 325–332

Miller MJ and Gennis RB (1985) The cytochrome *d* complex is a coupling site in the aerobic respiratory chain of *Escherichia coli.* J Biol Chem 260: 14003–14008

Miller MJ, Hermodson M and Gennis RB (1988) The active form of the cytochrome *d* terminal oxidase complex of *Escherichia coli* is a heterodimer containing one copy of each of the two subunits. J Biol Chem 263: 5235–5240

Mitchell HM (2001) Epidemiology of infection. In: Mobley HLT, Mendz GL, and Hazell SL (ed) *Helicobacter pylori*: Physiology and Genetics, pp 7–18. ASM Press, Washington, D. C.

Nagata K, Tsukita S, Tamura T and Sone N (1996) A *cb*-type cytochrome-*c* oxidase terminates the respiratory chain in *Helicobacter pylori*. Microbiology 142: 1757–1763

Nagata K, Nagata Y, Sato T, Fujino MA, Nakajima K and Tamura T (2003) L-serine, D- and L-proline and alanine as respiratory substrates of *Helicobacter pylori*: Correlation between in vitro and in vivo amino-acid levels. Microbiology 149: 2023–2030

Nedenskov P (1994) Nutritional requirements for growth of *Helicobacter pylori*. J Appl Env Microbiol 60: 3450–3453

Ødum L and Anderson LP (1995) Investigation of *Helicobacter pylori* ascorbic acid oxidising activity. FEMS Immunol Med Microbiol 10: 289–294

Olivieri R, Bugnoli M, Armellini D, Bianciardi S, Rappuoli R, Bayeli PF, Abate L, Esposito E, de Gregorio L, Aziz J, Basagni C and Figura N (1993) Growth of *Helicobacter pylori* in media containing cyclodextrins. J Clin Microbiol 31: 160–162

Olson JW, Mehta NS and Maier RJ (2001) Requirement of nickel metabolism proteins HypA and HypB for full activity of both hydrogenase and urease in *Helicobacter pylori*. Mol Microbiol 39: 176–182

Olson JW and Maier RJ (2002) Molecular hydrogen as an energy source for *Helicobacter pylori*. Science 298: 1788–1790.

Parkhill J, Wren BW, Mungall K, Ketley JM, Churcher C, Basham D, Chillingworth T, Davies RM, Feltwell T, Holroyd S, Jagels K, Karlyshev AV, Moule S, Pallen MJ, Penn CW, Quail MA, Rajandream M-A, Rutherford KM, van Vliet AHM, Whitehead S and Barrell BG (2000) The genome sequence of the food-borne pathogen *Campylobacter jejuni* reveals hypervariable sequences. Nature 43: 665–668

Pitson S, Mendz GL, Srinivasan S and Hazell SL (1999) The tricarboxylic acid cycle of *Helicobacter pylori*. Eur J Biochem 260: 258–267

Poole RK and Cook GM (2000) Redundancy of aerobic respiratory chains in bacteria: Routes, reasons and regulation. Adv Microb Physiol 42: 165–224.

Potter L, Angove H, Richardson D and Cole J (2001) Nitrate reduction in the periplasm of Gram-negative bacteria. Adv Microb Physiol 45: 51–112

Potter LC and Cole JA (1999) Essential roles for the products of the *napABCD* genes, but not *napFGH*, in periplasmic nitrate reduction by *Escherichia coli* K-12. Biochem J 344: 69–67

Reynolds DJ and Penn CW (1994) Characteristics of *Helicobacter pylori* growth in a defined medium and determination of its amino acid requirements. Microbiology 140: 2649–2656

Richardson DJ, Berks BC, Russell DA, Spiro S and Taylor CJ (2001) Functional, biochemical and genetic diversity of prokaryotic nitrate reductases. Cell Mol Life Sci 58: 165–178

Rönnberg M, Kalkkinen N and Ellfolk N (1989) The primary structure of *Pseudomonas* cytochrome *c* peroxidase. FEBS Lett 250: 175–178

Sellars MJ, Hall, SJ and Kelly DJ (2002) Growth of *Campylobacter jejuni* supported by respiration of fumarate, nitrate, nitrite, trimethylamine-N-oxide or dimethylsulfoxide requires oxygen. J Bacteriol 184: 4187–4196.

Simon J, Gross R, Einsle O, Kroneck PMH, Kröger A and Klimmek O (2000) A Nap/NirT-type cytochrome *c* (NrfH) is the mediator between the quinone pool and the cytochrome *c* nitrite reductase of *Wolinella succinogenes.* Mol Microbiol 35: 686–696

Skirrow MB and Blaser MJ (2000) Clinical aspects of *Campylobacter* infection. In: Nachamkin I and Blaser MJ (ed) *Campylobacter,* 2nd Ed, pp 69–88, ASM Press, Washington D.C.

Smith MA, Finel M, Korolik V and Mendz GL (2000) Characteristics of the aerobic respiratory chains of the microaerophiles *Campylobacter jejuni* and *Helicobacter pylori*. Arch Microbiol

174: 1–10

Spiro S and Guest JR (1991) Adaptive responses to oxygen limitation in *Escherichia coli.* Trends Biochem Sci 16: 310–314

Thöny-Meyer L (1997) Biogenesis of Respiratory cytochromes in bacteria. Microbiol Mol Biol Rev 61: 337–376

Tomb J-F, White O, Kerlavage AR, Clayton RA, Sutton GG, Fleishmann RD, Ketchum KA, Klenk HP, Gill S, Dougherty B, Nelson K, Quakenbush J, Zhou L, Kirkness EF, Peterson S, Loftus B, Richardson D, Dodson R, Khalak HG, Glodek A, McKenney K, Fitzegerald LM, Lee N, Adams MD, Hickey E, Berg DE, Gocayne JD, Utterback TR, Peterson JD, Kelley JM, Cotton MD, Weidman JM, Fujii C, Bowman C, Watthey L, Wallin E, Hayes WS, Borodovsky M, Karp PD, Smith HO, Fraser CM and Venter JC (1997) The complete genome sequence of the gastric pathogen *Helicobacter pylori.* Nature 388: 539–547

Trumpower BL (1990) Cytochrome bc_1 complexes of microorganisms. Microbiol Rev 54: 101–129

Tsukita S, Koyanagi S, Nagata K, Koizuka H, Akashi H, Shimoyama T, Tamura T and Sone N (1999) Characterization of a *cb*-type cytochrome *c* oxidase from *Helicobacter pylori.* J Biochem 125:194–201

van Vliet AHM, Baillon MLA, Penn C and Ketley JM (1999) *Campylobacter jejuni* contains two Fur homologues: Characterisation of iron-responsive regulation of peroxide stress defense genes by the PerR repressor. J Bacteriol 181: 6371–6376

Veron M, Lenvoise-Furet A and Beaune P (1981) Anaerobic respiration of fumarate as a differential test between *Campylobacter fetus* and *Campylobacter jejuni.* Curr Microbiol 6: 349–354

Weber I, Fritz C, Ruttkowski S, Kreft A and Bange FC (2000) Anaerobic nitrate reductase (narGHJI) activity of *Mycobacterium bovis* BCG in vitro and its contribution to virulence in immunodeficient mice. Mol Microbiol 35: 1017–1025

Weiner JH, MacIsaac DP, Bishop RE and Bilous PT (1988) Purification and properties of *Escherichia coli* DMSO reductase, an iron-sulfur molybdoenzyme with broad substrate specificity. J Bacteriol 170: 1505–1510

Yagi T, Yano T, Di Bernardo S and Matsuno-Yagi A (1998) Prokaryotic complex I (NDH-1), an overview. Biochim Biophys Acta 1364: 125–133

Chapter 4

Respiratory Chains in Acetic Acid Bacteria: Membrane-bound Periplasmic Sugar and Alcohol Respirations

Kazunobu Matsushita*, Hirohide Toyama, and Osao Adachi

Department of Biologichal Chemistry, Faculty of Agriculture, Yamaguchi University, Yamaguchi, Yamaguchi 753-8515, Japan

Summary

Acetic acid bacteria are obligate aerobes, and well known to have a strong ability to oxidize ethanol, sugars or sugar alcohols to produce the corresponding sugar acids. These oxidation reactions of sugars and alcohols in acetic acid bacteria are uniquely carried out by primary dehydrogenases located in the periplasmic side of the cytoplasmic membrane and linked to the terminal ubiquinol oxidase(s) via ubiquinone in the respiratory chain. The terminal ubiquinol oxidases working in the respiratory chain of acetic acid bacteria could be classified into four different types, cytochrome *o*, cytochrome a_1, cytochrome *d*, and CN-resistant bypass oxidase, which are not unique but found in other bacterial species; whereas the primary dehydrogenases working in the periplasmic sugar and alcohol respirations include many unique quinoproteins and quinoprotein-cytochrome *c* complexes, and flavoprotein-cytochrome *c* complexes. Such periplasmic sugar and alcohol respirations of acetic acid bacteria are involved in the accumulation of the oxidized products of sugars or alcohols, but not in the assimilation of these substrates, at least in their early growth phase. These sugar and alcohol respirations seem

*Author for correspondence, email: kazunobu@agr.yamaguchi-u.ac.jp

Davide Zannoni (ed): Respiration in Archaea and Bacteria. Vol 2: Diversity of Prokaryotic Respiratory Systems, pp. 81–99.

not to generate so much energy, but instead to work for rapid oxidation to produce a large amount of oxidized products, which are usually very hard to be utilized by other organisms, or confer very harmful conditions for other organisms to keep alive. Thus, the specific respirations of acetic acid bacteria seem to have evolved to their own living strategy for competing with other microorganisms.

I. Introduction

Acetic acid bacteria are obligate aerobes and are well known to have a strong ability to oxidize ethanol, sugars and alcohols to produce the corresponding sugar acids. Such oxidation reactions are traditionally called 'oxidative fermentation,' since they involve incomplete oxidation of these compounds. Bacteria capable of effecting such oxidative fermentation are called 'oxidative bacteria,' of which the most prominent ones are acetic acid bacteria. These oxidative bacteria do not oxidize such sugars or alcohols completely to carbon dioxide, at least in their early culture phase, and thus accumulate the corresponding incomplete oxidation products in large amounts in the growth medium.

These oxidation reactions of sugars and alcohols by acetic acid bacteria are uniquely carried out by membrane-bound dehydrogenases linked to the respiratory chain located in the periplasmic side of the cytoplasmic membrane of the organisms. Such unique periplasmic respirations are involved in only the partial oxidation reactions of these sugars and alcohols, but not in the complete oxidation and the assimilation of the substrates. Since the assimilation reaction can be carried out by cytoplasmic enzymes linked to the NADH respiration, the reaction products must be transported into the cytoplasm from the periplasmic reaction site. Thus, the periplasmic respiration of acetic acid bacteria seems reminiscent of iron or hydrogen respiration in chemolithotrophs, and thus has a unique position in heterotrophic energy metabolism.

In addition, since these periplasmic respiration systems are involved in the accumulation of a large amount of oxidation products such as acetic acid or L-sorbose in the culture media, acetic acid bacteria are also important for fermentation industries to produce useful biomaterials.

Acetic acid bacteria have been shown to belong to α-Proteobacteria, and the closest related species are *Rhodophila globiformis* and *Acidiphilium* species in the bacterial phylogenetic tree (Sievers et al., 1994, 1995). Acetic acid bacteria are classified into two genera, *Gluconobacter* and *Acetobacter* of the Family *Acetobacteraceae*. The classification has also been confirmed by phylogenetic analysis, in which all *Gluconobacter* species form a closely related cluster and are well separated from *Acetobacter* species; whereas *Acetobacter* species are relatively diverse, and are first separated into two groups, the first group including *Acetobacter aceti* and *Acetobacter pasteurianus*, and the other the remaining *Acetobacter* species. The remaining species seem to be further subdivided into two groups, the second group including *Acetobacter liquefaciens* and *Acetobacter diazotrophicus*, and the third group including *Acetobacter europaeus* and *Acetobacter xylinum*. Recently, the latter two groups were proposed to be included in the genus, *Gluconoacetobacter* (Yamada et al., 1997). Among these *Acetobacter* (and/or *Gluconoacetobacter*) species, *Acetobacter methanolicus* has a somewhat different location in between the first group and the second or third groups (Fig. 1). However, in the present chapter, all '*Acetobacter* species' are operationally defined as those included in the genus *Acetobacter*, *Gluconoacetobacter* and *Acidomonas*.

Such a phylogenetic relationship among acetic acid bacteria seems to be somewhat related to the sugar and alcohol respirations. *Gluconobacter* species exhibit highly active oxidation reactions on sugars or sugar alcohols such as D-glucose, D-gluconic acid, D-sorbitol, glycerol and so on, in addition to ethanol. In contrast, *Acetobacter* species have a highly active ethanol-oxidizing activity but fewer other sugar or sugar alcohol-oxidizing activities. Furthermore, *Acetobacter* species, unlike *Gluconobacter* species, are able to oxidize acetate, the oxidation product of ethanol, completely to CO_2 and water under some specific growth conditions. Thus, although the two genera exhibit a critical difference in their oxidizing ability, these sugar and alcohol respirations

---

Abbreviations: 2KGDH – 2-ketogluconate dehydrogenase; ADH– alcohol dehydrogenase; ALDH – aldehyde dehydrogenase; FDH – fructose dehydrogenase; GADH – gluconate dehydrogenase; GDH – glucose dehydrogenase; GLDH – glycerol dehydrogenase; PQQ – pyrroloquinoline quinone; SDH – sorbose dehydrogenase; SLDH – sorbitol dehydrogenase; SNDH – sorbosone dehydrogenase; UQ – ubiquinone; UQ_n – ubiquinone-n (e.g. ubiquinone-9)

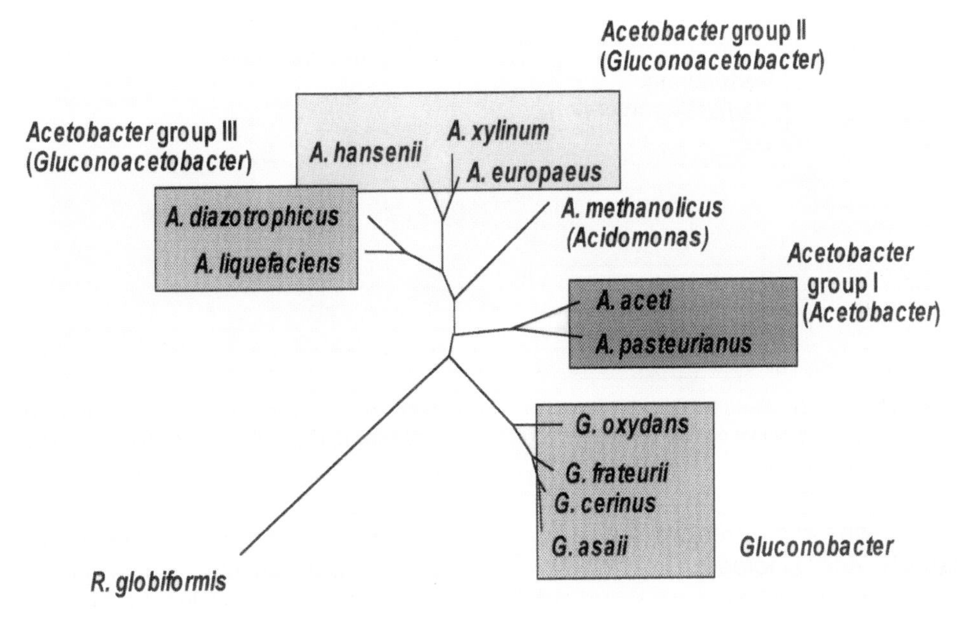

Fig. 1. Phylogenetic tree of acetic acid bacteria. This figure is redrawn based on Fig. 1 by Sievers et al. (1995).

in both genera seem to function by linking tightly to the aerobic respiratory chains of the organisms (Matsushita et al., 1994). Although a considerable amount of earlier work had been done on the sugar metabolism of acetic acid bacteria (Asai, 1968), little is known about the respiratory chain of acetic acid bacteria and of the relation with the sugar or alcohol oxidation reactions.

The respiratory chain of acetic acid bacteria not only has many common principles but also some diversity. In this chapter, the general aspects in the respiratory chains and also the characteristic periplasmic sugar and alcohol respirations of acetic acid bacteria are described.

II. Respiratory Chains of Acetic Acid Bacteria

A. General Aspects in the Respiratory Chain of Acetic Acid Bacteria

Bacterial respiratory chains have been classified into two categories based on the terminal oxidase; one is a cytochrome *c* oxidase and the other a ubiquinol oxidase. In this respect, the respiratory chain of acetic acid bacteria has been shown to be classified into one having ubiquinol oxidase at the terminal end of the respiratory chain. Thus, the respiratory chain of acetic acid bacteria is rather simple with respect to their arrangement of the respiratory components, like *Escherichia coli* (Anraku, 1988).

As shown in Fig. 2, the respiratory chain of acetic acid bacteria simply consists of primary dehydrogenases and terminal ubiquinol oxidase(s), both of which are connected by ubiquinone (UQ). The UQ present in the respiratory chain of acetic acid bacteria is either ubiquinone-9 (UQ_9) or ubiquinone-10 (UQ_{10}), which varies depending on the bacterial species and is mainly UQ_9 in *Acetobacter* species (strictly defined as *Acetobacter* species in this case) or UQ_{10} in *Gluconobacter* species and also in the recently defined *Gluconoacetobacter* and *Acidomonas* species. Although both the primary dehydrogenase and the terminal oxidase parts are very divergent in the respiratory chain of acetic acid bacteria, the diversity of the terminal oxidase is not unusual while the primary dehydrogenase is especially divergent in the respiratory chain. In acetic acid bacteria, many peculiar periplasmic dehydrogenases are working for the specific sugar, alcohol or sugar alcohol oxidation systems. The diversity and peculiarity of the periplasmic oxidation systems in the respiratory chain of acetic acid bacteria will be described in the next section.

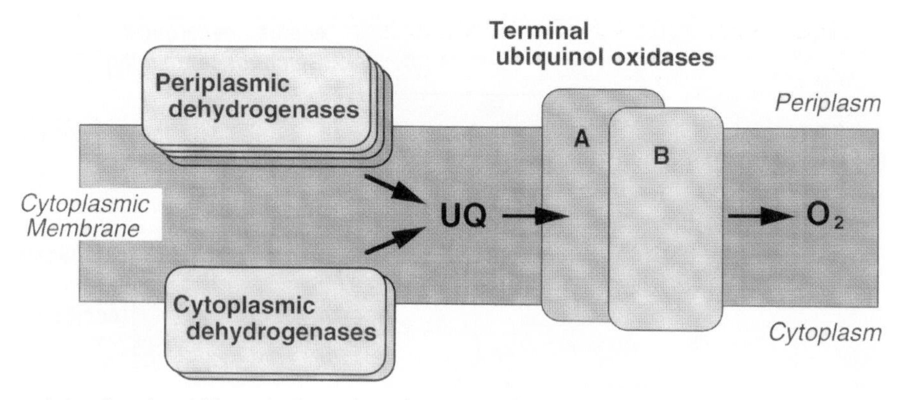

Fig. 2. Respiratory chain of acetic acid bacteria. It consists of many periplasmic dehydrogenases, some cytoplasmic dehydrogenases, UQ, and at least two sets of ubiquinol oxidase (A and B), all of which are bound tightly to the cytoplasmic membranes of acetic acid bacteria.

B. Terminal oxidases in the Respiratory Chains of Acetic Acid Bacteria

Several different species of ubiquinol oxidase are known in the respiratory chain of acetic acid bacteria, with respect to the cytochrome species, as summarized in Table 1. Originally, Bächi and Ettlinger (1974) have divided acetic acid bacteria into three groups based on the α-absorption peak in the longer wavelength region (over 570 nm) of cytochromes; (1) *Acetobacter* strains (*Peroxydans* group) exhibiting both peaks at ~590 nm and ~630 nm, (2) another *Acetobacter* species (*Oxydans* and *Mesoxydans* groups) having only the peak at ~590 nm, and (3) *Gluconobacter* strains, which do not show any corresponding peaks. The original classification can now be realized as the presence of the different species of the terminal oxidases in different classes of acetic acid bacteria. *Acetobacter* species exhibiting both α peaks at 590 and 630 nm seem to have cytochrome *d* (*bd*) oxidase, which is shown to exhibit both *b*-type (590 nm) and *d*-type (630 nm) absorption peaks due to the presence of heme *b* and heme *d*, respectively; whereas other *Acetobacter* species having only the 590 nm peak in the region are shown to have cytochrome a_1 (*ba*), which exhibits an α-absorption peak around 590 nm due to the presence of heme *a* in addition to a cytochrome *b* component. *Gluconobacter* species are shown to have only cytochrome *o* as a typical terminal oxidase, which does not have any typical α-peaks in such a long wavelength region. Thus, by adding a CN-resistant bypass oxidase, the component of which has not been identified, the terminal oxidases of acetic acid bacteria seem to be divided into four different types (Table 1). These terminal oxidases are changed mainly depending on the growth conditions, and two different ubiquinol oxidases are usually produced simultaneously (Fig. 3).

Unlike other acetic acid bacteria, *A. methanolicus* is an exceptional strain able to grow on methanol as well as other carbon sources. When grown on methanol, the strain has a methanol oxidase respiratory chain, where the terminal oxidase is a cytochrome *c* oxidase as in usual methylotrophs (Elliott and Anthony, 1988); whereas on ethanol- or glycerol-dependent growth, *A. methanolicus* has a ubiquinol oxidase as the terminal oxidase of the respiratory chain, as in the case of usual acetic acid bacteria. Cytochrome *c* oxidase and ubiquinol oxidase of both the respiratory chains have been shown to be a cytochrome *co* (seemingly cbb_3-type) and a cytochrome *o*, respectively (Table 1).

C. Cytochrome o as the General Major Terminal Oxidase in the Respiratory Chains of Acetic Acid Bacteria

Cytochrome *o* was found in *Gluconobacter suboxydans* in the 1950s, and then later characterized in 1970 and purified in 1987 from the same species. Although earlier studies had suggested that the organism contains cytochromes *c*, *b*, *o*, and *a* in the respiratory chain, more detailed analysis with a membrane preparation showed that the organism has only cytochromes *c* and *o* but no cytochrome *d* or *a*. Many aspects of the earlier observations have been confirmed and further improved by the purification and characterization of cytochrome *o* from *G. suboxydans* (Matsushita et al., 1994).

Cytochrome *o* of *G. suboxydans* is a ubiquinol oxidase which consists of four non-identical subunits

Table 1. Respiratory terminal oxi°dases of acetic acid bacteria

Type of terminal Oxidase	Prosthetic group	Strains	References
I. Cytochrome *o*	heme *b*	*G. suboxydans*	Matsushita et al. (1987)
(bo_3)	heme *o*		
	Cu^{++}		
		A. pateurianus	Williams and Poole (1988)
		A. aceti	Matushita et al. (1992a)
		A. methanolicus	Matsushita et al. (1992c)
II. Cytochrome a_1	heme *b*	*A. aceti*	Matsushita et al. (1990)
(ba_3)	heme *a*		Matsushita et al. (1992a)
	Cu^{++}	*A. diazotrophicus*	Flores-Encarnacòn et al. (1999)
III. Cytochrome *d*	heme *b*	*A. pasteurianus*	Williams and Poole (1987)
(*bd**)	heme *d*	*A. diazotrophicus*	Flores-Encarnacòn et al. (1999)
IV. CN-resistant bypass oxidase	??	*G. suboxydans*	Ameyama et al. (1987)
		A. diazotrophicus	Flores-Encarnacòn et al. (1999)
V. Cytochrome *co*	heme *b*	*A. methanolicus*	Matushita et al. (1992c)
(cbb_3*)	heme *c*		
	Cu^{++}		

*Not experimentally proved

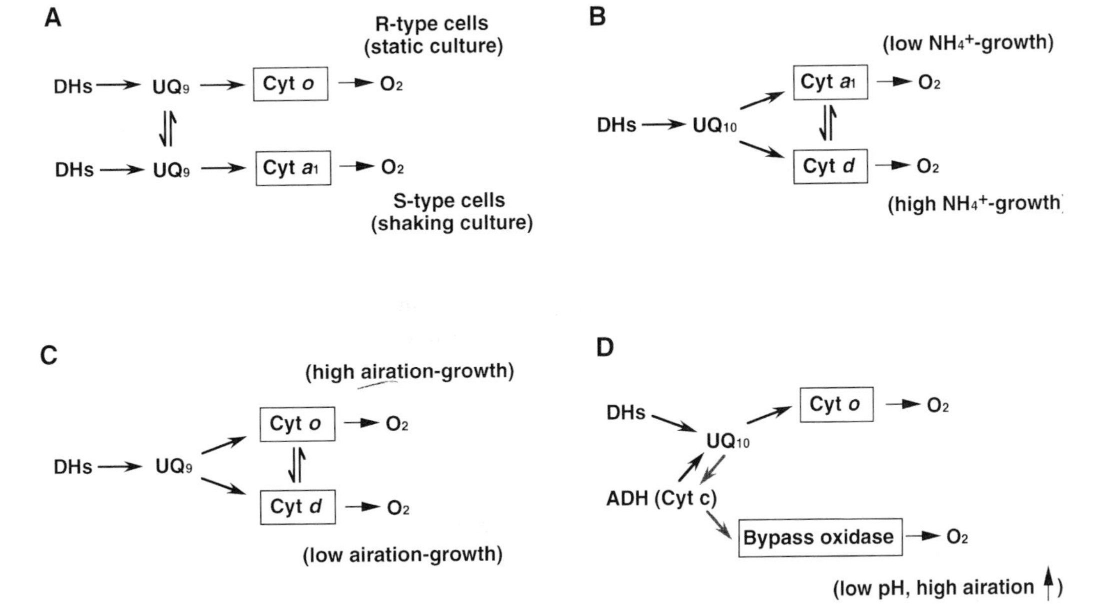

Fig. 3. Several typical respiratory chains of acetic acid bacteria, *Acetobacter aceti* (A), *Acetobacter diazotrophicus* (B), *Acetobacter pasteurianus* (C), and *Gluconobacter suboxydans* (D).

and contains heme *b*, heme *o*, and copper atom as the prosthetic group. Later, cytochrome *o* was found in one species of *Acetobacter aceti* R strain, which can grow on static culture by floating on the medium surface by producing pellicle polysaccharide (Matsushita et al., 1992b). The cytochrome *o* of *A. aceti* has also been shown to consist of one heme *b* and one heme *o* and 1 copper atom, and to be the same protein entity as cytochrome a_1 purified from *A. aceti* S strain, as described below. Cytochrome *o* (and also cytochrome a_1) has been shown to produce a proton-motive force by reconstituting into proteoliposomes (Matsushita et al., 1987, 1992d).

Cytochrome *o* might be present and function as at least one of the terminal oxidases in almost all acetic acid bacteria, since cytochrome *o* has also been detected in oxygen-sufficient *A. pasteurianus,* having cytochrome *d* as the terminal oxidase (Williams and Poole, 1988), and also in *A. methanolicus* under normal growth conditions having an ethanol oxidase respiratory chain (Matsushita et al., 1992c), in addition to *A. aceti* and *Gluconobacter* species.

D. Cytochrome a_1 as Another Major Terminal Oxidase in Acetobacter Species

Cytochrome a_1 was originally found in *Acetobacter* species by Warburg as a cytochrome exhibiting a weak absorption peak at 589 nm which is intensified in the presence of CN, and later identified as a terminal oxidase by Chance (see Matsushita et al., 1994). Although the presence of such a cytochrome oxidase was doubted later, a cytochrome a_1-like terminal oxidase was found in the membranes of *A. aceti* grown on shaking cultures (Matsushita et al., 1990). *A. aceti* produces either one of the two different terminal oxidases depending on the culture conditions (or the cell type) (Fig. 3A). Cells (S-type strain) producing cytochrome a_1 predominate on the shaking culture, while cytochrome *o* is predominantly present in cells (R-type strain) in static culture. The cytochrome a_1 was characterized as cytochrome *ba*-type ubiquinol oxidase consisting of four subunits and containing 1 mol each of heme *a*, heme *b*, and copper ion (Matsushita et al., 1990). Although no structural difference could be seen between cytochrome a_1 and cytochrome *o* of *A. aceti*, both enzymes have a difference in their oxygen affinity (4-times higher in cytochrome a_1) and CN sensitivity (2-times higher in cytochrome *o*) (Matsushita et al., 1992a). The genes (*cya* operon) encoding subunits of cytochrome a_1 were cloned from *A. aceti* chromosomes, and the deduced amino acid sequences of these subunits, especially subunit I, showed a great similarity to the cytochrome *o* of *E. coli* (Fukaya et al., 1993). Furthermore, Southern hybridization analysis showed that only one set of the gene exists for cytochrome a_1 and cytochrome *o* in *A. aceti*, and thus the two oxidases share an identical protein moiety (K. Matsushita et al., unpublished). Thus, the differences between two ubiquinol oxidases are solely due to a peripheral group of the heme at the binuclear center site, substitution of the methyl group at position 8 of heme *o* with a formyl group. Comparison of the binuclear site structure of both enzymes using resonance Raman, FT-IR, and EPR spectroscopies showed that the overall architecture and the electronic configurations of the binuclear center seem to be well conserved in their oxidized state, but that the binuclear center in cytochrome a_1 has conformational flexibility in the reduced state, which may induce a hydrogen bond network formation among the formyl group of heme *a*, water, and the surrounding amino acid residues (Tsubaki et al., 1997). This network is likely the origin of the higher affinity for oxygen and the lower sensitivity to CN of cytochrome a_1.

Cytochrome a_1 has been found in another *Acetobacter* species but not in *Gluconobacter* species. In addition to *A. aceti* R strain, *Acetobacter diazotrophicus* has recently been shown to produce cytochrome a_1 when the cells are grown under highly aerated and diazotrophic growth conditions (Fig. 3B), where a high energy state is required for the N_2 fixation (Flores-Encarnacón et al., 1999).

E. Cytochrome d as the Additional Terminal Oxidase in Acetobacter Species

Cytochrome *d* is one of the ubiquinol oxidases, and contains two hemes *b* and one heme *d* as the prosthetic groups, and the binuclear center for oxygen reaction is constituted of one of the hemes *b* and the heme *d* (Hill et al., 1993). Cytochrome *d* has been shown to be produced under limited oxygen conditions and to have low proton-motive force generating ability, different from cytochrome a_1 or *o*.

Cytochrome *d* was found in *Acetobacter peroxydans* by Bächi and Ettlinger (1974), and later in *A. pasteurianus* (seemingly the same strain as *A. peroxydans*) by Williams and Poole (1987). *A. pasteurianus* produces cytochrome *o* in air-sufficient growth conditions, while cytochrome *d* is produced, instead of cytochrome *o*, under low aeration conditions (Fig. 3C). Cytochrome *d* has also been detected in *A. diazotrophicus*, which have a unique ability to fix nitrogen from the air, as described above (Flores-Encarnacón et al., 1999). When the cells are grown in the presence of a high concentration of ammonium salts, which disturbs the diazotrophic growth, cytochrome *d* seems to become the major terminal oxidase in the respiratory chain, while the cells produce cytochrome a_1 as the terminal oxidase under diazotrophic growth conditions (Fig. 3B).

F. Cyanide-resistant Bypass Oxidase in Gluconobacter Species

G. suboxydans, especially when grown in a sugar-rich medium, exhibits extremely high oxidase activities for sugar or alcohol as well as NADH oxidase activity, in the respiratory chain. These sugar and alcohol respirations exhibit extremely high CN resistancy; more than 50% of the original oxidase activity is retained even in the presence of 10 mM CN. Thus, the respiratory chain of *G. suboxydans* seems to branch at the site of UQ, with CN-sensitive and resistant terminal oxidases, of which the CN-sensitive one is cytochrome *o*. However, component(s) involved in the CN-resistant terminal oxidase is not clear yet because the respiratory chain of *G. suboxydans* contains only two cytochrome components, a high level of *c*-type cytochrome and some *b*-type cytochrome, of which the former is mainly due to the triheme cytochrome *c* subunit of alcohol dehydrogenase (ADH) and the latter may be cytochrome *o*. Since depletion of the cytochrome *c* subunit from the membranes leads to the sugar and alcohol respirations to becoming CN-sensitive (Ameyama et al., 1987), the CN-resistant terminal oxidase system seems to contain ADH subunit(s), at least as one component of the system. This notion is also supported by the findings that *G. suboxydans* subspecies α-strains in which ADH, especially the cytochrome subunit, is deficient, have a relatively CN-sensitive respiratory chain, which can be changed to more CN-resistant by a biochemical reconstitution of the purified ADH into the membranes (Matsushita et al., 1991) and also by an in vivo reconstitution with the gene of ADH subunit II (Takeda et al., 1992). Thus, ADH, especially the cytochrome subunit, from *G. suboxydans* seems to be essential for the CN-resistant respiratory chain bypass. However, since ethanol oxidase respiratory activity was CN-sensitive when reconstituted into proteoliposomes with ADH, UQ, and cytochrome *o*, some other component(s) must be required to produce CN-resistant bypass oxidase in addition to the cytochrome subunit of ADH (Fig. 3D).

When the membranes of *G. suboxydans* are treated with a relatively low concentration of detergent, ADH can be solubilized while glucose dehydrogenase (GDH) is retained in the membrane residues. Although both GDH and ADH can reduce ferricyanide in the native membranes, GDH loses its ability to react with the artificial dye after the solubilization but ADH does not. The ferricyanide reductase activity of GDH can be reproduced in proteoliposomes reconstituted from GDH, ADH, and further UQ_{10}, all three components of which are indispensable (Shinagawa et al., 1990). Thus, the electron transfer from GDH to ferricyanide seems to be mediated by UQ and ADH in the membranes of *G. suboxydans*. Since GDH and ADH are able to react directly with UQ in the phospholipid bilayer, it is suggested that GDH donates electrons to UQ first and then the resultant ubiquinol reacts directly with ADH which subsequently reduces ferricyanide. Actually, ADH has been shown to have an additional function to accept electrons from ubiquinol (Matsushita et al., 1999), suggesting that ADH is able to mediate electron transfer from ubiquinol to ferricyanide in vitro, and also to the bypass oxidase system in vivo (Fig. 3D).

The same type of CN-resistant bypass oxidase involving some cytochrome *c* components has been shown in the respiratory chain of *A. diazotrophicus*, where most of the cytochrome *c* was solubilized with 0.2% Triton X-100 and the residual membrane retained a fully active glucose oxidase activity which is significantly more sensitive to CN (Flores-Encarnación et al., 1999). Thus, the CN-resistant bypass oxidase may work not only in *Gluconobacter* species but also in some *Acetobacter* species.

III. Periplasmic Sugar and Alcohol Respirations

A. Periplasmic Oxidase Systems

The periplasm and the periplasmic surface of the cytoplasmic membrane have been recognized as the important location for metabolism, especially for electron transport, of Gram-negative bacteria and many electron transport proteins have been assigned to work in this region. Such periplasmic primary dehydrogenases are coupled to the remaining part of the respiratory chain embedded in the membrane, constituting 'periplasmic oxidase systems' (Fig. 2); while NAD(P)-dependent dehydrogenases located in the cytoplasm are coupled via NADH dehydrogenase to the membrane-bound respiratory chain, and some flavoprotein dehydrogenases such as succinate and lactate dehydrogenases are also directly connected to it (Fig. 2). These flavoprotein dehydrogenases including NADH dehydrogenase are located at the inner surface of the cytoplasmic membrane, and thus constitute 'cytoplasmic oxidase systems,' which is in contrast to the periplasmic oxidase systems (Matsushita et al., 1994).

Of such periplasmic oxidase systems, some contain primary dehydrogenases freely soluble in the periplasmic space or loosely bound to the periplasmic side of the cytoplasmic membrane, which are linked to a membrane-bound terminal oxidase via cytochrome *c* or copper protein, a typical periplasmic electron transport protein. Such soluble dehydrogenases are found in the methanol- or methylamine- oxidizing system of methylotrophs, or in the alcohol- or amine-oxidizing system of some oxidative bacteria such as *Pseudomonas putida* or *Paracoccus denitrificans* (see C. Anthony, Chapter 10, Vol. 1); whereas the others have membrane-bound dehydrogenases linked to the respiratory chain via UQ. Many membrane-bound primary dehydrogenases are found in the periplasmic oxidase systems of acetic acid bacteria, and can be divided into quinoproteins and flavoproteins, which have pyrroloquinoline quinone (PQQ) and covalently-bound flavin adenine dinucleotide (FAD) as the prosthetic groups, respectively (Table 2). The flavoproteins are uniquely found in the sugar and alcohol respirations of acetic acid bacteria, except for the gluconate oxidase system also found in pseudomonads and some enteric bacteria; whereas in addition to the sugar and alcohol respirations of acetic acid bacteria, PQQ-quinoproteins are also found in the methanol oxidase system of methylotrophs, in the alcohol oxidase system of some oxidative bacteria, and also in the glucose

Table 2. Primary dehydrogenases functioning on sugar and alcohol respiration in acetic acid bacteria

Enzymes	Subunit structure and Prosthetic Group	Bacterial source (References)
Quinoproteins		
Glucose dehydrogenase (GDH)	80 Kd (PQQ)	*G. suboxydans* [1]
Glycerol dehydrogenase (GLDH)	80 Kd (PQQ), ~14 Kd	*G. industrius* [2], *G. suboxydans* [3,4]
(Glycerol dehydrogenase [2], Arabitol dehydrogenase[3] or Sorbitol dehydrogenase[4])		
Alcohol dehydrogenase (ADH)	72 Kd (PQQ/heme *c*), 48 Kd (3 hemes *c*), ~15 Kd	*G. suboxydans*[1], *A. aceti*[1], *A. polyoxygenus*[1], *A. pasteurianus*[5], *A. methanolicus*[6]
Flavoproteins		
Gluconate dehydrogenase (GADH)	64 Kd (FAD), 45 Kd (hemes *c*), ~21 Kd	*G. dioxyacetonicus* [1]
2-Ketogluconate dehydrogenase (2KGDH)	61 Kd (FAD), 47 Kd (hemes *c*), ~25 Kd	*G. melanogenus* [1]
Sorbitol dehydrogenase (SLDH)	63 Kd (FAD), 51 Kd (hemes *c*), ~17 Kd	*G. suboxydans* var. α[1]
Sorbose dehydrogenase (SDH)	58 Kd (FAD)	*G. oxydans* [1]
Other or unknown type enzymes		
Aledehyde dehydrogenase (ALDH)	84 Kd (Molybdopterin ?), 49 Kd [3] hemes c), ~17 Kd	*G. suboxydans* (1), *A. polyoxygenus* [1] *A. rancens* [1], *A. europaeus* [7]
Fructose dehydrogenase (FDH)	67 Kd (FAD or PQQ?), 51 Kd (hemes *c*), ~20 Kd	*G. industrius* [1]
Sorbosone dehydrogenase (SNDH)	48 Kd (?)	*A. liquefaciens*[8]

References: [1]see Matsushita et al., 1994, [2] Ameyama et al., 1985, [3] Adachi et al., 2001; [4] Sugisawa and Hoshino (2001), [5] Takemura et al., 1993; [6] Frebortova et al., 1997; [7] Thurner et al., 1997; [8] Shinjoh et al., 1995.

oxidase system of a variety of Proteobacteria (see C. Anthony, Chapter 10, Vol. 15).

Thus, acetic acid bacteria are unique bacteria that have extensively developed periplasmic oxidase systems, so-called 'oxidative fermentation' such as sugar and alcohol respirations, in which all the electron transport components including primary dehydrogenases are firmly bound to the periplasmic side of the cytoplasmic membrane. The most prominent 'oxidative fermentations' of acetic acid bacteria are ethanol, glucose and polyol (sugar alcohol) respirations, which are described below in more detail.

B. Ethanol Respiration (Acetic Acid Production)

The most typical 'oxidative fermentation' is ethanol respiration, which is a vinegar-producing system unique to acetic acid bacteria. NAD(P)-dependent alcohol dehydrogenase (ADH) and aldehyde dehydrogenase (ALDH) are found in the cytoplasm, besides the membrane-bound, NAD(P)-independent, ADH and ALDH (see Matsushita et al., 1994). In a mutant strain of *Acetobacter* species in which the membrane-bound ADH is defective, it has been shown that NAD-ADH and NADP-ALDH are dramatically increased and the cells grow very well in a medium containing ethanol without producing any acetic acid (Chinnawirotpisan et al., 2003). Thus, the membrane-bound ADH and ALDH are clearly shown to be involved in acetic acid production, while the cytoplasmic enzymes are extensively involved in ethanol assimilation (Fig. 4).

Membrane-bound ADH and ALDH, distributed among almost all the strains of both *Acetobacter* and *Gluconobacter* species, work by linking to the respiratory chain, constituting ethanol respiration, of acetic acid bacteria. ADH, catalyzing the first step of the respiration, is a quinohemoprotein-cytochrome *c* complex bound to the periplasmic side of the cytoplasmic membrane and functions as the primary dehydrogenase in the ethanol oxidase respiratory chain, where ADH oxidizes ethanol by transferring electrons to UQ embedded in the membrane phospho-

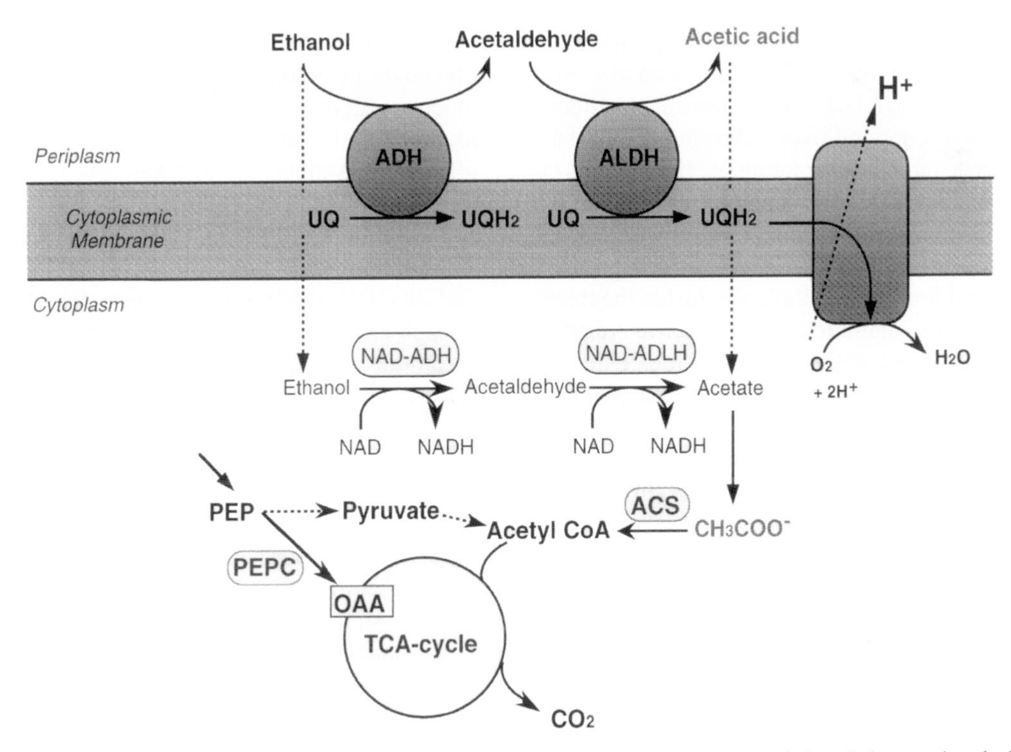

Fig. 4. Ethanol respiration of *Acetobacter* species consisting of alcohol and aldehyde respiratory chains. Quinoproteins alcohol dehydrogenase (ADH) and aldehyde dehydrogenase (ALDH) are located on the outer surface of the cytoplasmic membrane, while NAD-dependent alcohol dehydrogenase (NAD-ADH) and aldehyde dehydrogenase (NADP-ALDH) are in the cytoplasm. The ethanol respiration carried out by ADH and ALDH produces acetic acid which could be utilized by acetyl CoA synthase (ACS) and phosphoenolpyruvate carboxylase (PEPC)-dependent manner. When ADH is disrupted, ethanol is assimilated though NAD-ADH and NAD-ALDH.

lipids. ADH consists of three subunits, and subunit I contains PQQ and heme *c* moiety, and subunit II has 3 heme *c* moieties (Table 2; Matsushita et al., 1996). The periplasmic location of all the subunits (*adhAB* and *adhS*), the presence of PQQ and heme *c* in subunit I (*adhA*), and 3 hemes *c* in subunit II (*adhB*) were confirmed by genetic data (see Kondo and Horinouchi, 1997). ALDH, catalyzing the next step from aldehyde to acetic acid, is also a three subunit complex (Table 2), which is also supported by genetic data (Thurner et al., 1997). All three subunits (*aldFGH*) seem to have a signal sequence, which corresponds to the presence of ALDH in the periplasm. Although PQQ was proposed as a cofactor for this enzyme, the sequence data of the largest subunit (*aldH*) suggest that the subunit has no sequence homology to quinoprotein, but instead shows high homology to molybdopterin enzyme, which is consistent with the findings that a PQQ-deficient *Acetobacter* strain is still able to produce active ALDH (Takemura et al., 1994). The second subunit (*aldF*), similar to subunit II of ADH, is also suggested to have three heme *c* binding motifs.

ADH has been shown to donate electrons to UQ embedded in the membrane phospholipids, and then to the terminal oxidase (Matsushita and Adachi, 1993). Although no ADH subunits have a transmembranous domain, subunit II (cytochrome subunit) has been shown to have a UQ reacting site (Matsushita et al., 1996) and to have two amphiphilic α-helices as the possible membrane-anchor (K. Matsushita et al., unpublished). Since ADH has five prosthetic groups, one PQQ and four heme *c* moieties, inside the three subunit complex, an intramolecular electron transfer reaction should occur through these molecules to reduce UQ. From several pieces of circumstantial evidence, it has been suggested that electrons extracted from ethanol at the PQQ site could be transferred *via* the heme *c* site in subunit I to either of three heme *c* sites in the cytochrome subunit, and the electrons are passed to UQ through two of the hemes *c* (Matsushita et al., 1996; Frébortová et al., 1998). Thus, UQ reduction in ADH may occur via electron transfer from PQQ through three of the four heme *c* moieties present. Although no data have been presented in the case of ALDH, the enzyme also seems to bind to the membrane and thus to transfer the reducing equivalent to UQ via the cytochrome subunit.

An active ethanol oxidase respiratory chain could be reproduced in an artificial proteoliposome when both ADH and ubiquinol oxidase, either cytochrome a_1 or cytochrome *o*, were reconstituted into liposomes containing UQ_9 or UQ_{10} (Matsushita et al., 1992d). The reconstituted respiratory chain has a reasonable electron transfer turnover from ethanol to oxygen (~40% of the native respiratory chain) and also a membrane potential generation ability during the electron transfer (~110 mV inside negative). Since ALDH has UQ reductase activity similar to ADH, acetaldehyde may also be oxidized by terminal ubiquinol oxidase via UQ in the acetaldehyde respiratory chain. Thus, 'ethanol respiration' can be expected to occur in the respiratory chains consisting of two primary dehydrogenases, ADH and ALDH, and terminal ubiquinol oxidase, cytochrome a_1 or cytochrome *o*, both of which are connected via UQ_9 to UQ_{10}, as shown in Fig. 4.

C. Glucose Respiration

Figure 5 shows the possible dehydrogenases taking part in glucose metabolism in acetic acid bacteria. Although there are NAD(P)-dependent enzymes in the cytoplasm, such as D-glucose dehydrogenase and 2-keto-D-gluconate reductase, membrane-bound, NAD(P)-independent, enzymes such as GDH, gluconate dehydrogenase (GADH), and 2-keto-D-gluconate dehydrogenase (2KGDH) are extensively involved in the direct oxidation of D-glucose to D-gluconate, D-gluconate to 2-keto-D-gluconate, and further to 2,5-diketo-D-gluconate at the cell surface of *Gluconobacter* species (Matsushita et al., 1994). Thus, NAD(P)-independent enzymes, GDH, GADH, and 2KGDH, work as the primary dehydrogenases of each glucose, gluconate, and 2-keto-D-gluconate respiratory chain, respectively, constituting 'glucose respiration' of acetic acid bacteria. Of these primary dehydrogenases, GDH and GADH are not unique in acetic acid bacteria, but found in a wide variety of Proteobacteria including facultative anaerobes such as enteric bacteria and strictly aerobic bacteria such as pseudomonads; whereas 2KGDH is uniquely found only in acetic acid bacteria, especially *Gluconobacter* species. GDH is a well known quinoprotein having PQQ as the prosthetic group (see Chapter 10 in Volume 1), while GADH and 2KGDH have a covalently bound flavin, which is FAD bound via the methyl group at the C^8 site with the N^3 site of a histidine residue (McIntire et al., 1985).

GDH was purified, cloned, and then characterized from the membranes of *G. suboxydans* and *G. oxydans*. The structural and functional properties

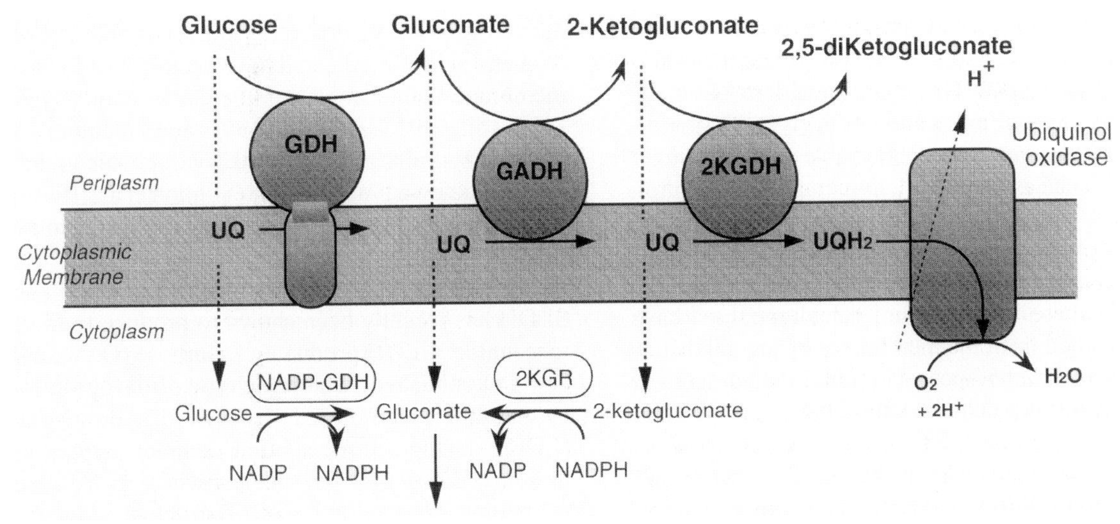

Fig. 5. Glucose respiration of *Gluconobacter* species consisting of glucose, gluconate and 2-ketogluconate respiratory chains. Quinoprotein D-glucose dehydrogenase (GDH), and flavoproteins D-gluconate dehydrogenase (GADH), 2-keto-D-gluconate dehydrogenase (2KGDH) are located on the outer surface of the cytoplasmic membrane, and work as the primary dehydrogenases of each respiratory chains. NAD-dependent glucose dehydrogenase (NAD-GDH) and NADP-dependent 2-ketogluconate reductase (2KGR) are working in the cytoplasm for partial assimilation of glucose, gluconate and 2-ketogluconate.

are shown to be almost the same as GDH from other bacterial strains such as *Pseudomonas* sp., *E. coli*, *Acinetobacter calcoaceticus* (Matsushita and Adachi, 1993). GDH is a single peptide quinoprotein (Table 2), of which the N-terminal region has a hydrophobic five membrane-spanning α-helix and the residual C-terminal part is a PQQ-containing catalytic domain located at the periplasmic side (Yamada et al., 1993). Although the enzyme is able to donate electrons to long-chain UQ, UQ_6 or UQ_9, as well as short-chain homologues, UQ_1 or UQ_2 (Matsushita et al, 1989b), only the C-terminal region has recently been shown to be sufficient to carry out the electron transfer to UQ (Elias et al., 2001). Unlike GDH, both GADH and 2KGDH are three subunit complexes including flavoprotein and cytochrome *c* (Table 2). Although some *Gluconobacter* strains have been reported to produce 5-keto-D-gluconate as well as 2-keto-D-gluconate (Shinagawa et al., 1999), GADH is only responsible for 2-keto-D-gluconate production, as described below. The GADH gene has not been cloned from acetic acid bacteria, but from *Erwinia cypripedii* (Yum et al., 1997). The data have shown that all the three subunits have a typical signal sequence, and that the largest subunit has a FAD binding motif and the second subunit has three heme *c* binding motifs and relatively high identity (34~40%) to subunit II of ADH of acetic acid bacteria.

As in the case of ADH, the glucose oxidase respiratory chain could be reconstituted in a proteoliposome having GDH, UQ_{10} and cytochrome *o* (Matsushita et al., 1989b), in which the electron transfer rate from glucose to oxygen is reasonably high (~60% of the native system), and able to generate a reasonable electrochemical proton gradient (–112 mV, 0.95 pH units). Although the UQ reductase activity was not examined in GADH and 2KGDH of *Gluconobacter* species, GADH of *Pseudomonas* species has also been shown to donate electrons to UQ_1, and also to reproduce gluconate oxidase activity only after adding UQ_2 or UQ_4 when the purified enzyme was added to the GADH-depleted membrane (Matsushita et al., 1979). Thus, GADH and 2KGDH are also expected to link to the rest of the respiratory chain via UQ, as in the case of GDH. As shown in Fig. 5, thus, 'glucose respiration' of acetic acid bacteria is effected by three different peripheral primary dehydrogenases, GDH, GADH and 2KGDH located on the periplasmic side of the membrane, and the terminal ubiquinol oxidase(s) present inside the membrane together with membranous UQ_{10}.

D. Sugar Alcohol Respiration and Other Sugar Respirations

Almost all *Gluconobacter* species are known to have

a high 'oxidative fermentation' ability against several different sugar alcohols, especially glycerol, besides alcohol and sugar. The most typical oxidation is dihydroxyacetone production from glycerol. In addition, *Gluconobacter* species are able to oxidize sugar alcohols such as D-sorbitol, D-mannitol, D-arabitol, D-ribitol or *meso*-erythritol to the corresponding sugars, L-sorbose, D-fructose, D-xylulose, D-ribulose, or L-erythrulose, respectively (Fig. 6). Recently, some of these sugar alcohol dehydrogenases have been purified from the membranes of several different *Gluconobacter* species; arabitol dehydrogenase and also sorbitol dehydrogenase have been purified from *G. suboxydans* IFO3257 and *G. suboxydans* IFO3255, respectively (Adachi et al., 2001; Sugisawa and Hoshino, 2002). Both enzymes exhibited similar molecular properties and similar broad substrate specificity, which was also found to be very similar to those of glycerol dehydrogenase previously purified partially from *Gluconobacter industrius* (Ameyama et al., 1985). Thus, the similarity of these sugar alcohol dehydrogenases were confirmed by a gene disruption based on the isolated gene (*sldAB*) of sorbitol dehydrogenase (Miyazaki et al., 2002), which showed that all these sugar alcohol dehydrogenases are the same enzyme based on the same genetic origin (Shinjoh et al., 2002). In addition, another oxidation reaction of D-gluconate to 5-keto-D-gluconate was also shown to be carried out by the same enzyme (Matsushita et al., 2003). Thus, as shown in Fig. 6, many different oxidation reactions have been shown to be done by glycerol dehydrogenase (GLDH), of which the name is selected because glycerol is the least reactive molecule. This enzyme has been shown biochemically to be a quinoprotein in both glycerol and arabitol dehydrogenases, which is also supported by the finding that the quinoprotein–type structure is well conserved in SldA. The enzyme has been shown to have UQ_2 reductase activity in arabitol dehydrogenase and to have a hydrophobic membrane spanning subunit (SldB) similar to the N-terminal membrane-spanning domain of GDH, suggesting that GLDH works as the primary dehydrogenase for sugar alcohol respiration together with UQ and the terminal ubiquinol oxidase, as in the case of GDH.

Other sugar- or sugar alcohol-oxidizing systems are also shown in Fig. 6, one of which is a different kind of alcohol-oxidizing system to oxidize L-sorbose dehydrogenase (SDH) and L-sorbosone dehydrogenase (SNDH) to produce 2-keto-L-gulonate, the precursor for Vitamin C. These alcohol and aldehyde dehydrogenases are different from ADH and ALDH of acetic acid bacteria, and have been shown to be a membrane-bound single peptide flavoprotein (Sugisawa et al., 1991; Saito et al., 1997) and a membrane-bound enzyme having N-terminal hydrophobic stretch but an unknown type cofactor (Shinjoh et al., 1995), respectively. In Fig. 6, another sorbitol dehydrogenase (SLDH) different from GLDH is shown, which is a flavoprotein-cytochrome *c* complex (Table 2). The SLDH has recently been shown to produce D-fructose, unlike GLDH producing L-sorbose (O. Adachi et al., unpublished). The same type of flavoprotein-cytochrome *c* complex, D-fructose dehydrogenase (FDH), having a three subunit complex is present in some *Gluconobacter* species, where D-fructose is oxidized further to 5-keto-D-fructose (Fig. 6). These sugar or sugar alcohol dehydrogenases are all membrane-bound, and seem to be present on the periplasmic side of the cytoplasmic membrane. Also, although there is no direct evidence, these oxidoreductases work as primary dehydrogenases by linking to the terminal ubiquinol oxidase via UQ, as in the case of other respirations.

IV. Physiological Function and Meanings of Sugar and Alcohol Respirations of Acetic Acid Bacteria

A. Energetic Aspect of CN-resistant Bypass Oxidase in Gluconobacter Respiratory Chains

As described above, since the respiratory chain of *Gluconobacter* species branches with a CN-sensitive cytochrome *o* and a CN-resistant bypass oxidase, the CN-resistant bypass oxidase may have some physiological meanings, especially from the energetic point of view. As shown in Fig. 7, the H^+/O ratio in the resting cells of *G. suboxydans* has been shown to vary from 1.1 to 2.3 depending on the respiratory substrates and on the extracellular pH of the growth medium (Matsushita et al., 1989a). CN-sensitive cytochrome *o* of *G. suboxydans* has been shown to generate an electrochemical proton gradient by the oxidation of ubiquinol (Matsushita et al., 1987). In general, cytochrome *o* catalyzes a scalar proton release from ubiquinol outside the membrane and a concomitant proton uptake inside the membrane to reduce oxygen ($\frac{1}{2}O_2$) to water (H_2O) where the scalar reaction produces a H^+/O ratio of 2 (Matsushita et

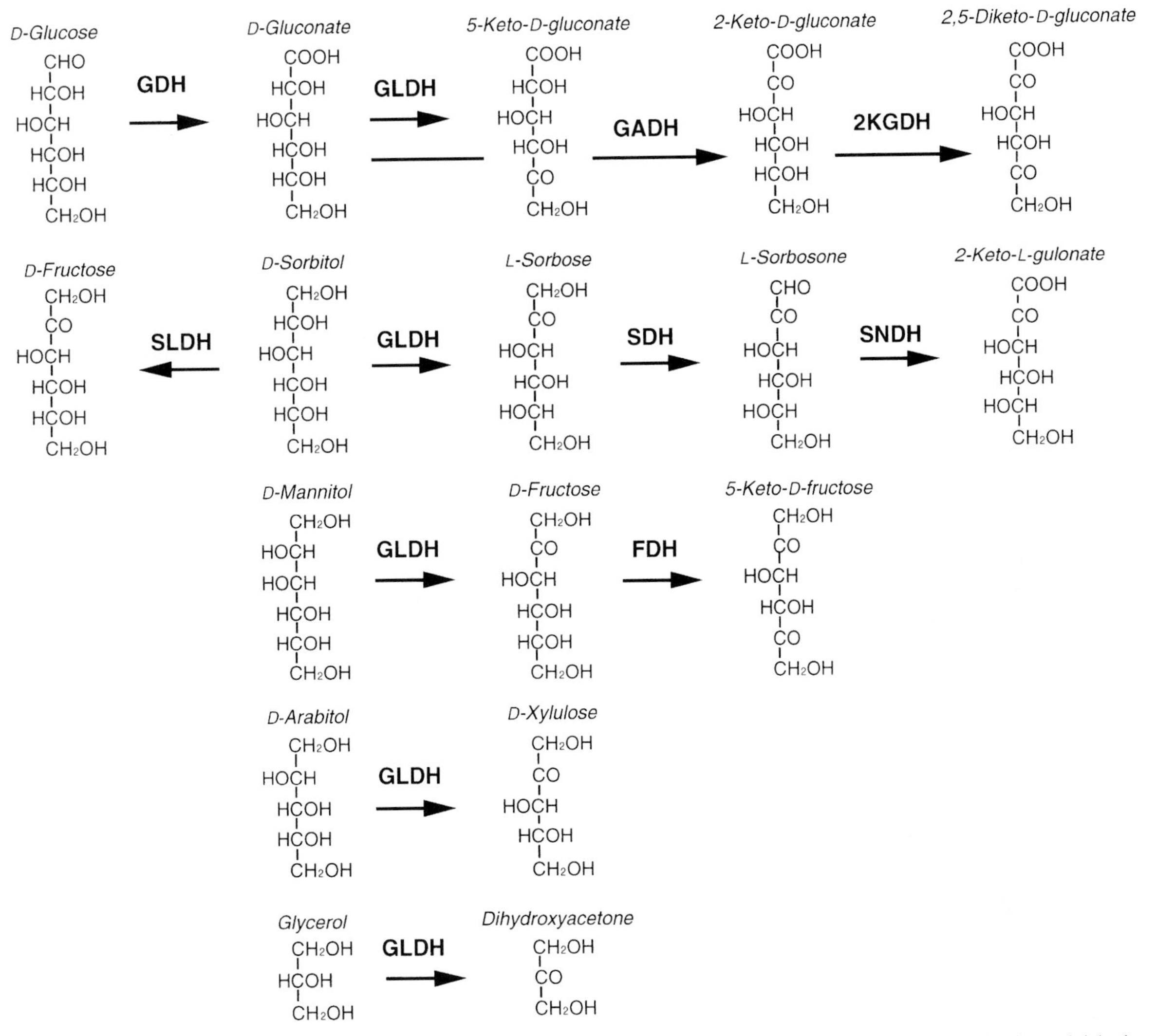

Fig. 6. Oxidation pathways of several sugar alcohols, sugars, and sugar acids in *Gluconobacter* species. Quinoprotein glycerol dehydrogenase (GLDH) is able to oxidize D-gluconate, D-sorbitol, D-mannitol, D-arabitol, glycerol to 5-ketogluconate, L-sorbose, D-fructose, D-xylulose, and dihydroxyacetone, respectively. D-Gluconate is also oxidized to 2-ketogluconate, then to 2,5-diketogluconate by a flavoprotein gluconate dehydrogenase (GADH) and 2-ketogluconate dehydrogenase (2KGDH), respectively. L-Sorbose is oxidized to L-sorbosone, then to 2-ketogulonate by a flavoprotein sorbose dehydrogenase (SDH) and an unidentified enzyme sorbosone dehydrogenase (SNDH), respectively. D-Fructose is oxidized to 5-ketofructose by a flavoprotein fructose dehydrogenase (FDH); whereas D-Sorbitol is oxidized to D-fructose by a flavoprotein sorbitol dehydrogenase (SLDH).

al., 1984), and further it has an additional ability to pump protons with a H^+/e^- ratio of 1 (Puustinen et al., 1991). Thus, although cytochrome *o* of *G. suboxydans* is expected to exhibit a H^+/O ratio of 4 (a H^+/e^- ratio of 2), the observed H^+/O ratio is lower than the value expected, with cytochrome *o* only working as the terminal oxidase. Furthermore, a membrane potential generated during the oxidation of glucose or ethanol has been shown to be dissipated completely with 1 mM KCN able to block cytochrome *o* completely in the membrane vesicles of *G. suboxydans* (Matsushita et al., 1989a). Thus, it is suggested that the CN-resistant bypass oxidase has no ability to produce an electrochemical proton gradient. This notion is highly consistent with the findings that the H^+/O ratio in cells grown at low extracellular pH, where CN-resistant bypass oxidase is increased, is lower than that in the cells grown at higher pH

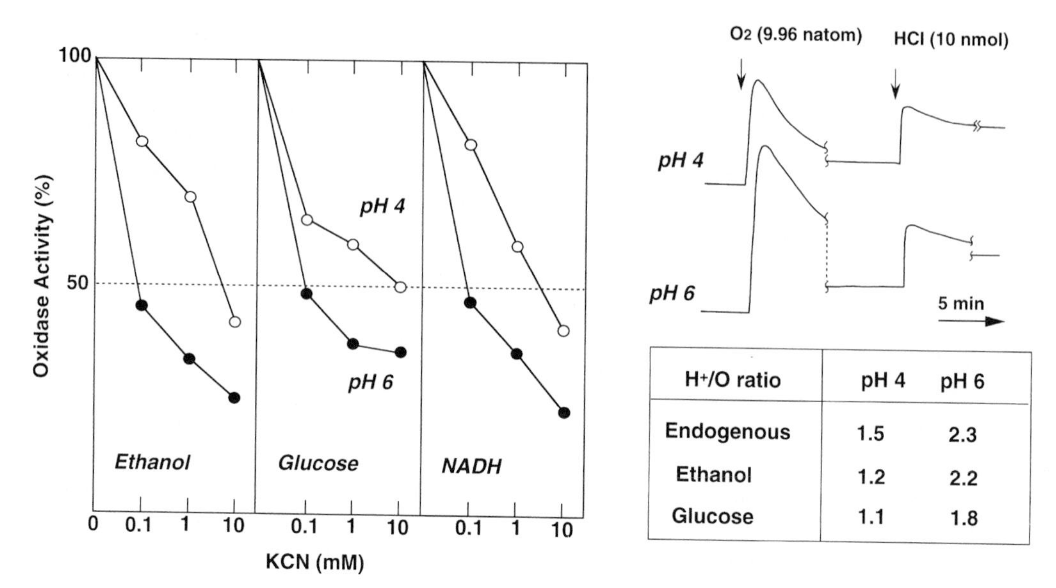

Fig. 7. Effect of growth pH on cyanide-sensitivity and H^+/O ratio of ethanol and glucose respirations of *Gluconobacter suboxydans*. Right panel shows the cyanide-sensitivity on ethanol, glucose, and NADH oxidase activities of the membranes prepared from the cells grown on pH 4 (○) and pH 6 (●). Left panel shows the measurement of endogenous respiration-dependent H^+/O ratio in the resting cells obtained from pH 4 and pH 6 cultures, and also the H^+/O values obtained from endogenous, ethanol, and glucose respirations.

(Fig. 7), but the membrane potential generation in the membrane vesicles of *G. suboxydans* is sensitive to CN regardless of the external pH during growth of the cells. Thus, non-energy-generating CN-resistant bypass oxidase may contribute 50% or more to the respiratory activity of *G. suboxydans* under the lower extracellular pH (pH 4.0).

Under the same acidic growth conditions, where CN-resistant bypass oxidase is increased, it was also found that the ADH content, and thus the cytochrome *c* content, in the membranes of *G. suboxydans* was largely increased but the activity did not change much, suggesting that such a condition produces an inactive form of ADH (inactive ADH) (Matsushita et al., 1995). Although the inactive ADH could not be distinguished from the active ADH with respect to structure, it exhibited ten-times lower UQ reductase activity than the active ADH. As described above, ADH is able to accept electrons from ubiquinol, and thus seems to mediate electron transfer from ubiquinol to the bypass oxidase system in vivo. Since inactive ADH was shown to exhibit higher ubiquinol oxidase activity than the active enzyme (Matsushita et al., 1999), inactive ADH might have a higher ability to donate electrons to the bypass oxidase. Thus, inactive ADH is produced under conditions of low pH and high aeration, where the bypass oxidase activity is highly elevated. Therefore, such an alternative electron flow in the respiratory chain might be generated by switching ADH from active to inactive. Since other membrane-bound dehydrogenases such as GADH or SLDH also have a triheme cytochrome *c* subunit homologous to subunit II of ADH, as described previously, such a specific electron transfer may also be carried out by the cytochrome subunit of these dehydrogenases.

Generation of inactive ADH and also the increase of the bypass oxidase activity occur under acidic pH growth conditions, which can be generated as a result of a large accumulation of oxidation products. Thus, it is reasonable to speculate that inactive ADH confers a higher non-energy-generating bypass oxidase activity on the respiratory chain, which keeps the oxidation reactions at a high level in any growth conditions.

B. Acetic Acid Resistance and Ethanol Respirations of Acetobacter Species

Acetic acid resistance is a crucial factor to stably produce large amounts of acetic acid by *Acetobacter* species, which are important strains for vinegar fermentation. In these *Acetobacter* species, spontaneous mutation is observed at high frequencies, and sometimes leads to simultaneous defects in both acetate-producing (ethanol-oxidizing) and acetic acid resistance abilities (Ohmori et al., 1982; Takemura et al., 1991). Since all these mutants also show a complete loss of ADH activity, ADH seems to be involved in acetic acid resistance. When acetic acid-sensitive mutants were isolated by NTG mutagenesis, however, only the genes, *aarABC* operon, related to the enzymes for acetate utilization including citrate synthase were obtained as the target gene, which could not complement the deficiency of acetate resistance in the spontaneous mutants described above (see Beppu, 1993). In ethanol culture of *A. aceti*, there are three growth phases (Fig. 8); *A. aceti* first grows by oxidizing ethanol completely to acetic acid (ethanol oxidation phase), then stops the growth and remains for a long time with the viable cell number being decreased (first stationary phase), and finally starts to grow again by utilizing the accumulated acetic acid, the phase of which is called as 'overoxidation of acetate.' It is thus conceivable that *Acetobacter* species have two different phases related to acetic acid resistance, the ethanol oxidation and the first stationary phases where the strains resist against acetic acid accumulated in the culture medium without utilizing the acetate, and the overoxidation phase where the strains just utilize the acetate for cell growth. Since it has been shown in the overoxidation phase that acetic acid utilization enzymes such as TCA cycle enzymes, acetyl CoA synthase, involved in the supply of acetyl CoA, and phosphoenolpyruvate carboxylase providing oxaloacetate are increased (Fig. 4; Saeki et al., 1999), the *aar* genes isolated as 'acetate resistance genes' are considered as one of the acetate utilization systems in the overoxidation phase; whereas these acetate-utilizing enzymes seem to be suppressed during the ethanol oxidation and the first stationary phases, in which another mechanism must be working to resist a high concentration of acetic acid produced by themselves. Such a resistant mechanism may be an energy-requiring process, of which the energy may be produced by ethanol respiration. Thus, since ADH deficiency causes the defect in ethanol respiration, cells having a mutation in ADH seem not to resist acetic acid. Thus, the ethanol respiration is always exposed to 'uncoupling' which may be induced by acetic acid itself or by the mechanism of acetate resistance, and thus the ethanol oxidation reaction by the respiratory chain seems to be very rapid but not to be disturbed by the electrochemical proton gradient generated by the respiration.

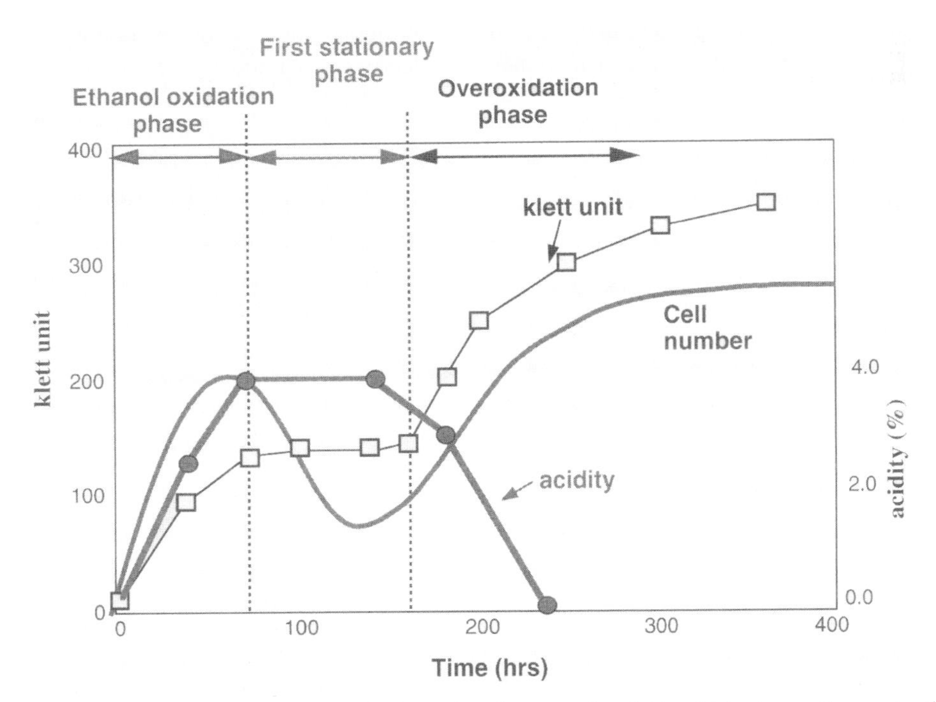

Fig. 8. Growth patterns of *Acetobacter* species in ethanol culture. *Acetobacter* strain exhibits a biphasic growth curve in ethanol culture, where the first phase has an ethanol oxidation to produce acetic acid, and the second phase an overoxidation of acetic acid (assimilation).

C. Ecological Aspects of Sugar and Alcohol Respiratory Processes in Acetic Acid Bacteria: Competition Mechanism

Acetic acid bacteria are able to accumulate large amounts of oxidation products outside the cells. The oxidation reactions are performed by sugar and alcohol respirations, of which the electron transfer generates an electrochemical proton gradient across the membranes that is useful for ATP generation or bioenergetic events for cell growth. As a general principle, however, the proton gradient thus generated may suppress the electron transfer by a feedback control, which may turn out to disturb these oxidation reactions. Thus, a rapid oxidation by such a respiratory chain disturbs the electron transfer by itself, which may be unfavorable for organisms that are required to produce large amounts of oxidation products by the respiratory chain. There seems to be two ways to overcome such a self-contradiction; one is that at least a part of the respiratory chain has a route with no energy generation, and the other is that an electron transfer and an energy generation in the respiratory chain are uncoupled in some way.

The former instance can be seen in *Gluconobacter* species where there is a non-energy-generating CN-insensitive bypass system in the respiratory chain and thus a rapid alcohol or sugar oxidation can be carried out with the generation of a little energy which may not interfere with the electron transfer so much. The latter instance, uncoupling of electron transfer and energy generation, could be expected in *Acetobacter* species which do not have a non-energy generating bypass oxidase unlike *Gluconobacter*. The organisms grow in an environment of a high concentration of acetic acid, which is known in general to be a powerful uncoupler and thus a well-known bactericidal agent, and also resist acetic acid by using a high level of energy. Therefore, the ethanol respiration and also other respirations of *Acetobacter* species may be affected by acetic acid which may uncouple energy generation and electron transfer of the respiratory chain and thus increase the rate of alcohol and sugar oxidation.

Acetic acid bacteria inhabit the surface of flowers (pistil), fruits and their fermented products such as vinegar, sake, wine or beer (Asai, 1968), and thus seem to be evolved in adapting to such a specific environment where high concentrations of sugars, alcohols, or sugar alcohols exist in highly aerobic conditions. In such a specific environment, acetic acid bacteria seem to arrange a specific respiratory chain to perform a rapid oxidation of high concentrations of sugars, alcohols, or sugar alcohols. Thus, *Gluconobacter* species are able to rapidly oxidize sugar alcohols to sugars, then sugars to sugar acids, and thus to finally accumulate sugar acids at a high concentration in their culture media, while *Acetobacter* species mainly oxidize ethanol to produce a high concentration of acetic acid. These reaction products, sugar acids including acetic acid, cause decreased pH and an even more bactericidal effect, and also may be hard to utilize by other organisms, so that these sugar acids may disturb the growth of other bacteria and yeasts living in the same habitat (Fig. 9). As shown in the phenomenon of 'acetic acid

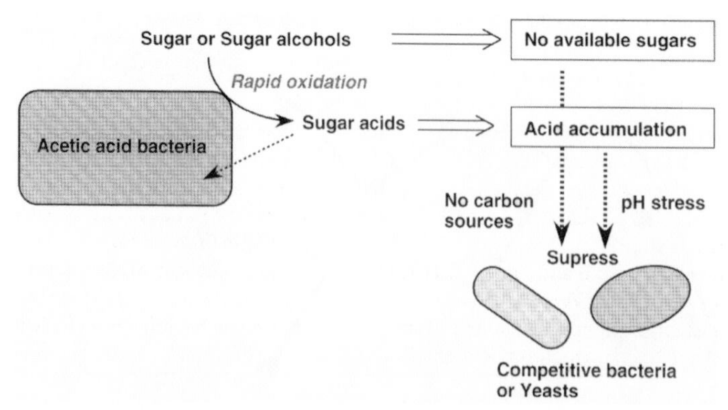

Fig. 9. Competition mechanism of acetic acid bacteria. Acetic acid bacteria are able to oxidize several sugars and sugar alcohols rapidly and then to accumulate a large amount of the corresponding sugar acids outside the cells. These oxidation reactions create acidic environment which is harmful for other competing bacteria or yeasts. At the same time, such a rapid oxidation of sugars or sugar alcohols to the sugar acids also disturb the growth of other competitive microorganisms because of the depletion of available sugars or sugar alcohols from their environment.

resistance' (Fig. 8), *A. aceti* strains do not utilize any acetic acid accumulated in the culture medium despite the cells dying, in the first stationary phase. Thereafter, when the cell number is decreased to some threshold level to be alive, the cells start to utilize the acetic acid to maintain their population. Thus, it is conceivable that *Acetobacter* species are patient for a long time in a high concentration of acetic acid, which is a more severe condition for other bacteria and microorganisms. Although the first stationary phase is not as long as in the case of *A. aceti* in ethanol culture, *Gluconobacter* species also have the same type of biphasic growth, where they can first accumulate the sugar acids at a high concentration, and utilize such sugar acids later. Thus, these sugar and alcohol respirations of acetic acid bacteria able to carry out rapid oxidation of sugar and alcohol seem to have evolved as a competition mechanism against other microorganisms, together with the delayed utilization strategy of such reaction products.

Acetic acid bacteria effect these specific sugar and alcohol respirations mainly for the oxidation of extracellular substrates to accumulate the reaction products outside of the cells, and partly for the energy generation. And, at least in the first stationary phase, these oxidation reactions seem not to be utilized for the assimilation of the substrates. Thus, such sugar and alcohol respirations of acetic acid bacteria seem to be related partly to the periplasmic oxidations of chemolithotrophs such as hydrogen or iron oxidation (see Chapters 9, 10, and 11 in this volume).

References

Adachi O, Fujii Y, Ghaly MF, Toyama H, Shinagawa E and Matsushita K (2001) Membrane-bound quinoprotein D-arabitol dehydrogenase of *Gluconobacter suboxydans* IFO 3257: A versatile enzyme for the oxidative fermentation of various ketoses. Biosci Biotechnol Biochem 65: 2755–2762

Ameyama M, Shinagawa E, Matsushita K and Adachi O (1985) Solubilization, purification and properties of membrane-bound glycerol dehydrogenase from *Gluconobacter industrius*. Agric Biol Chem 49: 1001–1010

Ameyama M, Matsushita K, Shinagawa E and Adachi O (1987) Sugar-oxidizing respiratory chain of *Gluconobacter suboxydans*. Evidence for a branched respiratory chain and characterization of respiratory chain-linked cytochromes. Agric Biol Chem 51: 2943–2950

Anraku Y (1988) Bacterial electron transport chains. Annu Rev Biochem 57: 101–132

Asai T (1968) Acetic Acid Bacteria. Classification and Biochemical Activities. Tokyo University Press, Tokyo

Bächi B and Ettlinger L (1974) Cytochrome difference spectra of acetic acid bacteria. Int J Syst Bacteriol 24: 215–220

Beppu T (1993) Genetic organization of *Acetobacter* for acetic acid fermentation. Antonie van Leeuwenhoek 64: 121–135

Chinnawirotpisan P, Matsushita K, Toyama H, Adachi O, Limtong A and Theeragool G (2003) Purification and characterization of two NAD-dependent alcohol dehydrogenases (ADHs) induced in the quinoprotein ADH-deficient mutant of *Scetobacter pasteurianus* SKU1108. Biosci Biotechnol Biochem 66:958–965

Elias M, Tanaka M, Sakai M, Toyama H, Matsushita K, Adachi O and Yamada M (2001) C-terminal periplasmic domain of *Escherichia coli* quinoprotein glucose dehydrogenase transfers electrons to ubiquinone. J Biol Chem 276: 48356–48361

Elliott EJ and Anthony C (1988) The interaction between methanol dehydrogenase and cytochrome *c* in the acidophilic methlotroph *Acetobacter methanolicus*. J Gen Microbiol 134: 369–377

Flores-Encarnacion M, Contreras-Zentella M, Soto-Urzua L, Aguilar GR, Baca BE and Escamilla JE (1999) The respiratory system and diazotrophic activity of *Acetobacter diazotrophicus* PAL5. J Bacteriol 181: 6987–6995

Frébortová J, Matsushita K, Yakushi T, Toyama H and Adachi O (1997) Quinoprotein alcohol dehydrogenase of acetic acid bacteria: Kinetic study on the enzyme purified from *Acetobacter methanolicus*. Biosci Biotech Biochem 61: 459–465

Frébortová J, Matsushita K, Arata H and Adachi O (1998) Intramolecular electron transport in quinoprotein alcohol dehydrogenase of *Acetobacter methanolicus*: A redox-titration study. Biochim Biophys Acta 1363: 24–34

Fukaya M, Tayama K, Tamaki T, Ebisuya H, Okumura H, Kawamura Y, Horinouchi S and Beppu T (1993) Characterization of a cytochrome a_1 that functions as a ubiquinol oxidase in *Acetobacter aceti*. J Bacteriol 175: 4307–4314

Hill JJ, Alben JO, and Gennis RB (1993) Spectroscopic evidence for a heme-heme binuclear center in the cytochrome *bd* ubiquinol oxidase from *Escherichia coli*. Proc Natl Acad Sci USA 90: 5863–5867

Kondo K and Horinouchi S (1997) Characterization of the genes encoding the three-component membrane-bound alcohol dehydrogenase from *Gluconobacter suboxydans* and their expression in *Acetobacter pasteurianus*. Appl Environ Microbiol 63: 1131–1138

Matsushita K and Adachi O (1993) Bacterial quinoproteins glucose dehydrogenase and alcohol dehydrogenase. In: Davidson VL (ed) Principles and Applications of Quinoproteins, pp 47–63. Marcel Dekker Inc, New York

Matsushita K, Shinagawa E, Adachi O and Ameyama M (1979) Membrane-bound D-gluconate dehydrogenase from *Pseudomonas aeruginosa*. Its kinetic properties and a reconstitution of gluconate oxidase. J Biochem 86: 249–256

Matsushita K, Patel L and Kaback HR (1984) Cytochrome *o* type oxidase from *Escherichia coli*. Characterization of the enzyme and mechanism of electrochemical proton gradient generation. Biochemistry 23: 4703–4714

Matsushita K, Shinagawa E, Adachi O and Ameyama M (1987) Purification, characterization and reconstitution of cytochrome *o*-type oxidase from *Gluconobacter suboxydans*. Biochim Biophys Acta 894: 304–312

Matsushita K, Nagatani Y, Shinagawa E, Adachi O and Ameyama M (1989a) Effect of extracellular pH on the respiratory chain and energetics of *Gluconobacter suboxydans*. Agric Biol Chem 53: 2895–2902

Matsushita K, Shinagawa E, Adachi O and Ameyama M (1989b) Reactivity with ubiquinone of quinoprotein D-glucose dehydrogenase from *Gluconobacter suboxydans*. J Biochem 105: 633–637

Matsushita K, Shinagawa E, Adachi O and Ameyama M (1990) Cytochrome a_1 of *Acetobacter aceti* is a cytochrome *ba* functioning as ubiquinol oxidase. Proc Natl Acad Sci USA 87: 9863–9867

Matsushita K, Nagatani Y, Shinagawa E, Adachi O and Ameyama M (1991) Reconstitution of the ethanol oxidase respiratory chain in membranes of quinoprotein alcohol dehydrogenase-deficient *Gluconobacter suboxydans* subsp. α strains. J Bacteriol 173: 3440–3445

Matsushita K, Ebisuya H and Adachi O (1992a) Homology in the structure and the prosthetic groups between two different terminal ubiquinol oxidases, cytochrome a_1 and cytochrome *o*, of *Acetobacter aceti*. J Biol Chem 267: 24748–24753

Matsushita K, Ebisuya H, Ameyama M and Adachi O (1992b) Change of the terminal oxidase from cytochrome a_1 in shaking cultures to cytochrome *o* in static cultures of *Acetobacter aceti*. J Bacteriol 174: 122–129

Matsushita K, Takahashi K, Takahashi M, Ameyama M and Adachi O (1992c) Methanol and ethanol oxidase respiratory chains of the methylotrophic acetic acid bacterium, *Acetobacter methanolicus*. J Biochem 111: 739–747

Matsushita K, Takaki Y, Shinagawa E, Ameyama M and Adachi O (1992d) Ethanol oxidase respiratory chain of acetic acid bacteria. Reactivity with ubiquinone of pyrroloquinoline quinone-dependent alcohol dehydrogenases purified from *Acetobacter aceti* and *Gluconobacter suboxydans*. Biosci Biotech Biochem 56: 304–310

Matsushita K, Toyama H and Adachi O (1994) Respiratory chain and bioenergetics of acetic acid bacteria. In: Rose AH and Tempest DW (ed) Advances in Microbial Physiology, Vol 36, pp 247–301. Academic Press, London

Matsushita K, Yakushi T, Takaki Y, Toyama H and Adachi O (1995) Generation mechanism and purification of an inactive form convertible in vivo to the active form of quinoprotein alcohol dehydrogenase in *Gluconobacter suboxydans*. J Bacteriol 177: 6552–6559

Matsushita K, Yakushi T, Toyama H, Shinagawa E and Adachi O (1996) Function of multiple heme *c* moieties in intramolecular electron transport and ubiquinone reduction in the quinohemoprotein alcohol dehydrogenase-cytochrome *c* complex of *Gluconobacter suboxydans*. J Biol Chem 271: 4850–4857

Matsushita K, Yakushi T, Toyama H, Adachi O, Miyoshi H, Tagami E and Sakamoto K (1999) The quinohemoprotein alcohol dehydrogenase of *Gluconobacter suboxydans* has ubiquinol oxidation activity at a site different from the ubiquinone reduction site. Biochim Biophys Acta 1409: 154–164

Matsushita K, Fujii Y, Ano Y, Toyama H, Shinjoh M, Tomiyama N, Miyazaki T, Sugisawa T, Hoshsino T and Adachi O (2003) 5-Keto-D-gluconate production is catalyzed by a quinoprotein glycerol dehydrogenase, major polyol dehydrogenase, in *Gluconobacter* sp. Appl Environm Microbiol 69:1959–1966

McIntire W, Singer TP, Ameyama M, Adachi O, Matsushita K and Shinagawa E (1985) Identification of the covalently bound flavins of D-gluconate dehydrogenase from *Pseudomonas aeruginosa* and *Pseudomonas fluorescens* and of 2-keto-D-gluconate dehydrogenase from *Gluconobacter melanogenus*. Biochem J 231: 651–654

Miyazaki T, Tomiyama N, Shinjo M and Hoshino T (2002) Molecular cloning and functional expression of D-sorbitol dehydrogenase from *Gluconobacter suboxydans* IFO3255, which requires pyrroloquinoline quinone and hydrophobic protein SldB for activity development in *E. coli*. Biosci Biotech Biochem 66: 262–270

Ohmori S, Uozumi T and Beppu T (1982) Loss of acetic acid resistance and ethanol oxidizing ability in an *Acetobacter* strain. Agric Biol Chem 46: 381–389

Puustinen A, Finel M, Haltia T, Gennis RB and Wikström M (1991) Properties of the two terminal oxidases of *Escherichia coli*. Biochemistry 30: 3936–3942

Saeki A, Matsushita K, Takeno S, Taniguchi M, Toyama H, Theeragool G, Lotong N and Adachi O (1999) Enzymes responsible for acetate oxidation by acetic acid bacteria. Biosci Biotech Biochem 63: 2102–2109

Saito Y, Ishii Y, Hayashi H, Imao Y, Akashi T, Yoshikawa K, Noguchi Y, Soeda S, Yoshida M, Niwa M, Hosoda J and Shimomura K (1997) Cloning of genes coding for L-sorbose and L-sorbosone dehydrogenases from *Gluconobacter oxydans* and microbial production of 2-keto-L-gulonate, a precursor of L-ascorbic acid, in a recombinant *Gluconobacter oxydans* strain. Appl Environ Microbiol 63: 454–460

Shinagawa E, Matsushita K, Adachi O and Ameyama M (1990) Evidence for electron transfer via ubiquinone between quinoproteins D-glucose dehydrogenase and alcohol dehydrogenase of *Gluconobacter suboxydans*. J Biochem 107:863–867

Shinagawa E, Matsushita K, Toyama H and Adachi O (1999) Production of 5-keto-D-gluconate by acetic acid bacteria is catalyzed by pyrroloquinoline quinone (PQQ)-dependent membrane-bound D-gluconate dehydrogenase. J Mol Catalysis B: Enzymatic 6: 341–350

Shinjo M, Tomiyama N, Asakura A and Hoshino T (1995) Cloning and nucleotide sequencing of the membrane-bound L-sorbosone dehydrogenase gene of *Acetobacter liquefaciens* IFO 12258 and its expression in *Gluconobacter oxydans*. Appl Environ Microbiol 61: 413–420

Shinjoh M, Tomiyama N, Miyazaki T and Hoshino T (2002) Main polyol dehydrogenase of *Gluconobacter suboxydans* IFO3255, membrane-bound D-sorbitol dehydrogenase, that needs product of upstream gene *sldB* for activity. Biosci Biotechnol Biochem 66:2314–2322

Sievers M, Ludwig W and Teuber M (1994) Phylogenetic position of *Acetobacter*, *Gluconobacter*, *Rhodophila* and *Acidiphilium* species as a branch of acidophilic bacteria in the α-subclass of *Proteobacteria* based on 16S ribosomal DNA sequences. Syst Appl Microbiol 17:189–196

Sievers M, Gaberthuel C, Boesch C, Ludwig W, Teuber M (1995) Phylogenetic position of *Gluconobacter* species as a coherent cluster separated from all *Acetobacter* species on the basis of 16S ribosomal RNA sequences. FEMS Microbiol Lett 126:123–126

Sugisawa T and Hoshino T (2002) Purification and properties of membrane-bound D-sorbitol dehydrogenase from *Gluconobacter suboxydans* IFO3255. Biosci Biotech Biochem 66: 57–64

Sugisawa T, Hoshino T, Nomura S and Fujiwara A (1991) Isolation and characterization of membrane-bound L-sorbose dehydrogenase from *Gluconobacter oxydans* UV10. Agric Biol Chem 55: 363–370

Takeda Y, Shimizu T, Matsushita K, Adachi O and Ameyama M

(1992) Role of cytochrome *c*-553 (CO), the second subunit of alcohol dehydrogenase, in the azide-insensitive respiratory chain and in oxidative fermentation of *Gluconobacter* species. J Ferment Bioeng 74: 209–213

Takemura H, Horinouchi S and Beppu T (1991) Novel insertion sequence IS1380 from *Acetobacter pasteurianus* is involved in loss of ethanol-oxidizing ability. J Bacteriol 173: 7070–7076

Takemura H, Tsuchida T, Yoshinaga F, Matsushita K and Adachi O (1994) Prosthetic group of aldehyde dehydrogenase in acetic acid bacteria not pyrroloquinoline quinone. Biosci Biotech Biochem 58: 2082–2083

Thurner C, Vela C, Thony-Meyer L, Meile L and Teuber M (1997) Biochemical and genetic characterization of the acetaldehyde dehydrogenase complex from *Acetobacter europaeus*. Arch Microbiol 168: 81–91

Tsubaki M, Matsushita K, Adachi O, Hirota O, Kitagawa T and Hori H (1997) Resonance Raman, Infrared, and EPR investigation on the binuclear site structure of the heme-copper ubiquinol oxidases from *Acetobacter aceti*: Effect of the heme peripheral formyl group substitution. Biochemistry 36: 13034–13042

Williams HD and Poole RK (1987) The cytochromes of *Acetobacter pasteurianus* NCIB 6428. Evidence of a role for a cytochrome a_1-like haemoprotein in electron transfer to cytochrome oxidase *d*. J Gen Microbiol 133: 2461–2472

Williams HD and Poole RK (1988) Cytochromes of *Acetobacter pasteurianus* NCIB 6428. Reaction with oxygen of cytochrome *o* in cells, membranes, and nonsedimentable subcellular fractions. Curr Microbiol 16: 277–280

Yamada M, Sumi K, Matsushita K and Adachi O (1993) Topological analysis of quinoprotein glucose dehydrogenase in *Escherichia coli* and its ubiquinone-binding site. J Biol Chem 268: 12812–12817

Yamada Y, Hoshino K and Ishikawa T (1997) The phylogeny of acetic acid bacteria based on the partial sequences of 16S ribosomal RNA: The elevation of subgenus *Gluconoacetobacter* to the generic level. Biosci Biotech Biochem 61: 1244–1251

Yum DY, Lee YP and Pan JG (1997) Cloning and expression of a gene cluster encoding three subunits of membrane-bound gluconate dehydrogenase from *Erwinia cypripedii* ATCC 29267 in *Escherichia coli*. J Bacteriol 179: 6566–6572

Chapter 5

Nitrogen Fixation and Respiration: Two Processes Linked by the Energetic Demands of Nitrogenase

Robert J. Maier*
815 Biology Science Bldg., University of Georgia, Athens, GA 30602 U.S.A.

Summary

Nitrogen fixation allows a diverse array of bacteria, either free-living in the environment or in symbiosis with plants, to grow in areas where fixed N is deficient. This confers them a large advantage over their non-N_2 fixing competitors. Nevertheless, N_2 fixation is an energy-demanding process, so that energy-generating respiration is oftentimes closely associated with efficient N_2 fixing systems. The root nodule bacteria have many (O_2-binding) terminal oxidases that differ in expression depending on whether the bacterium exists in free-living or in the symbiotic state. Many of the aerobic N_2 fixers contain a very high O_2 affinity terminal oxidase, like one that can function at free O_2 levels as low as 7nM when the bacteria are surrounded in leghemoglobin within legume root nodules. The previously described 'uncoupled' nature of some oxidases of N_2 fixing bacteria may not be the case, from more recent studies. The N_2 fixing enzyme, nitrogenase is labile to O_2 exposure; this creates a physiological problem for aerobic diazotrophs. The many strategies to overcome this problem must ensure a steady supply of ATP during nitrogenase function yet permit the enzyme to be protected from O_2 inactivation. Strategies include living in aggregates or within viscous slimy sheaths, or in O_2-restricted root nodule barriers or forming specialized cells lacking oxygenic O_2-evolving photosynthetic machinery. The temporal separation of N_2 fixation from O_2 evolving photosynthesis is a mechanism used by some marine filamentous

*Email: rmaier@arches.uga.edu

Davide Zannoni (ed): Respiration in Archaea and Bacteria. Vol 2: Diversity of Prokaryotic Respiratory Systems, pp. 101–120.

cyanobacteria. Employing vigorous respiratory activity at the membrane so that the inside of the cell is nearly anaerobic, or the use of a small redox-active iron-sulfur protein that binds to and protects nitrogenase from O_2 inactivation are well studied mechanisms. Most of the protection mechanisms, even the expression of terminal oxidase activities are generally co-regulated with N_2 fixation genes, and some of these protection systems may concomitantly combat oxidative stress. This additional stress is related to reactive oxygen species produced from N_2 fixation related proteins.

I. Introduction and Perspective

Oxygen-dependent respiration is a key process for the aerobic functioning of many bacteria, as the energetic advantages of respiratory metabolism make it beneficial for the survival of organisms having energetically demanding features. One-energy demanding bacterial process is nitrogen fixation, a key process for maintaining a balanced biogeochemical nitrogen cycle on earth. Nitrogen fixation is carried out by an ATP-utilizing N_2 reducing enzyme, nitrogenase. Energy-generating respiratory metabolism is therefore a requirement to fuel nitrogenase in many N_2 fixers. Indeed the bulk of N_2 fixing bacteria in nature are believed to be O_2-respiring ones (Mancinelli, 1996). The energy input into N_2 fixation via respiratory electron transport is well worth the cost, as the (N_2-fixing) process allows the organism to grow in environments where fixed nitrogen sources are lacking or deficient.

Nitrogen is frequently a limiting nutrient for growth of organisms, and only a limited number of bacteria have the ability to incorporate or 'fix' atmospheric N_2. Nevertheless, they are a highly diverse group of bacteria, with N_2 fixing members occupying all of the major taxonomic groups of prokaryotes (Fuchs, 1999). Furthermore, many of them have evolved complex interplay systems with higher organisms in the form of symbioses in order to inhabit a sanctuary from the more competitive environment of free living life. The need for nitrogen for all cells as a component of vital macromolecules means these bacteria can then grow where their (non-N_2 fixing) competitors cannot. Not only are the energetic demands of the N_2 reducing process great, but the O_2 lability of the N_2 fixing enzymes creates a need for a near-anaerobic environment within the cells of diazotrophic aerobes. Therefore the requirement for O_2 removal to permit N_2 fixation adds another valuable role to the overall importance of O_2 consuming respiration. Indeed, in some cases the basis for forming a symbiotic refuge is undoubtedly related to the need for a low level of oxygen supply to optimally fix N_2.

There is no disagreement that biological nitrogen fixation, the reduction of N_2 to form ammonium by organisms, is an energy-intensive process. The amount of ATP used per N_2 fixed can be as high as 42 according to some growth yield measurements (Stam et al., 1984), and under physiological conditions it is commonly reported as 20–30 ATP per dinitrogen molecule reduced (Burris, 1991). Similarly, N_2 fixing bacteria grow more slowly when fixing N_2 than when provided with a fixed nitrogen source. One reason for the considerable extra cost to the cell is due to the inefficient allocation of electrons to substrates by nitrogenase. First of all, nitrogenase allocates valuable reductant for the purpose of proton reduction in addition to passing electrons to carry out N_2 reduction. Further, nitrogenase does not efficiently allocate electrons to protons versus dinitrogen. For example, in some conditions the nitrogenase reaction in N_2 fixing root nodules can greatly favor the production of an undesirable product, namely H_2 (Maier and Triplett, 1996). Under physiological conditions, 40–60% of the energy input into N_2 fixation is lost as H_2 evolution from the root nodules (Schubert and Evans, 1976). Hydrogen production depletes the ATP and reducing equivalents normally used for the production of the desired end product, NH_3. In this way, ammonium formation from N_2 can cost considerably more ATP than the ideal N_2 reduction reaction shown below (Evans et al., 1985; Maier and Triplett, 1996). Under ideal conditions N_2 fixation via nitrogenase proceeds according to the following reaction:

$$N_2 + 8H^+ + 8e^- + 16\,ATP \rightarrow 2NH_3 + H_2 + 16\,(ADP + P_i)$$

Note that even under ideal conditions the energetic costs in terms of ATP is high. When one considers that

Abbreviations: AvI – component I of nitrogenase; Cyc*M* – cytochrome *c*M; NAC – nitrogen assimilation contro; PHB – poly-β-hydroxy butyrate; SOD – superoxide dismutase; FeSII – iron sulfur II protein

the low potential reductant used (the eight electrons in the reaction above) is usually poised at the ferredoxin or equivalent redox level, the overall energetic costs is very demanding to the cell. Consequently, microorganisms that fix N_2 oftentimes exhibit high carbon usage or high photosynthetic activities concomitantly with high respiratory rates. Other costs to the cell that cannot be readily measured would include the maintenance and synthesis of the nitrogenase structural, maturation, and the regulation-related proteins. These processes can be highly complex and multicomponent even for a single process within the overall mechanism of maturation or regulation (Yuvaniyama et al., 2000; Masepohl et al., 2002).

Nevertheless, N_2 fixation by nitrogenase accounts for millions of tons of N_2 fixed on earth annually, and it is a vitally important process for balanced nitrogen transformations on the planet. This chapter will review the physiologically linked processes of O_2 consuming respiration and N_2 fixation. It is appropriate that the two processes be considered as a single subject together, as the benefit in terms of energy production of one is tightly associated physiologically with the energy demands of the other. This chapter will focus primarily on the diazotrophic aerobes, especially *Rhizobium* and *Azotobacter*, as these are aerobic N_2 fixers for which we have a considerable amount of (respiratory metabolism) information. Their relationship with O_2 extends far beyond just the use of O_2 as a terminal electron acceptor for energy generation; it includes evolved responses to oxygen as both a hindrance to, and a requirement for, efficient N_2 fixation.

II. Nitrogen Fixation and Oxygen Tolerance

Nitrogenase is well documented to be an O_2-labile enzyme, and N_2 fixing bacteria display a wide variation in their tolerance to O_2 (Hill, 1988; Gallon, 1992). This is not surprising as they occupy a wide range of environments and they employ different mechanisms to ensure that nitrogenase remains stable. In spite of documented O_2 protection mechanisms, most N_2 fixers have optimum N_2 fixation rates in microaerobic rather than in atmospheric O_2 levels (Hill, 1988); thus the protection mechanisms may not be totally adequate in many environments commonly encountered. Avoidance of O_2 or fixing of N_2 only under fermentative conditions (Gallon, 1992) is sometimes considered to be strategies to protect nitrogenase.

The O_2 lability of the purified nitrogenase proteins does not vary much between organisms, and this is probably due to the highly conserved nature and structure of the nitrogenase components. Therefore the observed differences in O_2 tolerance among diazotrophs must be explained by physiological mechanism differences. A convenient example to demonstrate this physiological variability is the bacteroid response to O_2. Bacteroids (the term for the mature N_2 fixing bacteria in root nodules) from the nodules of different legumes show a considerably wide range of sensitivities to O_2. For example, *B. japonicum* bacteroids have no nitrogenase activity at 2 μM O_2 but *Azorhizobium caulinodans* has high activity at this O_2 tension (O'Brian and Maier, 1989). Also, lupin bacteroids have nitrogenase activity when isolated from nodules aerobically, whereas nitrogenase in soybean nodule bacteroids is abolished when the bacteroids are exposed to air. *B. japonicum* bacteroids have an O_2 optimum close to 100 nM for nitrogenase activity (Bergersen and Turner, 1975) while *R. leguminosarum* bacteroids have an O_2 optimum of 800 nM with regard to the greatest measured nitrogenase activities. Again, the explanation for this cannot lie in nitrogenase per se. Based on the observations, it seems that aerobic N_2 fixers must balance their relationship with O_2 to include aerobic life, as both hindrance to and a requirement for, efficient N_2 fixation. This balance must be precisely maintained for the organism to compete in environments lacking in fixed nitrogen.

In addition to preventing O_2 access to nitrogenase, diazotrophs need to keep metabolically active in terms of reductant and energy generation. Clear examples of loss of nitrogenase peptides by carbon starvation of cells are evident (Moshiri et al., 1994; Oelze, 2000), and this could be explained by a lack of electron flux to the nitrogenase peptides. It appears that the nitrogenase complex is more sensitive to oxygen inactivation upon carbon starvation of cells due to the lowered flux of electrons to the complex (Kuhla and Oelze, 1988). It is well known that the electronic nature of the metallocenter clusters of nitrogenase are altered by oxidation-reduction, and these metal centers (at least for the MoFe protein of nitrogenase) are the most O_2 labile areas too. Also, if ATP levels for nitrogenase are depleted, the enzymes reduction should be affected, as the transfer of electrons from the Fe protein to the MoFe protein requires ATP (Burris, 1991)

III. The N_2-fixing Rhizobium-legume Symbiosis

The N_2-fixing root nodule of leguminous plants is formed from the symbiosis of various *Rhizobium* bacteria with a particular plant host. These include *Bradyrhizobium*, *Mesorhizobium*, *Rhizobium,* and *Sinorhizobium*—all members within the family *Rhizobiaceae*. The nitrogenase enzyme that consumes ATP, and the associated reductant proteins, are located in the differentiated bacteroids within root nodules. These specialized bacterial cells are also the source of the respiratory machinery and ATP synthesis; the latter molecule drives the N_2 reducing reaction. The energy for nitrogen fixation is ultimately derived from plant host photosynthate, and high concentrations of carbohydrate, both as stored and immediate carbon sources, are found in the nodules (Poole and Allaway, 2000). The carbon-sources driving symbiotic nitrogen fixation and transported into the bacteroids are primarily C-4 dicarboxylic acids (Poole and Allaway, 2000). Results from bacteroid respiration experiments, physiological studies on mutants in dicarboxylic transport systems, and labeling studies have shown the importance of the C-4 dicarboxylic acids in intermediary metabolism for N_2 fixation energy input. In addition to providing carbon skeletons for synthesis of macromolecules, the catabolism of these carbon sources is linked to energy generation via respiration. The bacteroid is in essence carrying out rigorous respiratory metabolism to drive N_2 fixation. Its habitat (the root nodule) provides photosynthate carbon and a free O_2 level of less that 50 nM.

A. Nodules

1. Leghemoglobin

Although they function as aerobes, bacteroid metabolism within root nodules depends both on tissue structures and macromolecules surrounding the bacteroids that are designed to minimize the bacteroid's exposure to O_2. The important steps of bacteroid differentiation, and synthesis of nitrogenase and other enzymes all require a microaerobic environment (Batut and Boistard, 1994; Fisher, 1996). This microaerobic environment at the bacteroid surface is primarily achieved by an intriguing protein called leghemoglobin, a heme protein responsible for serving as an O_2 buffer found in the root nodules. Similar hemoglobins of plant origin have also been identified in the non-legume symbioses of *Bradyrhizobium* sp. and in the actinorhizal symbioses of *Frankia* with various plant hosts. The leghemoglobin apoprotein is a plant product and heme synthesis for the symbiosis is in part contributed by the bacterial symbiont (O'Brian et al., 1987a; Chauhan and O'Brian, 1993). The amino acid sequence of leghemoglobin shows it is homologous to animal globins in critical regions. This prompted Appleby to speculate that the origin of leghemoglobin, along with animal hemoglobin, are descended from a common ancestor (O'Brian and Maier, 1989). Interestingly, the engineered alteration of *B. japonicum* ALA dehydratase from a Mg^{2+} dependent enzyme to a Zn^{2+} dependent one does not alter the enzyme ability to supply heme in symbiosis (Chauhan and O'Brian, 1995).

Leghemoglobin found in root nodules facilitates the diffusion of O_2 to the bacteroids and buffers the free O_2 concentration in the nodule. The dissociation-binding constant of soybean leghemoglobin is 46 nM, so free O_2 levels in soybean nodules is 11 nM when the root nodule is at 20% partial oxygenation (Appleby, 1984). This value is in the range of the apparent dissociation binding constant estimated for the efficient bacteroid oxidase (Bergersen and Turner, 1975, 1980; Keefe and Maier, 1993; Preisig et al., 1996), thus efficient respiration can occur at the very low O_2 tension deep within the nodule. In other words, the O_2 can be poised at a low, but readily available concentration for respiration in nodules because leghemoglobin has a very high affinity for O_2. Furthermore, as the leghemoglobin concentration in soybean nodule cytosol is 3 mM (O'Brian and Maier, 1989) the free O_2 must be largely unavailable except to the most high O_2 affinity oxidase (Appleby, 1984); at least one such oxidase is present in bacteroids (see below). The high affinity of leghemoglobin for O_2 is due to a very fast association rate for O_2 (O'Brian and Maier, 1989) and a very slow O_2 dissociation rate. The O_2 dissociation rate constant is nevertheless fast enough for soybean leghemoglobin to be kinetically competent to carry O_2 to the bacteroids. This should be so even during vigorous respiratory metabolism, which is the case during high photosynthate supply.

Facilitated O_2 diffusion to permit bacteroid respiration (and ATP synthesis) was correlated to a function of leghemoglobin when Wittenberg et al., (1974) found that the addition of leghemoglobin to isolated bacteroids resulted in a ten-fold increase in nitrogenase activity with only a small increase in O_2 uptake activity. The properties of the root nodule with associated O_2 binding-proteins results in a remarkable O_2 buffering system; the low free O_2 concentration can be maintained during both rapid or sluggish bacteroid terminal oxidase activity. The increased local free O_2 gradient at the bacteroid surface, provided by leghemoglobin or other O_2 binding proteins, was postulated to stimulate O_2 uptake of the 'high affinity oxidase' (O'Brian and Maier, 1989). We now know this oxidase is primarily a cbb_3-type symbiotic specific oxidase (Preisig et al., 1993, 1996; see also Chapters 5 and 13, Vol. 15), and use of this oxidase presumably stimulates energy conserving electron flow.

2. Nodule Respiration and O_2 Permeability

Presumably to protect nitrogenase from O_2 damage, legume nodules also have mechanisms to regulate their permeability to O_2 (Witty et al., 1990; Hunt and Layzell, 1993). The inner cortex of the root nodule is an initial barrier to gas diffusion. Then, in the rhizobium-infected plant cells the O_2 tension is maintained at 5–50 nM compared with cells in equilibrium with air at 250 μM (King and Layzell, 1991; Denison et al., 1992). This is not solely attributed to leghemoglobin, but to interconnected gas-filled spaces between the cells. A physiological regulation of the volume of these (intercellular spaces) may help control gas permeability (Layzell et al., 1993). For intact nodules, nitrogenase activity and nodule respiration can be stimulated by gradually increasing the external partial pressure (Hunt et al., 1989) of O_2, so the capability for increased respiration and ATP synthesis is evident. Spectroscopic studies on soybean nodules indicates that the oxygen barrier can limit the nitrogenase activity (Layzell et al., 1990). When photosynthate supply is reduced experimentally (Denison et al., 1992; de Lima et al., 1994) nodule metabolism can be markedly increased by raising the external partial pressure of O_2. It is concluded (Kuzma et al., 1999) that nodule metabolism is continually limited by O_2 supply, and that nodules contain barrier mechanisms to decrease the O_2 supply to infected cells. Oxygen consumption can presumably continue even when photosynthate is depleted due to large carbon storage reservoirs in bacteroids.

Due to limitations on the O_2 availability to bacteroids, a presumably nodule-regulated process (Layzell et al., 1993), aerobic respiration is limited, so that ATP supply to nitrogenase is in turn limited (Werner, 1992; Kuzma et al., 1999). Due to the low K_M of (*B. japonicum*) bacteroids for O_2 the high affinity bacteroid respiration would limit the O_2 available for respiration by mitochondria of the nodule, so that ATP limitations would be 'seen' in the plant cells before the bacterial ones. The K_M of the terminal oxidases of nodule mitochondria of 50–100 nM (Rawsthorne and LaRue, 1986; Millar et al., 1995) is certainly greater than for the bacterial terminal oxidases. The actual site of O_2 limitation in nodules is therefore of interest. By measuring the adenylate pools as a measure of the hypoxic metabolic state of root nodules, Kuzma et al. (1999) concluded that the bacteroids are the site of O_2 limitation for nodule respiratory metabolism. Although researchers had predicted that bacteroids were the site of the O_2 limitation of nodule metabolism (Dilworth, 1974; Werner, 1992) the latest study (Kuzma et al., 1999) provides evidence that this is so within intact nodules. The mechanisms by which nitrogenase activity is limited by O_2 may be due to a limitation caused by the ATP needs of the nitrogenase reaction, or due to an inhibition of nitrogenase by accumulation of ADP (Kuzma et al., 1999).

B. Electron Transport in Rhizobia

1. Cultured Cells

Although some rhizobia are capable of nitrate respiration, the majority of respiration by rhizobia is thought to be O_2-dependent for both free-living and nodule-associated cells. The terminal oxidases used by rhizobia and other aerobes may differ in the reductants used, O_2 affinities, types of heme and metal requirements, but they are nevertheless related members of a heme-copper superfamily of oxidases (Garcia-Horsman et al., 1994; van der Oost et al ., 1994). The (characterized) terminal oxidases of *Rhizobium* are within this family. However the relatively large number of them (in *B. japonicum* in particular) is unusual. Initial spectral and inhibitor studies on membranes isolated from free-living and bacteroid forms of *B. japonicum* implicated the existence of a number of terminal oxidases (O'Brian and Maier, 1989). The spectral studies were (rightly) interpreted cautiously, as artifacts (of CO-binding components for example) can complicate clear interpretations (Appleby, 1969a, 1969b; Williams et al., 1990). The terminal oxidases clearly identified in *B. japonicum* include an aa_3 type cytochrome *c* oxidase, *a* heme *b* containing ubiquinol oxidase, an alternative heme-copper aerobically functioning cytochrome c oxidase, and a high O_2 affinity cbb_3-type symbiosis-specific functioning oxidase. Evidence for other possible oxidases that may not even contain cytochrome has been presented (O'Brian and Maier, 1989; Frustaci et al., 1991).

Electron transport in *Bradyrhizobium* has been studied most intensively in *B. japonicum*, and to a lesser extent in *Bradyrhizobium* sp. (*Lupinus)* and in *R. leguminosarum* (Delgado et al., 1998) and *R. etli* (Barquera et al., 1991a, 1991b). The complement of cytochromes expressed is significantly different in bacteroids compared with cells grown in culture. *B. japonicum* cultured cells contain *b*- and *c*-type cytochromes, and as terminal oxidases cytochromes aa_3 and *o*. Based on inhibitor as well as physiological studies of mutants, it is likely that other oxidases, perhaps the ones lacking heme peptides are probably expressed in cultured cells (Appleby, 1984; Frustaci et al., 1991). Cytochromes *o* and aa_3 form complexes with carbon monoxide in a CO-O_2 atmosphere (95: 5, v/v), but respiration is not completely inhibited under these conditions. Also, it must be kept in mind that heme *o* was not detected in *B. japonicum* membranes, so the spectral characteristics assigned to cytochrome *o* may actually be due to a bb_3 type terminal oxidase (Surpin et al., 1996). The aa_3-type terminal oxidase pathway exists in aerobically grown cells of many rhizobia. It is similar to that of mitochondria and many aerobic bacteria. The pathway is: dehydrogenase → Q → FeS protein/bc_1 complex → Cyt aa_3 → O_2.

A *B. japonicum* mutant that lacked Cyt *c* and Cyt aa_3 had respiratory properties that were like a mutant strain lacking only the Cyt aa_3 component (O'Brian et al., 1987); this is suggestive that cytochromes *c* and aa_3 participated in the same branch of a single terminal oxidase pathway. Other studies on oxidase deficient mutants also indicated that Cyt aa_3 and a membrane bound Cyt *c* shared a terminal respiratory branch to O_2 (Bott et al., 1991). Subsequently, a membrane-bound *c*-type cytochrome of 20 kDa called Cy*c*M was postulated to be the intermediary heme electron transfer component between the bc_1 complex and the Cyt aa_3 for both *B. japonicum* and *R. leguminosarum* biovar *viciae* (Delgado et al., 1998; Bott et al., 1991). The presence of Cy*c*M would help explain why the isolation of *B. japonicum* cytochrome mutants based on screening for a respiratory deficiency always yielded lesions in both the cytochrome *c* and aa_3 part of the electron transport system (O'Brian et al., 1987). The electron transport system of free-living *R. trifolii* is similar to that of *B. japonicum,* with the branch point at cytochrome *b*, and with aa_3- and *c*-type cytochromes forming a branch separate from that of cytochrome *o* (DeHollander and Stouthamer, 1980).

Like *B. japonicum, R. leguminosarum biovar viciae* contains CycM, and mutations in that gene also caused a lack of detectable Cyt aa_3 (Wu et al., 1996). The Cy*c*M of *R. leguminosarum* (Wu et al., 1996) is not necessary for a functioning symbiosis, and a function for Cy*c*M as a membrane anchor for another heme complex is indicated. Also, *R. leguminosarum* can apparently compensate for loss of the Cyt *c* and aa_3 pathway by increasing the level of an alternative oxidase, cytochrome *d* (Wu et al., 1996). Cytochromes *b*, *c*, aa_3, and *o* have been observed in many aerobically grown *Rhizobium* including *Rhizobium trifolii* (DeHollander and Stouthamer, 1980) and *R. leguminosarum* (Delgado et al., 1998). Both microorganisms and *R. etli* as well (Barquera et al., 1991a) express the unusual terminal oxidase cytochrome *d* when O_2-limited; interestingly, *B. japonicum* did not express a cytochrome *d* signal under the same conditions (Delgado et al., 1998). Microaerobic incubation conditions can mimic the root nodule environment and perhaps the lack of Cyt *d* in *B. japonicum* is related to the observed differences in optimal O_2 levels for *B. japonicum* versus *R. leguminosarum* bacteroids.

Rhizobium tropici mutants with elevated terminal oxidase activity (Marroqui et al., 2001) had increased levels of Cy*c*M, as well as of *c*-type cytochromes and Cyt aa_3. The strains had superior symbiotic performances on bean plants so this terminal oxidase pathway can be important to the symbiosis. The enhanced respiration and symbiotic performance were due to a mutation in *glgA* encoding a glycogen synthetase. These results underscore the importance to the cells of regulating cytochrome oxidase pathways in response to carbohydrate metabolism or availability. Such regulation is likely occurring continually in the bacteroid as photosynthate supply or stored carbon reserves fluctuate in level. Other cytochrome oxidase mutants of Rhizobium have been reported to have increased symbiotic performance as well (Soberon et al., 1989; Yurgel et al., 1998)

The adaptive responses needed for an organism to adjust to an oxygen tension change from 250 μM (air) to 11nM (the leghemoglobin-buffered concentration at the bacteroid surface) must be numerous. Not only must the new higher affinity oxidases be synthesized, but also it would be of benefit to the organism to down-regulate those oxidases that function at 200 μM

O_2 levels. Constitutive synthesis of the latter would deplete valuable metals and amino acids needed for other synthesis. Oxygen-dependent regulation of terminal oxidase (specifically Cyt aa_3) expression was demonstrated for *B. japonicum*; the amount of *coxA* transcription (based on probing RNA samples with *coxA*) by cells grown in 1% partial pressure O_2 was 6-fold less than in fully aerobic (air-grown) cells. Some *coxA* message was detected in bacteroids too, but it was only about 20% of the level observed for even the low O_2-adapted free living cultures (Gabel and Maier, 1993). Perhaps this residual message is due to the yet undifferentiated cells in the developing root nodule still expressing some coxA message. In contrast to *B. japonicum*, the same authors observed no O_2 dependent regulation of coxA message or Cyt aa_3 activity of *R. tropici*, and bean nodule bacteroids from *R. tropici* contained 65% of the coxA message as fully aerobic grown cells (Gabel et al., 1994). It is likely that the *R. tropici* bean nodule bacteroids use Cyt aa_3 for respiration, as the Cyt aa_3 levels determined either by difference spectra or by enzyme activity (as well as the level of the transcript) were all highly significant in amount. In this regard it is interesting to note that the leghemoglobin buffering capacity of bean nodules is less than for soybean, so that the free O_2 level will be higher at the bacteroid surface of bean nodules.

For free-living *B. japonicum*, *coxA* message was not influenced by copper addition to the medium, even when compared to medium treated to remove copper (Gabel et al., 1994). This was observed for both *B. japoncium* and *R. tropici* but a post-translational affect of copper addition for both of the free-living bacteria was observed. *B. japonicum* lacked spectrally-detectable Cyt aa_3 in medium nearly devoid of copper (less than 0.5 μg per l of copper), and *R. tropici* retained about 50% of the normal (copper supplemented level) cytochrome aa_3 detected by difference spectra of membranes. Whether this difference between the two bacteria is due to a specific affect on provision of copper to the terminal oxidase or to differences in the efficiency of copper sequestering/transport at the cell wall or membrane is not known.

The symbiotic development also results in an increased overall demand for cytochrome synthesis by *B. japonicum*. Accordingly, O_2 limitation has been shown to up-regulate the heme B gene (for ALA dehydratase) in *B. japonicum* (Chauhan and O'Brian, 1997). Such regulation is connected to global regulators that control many other genes related to bacterial differentiation and a functional symbiosis. This result illustrates the key role played by heme synthesis in the overall function of the (vigorously respiring) bacteroid.

Symbiotic expression of cytochrome aa_3 in some effective *B. japonicum* strains was reported, and perhaps the terminal oxidase is regulated differently in the different rhizobial strains (Marroqui et al., 2001; Hirsch et al., 2001). Oxygen may not be the sole regulator of the expression of this oxidase, as the different strains must presumably exist at the same leghemoglobin-buffered O_2 tension as the strains which do not express cytochrome aa_3 symbiotically. Functional cytochromes aa_3 has not been demonstrated in soybean nodules, and a cytochrome aa_3-deficient mutant, as well as a *coxMNOP* (the alternative heme-copper Cyt *c* oxidase) deficient mutant of *B. japonicum* strain still produce effective nodules. Also, mutations in cycM, the proposed electron donor to Cyt aa_3, were unaffected in symbiosis (Bott et al., 1991). Therefore it can be assumed that other oxidases must function in the various stages of the symbiosis (see Section III.B.2).

An integrated picture of *B. japonicum* symbiotic respiration must account for the roles of the multiple terminal oxidase complexes. Thus far we can assign one (*B. japonicum*) oxidase solely to a role in the symbiosis, and two oxidases specifically to the free-living condition. The oxidases identified thus far (as gene clusters) are *fix NOQP* (Preisig et al., 1993), *coxMNOP* (Bott et al., 1992), *coxBA* (Bott et al., 1990, Gabel and Maier, 1990), and *coxWXYZ* (Surpin et al., 1994, 1996). The roles of these based on analysis of mutants are given in the table. *coxBA* encodes two subunits of the cytochrome aa_3 oxidase complex and is expressed only under conditions of high aeration (Gabel and Maier, 1993; Gabel et al., 1994).

The oxidases that have been shown to play a role in free-living but low O_2 dependent growth (less than 10 μM O_2) with H_2 as substrate, are the products of genes *coxMNOP* and *coxWXYZ* (Surpin and Maier, 1998). These oxidase genes encode complexes with similarities to Cu_A-containing cytochrome *c* oxidases (Bott et al., 1992) and *b*-type ubiquinol oxidases (Surpin et al., 1996), respectively. Based on the predicted properties of CoxWXYZ as well as the lack of detectable heme *o* in *B. japonicum* membranes, it was concluded that CoxWXYZ is probably a bb_3 type ubiquinol oxidase. From cyanide inhibition titration experiments (Surpin and Maier, 1998) it was concluded that the oxidase is expressed microaerobically.

The roles played by CoxWXYZ and CoxMNOP in microaerobic free-living growth with H_2 as the electron donor was established by studying mutants in the two oxidases. A mutant in each of the oxidases was still able to grow via microaerobic respiration, but a double mutant was deficient in this growth mode, and in H_2/O_2 respiration. Apparently, each oxidase could substitute for the other.

The putative *B. japonicum* ubiquinol oxidase (CoxWXYZ) was still found to be important in symbiosis; symbiotic nitrogen fixation activities of the coxX mutant were 28 to 34% less than for the wild type (Surpin and Maier, 1999). The viable number of bacteria that could be re-cultured from crushed root nodules was less for the coxX mutant than for the wild type, so a role for this oxidase in survival of the bacteroids is implicated. It is possible that this oxidase is expressed during early stages of the symbiosis, when O_2 concentrations around the developing bacterial cells are high relative to leghemoglobin-bathed cells. Interestingly, free-living *R. etli* expresses a flavoprotein oxidase (Barquera et al., 1991b). It is not expressed in bacteroids and has a low O_2 affinity. Table 1 shows the terminal oxidases present in some of the *Rhizobia*. Their known roles, along with their branch point in the terminal oxidase specific pathway is not noted as well.

2. Bacteroids

The root nodule bacteria (*Bradyrhizobium, Mesorhizobium, Rhizobium* and *Sinorhizobium*) within the family *Rhizobiaceae* have the unique ability to form a nitrogen fixing symbiosis on legume roots. Due to the synthesis of both bacterial and plant oxygen- binding proteins within the nodule, a highly unique and varying oxygen environment must be considered when addressing rhizobial respiratory metabolism. Indeed, the terminal oxidases (with varying O_2 affinities) of the bacterial partner are likely to be expressed at different times during the progression of the symbiosis. The high affinity oxidase of the cbb_3 type with a K_M for O_2 of about 10 nM apparently accounts for a large portion of respiratory activity of bacteroids that function in very low free O_2 levels.

In *Azotobacter* species a small redox active Fe/S protein protects nitrogenase from O_2 damage in conditions when the O_2 tension becomes elevated due to insufficient respiration (caused by insufficient oxidizable carbon source availability). Interestingly, brief treatment of intact soybean roots with 1 atm O_2 diminishes activity of nitrogenase, but this activity is partially restored upon exposure to air (Patterson et al., 1983). From these and other studies (Appleby, 1984), the possibility that an O_2 protection mechanism involving one or more proteins binding to nitrogenase, has been considered for the root nodule symbioses. However, using the cloned *A. vinelandii* protective protein as a probe, even using low stringency hybridization conditions, Moshiri et al. (1994) found no evidence for the existence of the same (protective-type) FeSII protein in *B. japonicum*.

Bradyrhizobium japonicum bacteroids expressed carbon monoxide-reactive cytochromes *c*-552, *c*-554, *P*-420, and *P*-450; they are possibly putative oxidases (Appleby, 1969a,1984; O'Brian and Maier, 1989; Delgado et al., 1998). However, genetic and biochemical studies showed that most of these components are not too important to the symbiosis, as mutations in these genes have no effect of nitrogen fixation. Similarly a soluble cytochrome *c*550, a non CO-reactive cytochrome, was shown not to be important for the symbiosis but for NO_3 dependent respiration (Tully et al., 1991; Bott et al., 1995; Delgado et al., 1998). *Bradyrhizobium* sp. (*Lupinus*) bacteroids also express cytochrome *P*-450 as do, to a lesser extent, *R. leguminosarum* strain PRE and cultured cells of *Bradyrhizobium* sp. Strain 32H1 induced for nitrogenase activity (Appleby, 1984). This heme-containing oxidase could be a terminal oxidase but more likely function as an oxidase for other metabolic purposes. More recent biochemical and genetic evidence for some of the variable O_2-affinity oxidases, including a very high O_2 affinity one that was postulated to exist many years ago has been presented. This is the oxidase complex encoded by the gene cluster *fix NOQP* (Preisig et al., 1993 see below).

The identification of the gene clusters for terminal oxidases and the generation of the corresponding mutants do not mean we are finished identifying the specific components of the complex terminal respiratory system. Bergersen and Turner (1975,1980) originally identified four O_2 affinity states of *B. japonicum* bacteroid respiration, which presumably correspond to four oxidases or to four O_2 binding sites with different affinities for O_2. The highest O_2 affinity state showed an apparent K_s value of 6 nM (Bergersen and Turner, 1980); this free O_2 is near that estimated for the O_2 concentration within a soybean nodule (Appleby, 1984). The fact that the bacteroid FixNOQP oxidase has a K_M of 7 nM bolsters the original observations of Bergersen and Turner. There-

Table 1. Respiratory terminal oxidases in selected nitrogen fixing bacteria

Organism	Terminal oxidase	Branch point	Role
B. japonicum	Cyt aa_3	FeS/bc_1	Aerobic free-living
	CoxMNDP	FeS/bc_1	Aerobic free-living
	CoxWXYZ	Q	Microaerobic free-living
	(bb_3-type)	and symbiotic[#]	
	Cyt cbb_3	FeS/bc_1	Microaerobic, Symbiotic-specific[#]
R. leguminosarum	Cyt *d*	Q	Free living microaerobic
bv. *viciae*			
	Cyt cbb_3	FeS/*bc*1	Symbiosis
	Cyt aa_3	FeS/*bc*1	Aerobic
A. vinelandii	Cyt c_5o	Q-8	Aerobic, up-regulated with N_2 fixation
	Cyt c_4o	Q-8	Aerobic, up-regulated with N_2 fixation
	Cyt *b*/*d*	Q	Protection of nitrogenase
A. caulinodans	Cyt *b*/*d*	Q	Microaerobic, symbiosis[§]
	Cyt cbb_3	FeS/bc_1	Microaerobic, more so than
	Cyt *b*/*d*, symbiosis[§]		
	Cyt aa_3	FeS/bc_1	Aerobic free-living
	Cyt *b*/*o*	Q	Unknown

[§]The individual Cyt *b*/*d* and Cyt cbb_3 of *Azorhizobium caulinodans* knock-out mutants were both 50% reduced in symbiotic N_2 fixation activities (Kaminiski et al., 1996). [#]*Bradyrhizobium japonicum* knock-out mutants in *coxWXYZ* and in cbb_3 (*fixNOQO*) were reduced in symbiotic N_2 fixation by about 30% and 100%, respectively. The 'branch point' represents the electron carrier prior to the indicated terminal oxidase that initiates an electron flow pathway unique to that oxidase. Abbreviations: bv., biovar; Q-8, ubiquinone-8; FeS, iron sulfur protein. For a discussion on the features and roles, see text.

fore other O_2-affinity states may also be relevant to and correlated to identification of specific oxidases. *B. japonicum* bacteroid membranes were shown to contain three cyanide-binding sites (O'Brian and Maier, 1983), supporting the CO-binding data and the respiration kinetics experiments for multiple oxidases. NADH dependent O_2 uptake by *B. japonicum* that had been incubated microaerobically (maximum 1% partial pressure O_2 and corresponding to less than 15 μM dissolved O_2) had three CN^- inhibition phases (Surpin and Maier, 1998). One of these was attributed to Cyt aa_3 and another inhibition area was common to both CoxMNOP and CoxWXYZ. Still, at least one other CN^- inhibited but unidentified O_2 uptake activity remained (Surpin and Maier, 1998).

Studies on respiration of whole cells of the *B. japonicum* Cox mutants (microaerobically incubated) indicate that at least one more oxidase needs to be identified. The double mutant (*coxWXYZcoxMNOP*) was capable of significant respiration when tested in 50 μM O_2 levels, where Cyt aa_3 would be expected to be non-functional. The remaining oxidase activity could be symbiosis related, as the cells were incubated initially in very low O_2 conditions (less than 10 μM O_2 and with H_2 as the reductant) for 14 days (Surpin and Maier, 1998). Even anaerobically grown *B. japonicum* exhibited two distinct binding phases for O_2, and only one was attributed to the cbb_3 type (FixNOQP) oxidase (Preisig et al., 1996).

a. Fix NOQP

Bradyrhizobium japonicum genes related to a *fixN* region linked to nitrogenase genes of *R. meliloti* were characterized (Batut et al., 1989; Preisig et al., 1993). That this encoded an oxidase designed for symbiotic respiration was indicated by the following: *B. japonicum* mutants with insertions in *fixNOQP* were N_2 fixation negative, and the operon was induced in free-living culture under microaerobic conditions, (or anaerobically with NO_3^- as terminal acceptor). Two *c*-type cytochrome components of the complex are presumably FixO and FixP. FixQ is likely a hydrophobic membrane anchor, and FixN is a subunit I-type heme-copper oxidase. The oxidase falls into a subfamily of the heme-copper oxidase superfamily (van der Oost et al., 1994; Garcia-Horstman et al., 1994). It is present in all rhizobia thus far examined (Delgado et al., 1998). The (cbb_3) complex encoded by a *fixNOQP* operon is well suited to function microaerobically in the root nodule, and it accepts electrons from FeS/bc_1 (Zufferey et al., 1996). Not all of the Cyt *c* oxidase activity of membranes (from anaerobically grown cells) can be attributed to this oxidase (Preisig et al., 1996), so other oxidases are probably functioning. One candidate for the remaining activity is *CoxWXYZ*, a mutation in which was shown to adversely affect the symbiosis (Surpin and Maier, 1999).

A Cyt *c* oxidase with 7 to 8 subunits containing CO-reactive heme *c* was purified from *B. japonicum* bacteroid membranes (Keefe and Maier, 1993). This complex undoubtedly corresponds to the (cbb_3-type) oxidase encoded by the *fixNOQP* gene cluster, as some of the identified subunits as well as the heme content match with the cbb_3 oxidase. The latter oxidase complex was purified (from anaerobically-grown *B. japonicum*) using solubilization and purification procedures (Preisig et al., 1996) similar to those used to first purify the Cyt *b* and *c*-containing oxidase complex from the (*B. japonicum*) bacteroids (Keefe and Maier, 1993). The complex corresponding to *fixNOQP* that was purified from *B. japonicum* bacteroids (Keefe and Maier, 1993) also contained some of the bc_1 complex subunits, indicating a tight association of the terminal oxidase complex with the intermediary electron transfer complex. The K_M derived for O_2 of the cbb_3 complex (corresponding to *fixNOQP*) was 7 nM, and consistent with an oxidase that functions in the symbiosis.

The *fixNOQP* operons of *R. etli* are highly regulated by Fnr family transcriptional activators, and are poised for high expression in microaerobic conditions (and therefore for the symbiosis) (Lopez et al., 2001). In *Azospirillum brasilense*, a microaerobically-induced Cyt cbb_3 was identified (Marchal et al., 1998). It may not be related to N_2 fixation, as a Cyt N mutant was still capable of N_2 fixation rates almost equal to those of the wild type. Regulation experiments are badly needed for all the symbionts, in order to understand the environmental regulatory factors for expression of the various oxidases. For example, understanding the affect of a series of concentrations of various regulators like O_2, H_2, fixed N, and copper on (all) the oxidases expression would be useful.

C. Factors affecting nodule and bacteroid respiration

Studies of *R. leguminosarum* bacteroids (Haaker et al., 1996) showed that respiration rates may not only be determined by the affinities of the terminal oxidases, but by respiratory control, in effect by the adenylate pools. In this way, by generating fluctuating adenylate pools and requiring ATP nitrogenase in effect can regulate itself. The addition of leghemoglobin to bacteroid suspensions resulted in large increases in nitrogenase activity with only modest increases in O_2 consumption (Wittenberg et al., 1974). This indicated that ATP production is greatest at the lowest O_2 levels, but other explanations are plausible (see below).

From the early studies on bacteroids it was predicted that efficient respiration occurred at the lowest O_2 tensions, and that ATP synthesis is highest in *B. japonicum* bacteroids at leghemoglobin-buffered O_2 concentrations of 0.02–0.1 μM (Bergersen and Turner, 1975). It appeared that respiration became uncoupled from ATP synthesis at levels of oxygen above 0.1 μM (Bergersen and Turner, 1975). It was widely believed that uncoupled electron transport acted as a ‘protective respiration mechanism’ to maintain a low intracellular O_2 concentration at the expense of high-energy electrons to prevent inactivation of the O_2-labile nitrogenase enzyme (Maier et al., 1990). However, the multiple oxygen binding phases observed in bacteroids can be due to artifacts of the polarographic method (Haaker et al., 1996), and simply a change in the type of reductant used by bacteroids could lower the ATP levels at the high O_2 concentrations. This would make it appear as if high respiratory rates were less coupled to energy production. For example, if respiration was altered so that it is driven by electron donors entering at the level of Q rather than use of lower potential substrates (such as nitrogenase-produced H_2) a lower ATP/ADP ratio, but higher O_2 use would be observed. The best way to determine the function of the different oxidases (see Table 1) would be to reconstitute each into liposomes and measure the coupling efficiency ($\Delta\Psi$ and ΔpH) of each as a function of O_2 level.

The efficiencies of ATP production and thus nitrogenase activities by bacteroids were hypothesized to be determined by the O_2 affinity of the efficient versus inefficient terminal oxidase pathways (Bergersen and Turner, 1980). However, since nitrogenase in bacteroids can be controlled by fluctuating adenylate-pools and other factors that are in turn controlled by O_2 concentration (see Haaker et al., 1996) the efficient versus inefficient hypothesis is not needed to explain the early results. The 'uncoupled oxidase' hypothesis is therefore open to scrutiny. Haaker et al., (1996) observed that nitrogenase inhibition by excess O_2 was associated with high ATP/ADP ratios, and that the proton motive force controls respiratory activity and thus nitrogenase too. ATP hydrolysis by nitrogenase may keep the ATP/ADP ratio sufficiently low to relieve inhibition of respiration by respiratory control mechanisms (O'Brian and Maier, 1989; Haaker et al., 1996). At low (nM) levels of O_2, respiration would continue and nitrogenase would be active, regardless of the variability in coupling efficiency of the different electron transport pathways.

In a recent study, *B. japonicum* bacteroid respiration, was increased by increasing the O_2 supply, including by adding oxyleghemoglobin (Li et al., 2001). It was found that malate transport into the cells was stimulated along with respiration, and that the free O_2 level had little influence on the (malate) uptake rates. The study underscores the importance of performing O_2 use experiments via oxyleghemoglobin by bacteroids. Such conditions are needed in order to mimic the near-anaerobic nodule situation to relate the findings to in situ bacteriaroid metabolism.

IV. The FeSII Protein and Conformational Protection

A. Shethna Proteins

A problem inherent to aerobic nitrogen-fixing organisms is that O_2 is required for ATP synthesis via respiratory metabolism yet the nitrogenase enzyme is O_2-labile. Nitrogen fixing bacteria have developed a number of creative strategies to deal with this paradox. The strategies must ensure a steady supply of ATP during nitrogenase functioning, yet permit nitrogenase to be protected from O_2 inactivation. The *A. vinelandii* FeSII protein ('iron-sulfur II protein') was originally isolated as a pink protein present in relatively large amounts in nitrogen fixing cells (Shethna et al., 1968). Spectroscopic as well as direct metal analysis indicated it contained two 2Fe-2S clusters (4Fe/dimer) (Laane et al., 1980; Sherings et al., 1983). Bulen and LeCompte (1972) showed that the Fe-S protein was intimately associated with nitrogenase. The pure nitrogenase complex was composed of AvI (component 1 of nitrogenase, AvII is component 2 of nitrogenase), and the FeSII protein; as such nitrogenase was relatively stable to air exposure. The association of the *A. chroococcum* FeSII protein with the Mo-Fe protein and the Fe-protein of nitrogenase in the form of an isolatable oxygen-stable complex was rigorously demonstrated by Robson (Robson, 1979). The FeSII protein was found to co-purify with the AvI and AvII through a sucrose density gradient step, yielding a preparation, which retained a relatively high tolerance towards oxygen inactivation. The FeSII protein cannot protect either of the nitrogenase component proteins in the absence of the other component protein. Possible clues as to a redox-based mechanism of the FeSII protein have come from experiments which indicate that the FeSII protein can accept electrons from the Fe-protein (Laane et al., 1980). Despite evidence of an important role for the association of the FeSII protein with aerobic nitrogen fixation in *Azotobacter*, the role of this protein was essentially ignored for almost two decades (see Moshiri et al., 1994).

The nitrogenase-protective FeSII protein was first named the Shethna protein, but it is important to note that another Shethna protein (recently termed Shethna protein I) (Chatelet and Meyer, 1999) is also a 2Fe-2S protein but it is a different protein than the protective Shethna protein. The Shethna protein I, usually termed FeSI, is encoded within the *nif* gene cluster and has homologues in many N_2 fixing bacteria (Chatelet and Meyer, 1999). *Azotobacter vinelandii* synthesizes at least 12 different small ferredoxin-like proteins, and one recently shown to be involved in FeS cluster assembly was characterized and termed FeS IV (Jung et al., 1999). The structural gene encoding the *A. vinelandii* FeSII protein was used for the construction of strains bearing defined deletions of the *fesII* gene (Moshiri et al., 1994). A role for FeSII in protection of nitrogenase was obtained from experiments involving 'adding-back' the pure FeSII protein to extracts from the FeSII mutant strain and assaying the loss of nitrogenase activity in O_2 over a 10 or 15 min period (Moshiri et al., 1994, 1995).

The *A. vinelandii* FeSII protein was over expressed in *E. coli*, and characterized spectrally, and crystallized (Moshiri et al., 1995). The spectroscopic studies using UV-visible absorption, CD, and variable temperature MCD, EPR and Raman approaches showed the (2Fe-2S cluster) is coordinated by cysteine residues only. The data was unambiguous in assigning one (2Fe-2S) cluster to each subunit of the homodimer. The EPR properties showed the cluster is like the plant-type chloroplast ferredoxins, but the other spectra justified classification of the (FeSII protein) cluster into the hydroxylase type ferredoxins. The FeSII protein is therefore unusual in that it has characteristics of both classes of 2Fe-2S cluster-containing proteins. The vibrational properties of the cluster indicated it is not exposed to solvent, but it may be buried within the structure.

The nature of the trigger that facilitates association/dissociation of the FeSII protein with nitrogenase is unknown, but an attractive hypothesis is that this is determined by the oxidation state of the cluster, so that the cluster acts as a sensor of the intracellular redox potential. Therefore, the redox potential of the cluster is a crucial determinant to its function. That potential was determined to be –262 mV, a value substantially higher than for the chloroplast types of Fe-S clusters. The redox potential is in the range of values one would expect for a sensor that responds to the overall potential within an aerobic cell (Moshiri et al., 1995). Site directed mutants of the FeSII protein were created to test the importance of certain residues near the cluster in maintaining the integrity of the redox potential (Lou et al., 1999). The mutant versions were purified and shown to retain normal (like wild type) EPR and NMR spectroscopic properties. Mutation of a histidine residue near a cluster-liganding cysteine was shown create a FeSII protein with little nitrogenase-protective ability. A pK increase of that histidine upon 2Fe-2S cluster oxidation is indicative of ionization of another group near the cluster, which could modulate the FeSII proteins affinity for nitrogenase in a redox-dependent manner. Also, mutation (replacement with alanines) of two lysine residues of FeSII created an FeSII protein that permitted rapid degradation of nitrogenase in carbon-starved cells, and this (double lysine) mutant version provided almost no protective ability when added to nitrogenase-containing cell extracts. These lysine residues may be associated with the surface of FeSII, and may play a role in the initial steps of recognizing the nitrogenase component proteins (Lou et al., 1999).

B. Oxygen and Nitrogenase-Mediated Cell Death

Based on analysis of mutants it was concluded that the FeSII protein is most important to *A. vinelandii* when the cells are carbon substrate depleted (i.e. when respiratory protection is inoperable). We cultured cells in a low-carbon (10 mM sucrose rather than the normal 60 mM) medium to address the effects of the presence of the FeSII protein on the survivability of *A. vinelandii* (Maier and Moshiri, 2000). This effect was assessed by plating for viable cell number at 1.5 hr intervals, after the cells terminated their exponential growth phase. From the results, it was concluded the FeSII protein is important for aerobic survival of the cells, and that minimizing oxidative stress associated with nitrogenase inactivation may be a novel role of FeSII (Maier and Moshiri, 2000).

That aerobic respiratory metabolism is responsible for the production of harmful partially-reduced oxygen species is well-documented. That metalloenzymes not involved in respiratory electron transport can be the source of these oxygen radicals *in vivo* has only relatively recently been realized (Gaudu et al., 1994; Imlay, 1995; Storz and Imlay, 1999). The binding of oxygen to oxygen-sensitive bacterial redox proteins can lead to superoxide formation when anaerobically-grown cells are exposed to oxygen (Storz and Imlay, 1999). The accumulation of hydrogen peroxide and superoxide can lead to highly damaging hydroxyl radicals (via Fenton chemistry) in the presence of transition metal ions (Gallon, 1992; Storz and Imlay, 1999). Such radicals can damage a number of cellular components including DNA and the components of membranes, as $^{\bullet}OH$ radicals will attack virtually every molecule within the cell.

Although aerobic N_2-fixing bacteria have inherent physiological mechanisms to reduce contact of nitrogenase with O_2, nitrogenase can reduce low levels of O_2 to H_2O_2 without becoming inactivated (Thorneley and Ashby, 1989). Upon exposure to high levels of O_2, however, nitrogenase will be inactivated during O_2-reduction, through a proposed reaction involving the Fe protein and superoxide or hydroxyl radicals (Thornely and Ashby, 1989; Gallon, 1992). It is thought that nitrogenase even in vivo, can generate superoxide, $^{\bullet}OH$, and H_2O_2 by reducing O_2 (Gallon,

1992). It was determined that SOD levels could be varied, by varying the supply of iron to *E. coli*, so experiments were done to measure SOD levels for *A. vinelandii* as a function of the level of iron in the medium (Maier and Moshiri, 2000). It was found that SOD activities (for both the wild type and the FeSII strain) were considerably lower in *A. vinelandii* grown in Fe-limiting conditions than in cells with sufficient iron. The viability as percent of the t = 0 time point, (t = 0 is the point at which time the carbon supply is limited and exponential growth terminates), was determined at two times after the t = 0 time points. In SOD-lowered cultures, the survival of the FeSII mutant was affected much more than the wild type, in the time points after carbon exhaustion. The results related the loss of viability by lack of the FeSII protein to lowered SOD levels or ability of the cells to dissipate superoxide (Maier and Moshiri, 2000). Thus the cell death observed by lack of FeSII may be related to increases in toxic superoxide levels. N_2 fixation related proteins may generate the reactive oxygen species.

Oxidative stresses due to O_2 exposure, is a problem for other N_2 fixers as well (see Gallon, 1992). The powerful reductants associated with nitrogenase, like flavodoxin, ferredoxin, and nitrogenase itself could generate O_2^- and H_2O_2 by partially reducing O_2. Reactive oxygen species are produced as an early response by alfalfa to *Sinorhizobium meliloti* infection (Santos et al., 2001); it seems the rhizobia must combat the same barriers put forth by the legume to deter infection by other microorganisms. Rhizobia in mature root nodules are thought to be subject to toxic oxidative products of leghemoglobin reactions (Gallon, 1992). In legume nodules, the maintenance of leghemoglobin in an 'O_2-carrying state' as Fe^{++} by ferric leghemoglobin reductase may be responsible for H_2O_2 generation (see Gallon, 1992). Hydrogen peroxide in turn can degrade leghemoglobin liberating free Fe^{++}, which can give rise to even more damaging $^{\bullet}OH$ through Fenton reactions (Puppo and Halliwell, 1988).

C. Homologues of FeSII

In comparing the nucleic acid and deduced amino acid sequence of the FeSII protein against databases, the sequences with the greatest homology with the FeSII protein was found to be the product of the *R. capsulatus, fdxD* gene (Moshiri et al., 1994). Less homology was seen between the FeSII protein and the FdxH proteins, encoding heterocyst specific ferredoxins from the nitrogen fixing blue-green algae, *Anabaena* sp and *Fremyella diplosiphon* (Moshiri et al., 1994). The homologous area of these ferredoxins and FeSII was mainly in the areas around the cysteine residues, which ligate the FeS cluster. The *fdxD* gene (*R. capsulatus*) encodes a protein of 123 amino acids (including the initiator methionine), and the deduced amino acid sequence of its product is 50% identical (with an additional 15% similarity) to the FeSII protein, without any gaps in their alignment. Notably, all four of the 4 cysteine residues which are presumed to be the ligand for the [2Fe-2S] cluster of the FeSII protein are present, in the same position, in the FdxD sequence. In the other sequences which were found to be homologous with FeSII, such as the *Anabaena* FdxH protein, the three closely spaced cysteines (i.e. $C\text{-}X_4\text{-}C\text{-}X_2\text{-}C$ could also be aligned, but the fourth cysteine was present in rather variable positions in the remaining sequence. Interestingly, there is a di-lysine motif, which is also present close to the N-terminus of FdxD (at positions 12-13).

The *R. capsulatus* FdxD protein was expressed in *E. coli* and the purified protein was partially characterized (Armengaud et al., 1994). The expressed protein, called ferredoxin V, was shown to be a [2Fe-2S] protein which was unable to function as an electron donor to nitrogenase in vitro, a property consistent with its relatively high mid-point potential of –220 mV. Unaware of the sequence homologies between FdxD and FeSII, the French researchers (Armengaud et al., 1994; Jouanneau et al., 1995) noted the striking similarities between the spectroscopic properties and the redox potential of FdxD and FeSII, but still a function for FdxD could not be proposed. It was nevertheless concluded that FdxD may play a role in N_2 fixation in *R. capsulatus* (Jouanneau et al., 2000).

D. Model for Protective Protein Function Mechanism

The working model for the oxygen-protective function of the FeSII protein in the aerobic nitrogen fixing physiology of *Azotobacter* is as follows: during periods of oxygen stress, the FeSII protein by forming the oxygen-stable complex with nitrogenase allows the complex to assume a reversibly inactive state. We predict that the nitrogenase components of the complex undergo one final redox turnover, whereby the MoFe and Fe protein attain an oxidized form, and the FeSII protein is reduced. The reduced FeSII protein is auto-oxidized by its reaction with oxygen, and all three components remain associated. A number of redox titration experiments performed using different redox mediators and varying oxygen levels, have demonstrated that both the MoFe protein as well as the Fe protein can exist in a number of reversible, intermediate reduced, and oxidized states (Wang et al., 1985). It is likely that in the initial formation of the oxygen-stable complex, the FeSII protein by virtue of its intermediate redox potential, can act as a redox mediator between the MoFe and Fe proteins and oxygen, and maintain these components in a stable oxidized state. In the absence of the FeSII protein, however, oxidation of the nitrogenase components rapidly proceeds to an irreversibly oxidized-inactivated state (Wang et al., 1985), and toxic O_2-derived products may be generated as well. Perhaps with the formation of the tri-partite complex, production of these toxic products is eliminated or minimized. However, our in vitro experiments using spin-trap reagents in conjunction with EPR have been unsuccessful at providing evidence of greater nitrogenase-produced partial O_2 reduction products in extracts lacking the FeSII protein compared to extracts that contained the FeSII protein (unpublished observation). Nevertheless, the in vivo results of cell viability loss point to a link between a FeSII-nitrogenase complex and minimizing O_2 toxicity, and a probable role for the FeSII protein in preventing such toxicity problems.

V. Other O_2 Protection Mechanisms

A. Blocking the Accessibility of Molecular Oxygen

A recent study implicated an alginate capsule on the surface of *A. vinelandii* in playing a role in protection from O_2 (Sabra et al., 2000) by preventing O_2 diffusion across the cell wall. This is a mechanism that apparently is unique to phosphate-limited cultures. Nevertheless, growth in the presence of ammonia inhibits alginate formation, as expected for a protection system for nitrogenase. Interestingly, growth of cells in increasing O_2 tension led to concomitant formation of higher MW alginate and with a higher glucuronic acid content. This made the capsule thicker, furthering its function as an effective O_2 barrier.

Carbon starvation conditions (for example, for *A. vinelandii*) cause the cells to be particularly vulnerable to O_2 damage of nitrogenase (Moshiri et al., 1994). This may be due to either a deficiency in respiratory protection (blocking access of O_2 to the interior of the cell) or via a lack of reductant flux to nitrogenase. Also, an *A. vinelandii* mutant that has a limited ability to accumulate poly-β-hydroxy butyrate (a *PtsP* mutation) may be concomitantly limited in its ability to protect nitrogenase from O_2 damage (Segura and Espin, 1998). The inability of the PtsP mutant to grow diazotrophically was exacerbated by carbon starvation, and was relieved by supplying either ammonium or glucose. The results indicate a respiratory protection problem for the mutant, but it was acknowledged (Segura and Espin, 1998) that an unrealized interaction between FeSII (the protective protein) and PHB accumulation may exist to account for the results.

The marine filamentous N_2 fixing cyanobacteria *Trichodesmium* spp., a major player in N_2 fixed in the oceans, does not form heterocysts yet fixes N_2 while carrying out O_2-evolving photosynthesis (Zehr et al., 1997; Chen et al., 1999). This strategy is unusual as most cyanobacteria separate O_2-sensitive N_2 fixation from O_2 evolving photosynthesis by fixing N_2 in the dark, or in the specialized cells where photosystem II is absent. Still, *Trichodesmium's* nitrogenase proteins are sensitive to O_2 inactivation, and the protection mechanism proposed involves both spatial and temporal separation of N_2 fixation and photosynthesis (O_2 evolving type). The protection mechanism used by *Trichodesmium* spp. to fix N_2 (Berman-Frank et al., 2001) involves a brief (several hr) period within the entire photoperiod in which N_2 fixation can take place. The N_2 fixation related events can take place only when PSII is down regulated and stored carbohydrate is used, so that O_2 consumption via respiration exceeds the rate of O_2 evolution. At this brief time interval, the environment for nitrogenase is essentially anaerobic and nitrogenase is active. From this interesting work it is proposed that nitrogenase once served as an electron acceptor for anaerobic heterotrophic metabolism during early evolution of oxygenic photosynthesis. Another abundant diazotroph in the oceans are the cyanobacterial symbionts of diatoms (Zehr and Ward, 2002). They represent a significant source of fixed N in many parts of the ocean. The diatoms containing the heterocyst forming cyanobacteria form large aggregates, which may be another way to reduce damaging O_2 levels.

The observed behavior of *Azospirillum brasilense* when exposed to increased O_2 concentrations while incubated in soft agar indicates the bacteria needs to balance its O_2 exposure (Zhulin et al., 1997) between that needed for respiration and that so high as to be detrimental. This is probably a reflection of the O_2 level required for efficient O_2 respiration coupled with the O_2 level that will be tolerated by its nitrogenase. Aggregation is apparently used by many cells (including *Azotobacter vinelandii*) to limit O_2 to (the inner) cells, so that N_2 fixation can occur in the microenvironment in the center of the aggregate (see Gallon, 1992). Diazotrophs can separate the sites of N_2 reduction from high O_2 areas by a spatial separation of the processes. The formation of heterocysts (by filamentous *cyanobacteria*) as the differentiated and specialized N_2 fixation site to avoid O_2 exposure problems (Paerl, 1990), or forming actinorhizal vesicles (*Frankia*) to permit aero-tolerant N_2 fixation within an O_2 barrier (Lundquist, 2000) are highly interesting mechanisms.

The *Frankia-Alnus* root nodules responded to supra-atmospheric O_2 by decreasing nitrogenase activity, and a fixed O_2 diffusion barrier by the nodules was proposed (Lundquist, 2000). That vesicles of free-living *Frankia* produce thicker cell walls when exposed to higher O_2 levels (Harris and Silvester, 1992) indicate the vesicle structure may be regulated for N_2 fixation tolerance from O_2. Like heterocyst-formation, another mechanism involving spatial separation of processes would include the production of polysaccharide sheathes in slime-producing consortia. Physical viscous barriers to O_2, such as polysaccharide capsule production, may be a common mechanism in nature to prevent O_2 penetration or diffusion into cells or cell aggregates for many nitrogen fixing associations (see Gallon, 1992). It is proposed that the O_2-depleted internal regions deep within the slime layer of marine consortia support the highest N_2 fixation rates (Stal, 1995).

Non-heterocyst-forming cyanobacteria such as unicellular *Cloeothece* and filamentous types (*Oscillatoria* and *Microcoleous*) can temporally separate N_2 fixation (occurs at night) from oxygenic photosynthesis (occurs in daylight). Therefore, the mechanisms for occluding O_2 access to the nitrogenase peptides are numerous. It would be most interesting to study the expression of each of these mechanisms in concert with N_2 fixation genes.

The N_2 fixing endophyte of sugarcane, *Gluconacetobacter diazotrophicus* can adapt to a variety of O_2 tensions, apparently to maintain nitrogenase activity (Pan and Vessey 2001). Respiratory rates of colonies varied little in response to O_2 tension variations and nitrogenase activity returned to normal within 10 min after a high O_2 exposure. Therefore a switch on/off of nitrogenase in response to O_2 flux was proposed, a mechanism that will presumably involve other proteins or molecules that interact with nitrogenase.

B. Respiratory Protection

The energetic demands associated with nitrogen fixation require that efficient electron transport pathways' function in aerobic diazotrophs. However, respiratory pathways that play protective roles, by scavenging oxygen are also proposed to exist. The respiratory chain of *A. vinelandii* is terminated with either a cytochrome *o* type or a *b*/*d* type oxidase (Haddock and Jones, 1977; Maier et al., 1990). The heme moiety associated with the cytochrome *o* may be either a heme B or a heme O. The branch point, at the quinone level therefore yields $b \rightarrow d$ and $c \rightarrow o$ terminal pathways. Instead of the combined cytochrome $c_4c_5 \rightarrow o$ pathway that was proposed initially based on inhibitor experiments (Haddock and Jones, 1977; Yates, 1988) separate $c_4 \rightarrow o$ and cytochrome $c_5 \rightarrow o$ pathways are consistent with the spectral studies on terminal oxidase mutants of *A. vinelandii* (Ng et al., 1995). Since cytochrome *o* was still reduced in a cytochrome c_4 mutant, separate pathways to cytochrome *o* from the individual *c*-type cytochromes is indicated. The terminal oxidase pathways of *A. vinelandii* therefore consist of three separate branches, and this flexibility undoubtedly relates to N_2 fixation physiology. The complement of terminal oxidase in *Azotobacter vinelandii* and *Azorhizobium caulinodans* are included in Table 1.

The cytochrome *b*/*d* terminal oxidase of *Azotobacter vinelandii* has been implicated as playing role in using oxygen specifically to protect nitrogenase from O_2 damage (see Kelly et al., 1990; Moshiri et al., 1991). The O_2 affinity of this oxidase would be sufficient for O_2 binding within a large range of O_2 levels (Kolonay et al., 1994) as would be expected for respiratory protection. The up-regulation of this oxidase with nitrogen fixation genes supports such a role (Moshiri et al., 1991). Also, the characteristics of a cytochrome *b*/*d* mutant (it cannot fix N_2 in air) is consistent with a respiratory protection role for the *b*/*d* oxidase pathway (Kelly et al., 1990). The cytochrome *b*/*d* of *A. vinelandii* is nevertheless a coupled oxidase as pH and potential gradients are well established by the oxidase incorporated into liposome (Kolonay and Maier, 1997). Analysis of *c*-type cytochrome mutants of *A. vinelandii* indicates they may also play roles in respiratory protection of nitrogen fixation. A gene directed mutant in both cytochromes c_4 and c_5 was unable to grow at O_2 tensions of 2.5% or less on N-free medium (Rey and Maier, 1997). Since the wild type growth was unaffected by the various O_2 tensions, the conclusion that one role of the *c*-dependent pathways is to provide for respiration to support N_2 fixation at intermediate (5–10%) and low (below 5%) O_2 concentrations was reached. Nevertheless, these pathways (cytochromes $c \rightarrow o$) are still important for respiration in the absence of nitrogen fixation, as the c_4c_5 double mutant was growth-retarded in N-containing medium as well. From monitoring growth of cells in experimental conditions employing both liquid and solid support media in closed gas systems, it appears that the Cyt *c*-dependent terminal pathways augment respiration to permit survival. In the intermediate O_2 levels of 5-10% partial pressure of the gas; eventually a lower threshold level of O_2 can be achieved (below 2.5% partial pressure) where the Cyt *b*/*d* path fails to function, so that the wild type bacterium relies only on the *c*-type cytochrome pathways. Without these alternative (*c*-type cytochrome dependent) pathways the cells fail to survive, and this is most evident when the energy (and respiratory) demands are the greatest (i.e. N_2 fixation conditions).

Oxygen levels did not influence the transcriptional regulation of the two *c*-type cytochromes (c_4 and c_5), but they were regulated by fixed N supplementation. The promoter activities of the cycA (Cyt c_4) and cycB (Cyt c_5) were up-regulated 2.8 and 7-10 fold by N starvation. The N-starvation condition used also caused nitrogenase to be expressed. Other genes known to play roles in *Azotobacter* oxygen protection mechanisms (Cyt *d* and the FeSII protein) are moderately upregulated by conversion of the cells to nitrogen fixation conditions (Moshiri et al., 1991). However the *c*-type cytochromes that pass electrons to cytochrome *o* appear to be more highly N-regulated in concert with *nif* genes than the other electron transport pathways or protective (FeSII-type) mechanisms (Moshiri et al., 1991). The Cyt c_4 and c_5 transcript upregulation that coincides with fixed N starvation could be due to a factor like the *Klebsiella* nitrogen assimilation control (NAC) protein that couples transcription of σ^{70} dependent promoters to regulation by the N-regulation system (Muse and Bender, 1998).

The support for the concept or dogma of respiratory protection of nitrogenase in *Azotobacter* species has recently been brought into question (Oelze, 2000). The hypothesis that has existed for 25 years is that O_2 consumption by the cell membrane protects *Azotobacter* nitrogenase from O_2 damage. One of the tenets of the concept is that as O_2 levels are increased, the respiratory system will become more uncoupled. This has been observed, but it is not always the case (Oelze, 2000). The arguments that contradict the importance of respiratory protection are based on the observations that the respiratory activity in *Azotobacter vinelandii* is controlled by the C/N ratio, (i.e. the consumption of carbon per N compound used). It was concluded (Oelze, 2000) that respiratory protection is only important at the lower end (below 70 μM) of physiological O_2 levels. Although the C/N ratio may be involved, there is considerable evidence that respiratory protection is an important mechanism, especially for *Azotobacter* species. This has been corroborated by studies on gene-directed mutants. That *Azotobacter* respiratory activity is stimulated under N_2 fixation conditions appears to be a physiological adaptation to O_2 stress (Robson, 1979). The up-regulation of terminal oxidase pathways (Rey and Maier, 1997) with onset of nitrogenase expression is one piece of evidence that respiratory pathways are closely correlated to N_2 fixation in aerobes.

From studies on chemostat-cultures of *A. vinelandii*, it was concluded that rather than O_2-consuming respiration as the factor to maintain nitrogenase activity in aerobic *Azotobacter vinelandii*, the ATP supply to nitrogenase is a key element (Linkerhagner and Oelze, 1997). It is important to note that N_2 fixation activities and not nitrogenase inactivation was measured, so that the above studies do not directly impact conclusions on respiratory protection. That ATP levels are also important is indisputable. The controlled growth conditions to influence nitrogenase activity are more likely a function of ATP and reductant supply to the enzyme rather than protection of nitrogenase from O_2 damage. In *Anabaena* PCC 7120 respiratory protection of nitrogenase may function in the heterocyst by specific oxidases designed for this purpose of providing ATP for N_2 fixation (Jones and Haselkorn, 2002).

In addition to terminal oxidases, other electron transport components on the reducing side of the oxidases may play protection roles. For example, an *A. vinelandii* mutant in a non-coupled NADH: ubiquinone reductase was severely impaired in respiratory activity (Bertsova et al., 2001). The mutant strain was able to grow fine at low aeration or in the presence of fixed N, but was impaired in N_2-dependent growth in high aeration conditions. The uncoupled oxidase was also shown to be the ubiquinone oxidoreductase solely responsible for NADPH oxidation. This would mean that catabolism of certain carbon sources that yield NADPH rather that NADH may be particularly useful substrates under N_2 fixing condition.

References

Appleby CA (1969a) Electron transport systems of *Rhizobium japonicum*. 1.Haemoprotein P 450, other CO-reactive pigments, cytochromes and oxidases in bacteroids from N_2-fixing root nodules. Biochim Biophys Acta 172: 71–87

Appleby CA (1969b) Electron transport systems of *Rhizobium japonicum*. 2. Rhizobium haemoglobins, cytochromes and oxidases in free-living (cultured) cells. Biochim Biophys Acta 172: 88–105

Appleby CA (1984) Leghemoglobin and *Rhizobium* respiration. Annu Rev Plant Physiol 35: 443–478

Armengaud J, Meyer C and Jouanneau Y (1994) Recombinant expression of the fdxD gene of *Rhodobacter capsulatus* and characterization of its product, a [2Fe-2S]. Biochem J 300: 413–418

Barquera B, Garcia-Horsman A and Escamilla JE (1991a) Cytochrome *d* expression and regulation pattern in free-living *Rhizobium phaseoli*. Arch Microbiol 155: 114–119

Barquera B, Garcia-Horsman A and Escamilla JE (1991b) An alternate non-cytochrome containing branch in the respiratory system of free-living *Rhizobium phaseoli*. Arch Microbiol 155: 428–435

Batut J and Boistard P (1994) Oxygen control in *Rhizobium*. Antonie van Leeuwenhoek 66: 129–150

Batut J, Daveran-Mingot ML, David M, Jacobs J, Garnerone AM and Kahn D (1989) *fixK*, a gene homologous with *fnr* and *crp* from *Escherichia coli* regulates nitrogen fixation genes both positively and negatively in *Rhizobium meliloti*. EMBO J 8: 1279–1286

Berman-Frank I, Lundgren P, Chen YB, Kupper H, Kolber Z, Bergman B and Falkowski P (2001) Segregation of nitrogen fixation and oxygenic photosynthesis in the marine cyanobacterium *Trichodesium*. Science 294: 1534–1537

Bergersen FJ and Turner GL (1975) Leghaemoglobin and the supply of O_2 to nitrogen-fixing root nodule bacteroids: Presence of two oxidase systems and ATP production at low free O_2 concentration. J Gen Microbiol 91: 345–354

Bergersen FJ and Turner GL (1980) Properties of terminal oxidase systems of bacteroids from root nodules of soybean and cowpea and of N2-fixing bacteria grown in continuous culture. J Gen Microbiol 118: 235–252

Bertsova YV, Bogachev AV and Skulachev VP (2001) Noncoupled NADH: Ubiquinone Oxidoreductase of *Azotobacter vinelandii* is required for diazotrophic growth at high oxygen concentrations. J Bacteriol 183: 6869–6874

Bott M, Bollinger M and Hennecke H (1990) Genetic analysis of the cytochrome *c-aa*$_3$ branch of the *Bradyrhizobium japonicum* respiratory chain. Mol Microbiol 4: 2147–2157

Bott M, Ritz D and Hennecke H (1991) The *Bradyrhizobium japonicum cycM* gene encodes a membrane-anchored homolog of mitochondrial cytochrome *c*. J Bacteriol 173: 6766–6772

Bott M, Preisig O and Hennecke H (1992) Genes for a second terminal oxidase in *Bradyrhizobium japonicum*. Arch Microbiol 158: 335–343

Bott M, Thony-Meyer L, Loferer H, Rossbach S, Tully RE, Keister D, Appleby CA and Hennecke H (1995) *Bradyrhizobium japonicum* cytochrome c_{550} is required for nitrate respiration but not for symbiotic nitrogen fixation. J Bacteriol 177: 2214–2217

Bulen WA and LeComte (1972) Nitrogenase complex and its components. In: San Pietro A (ed) Methods of Enzymology, XXIV, pp 456–470. Academic Press, New York

Burris RH (1991) Nitrogenases. J Biol Chem 266: 9339–9342

Chatelet C and Meyer J (1999) The [2Fe-2S] protein I (Shetna protein I) from *Azotobacter vinelandii* is homologous to the [2Fe-2S] ferredoxin from Clostridium pasteurianum. J Biol Inorg Chem 4: 311–317

Chauhan D and O'Brian MR (1993) *Bradyrhizobium japonicum* δ-aminolevulinic acid dehydratase is essential for symbiosis with soybean and contains a novel metal binding domain. J Bacteriol 175: 7222–7227

Chauhan D and O'Brian MR (1995) A mutant *Bradyrhizobium japonicum* δ-aminolevulinic acid dehydratase with an altered metal requirement functions in situ for tetrapyrrole synthesis in soybean root nodules. J Biol Chem 270: 19823–19827

Chauhan S and O'Brian MR (1997) Transcriptional regulation of delta-aminolevulinic acid dehydratase synthesis by oxygen in *Bradyrhizobium japonicum* and evidence for developmental control of the hemB gene. J Bacteriol 179: 3706–3710

Chen YB, Dominic B, Zani S, Mellon MT and Zehr JP (1999) Expression of photosynthesis genes in relation to nitrogen fixation in the diazotrophic filamentous nonheterocystous cyanobacterium *Trichodesmium* sp. IMS 101. Plant Mol Biol 41: 89–104

Delgado MJ, Bedmar EJ and Downie JA (1998) Genes involved in the formation and assembly of Rhizobial cytochromes and their role in symbiotic nitrogen fixation. Advances in Microbial Physiology 40: 191–231

Denison RF, Hunt S and Layzell DB (1992) Nitrogenase activity, nodule respiration and O_2 permeability following detopping of alfalfa and birdsfoot trefoil. Plant Physiol 98: 894–900

De Hollander JA and Stouthamer AH (1980) The electron transport chain of *Rhizobium trifolii*. Eur J Biochem 111: 473–478

De Lima M, Oresnik IJ, Fernando SM, Hunt S, Smith R, Turpin DH and Layzell DB (1994) The relationship between nodule adenylates and the regulation of nitrogenase activity by oxygen in soybean. Physiol Plant 91: 687–695

Dilworth MJ (1974) Dinitrogen fixation. Annu Rev Plant Physiol 22: 121–140

Evans HJ, Hanus FJ, Russell SA, Harker AR, Lambert GR and Dalton DA (1985) Biochemical characterization, evaluation, and genetics of H_2 recycling in Rhizobium. In: Ludden PW, Burris JE (eds) Nitrogen Fixation and CO_2 Metabolism, pp 3–12. Elsevier Science Publishing, Amsterdam

Fischer HM (1996) Environmental regulation of rhizobial symbiotic nitrogen fixation genes. Trend Microbiol 4: 317–320

Frustaci JM, Sangwan I and O'Brian MR (1991) Aerobic growth and respiration of [delta]-aminolevulinic acid synthase (hemA) mutant of *Bradyrhizobium japonicum*. J Bacteriol 173: 1145–1150

Fuchs G (1999) Diversity of Metabolic Pathways. In: Lengeler

JW, Drews G, Schlegel HG, (eds) Biology of the Prokaryotes, pp163–186. Blackwell Science, New York

Gabel C and Maier RJ (1990) Nucleotide sequence of the *coxA* gene encoding subunit I of cytochrome aa_3 of *Bradyrhizobium japonicum*. Nucleic Acids Res 18: 143–151

Gabel C and Maier RJ (1993) Oxygen-dependent transcriptional regulation of cytochrome aa_3 in *Bradyrhizobium japonicum*. J Bacteriol 175: 128–132

Gabel C, Bittenger MA and Maier RJ (1994) Cytochrome aa_3 gene regulation in members of the family Rhizobiaceae: Comparison of copper and oxygen effects in *Bradyrhizobium japonicum* and *Rhizobium tropici*. AEM 60: 141–148

Gallon JR (1992) Reconciling the incompatible: N_2 fixation and O_2. New Phytol 122: 571–609

Garcia-Horsman J, Barquera B, Rumbley J, Ma J and Gennis RB (1994) The superfamily of heme-copper respiratory oxidases. J Bacteriol 176: 5587–5600

Gaudu P, Touati D, Niviere V and Fontecave M (1994) The NAD(P)H: Flavin oxidoreductase from *E. coli* as a source of superoxide radicals. J Biol Chem 269: 8182–8188

Haaker H, Szafran M, Wassink H, Klerk H and Appels M (1996) Respiratory control determines respiration and nitrogenase activity of *Rhizobium leguminosarum* bacteroids. J Bacteriol 178: 4555–4562

Haddock BA and Jones CW (1977) Bacterial respiration. Bacteriol Rev 41: 47–99

Harris SJ and Silvester WB (1992) Oxygen controls the development of *Frankia* vesicles in continuous culture. New Phytologist 121: 43–48

Hill S (1988) How is nitrogenase regulated by oxygen? FEMS Microbiol Rev 34: 111–130

Hirsch AM, Lum MR and Downie JA (2001) What makes the rhizobia-legume symbiosis so special? Plant Physiol 127(4): 1484–1492

Hunt S, King BJ and Layzell DB (1989) Effects of gradual increases in oxygen on nodule activity in soybean. Plant Physiol 91: 315–321

Hunt S and Layzell DB (1993) Gas exchange of legume nodules and the regulation of nitrogenase activity. Annu Rev Plant Physiol Plant Mol Biol 44: 483–511

Imlay JA (1995) A metabolic enzyme that rapidly produces superoxide, fumarate reductase of *Escherichia coli*. J Biol Chem 270: 19767–19777

Jones KM and Haselkorn R (2002) Newly identified cytochrome *c* oxidase operon in the nitrogen-fixing cyanobacterium *Anabaena* sp. Strain PCC 7120 specifically induced in heterocysts. J Bacteriol 184: 2491–2499

Jouanneau Y, Meyer C, Naud I and Klipp W (1995) Characterization of an fdxN mutant of *Rhodobacter capsulatus* indicates that ferredoxin I serves as electron donor to nitrogenase. Biochim Biophys Acta 1232: 33–42

Jouanneau Y, Meyer C, Asso M, Guigliarelli B and Willison JC (2000) Characterization of a *nif*-regulated flavoprotein (FprA) from *Rhodobacter capsulatus*. Redox properties and molecular interaction with a [2Fe-2S] ferredoxin. Eur J Biochem 267: 780–787

Jung YS, Gao-Sheridan HS, Christiansen J, Dean DR and Burgess BK (1999) Purification and biophysical characterization of a new [2Fe-2S] ferredoxin from *Azotobacter vinelandii*, a putative [Fe-S] cluster assembly/repair protein. J Biol Chem 274: 32402–32410

Kaminiski PA, Kitts CL, Zimmerman Z and Ludwig RA (1996) *Azorhizobium caulinodans* uses both cytochrome *bd* (quinol) and cytochrome cbb_3 (cytochrome *c*) terminal oxidases for symbiotic N2 fixation. J Bacteriol 178: 5989–5994

Keefe RG and Maier RJ (1993) Purification and characterization of an O_2-utilizing cytochrome-*c* oxidase complex from *Bradyrhizobium japonicum* bacteroid membranes. Biochim Biophys Acta 1183: 91–104

Kelly MJS, Poole RK, Yates MG and Kennedy C (1990) Cloning and mutagenesis of genes encoding the cytochrome *bd* terminal oxidase complex in *Azotobacter vinelandii*: Mutants deficient in the cytochrome *d* complex are unable to fix nitrogen in air. J Bacteriol 172: 6010–6019

King BJ and Layzell DB (1991) Effect of increases in O_2 concentration during the argon-induced decline in nitrogenase activity in root nodules of soybean. Plant Physiol 96: 376–381

Kolonay JF Jr and Maier RJ (1997) Formation of pH and potential gradients by the reconstituted *Azotobacter vinelandii* cytochrome *bd* respiration protection oxidase. J Bacteriol 179: 3813–3817

Kolonay JF Jr, Moshiri F, Gennis RB, Kaysser TM and Maier RJ (1994) Purification and characterization of the cytochrome *bd* complex from *Azotobacter vinelandii* comparison to the complex from *Escherichia coli*. J Bacteriol 176: 4177–4181

Kuhla J and Oelze J (1988) Dependence of nitrogenase switch-off upon oxygen stress on the nitrogenase activity in *Azotobacter vinelandii*. J Bacteriol 170: 5325–5329

Kuzma MM, Winter H, Storer P, Oresnik I, Atkins CA and Layzell DB (1999) The Site of Oxygen Limitation in Soybean Nodules. Plant Physiol 119: 399–407

Laane C, Sherings G, Matz L, Haaker H and VanZeeland-Wolbers L (1980) Membrane energization and nitrogen fixation in *Azotobacter vinelandii* and *Rhizobium leguminosarum*. In: Nitrogen Fixation, Vol I, pp 111–137. University Press, Baltimore

Layzell DB, Hunt S and Palmer GR (1990) Mechanism of nitrogenase inhibition in soybean nodules. Pulse-modulated spectroscopy indicates that nitrogenase activity is limited by O_2. Plant Physiol 92: 1101–1107

Layzell DB, Diaz del Castillo L, Hunt S, Kuzma M, van Cauzenberhe O and Oresnik I (1993) The regulation of oxygen and its role in regulating nodule metabolism. In: Palacios R, Mora J and Newton WE (eds) New Horizons in Nitrogen Fixation, pp 393–398. Kluwer Academic Publishers, Dordrecht

Li Y, Green LS, Holtzappel R, Day DA and Bergersen FJ (2001) Supply of O_2 regulates demand for O_2 and uptake of malate by N2-fixing bacteroids from soybean nodules. Microbiology 147: 663–670

Linkerhagner K and Oelze J (1997) Nitrogenase activity and regeneration of the cellular ATP pool in *Azotobacter vinelandii* adapted to different oxygen concentrations. J Bacteriol 179: 1362–1367

Lopez O, Morera C, Miranda-Rios J, Girard L, Romero D and Soberon M (2001) Regulation of gene expression in response to oxygen in *Rhizobium etli*: role of *FnrN* in *fixNOQP* expression and is symbiotic nitrogen fixation. J Bacteriol 183: 6999–7006

Lou J, Moshiri F, Johnson MK, Lafferty ME, Sorkin DL, Miller A-F and Maier RJ (1999) Mutagenesis studies of the FeSII protein of *Azotobacter vinelandii*: Roles of histidine and lysine residues in the protection of nitrogenase from oxygen damage. Biochemistry 38: 5563–5571

Lundquist P-O (2000) Nitrogenase activity in *Alnus incana* root nodules, responses to O2 and short-term N_2 deprivation. Plant Physiology 122: 553–561

Maier RJ and Moshiri F (2000) Role of the *Azotobacter vinelandii* nitrogenase-protective shethna protein in preventing oxygen-mediated cell death. J Bacteriol 182: 3854–3857

Maier RJ and Triplett EW (1996) Toward more productive, efficient, and competitive nitrogen-fixing symbiotic bacteria. CRC Critical Reviews in Plant Sciences CRC Press 15: 191–234

Maier RJ, Moshiri F, Keefe RG and Gabel C (1990) Molecular analysis of terminal oxidases in electron transport pathways of *Bradyrhizobium japonicum* and *Azotobacter vinelandii*. In: Gresshoff PM, Roth LE, Stacey G and Newton WE (eds) Nitrogen Fixation: Achievements and Objectives, pp 301–308. Chapman and Hall, New York and London

Mancinelli RL (1996 The nature of nitrogen: an overview. Life Support Biosph Sci 3: 17–24

Marchal K, Sun J, Keijers V, Haaker H and Vanderleyden J (1998) A cytochrome cbb_3 (cytochrome *c*) terminal oxidase in *Azospirillum brasilense* Sp7 supports microaerobic growth. J Bacteriol 180: 5689–5696

Marroqui S, Zorreguieta A, Santamaria C, Temprano F, Soberon M, Megias M and Downie JA (2001) Enhanced symbiotic performance by *Rhizobium tropici* glycogen synthase mutants. J Bacteriol 183: 854–864

Masepohl B, Drepper T, Paschen A, Gross S, Pawlowski A, Raabe K, Riedel KU, and Klipp W (2002) Regulation of nitrogen fixation in the phototrophic purple bacterium *Rhodobacter capsulatus*. J Mol Microbiol Biotechnol 4: 243–248

Millar AH, Day DA and Bergersen FJ (1995) Microaerobic respiration and oxidative phosphorylation by soybean nodule mitochondria: Implications for nitrogen fixation. Plant Cell Environ 18: 715–726

Moshiri F, Smith EG, Taormino JP and Maier RJ (1991) Transcriptional regulation of cytochrome *d* in nitrogen-fixing *Azotobacter vinelandii*. Evidence that up-regulation during N_2 fixation is independent of *nifA* but dependent on *ntrA*. J Biol Chem 266: 23169–74

Moshiri F, Kim JW, Fu C and Maier RJ (1994) The FeSII protein of *Azotobacter vinelandii* is not essential for aerobic nitrogen fixation but confers significant protection to oxygen-mediated inactivation of nitrogenase in vitro and in vivo. Molec Microbiol 14: 101–114

Moshiri F, Crouse BR, Johnson MK and Maier RJ (1995) the FeSII protein of *Azotobacter vinelandii*: Overexpression, characterization and crystallization. Biochem 34: 12973–12982

Muse WB and Bender RA (1998) The *nac* (nitrogen assimilation control) gene from *Escherichia coli*. J Bacteriol 180: 1166–1173

Ng TCN, Laheri AN and Maier RJ (1995) Cloning, sequencing and mutagenesis of the cytochrome c_4 gene from *Azotobacter vinelandii*: Characterization of the mutant strain and a proposed new branch in the respiratory chain. Biochim Biophys Acta 1230: 119–129

O'Brian MR and Maier RJ (1983) Involvement of cytochromes and a flavoprotein in hydrogen oxidation in *Rhizobium japonicum* bacteroids. J Bacteriol 155: 481–487

O'Brian MR and Maier RJ (1989) Molecular aspects of the energetics of nitrogen fixation in *Rhizobium*-legume symbioses. Biochim Biophys Acta 974: 229–246

O'Brian MR, Kirshbom PM and Maier RJ (1987) Tn5-induiced cytochrome mutants of *Bradyrhizobium japonicum*: Effects of the mutations on cells grown symbiotically and in culture. J Bacteriol 169: 1089–1094

Oelze J (2000) Respiratory protection of nitrogenase in *Azotobacter* species: Is a widely held hypothesis unequivocally supported by experimental evidence? FEMS Microbiol Rev 24: 321–333

Paerl H (1990) Physiological ecology and regulation of N_2 fixation in natural water. Adv Microb Ecol 11: 305–344

Pan B and Vessey JK (2001) Response of the endophytic diazotroph *Gluconacetobacter diazotrophicus* on solid media to changes in atmospheric partial O_2 pressure. Appl Environ Microbiol 67: 4694–4700

Patterson TG, Petersen JB and LaRue TA (1983) Plant Physiol 70: 695–700

Poole P and Allaway D (2000) Carbon and nitrogen metabolism in *Rhizobium*. Adv Micro Physiol 43: 117–163

Preisig O, Anthamatten D and Hennecke H (1993) Genes for a microaerobically induced oxidase complex in *Bradyrhizobium japonicum* are essential for a nitrogen-fixing symbiosis. Proc Natl Acad Sci USA 90: 3309–3313

Preisig O, Zufferey R, Thöny-Meyer L, Appleby CA and Hennecke H (1996) A high-affinity cbb_3-type cytochrome oxidase terminates the symbiosis-specific respiratory chain of *Bradyrhizobium japonicum*. J Bacteriol 178(6): 1532–1538

Puppo A and Halliwell B (1988) Generation of hydroxyl radicals by soybean nodule leghaemoglobin. Planta 173: 405–410

Rawsthorne S and LaRue TA (1986) Preparation and properties of mitochondria from cowpea nodules. Plant Physiol 81: 1092–1096

Rey L and Maier RJ (1997) Cytochrome c terminal oxidase pathways of *Azotobacter vinelandii*: Analysis of cytochrome c_4 and c_5 mutants and up-regulation of cytochrome c-dependent pathways with N_2 fixation. J Bacteriol 179: 7191–7196

Robson RL (1979) Characterization of an oxygen-stable nitrogenase complex isolated from *Azotobacter chroococcum*. Biochem J 181: 569–575

Sabra W, Zeng A-P, Lunsdorf H and Deckwer W-D (2000) Effect of oxygen on formation and structure of *Azotobacter vinelandii* alginate and its role in protecting nitrogenase. AEM 66(9): 4037–4077

Santos R, Herouart D, Sigaud S, Touati D and Puppo A (2001) Oxidative burst in alfalfa-*Sinorhizobium meliloti* symbiotic interaction. Mol Plant Microbe Interact 14: 86–89

Segura D and Espin G (1998) Mutational Inactivation of a gene homologous to *Escherichia coli ptsP* affects poly-B-hydroxybutyrate accumulation and nitrogen fixation in *Azotobacter vinelandii*. J Bacteriol 180: 4790–4798

Scherings G, Haaker H, Wassink H and Veeger C (1983) On the formation of an oxygen-tolerant three component nitrogenase complex from *A vinelandii*. Eur J Biochem 135: 591–599

Schubert KR and Evans HJ (1976) Hydrogen evolution: A major factor affecting the efficiency of nitrogen fixation in nodulated symbionts. PNAS 73: 1207–1211

Shethna YI, DerVartanian DV and Beinert H (1968) Non heme (iron-sulphur) proteins of *Azotobacter vinelandii*. Biochem Biophys Res Comm 31: 862–867

Soberon M, Williams HD, Poole RK and Escamilla E (1989) Isolation of a *Rhizobium phaseoli* cytochrome mutant with enhanced respiration and symbiotic nitrogen fixation. J Bacteriol 171: 465–472

Stal LJ (1995) Physiological ecology of cyanobacteria in microbial mats and other communities. New Phytol 131: 1–32

Stam H, Van Verseveld WH, deVries W and Stouthamer AH (1984) Hydrogen oxidation and efficiency of nitrogen fixation in succinate-limited cultures of *Rhizobium* ORS571. Arch Microbiol 139: 53–60

Storz G and Imlay JA (1999) Oxidative stress. Curr Opinion Microbiol 2: 188–194

Surpin MA and Maier RJ (1998) Roles of the *Bradyrhizobium japonicum* terminal oxidase complexes in microaerobic H_2-dependent growth. BBA 1364: 37–45

Surpin MA and Maier RJ (1999) Symbiotic deficiencies associated with a coxWXYZ mutant of *Bradyrhizobium japonicum*. AEM 65: 339–341

Surpin MA, Moshiri F, Murphy AM and Maier RJ (1994) Genetic evidence for a fourth terminal oxidase in *Bradyrhizobium japonicum*. Gene 143: 73–77

Surpin MA, Lubben M and Maier RJ (1996) The *Bradyrhizobium japonicum coxWXYZ* gene cluster encodes a bb_3-type ubiquinol oxidase. Gene 183: 201–206

Thorneley RNF and Ashby GA (1989) Oxidation of nitrogenase iron protein by dioxygen without inactivation could contribute to high respiration rates of *Azotobacter* species and facilitate nitrogen fixation in other aerobic environments. Biochem J 261: 181–187

Tully RE, Sadowsky RJ and Keister DL (1991) Characterization of cytochrome c_{550} and c_{555} from *Bradyrhizobium japonicum*: cloning, mutagenesis, and sequencing of the c_{555} gene (*cycC*). J Bacteriol 173: 7887–7895

Van der Oost J, de Boer APN, de Grier JW, Zumft WG, Stouthamer AH and van Spanning RJM (1994) The haem-copper oxidase family consists of three distinct types of terminal oxidases and is related to nitric oxide reductase. FEMS Microbiol Lett 121: 1–10

Wang Z-C, Burns A and Watt GD (1985) Complex formation and O_2 sensitivity of *Azotobacter vinelandii* nitrogenase and its component proteins. Biochemistry 24: 214–221

Werner D (1992) Physiology of nitrogen fixing legume nodules: Compartments and function. In: Stacey G, Burris RH and Evans HJ (eds) Biological Nitrogen Fixation, pp 391–431. Chapman and Hall, New York

Williams HD, Appleby CA and Poole RK (1990) The unusual behavior of the putative terminal oxidases of *Bradyrhizobium japonicum* bacteroids revealed by low-temperature photodissociation studies. Biochim Biophys Acta 1019: 225–232

Wittenberg BA, Wittenberg JB, Appleby CA and Turner GL (1974) the role of leghemoglobin in nitrogen fixation by bacteroids isolated from soybean root nodules. B Biol Chem 249: 4057–4066

Witty JF and Minchin FR (1990) Oxygen diffusion in the legume root nodule. In: Nitrogen Fixation: Achievements and Objectives, pp 285–292. Chapman and Hall, New York

Wu G, Delgado MJ, Vargas C, Davies AE, Poole RK and Downie JA (1996) The cytochrome bc_1 complex but not *CycM* is necessary for symbiotic nitrogen fixation by *Rhizobium leguminosarum*. Microbiology 142: 3381–3388

Yates MG (1988) The role of oxygen and hydrogen in nitrogen fixation. In: Cole JA and Ferguson (eds) The Nitrogen and Sulphur Cycles, pp. 383–416. Cambridge University Press, United Kingdom

Yurgel SN, Soberon M, Sharypova LA, Miranda J, Morera C and Simarov BV (1998) Isolation of *Sinorhizobium meliloti* Tn5 mutants with altered cytochrome terminal oxidase expression and improved symbiotic performance. FEMS Microbiol Lett 165: 167–173

Yuvaniyama P, Agar JN, Cash VL, Johnson MK and Dean DR (2000) *NifS*-directed assembly of a transient [2Fe-2S] cluster within the NifU protein. Proc Natl Acad Sci 97(2): 599–604

Zehr JP and Ward BB (2002) Nitrogen cycling in the ocean: New perspectives in processes and paradigms. Appl Environ Microbiol 68: 1015–1024

Zehr JP, Harris D and Salerno DB (1997) Structural analysis of the Trichodesmium nitrogenase iron protein: Implications for aerobic nitrogen fixation activity. FEMS Microbiol Lett 153: 303–309

Zhulin IB, Johnson MS and Taylor BL (1997) How do bacteria avoid high oxygen concentrations? Biosci Rep 17(3): 335–342

Zufferey R, Preisig O, Hennecke H and Thöny-Meyer L (1996) Assembly and function of the cytochrome cbb_3 oxidase subunits in *Bradyrhizobium japonicum*. J Biol Chem 271: 9114–9119

Chapter 6

The Oxidation of Ammonia as an Energy Source in Bacteria

Alan B. Hooper*[1], David Arciero[1], David Bergmann[2] and Michael P. Hendrich[3]

[1]*Department of Biochemistry, Molecular Biology and Biophysics, University of Minnesota, St. Paul, MN 55108, U.S.A.* [2]*College of Arts and Sciences, Black Hills State University, Spearfish, SD 57799-9114 U.S.A.* [3]*Department of Chemistry, Carnegie Mellon University, Pittsburgh, PA 15213 U.S.A.*

Author for correspondence, email: hooper@cbs.umn.edu

Davide Zannoni (ed): Respiration in Archaea and Bacteria. Vol 2: Diversity of Prokaryotic Respiratory Systems, pp. 121–147.

Summary

This chapter deals with the oxidation of ammonia ($NH_3 + 1.5\ O_2 \rightarrow HNO_2 + H_2O$) as a source of reducing power in the chemolithotrophic bacterium *Nitrosomonas europaea*. Direct knowledge of the enzymes involved together with the sequence of the genome reveal core elements of a redox system unique to oxidation of ammonia to nitrite which feeds into a more traditional bacterial electron transport/terminal oxidase system. The apparently low stoichiometry of protons translocated per ammonia oxidized hints at the basis of the low growth yields of this bacterium. Remarkably, the putative complex of hydroxylamine oxidoreductase (HAO), cytochrome c_{554} (Cyt c_{554}) and the membrane cytochrome c_{M552} (Cyt c_{M552}), which catalyzes the oxidation of a molecule of hydroxylamine and transfer of four electrons to membrane ubiquinone would involve 16 c-hemes per catalytic site or 48 hemes for the hypothetical aggregate containing the trimeric HAO. The dehydrogenation catalyzed at the novel catalytic heme (heme P460) is unique by comparison with other known catalytic hemes which bind substrate to the iron; in all others electrons enter the system and reduce the substrate whereas the reverse is true with HAO. This mode of catalysis may be functionally related to a cross link which is found only in HAO; a covalent bond between a methyne carbon of heme P460 and a ring carbon of a peptide tyrosine. The dramatic crystal structures of HAO and Cyt c_{554} have provided insights into catalysis and electron transfer as well as illustrating evolutionary relationships which are not reflected in homology of amino acid sequence. Considering their relative spatial arrangement. the 4 hemes of Cyt c_{554} can be precisely superimposed with 4 of the hemes of HAO. Evidence suggests that they have a common ancestor and have preserved heme configurations even when sequence homology had been lost. The novel anaerobic oxidation of ammonia ($NH_3 + HNO_2 \rightarrow N_2 + H_2O$) by a planctomycete bacterium and the oxidation of ammonia to nitrite in heterotrophic bacteria are described more briefly.

Abbreviations: AMO – ammonia monooxygenase; ANAMMOX – anaerobic ammonia oxidation; CCP – cytochrome c peroxidase; Cu NiR – copper-containing nitrite reductase; Cyt – cytochrome; Cyt c_{M552} – membrane cytochrome c_{M552} of *Nitrosomonas;* Cyt aa_3 complex – heme aa_3-containing terminal oxidase; Cyt bc_1 complex – ubiquinol cytochrome *c* oxido reductase; EPR – electron paramagnetic resonance; HAO – hydroxylamine oxidoreductase; NHE – normal hydrogen electrode; NiR – penta-heme cytochrome *c*-type nitrite reductase; N-side – negatively-charged region, cytoplasmic side of the plasma membrane; ORF – open reading frame; P-side – positively charged region, periplasmic side of the plasma membrane

I. Introduction: Metabolism of *Nitrosomonas*

This chapter deals with the biochemistry of the oxidation of ammonia as a source of reducing power for energy transduction and biosynthesis in bacteria. It will focus on the aerobic oxidation of ammonia to nitrite ($NH_3 + 1.5\ O_2 \rightarrow HNO_2 + H_2O$) in the chemolithotrophic bacterium *Nitrosomonas euro-*

paea, since that process has been most extensively characterized. The bacterial anaerobic oxidation of ammonia (ANAMMOX) pathway ($NH_3 + HNO_2 \rightarrow N_2 + H_2O$) and the oxidation of ammonia to nitrite in heterotrophic bacteria will be reviewed briefly.

The obligate aerobic chemolithotrophic ammonia-oxidizing bacteria, thought to account for the greater fraction of ammonia oxidation in nature, are found in the β and γ subdivisions of the proteobacteria (Teske et al., 1994). The species in the β subdivision, including *Nitrosomonas europaea*, are closely related and found in marine and fresh waters and in soils. The γ subdivision is represented by the obligate marine form *Nitrosococcus oceanus*. Although unable to reduce nitrate to nitrite or N_2O to N_2, at low concentrations of O_2 these organisms catalyze dissimilatory reduction of nitrite to NO and NO to N_2O and may be a major source of these gases in nature (Goreau et al., 1980; Anderson and Levine, 1986; Conrad, 1996).

The oxidation of ammonia to nitrite by *Nitrosomonas* is the sole source of reductant for biosynthesis and for creation of a proton gradient for energy transduction. Growth occurs in media containing only NH_3 as reductant and N-source, CO_2 as C-source, SO_4^{2-} as S-source, PO_4^{3-} and metal ions. In laboratory culture the doubling time is approximately 12 hours. During growth the media is acidified by nitrous acid. The population density can approach 10^8 cells/ml in a chemostat with pH control (Logan et al., 1995). The genome of *N. europaea* has been sequenced (Chain et al., 2003) and reveals that the organism is highly specialized for an autotrophic mode of life requiring genes encoding the N-oxidizing systems as well as biogenesis of all cellular components from inorganic N-, C- and S- but almost none of the pathways for the oxidation of organic molecules. Based on information from the genome, fixation of CO_2 is by the Calvin-Benson cycle and the citric acid cycle is present. NH_3 is assimilated by way of glutamate dehydrogenase. *Nitrosomonas* is able to reductively assimilate nitrite-N but not nitrate-N into protein (Wallace and Nicholas, 1968). This review deals with pathways for the oxidation of ammonia and production of reducing power (NADH) and ATP for biosynthesis. Previous reviews focusing on the biochemistry of the oxidation of ammonia and related electron transfer and energy transduction in *Nitrosomonas* have appeared (Wood, 1986; Vannelli et al., 1996; Hooper et al., 1997; Whittaker et al., 2000).

II. Summary of the Oxidative Pathways: Ammonia Monooxygenase, Hydroxylamine Oxidoreductase and the Passage of Electrons to Ubiquinone

A. An Overview of the Process

Ammonia is oxidized to hydroxylamine ($NH_3 + 2e^- + O_2 + 2H^+ \rightarrow NH_2OH$) (Hollocher, 1981) by the enzyme ammonia monooxygenase, AMO. Hydroxylamine is oxidized to nitrite ($NH_2OH + H_2O \rightarrow HNO_2 + 4e^- + 4H^+$) (Andersson and Hooper, 1983) by the enzyme hydroxylamine oxidoreductase, HAO (Fig. 1). The oxidation of hydroxylamine may be thought of as a dehydrogenation of substrate since electrons (and protons) are the product. The reaction is the source of all electrons required for reductive reactions in the organism. For each turnover of HAO the equivalent of two electrons removed from hydroxylamine must be expended in the AMO reaction. The AMO reaction yields no electrons and there is no evidence to date that, as a membrane redox enzyme, it can contribute to a chemiosmotic gradient during turnover. Thus the role of AMO is the production of the primary energy-generating molecule, hydroxylamine. Only two of the four electrons from the turnover of a molecule of HAO are thus available for the reductive reactions of the cell; they might be called 'usable' electrons. Of the two theoretical 'usable' electrons, an estimated 0.35 (approximately two electrons every six turnovers) pass to NAD for use in the reductive reactions of biosynthesis (Wood, 1986). It is assumed that NADH dehydrogenase, which is found to be encoded in the genome (Chain et al., 2003), catalyzes energy-dependent 'reverse electron transfer' from ubiquinol to NAD. The remaining 1.65 electron-equivalents (the usable pair of electrons in the other five turnovers of HAO) will ultimately pass to terminal electron acceptors including only O_2, nitrite or NO (catalyzed by a terminal oxidase, nitrite-reductase or NO-reductase, respectively) unless they serve as reductant in detoxification of potentially toxic forms of O_2 such as H_2O_2, the assimilatory reduction of sulfate or nitrite or the (physiologically futile) oxygenation of organic molecules by AMO (see below).

The electron transfer reactions related to N-oxidation and energy transduction in *Nitrosomonas* consist of a pathway unique to ammonia-oxidizing nitrifiers in which electrons from hydroxylamine are fed into membrane ubiquinone and then either returned to

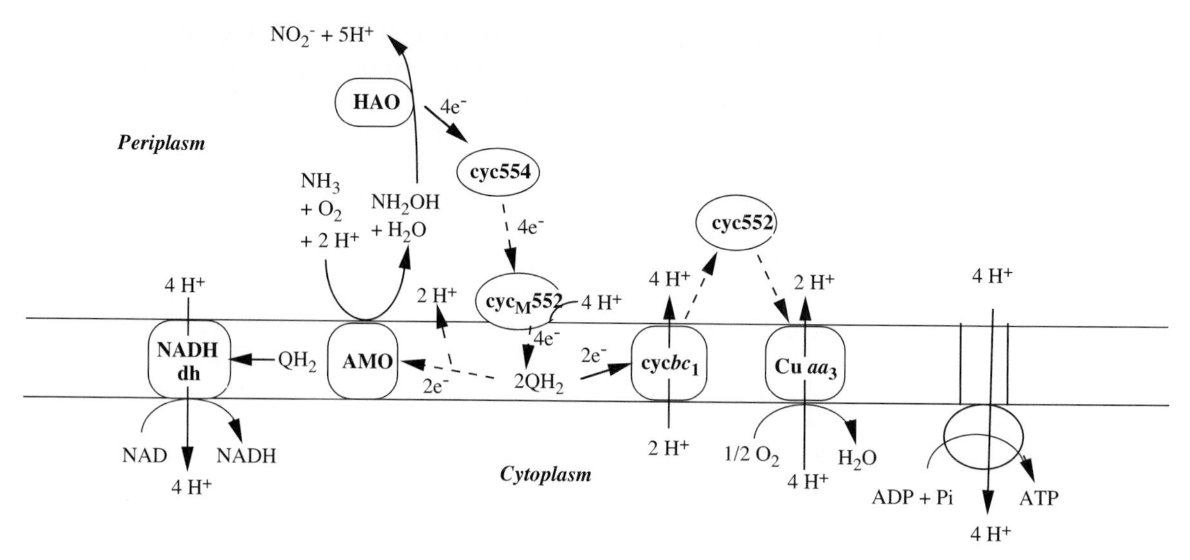

Fig. 1. The core electron-transport, proton translocating and energy transducing pathways of *Nitrosomonas*. Dashed lines indicate electron-transfer steps which have not been conclusively established by experiment.

AMO or channeled into a second, 'traditional', redox and energy transducing pathway starting with ubiquinol.

B. Membrane Systems of Nitrosomonas

Nitrosomonas is gram negative and has an extensive internal membrane system produced by invagination of the plasma membrane (Murray and Watson, 1965) giving rise to a large surface area of membrane and an extensive periplasmic space. It is likely that the enzymes of energy transduction in *Nitrosomonas* are in the internal regions of the membrane and periplasm (Hooper et al., 1984a; DiSpirito et al., 1985a) as is found with the analogous enzymes of methane oxidation in *Methylomicrobium album* (Brantner et al. 2002).

C. The N-oxidation Pathway

The majority of enzymes and electron-transfer carriers unique to N-oxidation in *Nitrosomonas* are now known and many have been extensively characterized. HAO is a trimer, each subunit contains eight *c*-hemes (hemes crosslinked to the protein by two thioether linkages to peptide cysteines). One of the *c*-hemes, the catalytic heme P460, has a third covalent bond; a crosslink to a tyrosine of the protein. Each HAO subunit has a binding site for its electron acceptor, the tetraheme cytochrome c_{554} (Cyt c_{554}). Electrons may then pass to the tetraheme membrane cytochrome c_{M552} (Cyt c_{M552}) and then to ubiquinone-8. The putative HAO-to-ubiquinone electron transfer pathway starting with a molecule of hydroxylamine in each of the three active sites of HAO includes 48 *c*-hemes and can, in theory reduce 12 molecules of ubiquinone to ubiquinol per turnover. It is thought that ubiquinol is the electron donor to AMO. Because it is not active in extracts, AMO has not been well characterized. Based on analysis of the sequence of a gene encoding a protein derivatized by acetylene (a mechanism-based inhibitor of AMO) and adjoining gene cluster, AMO appears to have three subunits. The translated proteins are predicted to have a large number of transmembrane domains. There are several copies of several of the genes involved in ammonia oxidation (McTavish et al.,1993b). The gene clusters containing AMO subunits, HAO, cytochrome c_{554} and cytochrome c_{M552} do not seem to have additional ORFs (Chain et al., 2003). Hence, all or most of the proteins involved with the oxidation of ammonia may have been identified.

D. The 'Traditional' Redox and Energy-Transducing Pathways of Nitrosomonas

The components of the 'traditional' or 'underlying' (Wood, 1986) energy-transducing pathway in *Nitrosomonas* have properties very similar to the equivalent enzymes of heterotrophic bacteria. As in many bacterial electron transport pathways, ubiquinol is a mobile electron carrier and a branching point to the AMO system, to the NAD reductase and to the terminal electron acceptor pathways.

1. Cytochrome bc_1

Genes for all proteins of a proton-dependent ubiquinol-cytochrome *c* reductase (cytochrome bc_1 complex) are found in a single gene cluster of the genome (Chain et al., 2003). The aerobic oxidation of either hydroxylamine or ammonia by cells of *Nitrosomonas* is inhibited by classic inhibitors of the ubiquinol-cytochrome c reductase such as antimycin A, myxothiazol, or stigmatellin indicating that the bc_1 functions 'downstream' from HAO in the electron transport pathway (Whittaker et al., 2000).

2. Cytochrome c_{552}

Cytochrome c_{552} (Yamanaka and Shinra, 1974) is a periplasmic mono-heme *c*-cytochrome which is homologous to the *c*-551 family of bacterial cytochromes and mitochondrial c_{550}. Its gene is present in a single copy which is not in any of the three gene clusters containing HAO, Cyt c_{554} and Cyt c_{M552}. Cyt c_{552} is thought to mediate electron transfer between the Cyt bc_1 complex and the aa_3 terminal oxidase (DiSpirito et al., 1986). Genetic or kinetic evidence has not established the role for cytochrome c_{552} although the value of reduction potential (+250 mV, Yamanaka and Shinra, 1974) is in keeping with this function. Cytochrome c_{552} is also able to rapidly reduce cytochrome *c* peroxidase (Arciero and Hooper, 1994) and slowly reduce the copper-containing nitrite reductase/oxidase (DiSpirito et al., 1985b) in vitro. Thus, distribution of electrons to alternate carriers may also be a function of Cyt c_{552}.

3. CuA Terminal Oxidase

A CuA-type cytochrome aa_3 oxidase has been isolated (DiSpirito et al., 1986) and genes encoding all subunits have been identified in the genome.

4. Control and Integration of Electron Transport Pathways

Protonophore uncouplers of oxidative phosphorylation such as 2,4 dinitrophenol or m-chlorocarbonylcyanidephenylhydrazone inhibit the oxidation of ammonia to nitrite and yet stimulate the oxidation of hydroxylamine to nitrite as catalyzed by intact cells (Hooper and Terry, 1973). This strongly suggests that, when the flux of electrons from ubiquinone through the Cyt bc_1 complex and the Cyt aa_3 oxidase complex is not retarded by the steady-state chemiosmotic gradient, all or most of electrons from ubiquinol go to the terminal oxidase pathway rather than to AMO (Whittaker et al., 2000). The results indicate that, in the absence of uncoupler, AMO catalyses the rate-limiting step in the conversion of ammonia to nitrite. The presence of artificial electron acceptors (e.g. phenazine methosulfate or methylene blue) (Hooper and Terry, 1973) also selectively inhibit ammonia oxidation and stimulate the rate of oxidation of hydroxylamine in whole cells. This effect is also probably caused by depriving AMO of a flux of electrons.

It will be interesting to know how the cell regulates NADH dehydrogenase activity. Given the large expenditure of energy required to reduce NAD (noted below) its re-oxidation by the electron transport chain and terminal oxidase would result in a net loss to the organism.

Genes are found which encode all enzymes of the citric acid cycle (Chain et al., 2003). No activity of α-ketoglutarate dehydrogenase was found in extracts (Hooper, 1969). This apparent absence was once related to a possible biochemical basis of obligate autotrophy which still allowed synthesis of amino acids dependent on the other enzymes of the cycle. It remains to be seen whether all genes are expressed. It is likely that membranes contain succinate dehydrogenase.

E. Minor Electron Transfer Pathways to Terminal Electron Acceptors

1. Oxidation of Alternate Substrates by AMO

In vivo, ammonia monooxygenase of *Nitrosomonas* is able to oxidize many organic compounds (Hooper et al., 1997). Note that two of the 'usable' electrons are used per turnover of AMO. There is no example where the product of the reaction can be utilized for energy transduction; hence this is an energetically unproductive activity Z(Fig. 2).

2. Alternate Oxidases

It is notable that *Nitrosomonas* has so few optional

terminal oxidases. Genes encoding no other members of the membrane-bound super-family of heme-copper respiratory oxidases, are found. By HPLC analysis heme *o* is not present. The absence of a quinol oxidase in this family suggests that the contribution to the proton gradient of the turnover of the Cyt bc_1 complex is very important. *Nitrosomonas* has a soluble, periplasmic copper protein which catalyses oxidation of dyes by oxygen or nitrite (Miller and Wood, 1983; DiSpirito et al., 1985b). Its possible role in reduction of dioxygen as a terminal electron acceptor is under investigation.

3. Denitrification

The NO and N_2O produced during oxidation of ammonia by *Nitrosomonas* at low concentrations of oxygen can account for 10% or more of the NH_3–N oxidized (Lipschultz et al., 1981). Production of NO or N_2O in vivo appears to occur by the reduction of nitrite produced by HAO rather than the incomplete oxidation of hydroxylamine by HAO (Poth and Focht, 1985; Hooper et al., 1990; Remde and Conrad, 1990). A soluble enzyme with nitrite reductase and dye oxidase activity has been isolated (DiSpirito et al., 1985 b). Interestingly, the gene for this blue copper oxidase/nitrite reductase is immediately upstream of genes for a monoheme *c*-cytochrome and a diheme *c*-cytochrome (both of which are also expressed and soluble) and then a *nir*K-type gene for a copper-containing nitrite reductase. The role of the proteins encoded by this '*nir*K' gene cluster is under investigation (Beaumont et al., 2002); current speculation is that both enzymes can contribute to denitrification. In keeping with physiological observations, genes encoding a NO reductase are in the genome whereas genes for nitrate- or N_2O- reductase are not (Chain et al., 2003).

4. Possible Bypass of the Cyt bc_1 Complex

In vivo, inhibition of the oxidation of ammonia or hydroxylamine by antimycin A, myxothiazol or stigmatellin is not complete. Hence, under certain circumstances, a small fraction of the flux of electrons from HAO can possibly bypass the Cyt bc_1 complex. Although the X-ray structure of cytochrome c_{554} shows a docking site for HAO but not for cytochrome c_{552} (Iverson et al., 1998), in vitro electron flow from hydroxylamine to cytochrome c_{552} is extremely rapid in the presence of catalytic amounts of HAO and cytochrome c_{554} (Yamanaka and Shinra, 1974). This could, in theory provide a bypass of the bc_1 complex. However the response of cells to inhibitors of the bc_1 complex indicates that this activity is not nearly as rapid in vivo as in vitro. A second candidate for a bypass of the Cyt bc_1 complex would be electron transfer through a cytochrome c_4-like protein which is present in membranes of *Nitrosomonas* at 5% on a molar basis relative to the core proteins of N-oxidation (Whittaker et al., 2000). Examples of cytochrome c_4 in other organisms are thought to participate in electron transfer from ubiquinol to cytochromes as part of a path leading to a terminal oxidase.

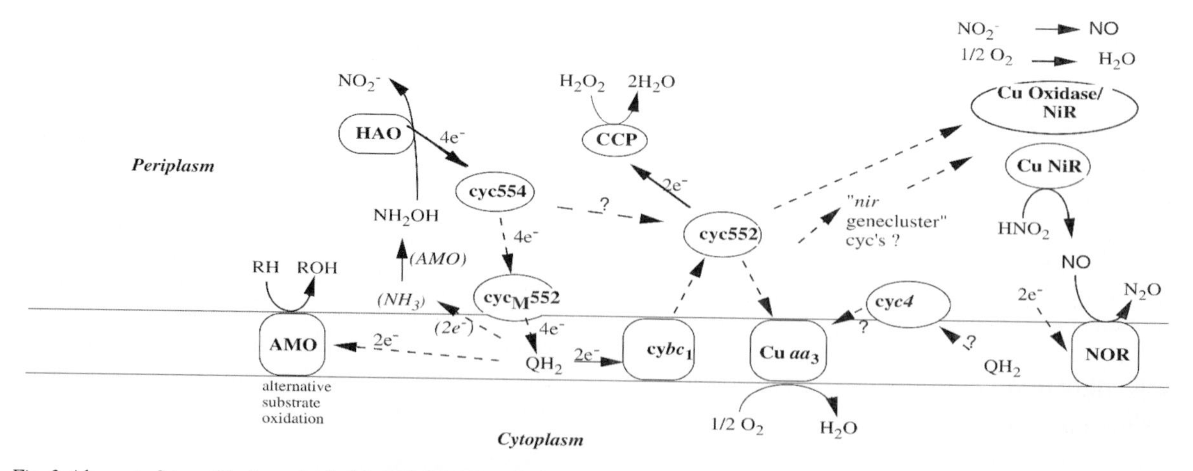

Fig. 2. Alternate fates of the hypothetical 'usable' two electrons from the HAO reaction. 'Usable' refers to the 2 of 4 electron-equivalents per catalytic cycle of HAO which are not obligatorily consumed by the AMO reaction. Dashed lines indicate electron-transfer steps which have not been conclusively established by experiment.

5. Molar Ratio of Redox Components

The relative amounts of electron transfer components are approximately 1.0 HAO subunit : 1.3 cytochrome c_{554}:0.6 cytochrome c_{M552}: 13 ubiquinone-8 : 0.78, Cyt bc_1 complex (together with NO reductase) : 3.5 cytochrome c_{552}:0.78 cytochrome aa_3 (Whittaker et al., 2000). It is intriguing that there are roughly equal amounts of cytochrome c_{554}, cytochrome c_{M552}, Cyt bc_1 complex and cytochrome aa_3 per subunit of HAO. In contrast the mobile electron carriers, membrane ubiquinone-8 and the soluble periplasmic cytochrome c_{552} are in considerable excess. The equal amounts of cytochrome c_{554} and cytochrome c_{M552} per subunit of HAO prompt speculation that these components might form a stoichiometric aggregate radiating from the HAO trimer.

F. Other redox proteins

1. Nitrosocyanin

Nitrosocyanin is a novel soluble, red, copper protein located in the periplasm of *Nitrosomonas* (Arciero et al., 2002). It is found in quantities equivalent to HAO and so is thought to possibly have a role in a central N-oxidation pathways. Based on its primary and tertiary structure it belongs to the cupredoxin family of redox proteins (Lieberman et al., 2001). However it differs from other cupredoxins in binding copper in a 'type 2' site having an open copper coordination site. Thus the possibility of either an electron transport or catalytic role is being considered.

2. Cytochrome P460

Cytochrome P460, a soluble, periplasmic monoheme *c*-cytochrome with hydroxylamine oxidizing activity is also found in small quantity (Erickson and Hooper, 1972; Numata et al., 1990; Bergmann and Hooper,1994b). It is characterized by an optical spectrum with a broad Soret band at 435 nm in the ferric form and 463 nm in the ferrous form. α and β absorption bands of typical *c*-hemes are not seen, whereas the heme P460 of HAO is a c-heme also covalently bound to a peptide tyrosine, the c-heme of cytochrome P-460 is cross-linked to a peptide lysine (Arciero and Hooper, 1997). Its physiological role is not known. Recent work suggests that its role is in the detoxification of NO or related compounds. Its sequence is homologous to that of a cytochrome P460 with similar optical properties and reactivity with N-oxides found in the methane oxidizing bacteria (Bergmann et al., 1998). Significantly, the latter also has great sequence homology with cytochrome c′ from the same organism, a protein whose role is thought to be sequestration of NO. Further, replacement of the crosslinked lysine with Arg, Ala or Tyr results in heterologously-expressed mutant protein with ligand binding ability, without catalytic activity and with optical absorption properties dissimilar to the wild type protein but resembling those of cytochrome *c*′ (Bergmann and Hooper, 2003).

3. Diheme Cytochrome c Peroxidase

Nitrosomonas contains a periplasmic diheme cytochrome *c* peroxidase which may scavenge hydrogen peroxide (Arciero and Hooper, 1994). Cyt c_{552} serves as an effective electron donor. The di-ferric form of the di-heme cytochrome *c* peroxidase of *Nitrosomonas* has the novel ability to react with H_2O_2 to form a stable ferryl-oxygen derivative (compound I) and H_2O. This contrasts with the more commonly observed type of diheme cytochrome *c* peroxidase from *P. aeruginosa* which requires reduction of the high potential electron-transfer heme before the catalytic heme becomes reactive with H_2O_2 and can then form compound I. Comparison of the crystal-derived structures of the two proteins (Fig.3) indicates that a histidine is the sixth axial ligand of the catalytic heme of the enzyme from *P. aeruginosa* whereas in the enzyme from *Nitrosomonas* the loop containing the homologous histidine is held away from the heme; thus providing access of peroxide to the reactive iron (Shimizu et al., 2001). This histidine-containing loop of the enzyme from *P. aeruginosa* is presumed to fold away upon reduction of the electron transfer heme to allow catalysis to commence. It is possible that this property of the enzyme from *Nitrosomonas* allows sequestration of H_2O_2 at relatively high steady-state redox potentials within the cell. Full reduction of the second oxygen and completion of turnover of CCP of *Nitrosomonas* still requires the entry of two additional electrons.

G. Proton Gradient

1. Generation of the Gradient

The stoichiometry of proton-gradient generated per electron-pair from hydroxylamine is hard to predict.

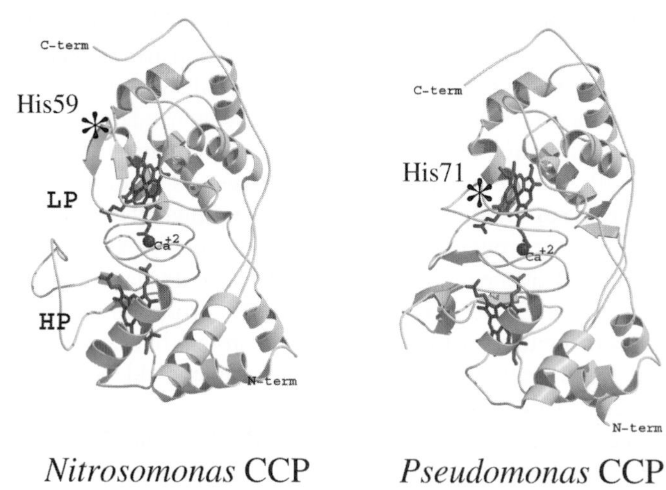

Fig. 3. Comparison of the structures of the resting, oxidized diheme cytochrome *c* peroxidases (CCP) of *Nitrosomonas europaea* and *Pseudomonas aeruginosa*. The two enzymes have similar domains enclosing a low potential (LP), catalytic c-heme and high potential (HP), electron-transfering *c*-heme. His 71* of the *Pseudomonas* enzyme is the sixth axial ligand of the ferric LP heme whereas the loop containing the homologous His 59* of the *Nitrosomonas* enzyme is relatively displaced resulting in an open catalytic site and penta-coordinate catalytic c-heme.

It is fair to assume that oxidation of ubiquinol by the cytochrome bc_1 complex and cytochrome oxidase would yield the net translocation of six protons from N to P side (Fig. 1).

The reaction of HAO releases four protons on the P side (two per ubiquinone reduced). The ionization of nitrite might be considered as contributing a fifth proton. However the acidification of the media during growth is approximately stoichiometric with nitrite production indicating that this should not be considered an energetically-useful proton; i.e. one which could be translocated to N-side during turnover of an enzyme such as ATP synthase. The net translocation of protons in the reduction of ubiquinone by electrons from hydroxylamine would depend on the membrane location of the site of reduction of- and uptake of protons by- ubiquinone. Cytochrome c_{M552} is attached to the P-side of the membrane by an N-terminal transmembrane α-helix. If the globular tetraheme region lies on the membrane surface and reduces ubiquinone during catalysis, 2 protons would be absorbed into membrane quinol on the P-side. In this case the net proton translocation during the reduction of ubiquinone by 2 electrons from hydroxylamine would be 0. Interestingly, this orientation of Cyt c_{M552} could allow the reduction of ubiquinone (–90 mV) by the +47 mV hemes of cytochrome c_{554} (Arciero et al., 1991b) to be facilitated by the steady-state transmembrane charge differential. If, on the other hand, cytochrome c_{M552} projects into the membrane far enough to reduce ubiquinone on the N-side, two N-side protons could be taken up. In that case, the reduction of one molecule of ubiquinone by two electrons from HAO would result in the equivalent of the translocation of two protons from N to P side.

It is not known whether net translocation of protons occurs in the AMO reaction. It depends on whether the release of two protons produced in the oxidation of ubiquinol (ubiquinol → ubiquinone + $2e^-$ + $2H^+$) and the uptake of the 2 substrate protons of the monooxygenase reaction ($NH_3 + O_2 + 2e^- + 2H^{+}$) take place on the P or N side. The large number of transmembrane α-helices predicted in AMO (see *AMO*, below) have the potential to provide the scaffolding to support a system for moving either protons or electrons across the membrane and could theoretically support several scenarios. Release of two protons on the P side and uptake of two protons on the N side would result in the equivalent of the translocation of two protons to the P side whereas the reverse orientation of the reactions would result in a net translocation of two protons to the N side. Interestingly, the latter orientation of reactions would allow the AMO reaction to be driven by the steady state transmembrane charge differential. If this were so it could account for the great sensitivity of the AMO reaction to uncouplers. If both the proton-yielding and proton utilizing reactions of AMO are on the same side of the membrane, the net translocation of protons would be zero.

Until more is known regarding the hydroxylamine-to-ubiquinone steps and the AMO reaction, one can only guess the number of protons which are translocated to the P side per transfer of four electrons from NH_2OH; two of which then reduce O_2 and two of which reduce AMO to replace the oxidized NH_2OH. The equivalent of six protons are translocated by Cyt bc_1 and Cyt aa_3 in the reoxidation of a molecule of ubiquinol. The high and low values of the possible contribution of the 4-electron hydroxylamine-to-ubiquinone step are +4 and 0 protons. The high and low values of the contribution of the two-electron AMO reaction are +2 and –2. Hence the theoretical overall yield ranges from +12 (+4 by HAO/quinone, +4 by Cyt bc_1/Cyt aa_3 and +4 by AMO) to +4 (0 by HAO/quinone, +6 Cyt bc_1 and Cyt aa_3 and –2 from AMO).

2. Utilization of the Gradient

a. ATP Synthase

ATP synthase is driven by a proton gradient which can be experimentally demonstrated during oxidation of hydroxylamine (Kumar and Nicholas, 1982, 1983). A steady-state value of Δp of 173 mV (inside negative) has been estimated for *Nitrosomonas* (Hollocher et al., 1982). All genes for a proton-dependent ATP synthase are present (Chain et al., 2003). It would presumably translocate approximately four protons to the N side per ATP formed. Note that this may correspond to all or more than the proton gradient yield of the oxidation of one molecule of ammonia.

b. Reduction of Pyridine Nucleotides

A gene cluster encodes all subunits of a proton-dependent NAD dehydrogenase (complex I) which is presumed to function predominantly or only as a NAD reductase in this organism. Interestingly, each turnover of this enzyme may require AMO and HAO to turnover twice; once to reduce ubiquinone (and then NAD) and once to replenish the 4 H^+ gradient equivalents utilized in the reduction of NAD by ubiquinol. Another gene cluster encodes a proton gradient-dependent NAD/NADP transhydrogenase, presumably for the generation of NADPH. Hence NADPH is very energetically expensive, possibly requiring three turnovers of the AMO/HAO system. The genome encodes a Na^+-dependent NAD reductase and Na^+/H^+ antiporter which might have roles in sodium symport or antiport.

III. Hydroxylamine Oxidoreductase

A. Overview of Reaction

The reaction catalyzed by HAO can be thought of as a dehydrogenation of hydroxylamine to nitrosonium,

$$NH_2OH \rightarrow NO^+ + 4\,e^- + 3\,H^+ \quad (1)$$

followed by hydrolysis to nitrous acid

$$NO^+ + H_2O \rightarrow HNO_2 + H^+ \quad (2)$$

HAO will also rapidly oxidize hydrazine to dinitrogen,

$$NH_2\,NH_2 \rightarrow N_2 + 4\,e^- + 4\,H^+ \quad (3)$$

The oxidation of hydrazine by HAO has been experimentally useful but has no known function in *Nitrosomonas*. The reaction as catalyzed by an HAO with similar catalytic and molecular properties has an important role in the bacterial anaerobic transformation of ammonium and nitrite to dinitrogen discussed later (Schalk et al., 2000). The turnover of the enzyme is commonly measured in the presence of an artificial electron acceptor such as phenazine methosulfate (catalytic in the presence of O_2), dichlorophenolindophenol, mammalian cytochrome c or cytochromes of *Nitrosomonas* (e.g. cytochrome c_{552} in the presence of catalytic amounts of cytochrome c_{554}, Yamanaka and Shinra, 1974).

Several elements of the possible catalytic mechanism may be suggested. Catalysis must involve selective binding of substrate. N-intermediates must be stabilized to prevent the release of such possible compounds as HNO or NO. Activation of substrate and N-H intermediates may involve conversion to electron rich forms by deprotonation. Water may also be activated by deprotonation. Electrons may be removed in single steps or pairs (possibly as a hydride) or a combination (e.g. as a hydride in one step followed by two one-electron steps). Electrons must be withdrawn and also effectively removed from the site of catalysis to high potential redox centers so as to prevent 'back-flow' of electrons to possible oxidants such as dioxygen or hydrogen peroxide or to prevent reversal of the N-oxidation reaction. Removal of protons could occur by way of an intra-protein relay or a channel. The product, nitrous acid, may exit through a channel or through separate nitrite channel and proton path.

Substrate binds with high affinity; HAO has a K_m for hydroxylamine or hydrazine in the low μM range (Hooper and Nason, 1965). The measured rate of transfer of electrons from substrate to hemes of the enzyme is at least 30 S^{-1} (Hooper et al., 1984b; Arciero et al., 1991a). Under some reaction conditions, the isolated enzyme produces some N_2O (possibly the product of the dimerization and hydrolysis of two nitroxyls, HNO); hence, the first step may be removal of two electrons or a hydride (Hooper, 1968). NO can also be a product indicating that the second stage can sometimes occur in vitro in single electron steps (Hooper et al., 1977; Hooper and Terry, 1979). NO will not serve as substrate for the production of nitrite (Hendrich et al., 2001), however NO will re-oxidize ferrous *c*-hemes Release of intermediates does not occur in vivo but can occur in vitro depending on the nature of the artificial electron acceptor and the assay conditions.

B. Overview of the Structure of HAO

1. Chemical Characterization

Purified HAO is a rich red in color due to its high content of heme (Hooper et al., 1978). In fact cells have a striking brownish red color due to the presence of large amounts of HAO and other *c*-type cytochromes. Several years of rigorous protein chemistry revealed HAO to be an oligomer in which each monomer contained seven *c*-hemes (each covalently attached by two thioether bonds from vinyl groups to cysteines) and an unusual c-heme derivative, heme P460. This was determined by proteolysis and isolation of all heme containing peptides (Arciero and Hooper, 1993). As revealed by analysis of an unusual crosslinked dipeptide (recognized by having two N-terminal amino acid residues), heme P460 has an additional covalent bond from a ring carbon of a peptide tyrosine to a methyne carbon of a heme (Arciero et al., 1993). Treatment of ferric HAO with hydrogen peroxide specifically and irreversibly alters the enzyme so that the 460 nm optical band is no longer seen in the ferrous enzyme. In addition the turnover of the enzyme and the reducibility of *c*-hemes by substrate is lost. Hence heme P460 is critical to catalysis (Hooper and Terry, 1977). All evidence indicates that, in fact, heme P460 binds to substrate. The sequence deduced from the cloned gene resulted in numbering of hemes # 1–8 from the N-terminus (Sayavedra-Soto et al., 1994).

The optical spectrum of ferrous HAO consists of a large Soret band, a shoulder at 460 nm of heme P460 and α- and β-bands at 553 and 559 nm due to c-hemes (Fig.4). The ferric spectrum contains only the broad Soret band and α- and β-bands of the *c*-hemes and a weak broad band at 740 nm due to heme P460 (Collins et al., 1993). Optical spectro-electrochemistry (Fig.4 inset) revealed the remarkable range of midpoint potentials (Table 1) to be +288 mV to −412 mV. The reduction potential of heme P460 decreased by 60 mV per pH unit, i.e. protonation of a moiety on the enzyme promotes uptake of electrons by the catalytic iron. The hemes with potentials at +288, +11 and −10 mV are reduced either at equilibrium in the presence of hydroxylamine or in the steady state in the presence of hydroxylamine and electron acceptors.

2. The Crystal Structure of HAO

The determination of the crystal structure of HAO has advanced the field remarkably (Igarashi, 1997). A side view of a monomer (Fig. 5A) reveals c-hemes concentrated in the lower portion of the molecule. The molecule is wide at the base and a spine of long α-helical domains rise at the back. Fig. 5B more clearly shows the arrangement in space of the 8 hemes of a monomer. In this view, heme 4 (heme P460 as identified by its cross-linking to tyrosine) is the highest heme. By their spacing the hemes of each subunit appear to group into four clusters; a trimer of hemes 4,6 and 7, dimers of hemes 3 and 5 and 1 and 2 and a monomeric heme 8 (Igarashi, et al., 1997). The trimer of HAO has a three clove 'garlic' shape. As seen from the side view (Fig.6), the hemes are concentrated in the wide bottom half of the molecule around the tower of long α-helices. A shallow cavity projects into the base of the trimer. The three catalytic heme P460s are at the top of the cloud of c-hemes in the HAO trimer. Remarkably, the heme crosslink of heme P460 is to a tyrosine in the adjacent subunit. Hence HAO is a covalently linked entity containing 24 hemes. From the top view (Fig. 7) the hemes appear in a ring with a heme 1 projecting radially from each subunit.

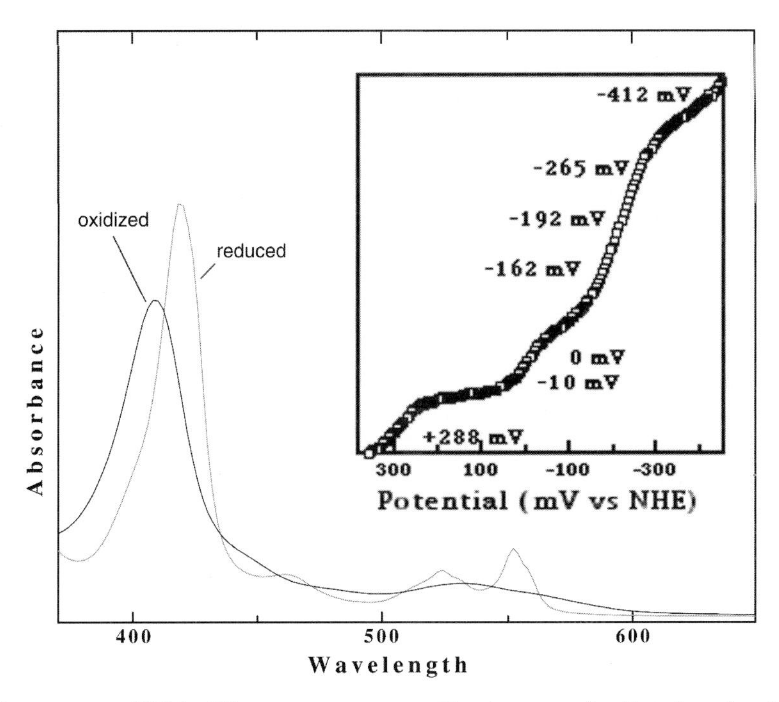

Fig. 4. Optical absorption spectrum of HAO of *Nitrosomonas europaea*. Absorbance as a function of wavelength. Darker line; oxidized enzyme. Lighter line; reduced enzyme. Inset; redox titration of the seven c-hemes of HAO (data for heme P460 not shown). Absorbance as a function of potential (□). Data based on Collins et al. (1993).

Table 1. Reduction potentials for the hemes of HAO (Collins et al., 1993)

$E'_{m7.0}$ (NHE)	+288	+11	–10	–162	–192	–260	–265	–412
Absorption Band	553	559	553	553	553	460	553	553

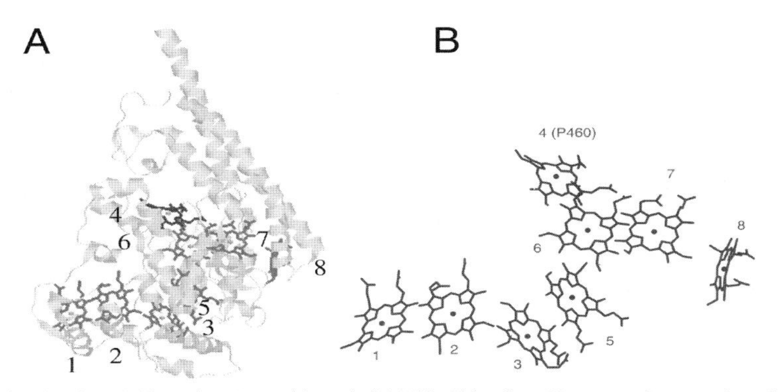

Fig. 5. Structure of a subunit of HAO [based on Igarashi et al. (1997)]. Side view. Heme numbers are based on the primary structure and begin at the N-terminal. A. hemes and ribbon view of protein. B. Hemes alone.

3. Possible Path of Electrons from Heme P460

Access to solvent by the hemes is limited to a substrate-accessible channel leading to an open axial position of the catalytic heme P 460 (heme 4), to the porphyrin edge of heme 1 (most distal from the center of the 24-heme cluster), and to a region in the base cavity near hemes 3 and 5. On this basis, heme 1 was predicted to be the point of exit of electrons from HAO (Igarashi et al., 1997). The irons of all hemes are within 10 Å from the nearest heme iron. Hence there is no apparent kinetic barrier to electron transfer.

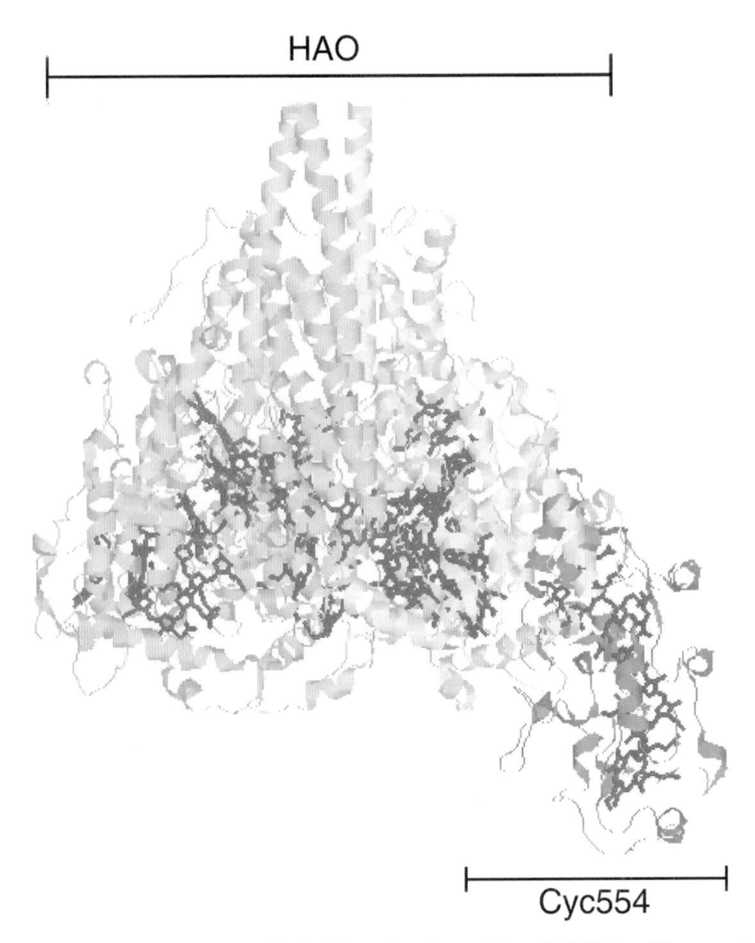

Fig. 6. Structure of trimer of HAO with one Cyt c_{554} docked [based on Iagarashi et al. (1997) and Iverson et al. (1998)]. Side view.

Hemes 8 are also close enough to hemes 2 of the adjacent subunit to allow rapid electron transfer. Hence, a low resistance electron path could begin at heme P460 of a given subunit and, at heme 6, bifurcate to a path through heme 1 of that subunit and a second path through heme 7 and 8 and hemes 2 and 1 of the adjacent subunit (interestingly, the subunit to which heme P460 is covalently bound) (Fig. 8).

4. Electronic and Redox Properties of the Hemes

The ferric enzyme analyzed by perpendicular mode EPR is seen to contain a variety of low spin axial bis-His heme irons with g values of approximately 3 and pairs of hemes exhibiting electronic exchange (Fig. 9). One of the electronically coupled pairs exhibits signals at g = 3.4, 2.7 and 1.66. An EPR signal for the second pair is not seen except by parallel mode EPR by which an integer-spin signal at g = 7.7 is seen (Hendrich et al., 1994). Reduction of the enzyme with substrate (Lipscomb and Hooper, 1982) or redox titration of the enzyme (Prince and Hooper, 1984) results in an array of shifting EPR signals reflecting interaction between hemes. An exhaustive series of studies have been carried in which EPR spectra were measured on poised samples of HAO. Measurement of EPR spectra at several frequencies and analysis of the heme geometries from the crystal structure allowed assignment of EPR properties and redox potentials to several hemes (Hendrich et al., 2001). The catalytic heme P460 (–260 mV) is exchange-coupled to the neighboring heme 6. The values of midpoint reduction potential of –260 mV and –190 mV were assigned to heme P460 and heme 6, respectively. These low values of midpoint potential are in keeping with the inability to experimentally observe reduction of heme P460 by substrate in the steady

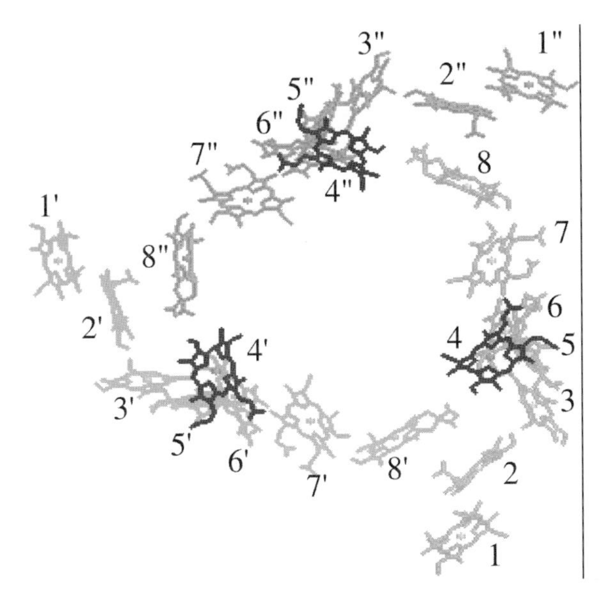

Fig. 7. Spatial arrangement of hemes of trimer of HAO [based on Igarashi et al. (1997)]. Top view. Hemes #4 (P460) of the three subunits are numbered 4, 4′ and 4″ and are darker.

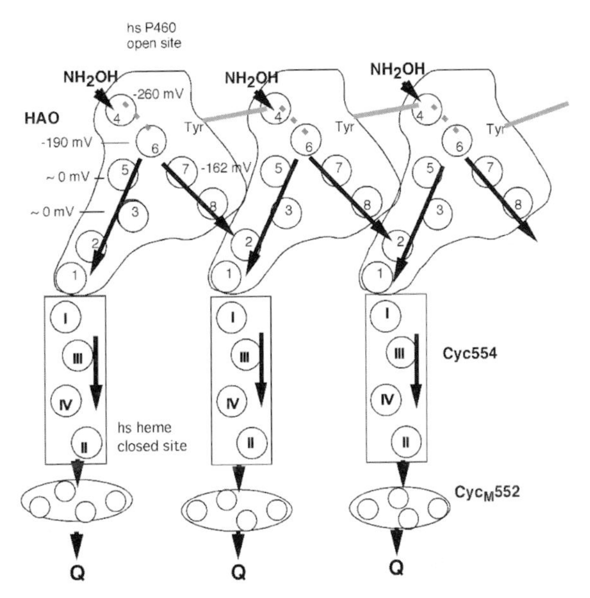

Fig. 8. Diagram of possible electron paths from the three active sites of the HAO trimer through hemes of HAO, Cyt c_{554} and Cyt c_{M552}. Hemes are shown as circles.

state or by stopped-flow measurements in the msec range (Hooper et al., 1984b). In the ferric enzyme the coupling between these two hemes can be disrupted by binding of substrate or cyanide to heme P460 (Hendrich et al., 1994). Cyanide is a ligand to heme P460 and an inhibitor of the enzyme activity (Table 2, Logan et al., 1995).

The two hemes with midpoint potentials near 0 are also exchange coupled and have been identified as hemes 3 and 5. Notably these are two of the hemes that are reduced during reaction of HAO with hydroxylamine or hydrazine. The electronic coupling between hemes in these two pairings implies that electrons equilibrate rapidly between the two heme irons. This property might make them especially suitable as 2-electron redox mediators or sinks. Heme 7 has a midpoint potential of –150 mV. The other three hemes, 1,2 and 8 have not yet been matched with the unassigned values of midpoint potential (+288, –162 and –412).

Interestingly, because the electron paths in the enzyme divide and rejoin (Fig.8), the 9 highest potential hemes of the HAO trimer (six hemes of ~ 0 mV and three of +288 mV) can, in theory, provide a reservoir for at least 4 electrons and thus allow complete turnover and release of nitrite once a substrate molecule is bound.

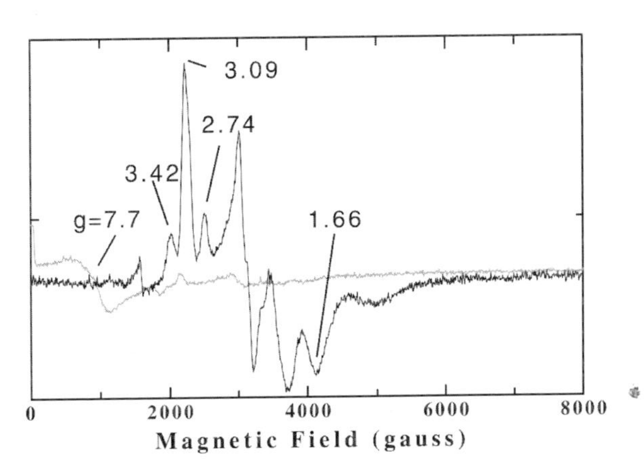

Fig. 9. EPR spectrum of HAO of *Nitrosomonas*. Conditions are as described in Hendrich et al. (1994, 2001). Darker line, perpendicular-mode EPR. Lighter line, parallel-mode EPR.

5. Speculation Regarding the Mechanism of Catalysis

a. Binding and Deprotonation of Substrate

Figure 10 shows the orientation of amino acid residues of the active site as determined by crystallography (Igarashi et al., 1997), modeled with a molecule of hydroxylamine bound by its nitrogen to the iron. Atoms of several residues project into the active

Table 2. Reactants with heme P460 of HAO

	Fe^{III}	reference	Fe^{II}	reference
Reductant/	NH_2OH	(Hooper and Nason, 1965)	O_2	(Hooper and Balny,1982)
Oxidant	N_2H_4	(Hooper and Nason, 1965)	H_2O_2	(Hooper et al., 1983)
	CH_3NHOH	(Ritchie and Nicholas, 1974)	HNO_2,	(Yamanaka and Shinra, 1974)
	NO	(Hendrich et al., 2002)	NO	(Hendrich et al., 2002)
Suicide-	phenyl-N_2H_4,	(Logan and Hooper, 1995)		
substrates	CH_3-N_2H_4,			
	HOC_2H_4-N_2H_4			
	H_2O_2	(Hooper and Terry, 1977)		
Ligand	CN^-	(Logan et al., 1995)		
	NO	(Hendrich et al., 2002)	CO	(Hooper et al., 1983)
Non-reactive	NH_2OCH_3	(Ritchie and Nicholas, 1974)		
compounds	N_3^-, SCN^-, OCN^-,	(Logan et al., 1995)		
	F^-,Cl^-,Br^-			

site within hydrogen bonding distances (2–3.3Å) of hydrogens of hydroxylamine and are thus potentially capable of facilitating selective binding of substrate, stabilization of intermediates and activation of the substrate by deprotonation. They include the crosslinked tyrosine 467 as well as histidine 268, aspartate 267 and tyrosine 334 which are above the substrate. H-bonds could extend from the hydroxyl-H of substrate to the epsilon N of histidine 268, from one substrate amino-H to the ring oxygen of tyrosine 467, from the other substrate amino-H to the carboxylate-O of aspartate 267. Hydroxylamine and hydrazine reduce the *c*-hemes of the enzyme. N-methylhydroxylamine will also reduce *c*-hemes of HAO but O-methyl-hydroxylamine is not reactive (Ritchie and Nicholas, 1974), emphasizing the possible importance of being able to form an oxyanion during oxidation of hydroxylamine.

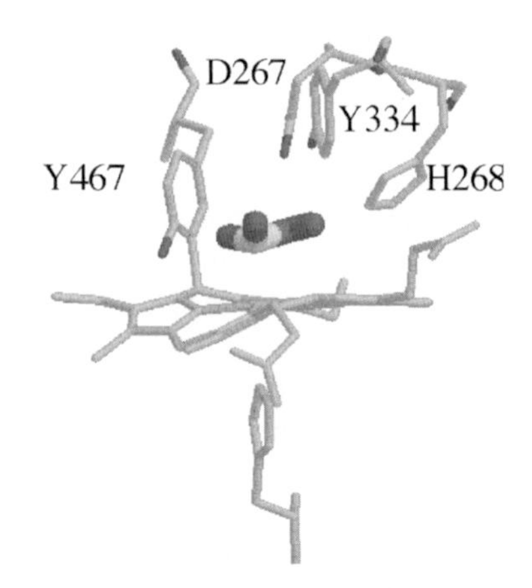

Fig. 10. Heme P460 of HAO showing nearby peptide residues and putative mode of binding of substrate. Data based on Igarashi et al. (1997). Shown are the crosslinked tyrosine 467 and three residues that project into the active site; aspartate 267, tyrosine 334 and histidine 268. Hydroxylamine is modeled as having an N-Fe bond to the heme. The axial histidine appears below the heme.

b. Electron Removal

The electronic coupling of the catalytic heme pair (4 and 6) make it a reasonable hypothesis that the reaction occurs in two-electron steps. In the first step, deprotonation followed either by removal of a hydride or a second deprotonation and removal of two electrons would produce iron-HNO. In the second step hydride transfer or deprotonation and removal of two electrons would produce iron-nitrosonium, (NO^+).

The low values of redox potential for the catalytic heme pair make it likely that electrons quickly pass to the 0 mV pair (hemes 3 and 5) in keeping with the inability to experimentally demonstrate reduction of heme P460 during catalysis.

c. The Reaction of HAO is Unique for a Catalytic Heme

In the reaction catalyzed by HAO, substrate first binds to the iron of a heme. In the course of the reaction electrons are removed from substrate and are moved out of the system through the other c-hemes of the protein. This is the only reaction of this type in nature. All other redox reactions of substrates bound to the iron of a heme involve transfer of electrons into the catalytic system and then to the iron-bound substrate (Table 3). This describes the reactions of the siroheme-containing nitrite or sulfite reductase, the cytochrome cd-nitrite reductase the penta-heme-nitrite reductase (*NiR)* and the nitric oxide reductase as well as reduction of oxygen by cytochrome oxidase or cytochrome *c* peroxidase. In the case of cytochrome oxidase, electrons may be thought of as entering the catalytic system and reducing one of the oxygen atoms to water. At the same time the iron and porphyrin ring donate electrons as the ferryl-oxygen intermediate is formed. The latter is reduced to water as two additional electrons enter the system. In the case of cytochrome P450 monooxygenase, electrons may also be thought of as entering the catalytic system and reducing one oxygen to water; electrons are also withdrawn from the iron and porphyrin ring as the ferryl-oxygen intermediate is formed. The latter decays as substrate is hydroxylated. In the reaction of catalase with a molecule of hydrogen peroxide; the porphyrin and iron donate electrons as the ferryl-oxygen intermediate and water are formed. This is followed by reaction of the oxygen with a second molecule of hydrogen peroxide.

Since the trait of electron withdrawal from substrate which is bound to the iron of heme P460 is so novel, we strongly suspect that it is related to the most novel aspect of the structure of HAO; the tyrosine cross-linked to the catalytic heme. This is reinforced by a comparison of HAO with the penta-*c*-heme nitrite reductase. In both enzymes catalysis takes place at the iron of a *c*-heme and electrons are either withdrawn to- (HAO) or donated by- (NiR) auxiliary c-hemes of the enzyme. Remarkably, although significant homology of amino acid sequence is not observed, the tertiary structure of subunits of HAO and NiR are very similar and the 5 hemes of NiR have exactly the same orientation in space relative to one another as hemes 4–8 of HAO (Einsle et al., 1999, 2000). The catalytic heme-iron of NiR differs from heme P460 in having a lysine rather than histidine as the axial ligand. Most notably, however, the catalytic heme of NiR lacks the tyrosine cross-link found in HAO. With this in mind, determination of the role of the tyrosine cross-link in HAO is a high priority for future work.

By showing that the plane of the tyrosine ring is perpendicular to the porphyrin plane (Fig. 7) rather than in the same plane, the crystal structure of HAO eliminated mechanisms dependent on a continuous conjugated π electron system between porphyrin and tyrosine rings. For example, one hypothetical mechanism had involved the oxidation of a Schiff's base which wuld have been previously formed by reaction of hydroxylamine with a carbonyl oxygen of tyrosine conjugated with the heme ring (Arciero and Hooper, 1998). In fact, the orientation of the tyrosine to porphyrin covalent bond indicates that the relevant porphyrin carbon is a tetrahedral center bonded to the two adjacent pyrrole carbons, the tyrosyl carbon and, presumably, hydrogen (i.e. the porphyrin ring could not be unsaturated at that point). This structure, a dihydroporphyrin, could theoretically participate in hydride removal from iron-bound hydroxylamine (Arciero and Hooper, 1998). In that model, the role of iron was the stabilization and activation of substrate rather than electron transfer. Other possible models which are under consideration call for a clearer determination of the nature of the tyrosyl-porphyrin structure (Arciero and Hooper, 1999).

d. Reactivity of the Active Site of HAO

Organohydrazines are suicide substrates of HAO (Logan and Hooper, 1995). It is hypothesized that organohydrazine is oxidized to dinitrogen and a radical cation which subsequently reacts with heme P460. As noted above, hydrogen peroxide reacts with and destroys heme P460 of HAO. Hydrogen peroxide is hypothesized to bind at the iron, undergo deprotonation and react with ferric heme P460 to form a ferryl-oxygen intermediate which may decay or dismutate to reactive forms of oxygen that open the porphyrin ring. This inactivation is apparently prevented in vivo.

Either O_2 or H_2O_2 are catalytically reduced to H_2O_2 or H_2O, respectively, at heme P460 of fully ferrous HAO or in the presence of an excess of reductant; i.e. the enzyme can act as an oxidase or peroxidase (Hooper and Balny, 1982; Hooper et al., 1983; Table 3). The reactions are rapid only during the oxidation of the low potential hemes of the enzyme. These reactions apparently do not occur in vivo and in fact, must be avoided. As expected for a ferrous heme which reacts with dioxygen, ferrous heme P460 binds CO causing the Soret band to shift to 450 nm. As determined by optical spectroscopy, ferric HAO binds cyanide but does not bind the anions azide, cyanate or thiocyanate, fluoride, chloride or Bromide (Logan et al., 1995). Cyanide inhibits enzyme activity. NO forms a derivative with ferric heme P460 in the resting enzyme and causes the reoxidation of ferrous hemes of the partially-reduced enzyme (Hendrich et al., 2002). Several lines of evidence indicate that ferrous HAO reduces nitrite. The stoichiometric conversion of hydroxylamine and nitrite to nitrous oxide in the presence of HAO possibly involves the intermediate production of HNO from nitrite (Hooper, 1968). Further, the reduced enzyme is reoxidized in the presence of nitrite (Yamanaka. et al., 1979; Schalk et al., 2000; Table 3).

IV. Transfer of Electrons from HAO to Ubiquinone

A. Cytochrome c_{554}

Electrons from HAO pass to Cyt c_{554} (Yamanaka and Shinra,1974). The c-hemes of this cytochrome have redox potentials of (E_m' vs. NHE) + 47, + 47, –147, and –276 mV (Arciero et al., 1991b). During turnover with HAO, the two hemes with midpoint potentials of +47 mV are reduced (Arciero et al., 1991 b). The magnetic properties of the 4 c-hemes have been partially characterized and exhibit` electronic interactions similar to those observed in HAO (Andersson et al., 1986). Most of Cyt c_{554} is soluble in extracts. However a significant fraction of Cyt c_{554} is membrane-associated and can be washed off with high concentrations of NaCl. The dissociated Cyt c_{554} re-binds in approximately stoichiometric amounts (McTavish et al., 1995). The binding capabilities of Cyt c_{554} in the initial supernatant, the salt wash and highest level of purification are the same. The amino acid sequence (Bergmann et al., 1994; Hommes et al., 1994) and the X-ray structure have been completed (Iverson et al., 1998, 2001a). A side view of the X-ray structure is shown in Fig. 11. Hemes are numbered I–IV starting from the N-terminus. Reduction potentials and spectroscopic properties and heme-heme interactions have been assigned to the 4 hemes of the protein (Upadhyay et al., 2003). The distribution of surface charge and the surface geometry of HAO and cyc-554 reveal a reasonable docking site which brings the exposed edge of heme 1 of HAO within 8 Å of the solvent exposed edge of heme I of Cyt c_{554} (Fig. 6) (Iverson et al., 1998). The

Table 3. Redox catalysis by hemes

Reduction of Fe-bound substrate		
Cyt *cd* nitrite reductase	$2H^+ + 2e^- + HNO_2$	$\rightarrow NO + H_2O$
Siroheme sulfite/nitrite reductase	$6H^+ + 6e^- + SO_3^=$	$\rightarrow S^= + 3H_2O$
	$6H^+ + 6e^- + HNO_2$	$\rightarrow NH_3 + 2H_2O$
Cyt c_{552} pentaheme nitrite recuctase	$6H^+ + 6e^- + HNO_2$	$\rightarrow NH_3 + 2H_2O$
NO reductase	$2H^+ + 2e^- + 2NO$	$\rightarrow N_2O + H_2O$
Cyt *c* oxidase	$2H^+ + 2e^- + Fe^{II} + O^2$	$\rightarrow Fe^{IV}{=}O + H_2O$
	$2H^+ + 2e^- + Fe^{IV}{=}O$	$\rightarrow Fe^{III} + H_2O$
Cyt *c* peroxidase	$Fe^{II} + H_2O_2$	$\rightarrow Fe^{IV}{=}O + H_2O$
	$2H^+ + 2e^- + Fe^{IV}{=}O$	$\rightarrow Fe^{III} + H_2O$
Cyt P450	$2H^+ + 2e^- + Fe^{II} + O_2$	$\rightarrow Fe^{IV}{=}O + H_2O$
	$XH_2 + Fe^{IV}{=}O$	$\rightarrow Fe^{III} + XHOH$
Catalase	$Fe^{III} + H_2O_2$	$\rightarrow Fe^{IV}{=}O + H_2O$
	$Fe^{IV}{=}O + H_2O$	$\rightarrow Fe^{III} + H_2O + O_2$
Oxidation of Fe-bound substrate		
Hydroxylamine oxidoreductase	$NH_2OH + H_2O$	$\rightarrow HNO_2 + 4e^- + 4H^+$

iron-iron distance between the two hemes is 20 Å. Within Cyt c_{554}, a possible electron transport chain through hemes I,III,IV and II would have iron-iron distances of 9–12 Å and allow rapid equilibrium of electrons among the hemes of Cyt c_{554} (and bound HAO). In the steady state or at equilibrium, electrons could accumulate in the two +47 mV hemes of Cyt c_{554} as well as the ~ 0 mV and +288mV hemes of HAO. Recalling the apparent capacity for electron transport between subunits, a single active site of HAO in a fully occupied [HAOCyt c_{554}]complex theoretically has an electron reservoir of 9 high potential c-hemes on HAO and 6 on Cyt c_{554} (Fig. 8). The utility of this arrangement may be the assurance of complete turnover once a substrate molecule has bound.

B. Cytochrome c_{M552}: A Putative Cytochrome c_{554}-Ubiquinone Reductase

In a gene cluster containing HAO (Sayavedra-Soto et al., 1994), the gene for Cyt c_{554} shares an operon and is co-transcribed with the gene encoding the membrane tetraheme Cyt c_{M552} (Bergmann and Hooper, 1994). The N-terminal sequence of the purified detergent-solubilized protein is the only transmembrane domain (Vannelli et al., 1996). Based on its gene sequence, this protein is a member of the 'NirT/NapC' family of tetraheme membrane-anchored cytochromes which are implicated in electron transfer from membrane quinols to periplasmic terminal electron acceptors. This family is exemplified by NirT, NapC and NrfH which direct electrons to the Cu-containing nitrite reductase (Jungst et al., 1991), the nitrate reductase (Roldan et al., 1998) or the penta-heme nitrite reductase (Simon et al 2000), respectively. Electron transfer from quinol to the appropriate periplasmic cytochrome has been experimentally demonstrated with NrfH. In all the members of this family, the subsequent periplasmic electron transfer step is mediated by a soluble periplasmic *c*-cytochrome or a fifth *c*-heme in an additional very large C-terminal domain of the quinol dehydrogenase.

It is proposed that Cyt c_{M552} of *Nitrosomonas* transfers electrons from the periplasmic oxidation of hydroxylamine by HAO/c_{554} to ubiquinone in the membrane. Although members of the NirT/NapC family have similar large central amino acid sequence domains containing the four *c*-heme binding motifs, they differ very significantly in the sequence of C-terminal domains. Since the family of proteins are all thought to react with membrane quinones, the variable

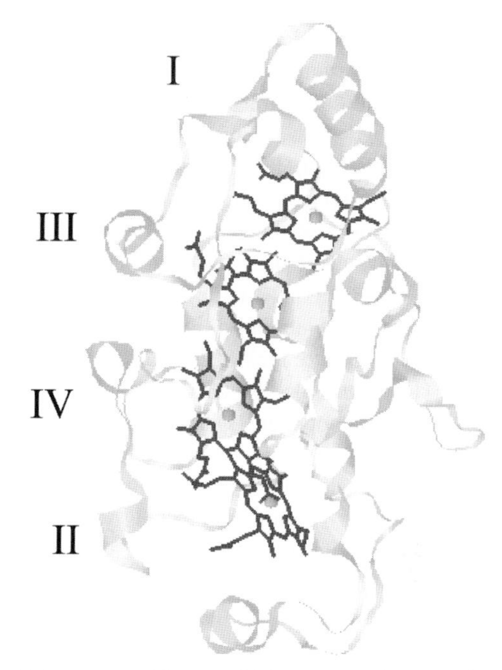

Fig. 11. Structure of Cyt c_{554} of *Nitrosomonas* [based on Iverson et al. (1998)]. Side view. The protein is arranged as if prepared to dock at the left side of the HAO trimer of Fig. 6. Heme numbers are based on the primary structure and begin at the N-terminal.

C-terminal domains may play a role in the specificity of periplasmic redox partner. The C-terminal domain of the *Nitrosomonas* Cyt c_{M552} is remarkably acidic; 24 of the 60 amino acids are aspartate or glutamate.

If the enzymatic role of Cyt c_{M552} is confirmed, the transfer of electrons from an HAO-bound NH_2OH to ubiquinone will contain 16 hemes per HAO subunit or 48 in the hypothetical fully-occupied [HAOCyt c_{554}Cyt c_{M552}] complex (as in Fig.8). Obviously one asks why so many hemes are required by this process. The answer might include rapid and irreversible removal of electrons to a safe distance.

V. Speculation on the Evolution of the Hydroxylamine-Oxidizing System

During analysis of the crystal structure of Cyt c_{554}, pairs of hemes within HAO and Cyt c_{554} were seen to have the same spatial orientation relative to one another; i.e. the pairs from the two proteins could be superimposed with high precision and were thus recognized as diheme 3-dimensional motifs (Iverson et al., 1998, 2001b). The motifs were seen to occur in several other multiheme cytochromes which had an electron-transfer role in the oxidation or reduction of inorganic nitrogen or sulfur compounds. Most dramatically, the entire tetraheme motif of Cyt c_{554} (hemes II, IV, III, I) was precisely superimposable on hemes 4,6, 5, 3 of HAO. This can be visualized by lifting the collective hemes of Cyt c_{554} from the protein in its docking site with HAO, turning them end-to-end and setting them on the 4 hemes of HAO (Fig. 12). The similarity extends to much of the secondary structure surrounding the hemes in spite of an almost complete lack of homology in primary structure (Iverson et al., 200lb). Given the functional interrelationship between these proteins and their presence in the same genome, the hypothesis of common ancestry is simpler and perhaps more likely than convergent evolution. Interestingly, heme II of Cyt c_{554} (which is homologous to the catalytic heme P460 (heme 4) of HAO as revealed in the alignment of the tetraheme motif) has the unusual property of being pentacoordinate and yet being protected from access to solvent by a protein fold. This property is correlated with its lack of reactivity with ligands and the seemingly paradoxical possession of a Soret optical band characteristic of a high spin EPR center. Cyt c_{554} could, in theory, have evolved from a primitive HAO by gene duplication followed by the loss of heme-binding CXYCH motifs and some of the protein domains for hemes 1,2, and 6, 7,8. This would have been accompanied by acquisition of a protein fold covering heme II, and loss of surface properties related to the trimer structure and the tyrosine crosslink. Alternatively the more primitive molecule may have been an enzyme resembling Cyt c_{554} which possessed a catalytic site open to solvent. This ancestral enzyme could have evolved to HAO and the electron transport mediator, Cyt c_{554} following gene duplication.

The completion of the crystal structure of the pentaheme cytochrome c-nitrite reductase from several species added another complex multiheme c-cytochrome to the family. It shares its 5-heme motif with hemes 4–8 of HAO and 3 hemes (#II–IV) of Cyt c_{554} (Einsle et al., 1999, 2000). These relationships are diagramed in Fig. 13. Remarkably, the three catalytic or pentacoordinate hemes (heme P460 of HAO, heme 1 of NiR and heme II of Cyt c_{554}) are homologous by this criterion of orientation in space. The relationship between HAO and NiR is especially interesting since the reactions involve the six electron reduction of nitrite by NiR or the production of nitrite by the extraction of four electrons from substrate by HAO, repectively. The subunit size, amino acid composition (Hooper et al., 1990) and much of the secondary structure is similar in the two enzymes though they have little homologous sequence. Further, NiR is a dimer and lacks the inter-molecular tyrosine crosslink to the catalytic heme. In fact steric considerations might make formation of two such crosslinks in a dimer of HAO subunits impossible. If so the trimeric HAO might be the smallest oligomer allow-

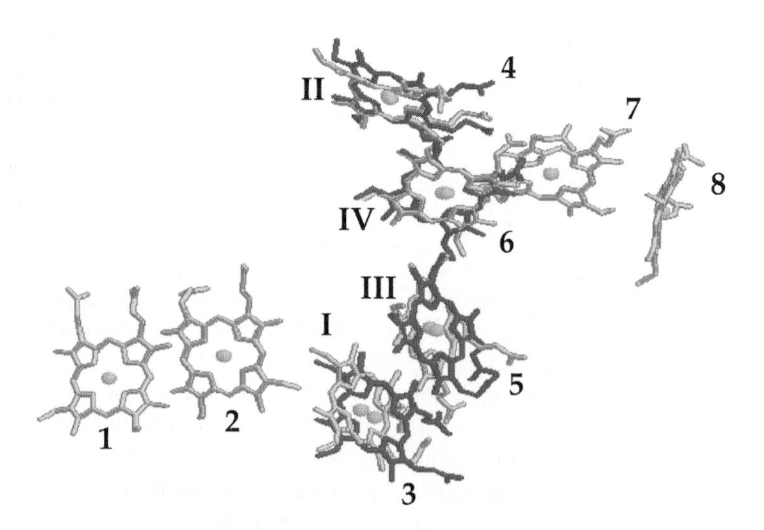

Fig. 12. Alignment of hemes of a subunit of HAO (as in Fig. 5B) and the hemes of Cyt c_{554} (as rotated 180° in the plane of the page from their arrangement in Fig.11). Data of Iverson et al. (1998, 2001a,b).

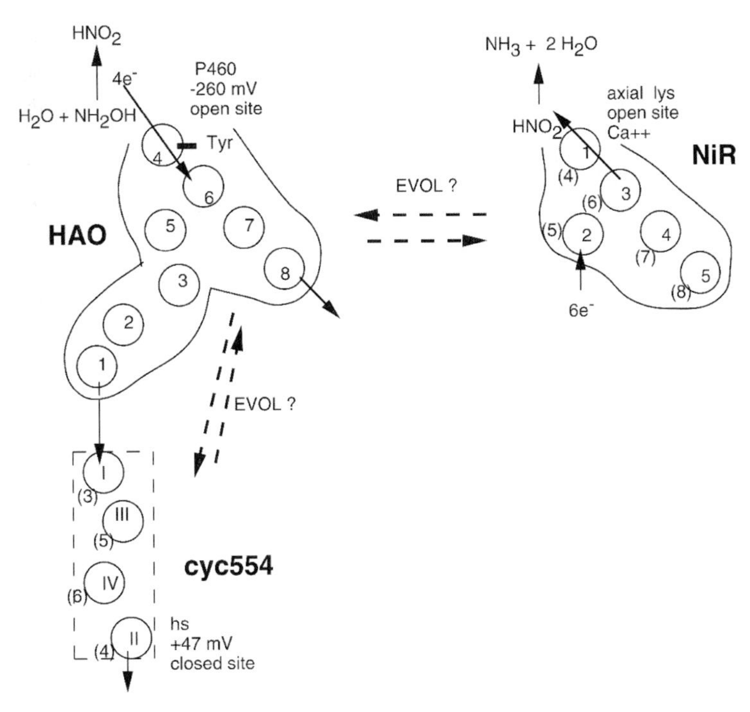

Fig. 13. Hypothetical evolutionary relationships between Cyt c_{554}, HAO and NiR based on 3D heme motifs. Based on data of Igarashi et al. (1997), Iverson, et al. (1998, 2001a,b) and Einsle et al (1999, 2000). The similarity of heme spatial arrangements for HAO and Cyt c_{554} is illustrated in Fig. 12. Numbers in parentheses are homologous hemes of HAO as categorized by orientation in space relative to other *c*-hemes in the same molecule.

ing equivalent crosslinks to all three catalytic hemes. Heme 1 of NiR is the first example of a c-heme with a –CXYCK- motif with the lysine as axial ligand to the iron. In the dimer of NiR, subunits associate at the base near heme 5 (equivalent to heme 8 of HAO). Heme 2 of NiR, which is proposed to be the point of *entry* of electrons into the enzyme (Einsle et al., 1999), corresponds to heme 5 of HAO. Interestingly, heme 5 in HAO is the apparent electron donor to hemes 3, 2 and 1 (whose homologues are not found in NiR) of the *exit* path from HAO. As with HAO and Cyt c_{554}, HAO and NiR could relatively easily have arisen from a common ancestor. This would require either acquisition or loss of the binding sites for hemes 1,2 and 3 of HAO and the heme crosslink and tuning of the catalytic site for deprotonation and oxidation or protonation and reduction, respectively, of N-oxides.

The evolutionary relationships between genes for these proteins are subject for future speculation. Looking at 3 examples of members of this family it is interesting to note that, because the order of spatially homologous hemes are co-linear with the amino acid sequence (Table 4), most evolutionary transformations suggested can occur with simple mutation of a heme ligation site and/or deletion/addition of terminal domains and do not require excision or insertion of domains internal to the gene. Importantly, the lack of sequence homology indicates that these are old enzymes which diverged from common ancestry long ago.

Caution must accompany the hypothesis that these particular heme motifs are evolutionarily related. It has been pointed out that, given the two thioether heme-to-peptide linkages, the histidine axial iron ligation from the –CXYCH- primary sequence motif and axial ligation of each heme with a second His, steric hindrance allows few orientations of two hemes if packed within a hydrophobic environment such that the hemes are less than 14Å apart (Brige et al., 2002). Further, the observed orientations within the diheme motifs of the characterized cytochromes are the one's predicted from these steric considerations.

Table 4. Order of hemes in primary sequence of Nir, HAO and Cyt c_{554}

NiR	1	2	3	4	5			
HAO	1	2	3	4	5	6	7	8
Cyt c_{554}	I	II	III	IV				

Numbers in rows refer to order of heme covalent binding motifs in primary structure of protein. Hemes in the same column are homologous based on the criterion of spatial orientation relative to other hemes in the same protein. Data: pentaheme nitrite reductase (Darwin et al., 1993); HAO (Sayavedra-Soto et al., 1994) Cyt c_{554} (Bergmann et al., 1994a; Hommes et al., 1994).

VI. Enzymology of Ammonia Monooxygenase

A. Structure of AMO Subunits

The reaction-based inactivator, acetylene, labels a 28 kDa membrane protein seen in denaturing gels (Hyman and Wood, 1985). An apparent complex consisting of the acetylene-derivatized protein (AMO-A) and a 40 kDa protein, AMO-B, was isolated. Based on the N-termini, an operon containing the genes *amo*A and *amo*B for the putative subunits AMO-A and –B was cloned and sequenced (McTavish et al 1993a; Bergmann and Hooper, 1994b). Homologous genes for subunits of the particulate methane monooxygenase (pMMO) were subsequently found in *Methylococcus capsulatus*, Bath (Nguyen et al., 1996) together with a third upstream gene, *pmmo*C. A homologous gene, *amo*C, was then found in the ammonia-oxidizing nitrifiers 200 b upstream from *amo*B (Klotz et al.,1997). There are no other open reading frames in the *amo* gene cluster of *Nitrosomonas* (Chain et al., 2003). Based on amino acid sequence, the subunits of AMO and pMMO are clearly in the same evolutionary family of enzymes and distinct from the soluble methane monooxygenase of the methylotrophic bacteria (Holmes et al., 1995). Hydrophobicity plots of deduced amino acid sequences of the three subunits predict a large number of transmembrane segments; 6 in both AMO-A and -C and 2 in AMO-B (Fig.14). Regions with significant identity for AMO and pMMO are found in the large N- and C- terminal periplasmic domains of subunit B as well as the most C-terminal periplasmic loop of subunit A of AMO. The latter loop is thought to be cytoplasmic in a theoretical sequence-based analysis of the homologous subunit in pMMO of *M. capsulatus* where seven transmembrane segments were predicted (Tukhvatullin et al., 2001).

B. Mechanism of Turnover of AMO

1. General Nature of the Enzyme

The electron donor to AMO has not been identified. Duroquinol serves as electron donor in crude AMO or purified pMMO systems (Shears and Wood, 1986; Zahn and DiSpirito, 1996). Thus the donor is hypothesized to be ubiquinol. Many observations suggest a role for Cu ion in AMO. With cells, the copper binding agents allylthiourea, KCN or diethyldithiocarbamate inhibit ammonia oxidation (Hooper and Terry, 1973). Copper ions stimulate ammonia oxidation in vitro (Ensign et al., 1993). Specific photo-inactivation of ammonia-oxidizing activity (but not hydroxylamine-oxidizing activity) in cells of *Nitrosomonas* is caused by 420 nm light (Hooper and Terry, 1974).

Enzyme activity is low in extracts. Hence the present limited understanding of the mechanism of AMO has come from studies with intact cells. In cells, AMO is reactive with a remarkable variety of non-polar substrates with a susceptible C-H bond or π electrons. The list includes NH_3, CO, methane, methanol, propylene, cyclohexane, benzene, phenol, ethylene, halogenated aliphatic compounds, halobenzenes, naphthalene and 2-chloro, 6 trichloromethylpyridine (Hyman and Wood, 1983; Vannelli and Hooper, 1992; Keener and Arp, 1994; reviewed in Hooper et al., 1997). Based on the pH optimum for oxidation, ammonia rather than ammonium may be the preferred substrate (Suzuki et al., 1974). Polar compounds in general are not good substrates suggesting that the active site is non-polar.

The broad range of suitable substrates suggests that turnover is initiated by the activation of oxygen rather than the activation of substrate. In theory, O-activation could involve binding and 2-electron reduction of O_2 by a metal center. The hydroxylating agent could be the resulting peroxyl derivative or a metal-O derivative produced by heterolytic cleavage

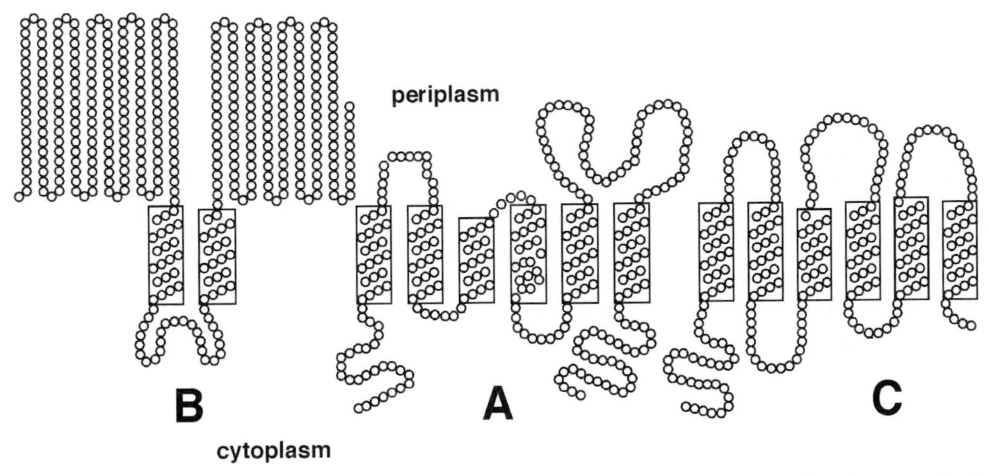

Fig. 14. Hypothetical transmembrane arrangement of subunits of AMO predicted by hydrophobicity plots. Sequence data is from Mc-Tavish et al., 1993a; Bergmann and Hooper, 1994a; Klotz et al., 1997. Prediction of structure was determined as described by Jones et al. (1994).

of peroxyl (with release of water).

$$M + 2e^- + 2H^+ + O_2 \rightarrow M{=}O + H_2O \quad (4)$$

A hydrogen may be extracted from substrate (forming a planar tricoordinate intermediate) by either the peroxide (releasing water) or the metal=O species to form a metal hydroxide and a substrate radical. The 'rebound' reaction of the metal hydroxyl and substrate radical would generate metal and product.

$$NH_3 + M{=}O \rightarrow [NH_2^{\cdot}] + M\text{-}OH \quad (5)$$

$$[NH_2^{\cdot}] + M\text{–}OH \rightarrow M + NH_2OH \quad (6)$$

AMO in cells of *Nitrosomonas* catalyzes two reactions in addition to monooxygenation which can be rationalized in terms of the these mechanisms. The dehydrogenation of ethylbenzene to form styrene (Keener and Arp, 1994) could involve formation of the M=O intermediate followed by two cycles of dehydrogenation of substrate and release of water. The anaerobic reductive dechlorination of 2-chloro-6-trichloromethyl-pyridine to 2-chloro-6-dichloromethyl-pyridine (Vannelli and Hooper, 1993) might have resulted from orientation of the pyridine in a way that the ring-N was reduced instead of dioxygen.

2. Speculation on the Geometry of the Active Site

Comparison of the reactivity of various substrates also suggests that the oxygenating site is deep in a somewhat flattened tubular pocket. For example, alkanes of 4 to 8 carbons in length are preferentially hydroxylated at the 1- or 2-postion (Hyman, et al., 1988). Further, the substituted aromatic compound, anisole, is oxidized preferentially at the substituent group, and when the ring is hydroxylated during the reaction, the process occurs preferentially at the p- rather than o- or m- positions (Keener and Arp, 1994) i.e. the substitute–d aromatic compounds enters the active site leading with either one end or the other.

A series of observations indicate that the aromatic ring of the substrate is immobilized in the active site so as to limit rotation of the benzyl moiety and to limit rotation around the C_1-benzyl bond. Further, the data suggest that the activated oxygen of the active site is positioned asymmetrically relative to the two C_1 Carbons (Vannelli et al., 1996). AMO exhibits strongly regioselective hydroxylation of C_1 in preference to C_2 of ethylbenzene i.e. the substrate is oriented so that the active oxygen is closer to C_1 (Fig. 15). Substitution of deuterium for hydrogen on C_1 changed the regioselectivity; C_2 rather than C_1 was preferentially hydroxylated. This deuterium effect indicated that a C-H bond is broken during catalysis. In terms of the catalytic model, this would be the abstraction of an H (or D) by the reactive oxygen. The result suggests that, although the orientation of the substrate by the active site placed C_1 deuterium closest to the reactive oxygen (as indicated by the regioselectivity of the non-deuterated substrate), the C_2 hydrogen was close enough that the latter was selected due to the greater stability of the C-D bond as compared with the C-H bond. Additional details of the geometry of the active site in the vicinity of the oxygen-activating center were suggested by the relative reactivity of the proR or proS carbons of the chiral substrates, R- or S-[1-^{2}H]-ethylbenzene. With S-[1-^{2}H]-ethylbenzene the pro-R position was hydroxylated approximately six-fold more frequently than the pro-S position. Aspects of the model are illustrated in Fig. 15. As suggested by the regioselectivity, the hydrogens of the methyl group are positioned further from the active oxygen than either hydrogen on C_1. As suggested by the enantiomeric selectivity, the R hydrogen is positioned closer to the active oxygen than the S hydrogen.

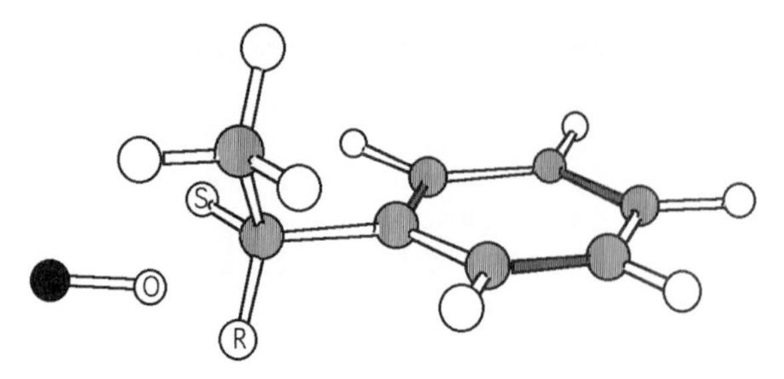

Fig. 15. Hypothetical orientation of ethylbenzene relative to the putative activated oxygen in the active site of AMO. Opening of the long active site is to the right. The active site binds substrate so as to immobilize the aromatic ring relative to the activated oxygen to the left. The putative O is closer to the R than the S hydrogen.

3. Evidence for a Radical Rebound Mechanism

The hypothetical mechanism involves formation of a substrate radical which rebounds to reaction with the metal hydroxyl. The observation that inversion of hydrogens occurs during hydroxylation of C_1 of ethylbenzene by AMO is consistent with the involvement of a tri-coordinate radical intermediate (Vannelli et al., 1996). In 14% of the cases where the proR hydrogen was extracted, inversion around the C_1-benzyl bond was found to occur. Following proS abstraction, inversion occurred in 28 or 57% of the cases, respectively, depending on whether the substrate was D or H in the proS position. A simple interpretation of this result is that substrate oxidation was initiated when H (or D) was abstracted by the activated oxygen and that the resulting substrate radical 'rebounded' to the metal-OH center with formation of product. According to this hypothesis the rate of the rebound process was slow enough to allow rotation of some fraction of the substrate radicals before attack by the hydroxyl.

4. Comparison of the Reactivity of pMMO and AMO with Ammonia and Methane

The rate of oxidation of ammonia by AMO is roughly ten-fold greater than the rate of oxidation of methane. This ratio of relative activity with the substrates is reversed with pMMO. Interestingly, the AMO of *Nitrosococcus oceanus*, which lies between the AMO of β-nitrifiers and pMMO of *M. capsulatus* in sequence homology, has similar activity with either ammonia or methane (Ward, 1987). *Nitrosomonas* and *Nitrosococcus* oxidize methane to CO_2 and the entire process is likely to be catalyzed by AMO (Jones and Morita, 1983; Voysey and Wood, 1987). It may be nutritionally significant that the organism also incorporates a significant fraction of the methane-C into biomass (Jones and Morita, 1983; Ward, 1987). AMO of *Nitrosomonas* is also very much more reactive with aromatic compounds than pMMO (Lontoh et al., 2000).

VII. Anaerobic Respiration by Autotrophic Nitrifiers

Very slow growth of *Nitrosomonas europaea* under anaerobic conditions has been reported with pyruvate as reductant and nitrite as terminal electron acceptor (Abielovich and Vonshak, 1992). The data in the genome reveal the presence of genes encoding many of the enzymes necessary for this process as mediated by pyruvate dehydrogenase, the citric acid cycle and nitrite reductase (Chain et al., 2003). Anaerobically, *Nitrosomonas eutropha* and *Nitrosomona europaea* are reported to oxidize hydrogen while reducing nitrite (Bock et al., 1995). A gene for hydrogenase is not seen in the chromosome of the strain of *N. europaea* which has been sequenced.

Nitrosomonas eutropha is also able to grow anaerobically while consuming equal amounts of NH_3 and NO_2 and producing a mixture of NO, NO_2^-, and N_2 (Schmidt et al., 2001). The initiating reaction is thought to involve NO_2 as oxygen donor to AMO resulting in the production of NH_2OH and NO.

VIII. Anaerobic Oxidation of Ammonia (ANAMMOX)

An autotrophic bacterium, *Candidatus* Brocadia anammoxidans, of the order planctomycetes grows anaerobically based on the overall energy-generating reaction:

$$NH_3 + HNO_2 \rightarrow N_2 + 2H_2O \quad (7)$$

(van de Graf et al., 1995; Strous et al., 1999).

Use of $^{15}NH_3$ with $H^{14}NO_2$ resulted in the production of only $^{14/15}N_2$. This elegant outcome indicated that a single active site must have simultaneously bound ammonia and nitrite or their respective intermediate products. Measurable amounts of hydrazine and hydroxylamine accumulated during the process. The hypothesis which is favored by these investigators involves three reactions: reduction of nitrite to hydroxyamine, reaction of ammonia with hydroxylamine to form hydrazine and, finally, the dehydrogenation of hydrazine:

$$HNO_2 + 4\ e^- + 4\ H^+ \rightarrow NH_2OH + H_2O \quad (8)$$

$$NH_3 + NH_2OH \rightarrow NH_2NH_2 + H_2O \quad (9)$$

$$NH_2NH_2 \rightarrow N_2 + 4\ e^- + 4\ H^+ \quad (10)$$

The reactions are proposed to be arranged in separate membrane compartments so as to drive formation of a chemiosmotic gradient. The organism has an unusual internal membrane system which is extremely difficult to analyze (Lindsay et al., 2001). It is described as consisting of a peripheral cytoplasmic membrane inside the cell wall. Within the enclosed 'cytoplasm' is a region called the 'riboplasm' which is enclosed by a second internal membrane (the intracytoplasmic membrane). The riboplasm contains ribosomes and a nucleoid. The extensive space between the intracytoplasmic membrane and cytoplasmic membrane, is termed 'paryphoplasm'. The latter is seen to form a balloon-like intrusion into the riboplasm and, perhaps, pinch off producing yet a third layer of concentric vacuoles.

The cells are red in color due to the presence of an HAO which is able to oxidize hydroxylamine or hydrazine (Schalk et al., 2000). Most of the molecular properties are very similar to those of HAO of *Nitrosomonas* and *Nitrosococcus*. The reduced and optical spectra of the HAO have major Soret and α and β bands attributable to the approximately 26 c-hemes in a trimer of 183 kDa. The ferrous spectrum has a weak Soret band at 468 nm which is not seen if the ferric enzyme had been treated with hydrogen peroxide (resulting in loss of enzyme activity). This appears to correspond to heme P460 of the *Nitrosomonas* enzyme. The EPR spectrum contains some signals from iron of low-spin c-hemes as well as atypical signals possibly representing electronically-coupled heme irons. The enzyme is thought to at least account for the oxidation of hydrazine to dinitrogen in the organism and possibly catalyze the reduction of nitrite to NO.

The heme composition and catalytic and EPR properties of the HAOs of ANAMMOX and of *Nitrosomonas* are similar enough to suggest that the hemes of the two proteins might ultimately be found to align in similar spatial configurations.

IX. Heterotrophic Nitrification

The oxidation of ammonia to nitrite also occurs in the heterotrophic bacteria. The rates of oxidation of ammonia per cell are lower than with the autotrophic nitrifiers. Biochemical characterization of the reactions involved has not been extensive. An ammonia monooxygenase from membranes of *Paracoccus denitrificans* has been solubilized and purified to a form containing subunits of 38 and 46 kDa (Moir et al., 1996). The assay employed duroquinol as reductant. The enzyme resembled AMO from *Nitrosomonas* in sensitivity to light and copper-binding agents. The oxidation of ammonia to hydroxylamine, nitrite and nitrate by *Pseudomonas putida* was seen to be inactivated by acetylene (Daum et al., 1998). An open reading frame identified by its hybridization with a probe for *amo*A contained sequence similarity to the gene in autotrophic nitrifiers. Thus it appears that AMO of the AMO/pMMO family is utilized by many organisms. In contrast, the hydroxylamine-oxidizing enzymes that have been isolated thus far from heterotrophic bacteria do not resemble HAO of the autotrophic nitrifiers. An enzyme from *Arthrobacter globiformis* (Kurokawa et al., 1985) has hydroxylamine-dye reductase activity or hydroxylamine-cytochrome c reductase activity but does not contain heme. A similar activity from *Paracoccus denitrificans* appears to be catalyzed by an iron sulfur protein (Moir et al., 1996).

Acknowledgments

The authors thank Mark Whittaker, Tom Poulus, Todd Vannelli and Paul Kluge for discussions and help in preparation of the manuscript. This work was supported by funds from the US National Science Foundation (# NSF/MCB 0093447) and US Department of Energy (# DE-FG02-95ER20191) to ABH and funds from NIH 49970 to MPH.

References

Abielovich A and Vonshak A (1992) Anaerobic metabolism of *Nitrosomonas europaea.* Arch Microbiol 158: 267–270

Anderson IC and Levine JS (1986) Relative rates of nitric oxide and nitrous oxide production by nitrifiers, denitrifiers, and nitrate respirers. Appl and Environ Microbiol 51: 938–945

Andersson KK and Hooper AB (1983) O_2 and H_2O are each the source of one O of HNO_2 produced from NH_3 by *Nitrosomonas*; ^{15}N-NMR evidence. FEBS Lett 164: 236–240

Andersson KK, Lipscomb JD, Velentine M, Munck E and Hooper AB (1986) Tetraheme cytochrome *c*-554 from *Nitrosomonas europaea*., heme-heme interactions and ligand binding. J Biol Chem 261: 1126–1138

Arciero DM and Hooper AB (1993a) Hydroxylamine oxidoreductase from *Nitrosomonas europaea* is a multimer of an octa-heme subunit. J Biol Chem 268: 14645–14654

Arciero DM and Hooper AB (1994) A di-heme cytochrome *c* peroxidase from *Nitrosomonas europaea* catalytically active in both the oxidized and half-reduced states. J Biol Chem 269: 11878–11886

Arciero DM and Hooper AB (1997) Evidence for a peptidyl lysine-to *c* heme crosslink in cytochrome P-460 of *Nitrosomonas europaea*. FEBS Lett 410: 457–460

Arciero DM and Hooper AB (1998) Consideration of a phlorin structure for haem P460 of hydroxylamine oxidoreductase and its implications regarding reaction mechanism. Biochem Soc Trans 26: 385–389

Arciero DM and Hooper AB (1999) Heme P460 of hydroxylamine oxidoreductase: Models for catalysis based on dihydroporphyrin and isoporphyrin ring systems. J Inorg Biochem 74: 67

Arciero D, Balny C and Hooper AB (1991a) Spectroscopic and rapid kinetic studies of reduction of cytochrome *c*554 by hydroxylamine oxidoreductase from *Nitrosomonas europaea*. Biochemistry 30: 11466–11472

Arciero DM, Collins M, Haladjian J, Bianco P and Hooper AB (1991b) Resolution of the four hemes of cytochrome c_{554} from *Nitrosomonas europaea* by redox potentiometry and optical spectroscopy. Biochemistry 30: 11459–11465

Arciero DM, Hooper AB, Cai M and Timkovich R (1993) Evidence for the structure of the active site heme P460 in hydroxylamine oxidoreductase of *Nitrosomonas*. Biochemistry 32: 9370–9378

Arciero DM, Pierce BS, Hendrich MP and Hooper AB (2002) Nitrosocyanin, a red cupredoxin-like protein from *Nitrosomonas europaea.* Biochemistry 41:1703–1709

Beaumont HEB, Hommes NG, Sayavedra-Soto LA, Arp DJ, Arciero DJ, Hooper AB, Westerhoff HV and Van Spanning RJM (2002) Nitrite reductase of *Nitrosomonas europaea* is not essential for production of gaseous nitrogen oxides and confers tolerance to nitrite. J Bacteriol 184: 2557–2560

Bergmann DJ and Hooper AB (1994a) The primary structure of cytochrome P460 of *Nitrosomonas europaea*: Presence of a *c*-heme binding motif. FEBS Lett 353: 324–326

Bergmann DJ and Hooper AB (1994b) Sequence of the gene *amo*B which encodes the 46 kDa polypeptide of ammonia monoxygenase of *Nitrosomonas europaea*. Biochem Biophys Res Commun 204: 759–762

Bergmann DJ and Hooper AB (2003) Cytochrome P460 of *Nitrosomonas europaea*: Formation of the heme-lysine crosslink in a heterologous host and mutagenic conversion to a non cross-linked cytochrome *c′*. Eur J Biochem 270: 1935–1941

Bergmann DJ, Arciero D and Hooper AB (1994) Organization of the HAO gene cluster of *Nitrosomonas europaea*: Genes for two tetraheme cytochromes. J Bacteriol 176: 3148–3153

Bergmann DJ, Zahn JA, Hooper AB and DiSpirito AA (1998) Cytochrome P460 genes from the methanotroph *Methylococcus capsulatus* bath. J Bacteriol 180: 6440–6445

Bock E, Schmidt I, Stüven R and Zart D (1995) Nitrogen loss caused by denitrifying *Nitrosomonas* cells using ammonium or hydrogen as electron donors and nitrite as electron acceptor. Arch Microbiol 163: 16–20

Brantner CA, Remsen CC, Owen HA, Buchholz LA and Collins MLP (2002) Intracellular localization of the particulate methane monooxygenase and methanol dehydrogenase in *Methylomicrobium album* BG8. Arch Microbiol 178: 59–64

Brige A, Leys D, Meyer TE, Cusanovich MA and Van Beeumen JJ (2002) The 1.25Å resolution structure of the diheme NapB subunit of soluble nitrate reductase reveals a novel cytochrome *c* fold with a stacked heme arrangement. Biochemistry 41: 4827–4836

Chain P, Lamerdin J, Lariner F, Regala W, Lao V, Land M, Hauser L, Hooper AB, Klotz M, Norton J, Sayavedra-Soto L, Arciero D, Hommes N, Whittaker M and Arp D (2003) Complete genome sequence of the ammonia oxidizing bacterium and obligate *chemolithoautotroph Nitrosomonas europaea*. J Bacteriol 185: 2759–2773

Collins M, Arciero DM and Hooper AB (1993) Optical spectrophotometric resolution of the hemes of hydroxylamine oxidoreductase. Heme quantitation and pH dependence of E_m. J Biol Chem 268: 14655–14662

Conrad R (1996) Soil microorganisms as controllers of atmospheric trace gases (H_2, CO, CH_4, OCS, N_2O, and NO). Microbiol Rev 60: 609–640

Darwin A, Hussain H, Griffiths L, Grove J, Sambongi, Y, Busby S, and Cole, J (1993) Regulation and sequence of the structural gene for cytochrome c552 from *Escherichia coli*: Not a hexahaem but a 50 kDa tetrahaem nitrite reductase MoleC Microbiol 9: 1255–1265

Daum M, Zimmer W, Papen H, Kloos K, Nawrath K and Bothe H (1998) Physiological and molecular biological characterization of ammonia oxidation of the heterotrophic nitrifier *Pseudomonas putida*. Current Microbiol 37: 281–288

DiSpirito AA, Taaffe LR and Hooper AB (1985a) Localization and concentration of hydroxylamine oxidoreductase and cytochromes *c*-552, *c*-554, c_M-553, c_M-552 and *a* in *Nitrosomonas europaea*. Biochim. Biophys. Acta 806: 320–330

DiSpirito AA, Taaffe LR, Lipscomb JD and Hooper AB (1985b) A 'blue' copper oxidase from *Nitrosomonas europaea* Biochim

Biophys Acta 827: 320–326

DiSpirito AA, Lipscomb JD and Hooper AB (1986) Cytochrome aa_3 from *Nitrosomonas europaea*. J Biol Chem 261: 17048–17056

Einsle O, Messerschmidt A, Stach P, Bourenkov GB, Bartunik HD, Huber R and Kroneck PMH (1999) Structure of cytochrome *c* nitrite reductase. Nature 400: 476–480

Einsle O, Stach P, Messerschmidt A, Simon J, Kroger A, Huber R and Kroneck PMH (2000) Cytochrome *c* nitrite reductase from *Wolinella succinogenes*. Structure at 1.6 Å resolution, inhibitor binding, and heme-packing motifs. J Biol Chem 275 (50): 39608–39616

Ensign SA, Hyman MR and Arp DJ (1993) In vitro activation of ammonia monoxygenase froM nitrosomonas by copper. J Bacteriol 175: 1971–1098

Erickson RH and Hooper AB (1972) Preliminary characterization of a variant CO-binding heme protein from *Nitrosomonas*. Biochim Biophys Acta 275: 231–244

Goreau TJ, Kaplan WA, Wofsy SC, McElroy MB, Valois FW and Watson SW (1980) Production of NO_2- and N_2O by nitrifying bacteria at reduced concentrations of oxygen. Appl Environ Microbiol 40: 526–532

Hendrich M, Logan MSP, Andersson KK, Arciero DM, Lipscomb JD and Hooper AB (1994) The active site of hydroxylamine oxidoreductase: Evidence from integer spin EPR. J Am Chem Soc 116: 11961–11968

Hendrich MP, Petasis D, Arciero DM and Hooper AB (2001) Correlations of structure and electronic properties from EPR spectroscopy of hydroxylamine oxidoreductase. J Am Chem Soc 123: 2997–3005

Hendrich MP, Upadhyay AK, Riga J, Arciero DM and Hooper AB (2002) Spectroscopic characterization of the NO adduct of hydroxylamine oxidoreductase. Biochemistry 41: 4603–4611

Hollocher TC, Kumar S and Nicholas DJD (1982) Respiration-dependent proton translocation in *Nitrosomonas europaea* and its apparent absence in *Nitrobacter agilis* during inorganic oxidations. J Bacteriol 149: 1013–1020

Holmes AJ, Costello A, Lidstrom ME and Murrell JC (1995) Evidence that particulate methane monoxygenase and ammonia monoxygenase may be evolutionarily related. FEMS Microbiol Lett 132: 203–208

Hommes NG, Sayavedra-Soto LA and Arp DJ (1994) Sequence of *hcy*, a gene encoding cytochrome *c-554* in *Nitrosomonas europaea*. Gene 146:87–89

Hooper AB (1968) A nitrite-reducing enzyme from *Nitrosomonas europaea* preliminary characterization with hydroxylamine as electron donor. Biochim Biophys Acta 162: 49–65

Hooper AB (1969) Biochemical basis of obligate autotrophy in *Nitrosomonas europaea*. J Bacteriol 97: 776–779

Hooper AB and Nason A (1965) Characterization of hydroxylamine-cytochrome *c* reductase from the chemoautotrophs *Nitrosomonas europaea* and *Nitrosocystis oceanus*. J Biol Chem 240: 4044–4057

Hooper AB and Terry KR (1973) Specific inhibitors of ammonia oxidation in *Nitrosomonas*. J Bacteriol 115: 480–485

Hooper AB and Terry KR (1974) Photoinactivation of ammonia oxidation in *Nitrosomonas*. J Bacteriol 119: 899–906

Hooper AB and Terry KR (1977) Hydroxylamine oxidoreductase of *Nitrosomonas*: Inactivation by hydrogen peroxide. Biochemistry 16: 455–459

Hooper AB and Terry KR (1979) Hydroxylamine oxidoreductase of *Nitrosomonas*: Production of nitric oxide from hydroxylamine. Biochim Biophys Acta 571: 12–20

Hooper AB and Balny C (1982) Reaction of oxygen with hydroxylamine oxidoreductase of *Nitrosomonas*: Fast kinetics. FEBS Lett 144: 299–303

Hooper AB, Terry KR and Maxwell PC (1977) Hydroxylamine oxidoreductase of *Nitrosomonas*: Oxidation of diethyldithiocarbamate and concomitant stimulation of nitrite synthesis. Biochim Biophys Acta 462: 141–152

Hooper AB, Maxwell PC and Terry KR (1978) Hydroxylamine oxidoreductase from *Nitrosomonas*: Absorption spectra and content of heme and metal. Biochemistry 17: 2984–2989

Hooper AB, Debey P, Andersson KK and Balny C (1983) Reaction of hydroxylamine oxidoreductase of *Nitrosomonas* with H_2O_2 and CO: Fast kinetic studies. Eur J Biochem 134: 83–87

Hooper AB, DiSpirito AA, Olson TC, Andersson KK, Cunningham W and Taafe LR (1984a) Generation of the proton gradient by a periplasmic dehydrogenase. In.: Crawford R and Hanson R (eds) Microbial Growth on C_1 Compounds, pp 53. American Society of Microbiology Press, Washington, DC

Hooper AB, Tran JM and Balny C (1984b) Kinetics of reduction by substrate or dithionite and heme-heme electron transfer in the multiheme hydroxylamine oxidoreductase. Eur J Biochem 141: 565–571

Hooper AB, Arciero DM, DiSpirito AA, Fuchs J, Johnson M, LaQuier F, Mundfrom G and McTavish H (1990) Production of nitrite and N_2O by the ammonia-oxidizing nitrifiers. In: Gresshof PM, Newton WE, Roth WE and Stacey G (eds) Nitrogen Fixation: Achievements and Objectives, pp 387–391. Chapman-Hall, New York

Hooper AB, Vannelli T, Bergmann DJ and Arciero DM (1997) Enzymology of the oxidation of ammonia to nitrite by bacteria. Antonie van Leeuwenhoek 71: 56–67

Hyman MR and Wood PM (1983) Methane oxidation by *Nitrosomonas europaea* Biochem J 212: 31–37

Hyman MR and Wood PM (1985) Suicidal inactivation and labelling of ammonia mono-oxygenase by acetylene. Biochem J 227: 779–725

Hyman MR Murton IB and Arp DJ (1988) Interaction of ammonia monooxygenase from *Nitrosomonas europaea* with alkanes, alkenes, and alkynes. Appl Envt Microbiol 54:3187–3190

Igarashi N, Moriyama H, Fujiwara T, Fukumori Y. and Tanaka N (1997) The 2.8Å structure of hydroxylamine oxidoreductase from a nitrifying chemoautotrophic bacterium, *Nitrosomonas europaea*. Nature Struct Biol 4: 276–284

Iverson TM, Arciero DM, Hsu BT, Logan MSP, Hooper AB and Rees DC (1998) Heme packing motifs revealed by the crystal structure of the tetraheme cytochrome *c*554 from *Nitrosomonas euroapea*. Nature Struct Biol 5: 1005–1012

Iverson T, Arciero DM, Hooper AB and Rees DC (2001a) High-resolution structures of cytochrome *c*554 from *Nitrosomonas europaea*. J Biol Inorg Chem 6: 390–397

Iverson, TM, Hendrich MP, Arciero DM, Hooper AB and Rees DC (2001b) Cytochrome *c*554. Messerschmidt A, Huber R, Poulos T and Weighardt K (eds). Handbook of Metalloproteins, pp 136–146. John Wiley and Sons LTD, Chichester

Jones RD and Morita RY (1983) Methane oxidation by *Nitrosococcus oceanus* and *Nitrosomonas europaea*. Appl Envt Microbiol 45.401–410

Jones JT, Taylor WR and Thornton JM (1994) A model recognition approach to the prediction of all-helical membrane protein

structure and topology. Biochemistry 33: 3038–3049
Jungst A, Wakayabayashi, S, Matsubara H and Zumft WG (1991) The *nicSTBM* region coding for cytochrome cd_1-dependent nitrite respiration of *Pseudomonas stutzeri* consists of a cluster of mono-, di-, and tetraheme proteins. FEBS Lett 279: 205–209
Keener WK and Arp DJ (1994) Transformations of aromatic compounds by *Nitrosomonas europaea*. Appl Envt Microbiol 60:1914–1920
Klotz MG, Alzerreca J and Norton JM (1997) A gene encoding a membrane protein exists upstream of the *amoA/amoB* genes in ammonia-oxidizing bacteria: a third member of the amo operon? FEMS Micro Lett 150: 65–73
Kumar S and Nicholas DJD (1982) A protonmotive force-dependent adenosine-5′ triphosphate synthesis in spheroplasts of *Nitrosomonas europaea*. FEMS Microbiol Lett 14: 21–25
Kumar S and Nicholas DJD (1983) Proton electrochemical gradients in washed cells of *Nitrosomonas europaea* and *Nitrobacter agilis*. J Bacteriol 154: 65–71
Kurokawa M, Fukumori Y and Yamanaka T (1985) A hydroxylamine-cytochrome *c* reductase occurs in the heterotrophic nitrifier *Arthrobacter globiformis*. Plant Cell Physiol 26: 1439–1442
Lieberman RL, Arciero DM, Hooper AB and Rosensweig AC (2001) Crystal structure of a novel red copper protein from *Nitrosomonas europaea*. Biochemistry 40: 5674–5681
Lindsay MR, Webb RI, Strous M, Jetten MSM, Butler MK, Forde RJ and Fuerst JA (2001) Cell compartmentalisation in planctomycetes: Novel types of structural organisation for the bacterial cell. Arch Microbiol 175: 413–429
Lipschultz F, Zafiriou OC, Wofsy SC, McElroy MB, Valois FW and Watson SW (1981) Production of NO and N_2O by soil nitrifying bacteria. Nature 294: 641–643
Lipscomb JL and Hooper AB (1982) Resolution of multiple heme centers of hydroxylamine oxidoreductase from *Nitrosomonas*. 1. Electron paramagnetic resonance spectroscopy. Biochemistry 21: 3965–3972
Logan MSP and Hooper AB (1995) Suicide inactivation of hydroxylamine oxidoreductase of *Nitrosomonas europaea* by organohydrazines. Biochemistry 34: 9257–9264
Logan MSP, Balny C and Hooper AB (1995) Reaction with cyanide of hydroxylamine oxidoreductase of *Nitrosomonas europaea*. Biochemistry 34: 9028–9037
Lontoh S, DiSpirito AA, Krema CL, Whittaker MR, Hooper AB and Semrau JD (2000) Differential inhibition in vivo of ammonia monooxygenase, soluble methane monooxygenase, and membrane-associated methane monooxygenase by phenylacetylene. Environ Microbiol 2: 485–494
McTavish H, Fuchs J and Hooper AB (1993a) Sequence of the gene for ammonia monoxygenase of *Nitrosomonas europaea*. J Bacteriol 175: 2436–2444
McTavish H, LaQuier F, Arciero D, Logan M, Mundfrom G, Fuchs J and Hooper AB (1993b) Multiple copies of genes for electron transport proteins in the bacterium *Nitrosomonas europaea*. J Bacteriol 175: 2445–2447
McTavish H, Arciero DM and Hooper AB (1995) Interaction with membranes of cytochrome *c*554 from *Nitrosomonas europaea*. Arch Biochem Biophys 324: 53–58
Miller DJ and Wood PM (1983) The soluble cytochrome oxidase of *Nitrosomonas europaea*. J Gen Microbiol 129: 1645–1650
Moir JWB, Wehrfritz J-M, Spiro S and Richardson DJ (1996) The biochemical characterization of a novel non-haem-iron hydroxylamine oxidase from *Paracoccus denitrificans* GB17. Biochem J 319: 823–827
Murray RGE and Watson SW (1965) Structure of *Nitrosocystis* and *oceanus* and comparison with *Nitrosomonas* and *Nitrobacter*. J Bacteriol 89: 1594–1609
Numata M, Saito T, Yamazaki T, Fukumori Y and Yamanaka T (1990) Cytochrome P-460 of *Nitrosomonas europaea*: Further purification and further characterization. J Biochem 108: 1016–1023
Nguyen HAT, Zhu M, Elliot SJ, Nakagawa KH, Hedman B, Costello AM, Peples TL, Wilkinson B, Morimoto H, Williams PG, Floss HG, Lidstrom ME, Hodgson KO and Chan S (1996) The biochemistry of the particulate methane monooxygenase. In: Lidstrom ME and Tabita FR (eds) Proceedings of the 8th International Symposium on Microbial Growth on C_1 Compounds, pp150–158, Kluwer Academic Publishers
Poth M and Focht DD (1985) ^{15}N kinetic analysis of N_2O production by *Nitrosomonas europaea*: An examination of nitrifier denitrification. Appl Environ Microbiol 49:1134–1141
Prince RC and Hooper AB (1987) Resolution of the hemes of hydroxylamine oxidoreductase by redox potentiometry and electron spin resonance spectroscopy. Biochemistry 26: 970–974
Remde A and Conrad R (1990) Production of nitric oxide in *Nitrosomonas europaea* by reduction of nitrite. Arch Microbiol 154:187–191
Ritchie GAF and Nicholas DJD (1974) The partial characterization of purified nitrite reductase and hydroxylamine oxidase from *Nitrosomonas europaea*. Biochem J 138: 471–480
Roldan MD, Sears HJ, Cheesman MR, Ferguson SJ, Thomson AJ, Berks BC and Richardson DJ (1998) Spectroscopic characterization of a novel multiheme *c*-type cytochrome widely implicated in bacterial electron transport. J Biol Chem 273: 28785–28790
Sayavedra-Sota LA, Hommes NG and Arp DJ (1994) Characterization of the gene encoding hydroxylamine oxidoreductase in *Nitrosomonas europaea*. J Bacteriol 176: 504–510
Schalk J, de Vries S, Kuenen JG and Jetten MSM (2000) Involvement of a novel hydroxylamine oxidoreductase in anaerobic ammonium oxidation. Biochemistry 39: 5405–5412
Schmidt I, Bock E and Jetten MSM (2001) Ammonia oxidation by *Nitrosomonas eutropha* with NO_2 as oxidant is not inhibited by acetylene. Microbiology 147: 2247–2253
Shears JH and Wood PM (1986) Tri- and tetra-methylhydroquinone as electron donors for ammonia monoxygenase in whole cells of *Nitrosomonas europaea*. FEMS Microbiol Lett 33: 281–284
Shimizu H, Schuler DJ, Lanzillotta WN, Sundaramoorthy M, Arciero D, Hooper AB and Poulos TL (2001) Crystal structure of *Nitrosomonas europaea* cytochrome c peroxidase and the structural basis for ligand switching in bacterial di-heme peroxidases. Biochemistry 40: 13483–13490
Simon J, Gross R, Einsle O, Kroneck PMH, Kroger A and Klimmek O (2000) A NapC/NirT-type cytochrome *c* (NrfH) is the mediator between the quinone pool and the cytochrome *c* nitrite reductase of *Wolinella succinogenes*. Mol Microbiol 35: 686–696
Strous M, Fuerst JA Kramer EHM, Logemann S, Muyzer G, van de Pas-Schoonen KT, Webb R, Kuenen JG and Jetten MSM (1999) Missing lithotroph identified as new planctomycete. Nature 400: 446–449

Suzuki I, Dular U and Kwok SC (1974) Ammonia and ammonium ion as substrate for oxidation by *Nitrosomonas* cells and extracts. J Bacteriol 120: 556–558

Teske A, Alm E, Regan JM, Toze S, Rittman BE and Stahl DA (1994) Evolutionary relationships among ammonia- and nitrite-oxidizing bacteria. J Bacteriol 176: 6623–6630

Tukhvatullin IA, Gvozdev RI and Andersson KK (2001) Chemistry of natural compounds, bioorganic, and biomolecular chemistry. Structural and functional model of methane hydroxylase of membrane-bound methane monooxygenase from *Methylococcus capsulatus* (Bath). Russian Chem Bull, Intl Ed 50: 1–10

Upadhyay AK, Petasis DT, Arciero DM, Hooper AB and Hendrich MP (2003) Spectroscopic characterization of assignment of reduction potentials in the tratheme cytochrome *c*554 from *Nitrosomonas europaea*. J Am Chem Soc 125: 1738–1747

van de Graaf AA, Mulder A, de Bruijn P, Jetten MSM and Robertson LA (1995) Anaerobic oxidation of ammonium is a biologically mediated process. Appl Envt Microbiol 61: 1246–1251

Vannelli T and Hooper AB (1992) Oxidation of nitrapyrin to 6-Chloropicolinic acid by the ammonia-oxidizing bacterium *Nitrosomonas europaea.* Appl Envt Microbiol 58: 2321–2325

Vannelli T and Hooper AB (1993) Reductive Dehalogenation of the Trichloromethyl Group of Nitrapyrin by the Ammonia-oxidizing Bacterium *Nitrosomonas europaea.* Appl Envt Microbiol 59: 3597–3601

Vannelli T, Bergmann D, Arciero DM and Hooper AB (1996) Mechanism of N-oxidation and electron transfer in the ammonia-oxidizing autotrophs. In: Lindström ME and Tabita FR (eds) Proceedings of the 8th International Symposium on Microbial Growth on C_1 Compounds, pp 80–87, Kluwer Academic Publishers, Dordrecht

Voysey PA and Wood PM (1987) Methanol and formaldehyde oxidation by an autotrophic nitrifying bacterium. J Gen Microbiol 33:283–290

Wallace W and Nicholas DJD (1968) Properties of some reductase enzymes in the nitrifying bacteria and their relationship to the oxidase systems. Biochem J 109: 763–773

Ward B (1987) Kinetic studies on ammonia and methane oxidation by *Nitrosococcus oceanus.* Arch Microbiol 147: 126–133

Whittaker M, Bergmann DJ, Arciero DM and Hooper AB (2000) Electron transfer during the oxidation of ammonia by the chemolithotrophic bacterium *Nitrosomonas europaea.* Biochim Biophys Acta 1459: 346–355

Wood PM (1986) Nitrification as a bacterial energy source. *In:* Prosser JL (ed.), Nitrification, Special Publications of the Society for General Microbiology, Vol 20, p 39–62. Society for General Microbiology and IRL Press, Oxford

Yamanaka T and Shinra M (1974) Cytochrome *c*-552 and cytochrome *c*-554 derived from *Nitrosomonas europaea.* Purification, properties and their function in hydroxylamine oxidation. J Biochem 75: 1265–1273

Yamanaka T, Shinra M, Takahashi K and Shibasaka M (1979) Highly purified hydroxylamine oxidoreductase derived from *Nitrosomonas europaea*. Some physiochemical and enzymatic properties. J Biochem 86: 1101–1108

Zahn JA and DiSpirito AA (1996) Membrane-associated methane monooxygenase from *Methylococcus capsulatus* (Bath). J Bacteriol 178:1018–1029

Chapter 7

Respiration in Methanotrophs

Alan A. DiSpirito*[1], Ryan C. Kunz[1], Don-Won Choi[1,2], and James A. Zahn[3]
[1]Department of Biochemistry, Biophysics and Molecular Biology; [2]Graduate Program in Microbiology, Iowa State University, Ames, Iowa 50011 U.S.A.; [3]Dow Agro-Sciences, Harbon Beach, MI 48441 U.S.A.

*Author for correspondence, email: aland@iastate.edu

Davide Zannoni (ed): Respiration in Archaea and Bacteria. Vol 2: Diversity of Prokaryotic Respiratory Systems, pp. 149–168.

Summary

Methanotrophs are a unique bacterial group characterized by the ability to utilize methane as a sole carbon and energy source. Methanotrophs oxidize methane to CO_2 via a series of two electron steps with methanol, formaldehyde, and formate as intermediates. In addition to the known growth substrates methanotrophs will oxidize a variety of multi-carbon compounds, molecular hydrogen, and ammonia. With the exception of methane, and in some cases, methanol, the oxidation of these other substrates do not support growth and have been termed co-oxidation because of the need to provide a second substrate, methane or methanol, to support growth. However, recent studies have demonstrated, the oxidation of some non-growth substrates provides a secondary source of reducing equivalents to the cells. This review focuses on the electron flow between the enzymes that catalyze the oxidation of both growth and non-growth substrates, and the respiratory chains in methanotrophs. The properties of the individual oxidative enzymes and respiratory components are reviewed and working models presented for electron flow during methane and ammonia oxidation.

I. Introduction

A. Organisms

Methanotrophs are Gram-negative bacteria characterized by the utilization of methane or methanol as a sole carbon and energy source. Two general categories of methanotrophs have been identified, Type I and Type II, based on several characteristics including the pattern of internal membranes, carbon assimilation pathway, and predominant fatty acid chain length (Whittenbury and Dalton, 1981; Anthony, 1982; Hanson and Hanson, 1996). One strain, *Methylococcus capsulatus* Bath has characteristics of both types and is classified as Type X. Methanotrophs play a key role in the global carbon cycle, and may be a significant sink for atmospheric methane (Kiene, 1991; Topp and Hanson, 1991; Oremland and Culbertson, 1992). In addition to their ecological significance, the potential use of these microorganisms in bioremediation and biotransformations processes has provided incentive for the biochemical characterization of methanotrophs (DiSpirito et al., 1992; Hanson and Hanson, 1996; Lontoh et al., 1999, 2000; Murrell et al., 2000).

Abbreviations: AMO – ammonia monooxygenase; CCP – cytochrome *c* peroxidase; Cu-cbc – copper binding compound; Cu-cbc – copper containing cbc; EPR – electron paramagnetic resonance; FalDH – formaldehyde dehydrogenase; FDH – formate dehydrogenase; H_2ase – hydrogenase; H_4F – tetrahydrofolate; HS – high spin; LS – low spin; MDH – methanol dehydrogenase; MMO – methane monooxygenase; NDH – NADH:Quinone oxidoreductase; pMMO – membrane-associated or particulate methane monooxygenase; PQQ – pyrroloquinoline quinone; Q_8 – ubiquinone-8; RuMP – ribulose monophosphate cycle; sMMO – soluble methane monooxygenase

B. Oxidation of Methane to Carbon Dioxide

The oxidation of methane to carbon dioxide by methanotrophs involves a series of two electron steps with methanol, formaldehyde, and formate as intermediates (Fig. 1) (Anthony, 1982; Hanson and Hanson, 1996). The first enzyme in this pathway, the methane monooxygenase (MMO) catalyzes the energy-dependent oxidation of methane to methanol. In some methanotrophs, methane is oxidized to methanol by two different methane monooxygenases (MMOs), a membrane-associated MMO or particulate MMO (pMMO) and a soluble MMO (sMMO) depending on the copper concentration during growth (Stanley et al., 1983; Dalton et al., 1984; Prior and Dalton, 1985). In cells cultured under a low copper-to-biomass ratio, the sMMO is predominately expressed, with low, but detectable levels of pMMO expression (Stanley et al., 1983; Nielsen et al., 1996, 1997; Zahn and DiSpirito 1996; Murrell et al., 2000; Choi et al., 2003). Cells cultured under higher copper-to-biomass ratios express the pMMO exclusively, with no detectable expression of sMMO (Nielsen et al., 1997; Murrell et al., 2000; Stolyar et al., 2001; Choi et al., 2003). The reductant for the first, energy dependent, step is supplied by either NADH in cells expressing the sMMO or by the respiratory chain in cells expressing the pMMO (Anthony, 1982; Stanley et al., 1983; Dalton et al., 1984; Nguyen et al., 1998; Zahn et al., 2001; Basu et al., 2002; Cook and Shiemke, 2002 b; Choi et al., 2003; Lieberman et al., 2003). The second two electron step, catalyzed by the methanol dehydrogenase (MDH), involves the oxidation of methanol to formaldehyde utilizing a *c*-type cytochrome as an electron acceptor (Anthony, 1975,1992a,b,1996) (see

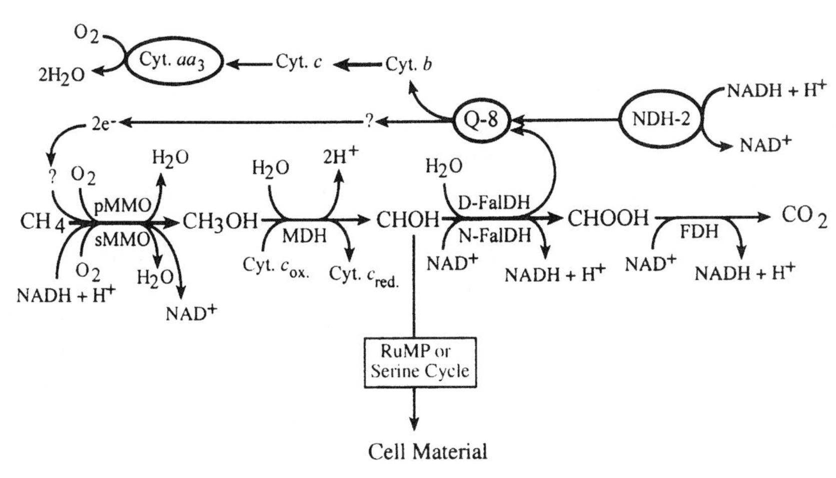

Fig. 1. Proposed pathways of methane oxidation in *M. capsulatus* Bath. Membrane associated proteins are shown above the carbon oxidation steps, soluble proteins are shown below the carbon oxidation steps. Abbreviations: Cyt, cytochrome; D-FalDH, dye-linked formaldehyde dehydrogenase; FDH, formate dehydrogenase; MDH, methanol dehydrogenase; N-FalDH, NAD(P)-linked formaldehyde dehydrogenase; pMMO, particulate methane monooxygenase; Q-8, ubiquinone-8; RuMP, ribulose monophosphate; sMMO, soluble methane monooxygenase.

also Chapter 10 by C. Anthony, Vol. 1 of this book). Formaldehyde is either assimilated via the serine or ribulose monophosphate cycle (RuMP) or oxidized to formate by either an NAD^+-linked, a dye-linked formaldehyde dehydrogenase, or by a tetrahydromethanopterin/methanofuran-mediated pathway (Stirling and Dalton, 1978; Marison and Attwood, 1980, 1982; Anthony, 1982; Speer et al., 1994; Chistoserdova et al., 1998; Vorholt et al., 1998a,b,1999; Tate and Dalton, 1999; Vorholt, 2002; Zahn et al., 2001). Lastly, formate is oxidized to carbon dioxide by an NAD^+-linked formate dehydrogenase (Anthony, 1982; Jollie and Lipscomb 1990, 1991; Yoch et al., 1990).

II. Oxidative Enzymes

A. Methane Monooxygenases (MMO)

1. Soluble Methane Monooxygenase (sMMO)

The sMMO is present in many, but not all, genera of methanotrophs (Anthony, 1982; Hanson and Hanson, 1996; Murrell et al., 2000). The sMMO has been the topic of recent reviews (Wallar and Lipscomb 1996; Murrell et al., 2000) and will only be considered briefly here. The sMMO is a three component enzyme (Colby et al., 1977; Colby and Dalton, 1978, 1979; Woodland and Dalton, 1984; Green and Dalton, 1985; Fox et al., 1989; Pilkington and Dalton, 1990; Wallar and Lipscomb, 1996). Protein A or the hydroxylase component is composed of three subunits with molecular masses of 54,000, 42–43,000 and 17–22,700 Da, with a subunit structure of $(\alpha\beta\gamma)_2$. High activity preparations of protein A contain 4 mol of non heme iron in 2 μ-oxo- or μ-hydroxo-bridged iron clusters associated with the α subunit (Woodland and Dalton 1984; Fox et al., 1988; Wallar and Lipscomb 1996). The reductase component or protein C contains both FAD and $[Fe_2S_2]$ cofactors (Colby and Dalton, 1978, 1979; Fox et al., 1989). This component oxidizes NADH and transfer the two electrons to the 2 μ-oxo-bridged iron centers of the hydroxylase component (Fox et al., 1989; Wallar and Lipscomb 1996). Protein B contains no metal or cofactors and is a small regulatory protein that binds to the α subunit of the hydroxylase component (Green and Dalton, 1985; Wallar and Lipscomb, 1996; Chang et al., 1999; Walters et al., 1999).

2. Membrane-Associated or Particulate Methane Monooxygenase (pMMO)

In contrast to the sMMO, little is known on the molecular properties of the pMMO. Purification of the pMMO has been reported from *M. capsulatus* Bath (Zahn and DiSpirito, 1996; Nguyen et al., 1998; Basu et al., 2002; Choi et al., 2003; Lieberman et al., 2003) and *Methylosinus trichosporium* OB3b (Takeguchi et al., 1998). As isolated, the pMMO is composed of three polypeptides with molecular masses of 45,000,

26,000, and 23,000 Da with a $(\alpha\beta\gamma)_2$ subunit structure (Nguyen et al., 1998; Takeguchi et al., 1998; Basu et al., 2002; Choi et al., 2003; Lieberman et al., 2003). Active pMMO preparations contain approximately 2–21 copper and 0–2 iron atoms depending on the reporting laboratory (Nguyen et al., 1998; Zahn and DiSpirito, 1996; Takeguchi et al., 1998). Researchers working on the pMMO all agree Cu is or is part of the catalytic center of the enzyme. However, there are two different theories on the number and type of Cu involved in catalysis. One theory states the two type II (Cu^{2+}) copper atoms are associated with the two larger subunits of the enzyme (Zahn and DiSpirito, 1996; Tukhvatullin et al., 2000, 2001; Basu et al., 2002; Choi et al., 2003; Lieberman et al., 2003). These two larger subunits are considered the catalytic center of the enzyme. The remaining 10 to 13 copper atoms are bound to a small, 1,218 Da copper binding peptide/compound (Cu-cbc) that is associated with the pMMO (Zahn et al., 1996; DiSpirito et al., 1998; Choi et al., 2003). The role of copper containing cbc (Cu-cbc) is not known but it may protect the enzyme from oxygen radicals, maintain a particular redox state, or act as a copper chaperone. The second theory coordinates the 15 to 21 copper atoms into 5–7 spin-coupled trinuclear copper clusters, of which 2–3 clusters are catalytic and 3–4 clusters are involved in electron transfer (Nguyen et al., 1994, 1996, 1998).

Spectral studies of membrane fractions and purified pMMO from *M. capsulatus* Bath and *M. trichosporium* OB3b are complicated by the presence of high concentration of Cu-cbc. Comparison of the electron paramagnetic resonance (EPR) spectra between membrane fractions of *M. capsulatus* Bath expressing the pMMO, as well as purified pMMO samples from *M. capsulatus* Bath and *M. trichosporium* OB3b, show a number of similarities to Cu-cbc (Nguyen et al., 1994, 1996, 1998; Zahn and DiSpirito, 1996; DiSpirito et al., 1998). The complex Cu-cbc spectra can account for much of the broad isotropic copper signals associated with purified preparations of pMMO and may be responsible for much of the data leading to the trinuclear copper cluster theory. The better resolved EPR spectra of whole cell and of the membrane fraction from *M. albus* BG8 can be attributed to the absence of the underlying signals from Cu-cbc (Yuan et al., 1997, 1998a,b, 1999; Lemos et al., 2000). The results from spectral characterization of the membrane fraction from *Methylomicrobium albus* BG8 have correlated two, type II (Cu^{+2}) copper centers with activity levels of pMMO (Yuan et al., 1997, 1998a,b, 1999; Lemos et al., 2000; Choi et al., 2003). The results also indicate that there are four nitrogens coordinated to the two copper centers and that three or four of these nitrogens are histidine imidazoles. The high expression levels of the pMMO, which can account for 5 to 10% of the total cell protein, makes spectral studies of the enzyme in whole cell or washed membrane samples possible. In addition, when considered in conjunction with studies using purified pMMO (Zahn and DiSpirito, 1996; Takeguchi et al., 1998, 1999a; Choi et al., 2003; Lieberman et al., 2003), or partially purified pMMO (Basu et al., 2002) as well as purified Cu-cbc (Zahn and DiSpirito, 1996; DiSpirito et al., 1998), and sequence analysis (Semrau et al., 1995; Murrell and Holmes, 1996; Gilbert et al., 2000; Tukhvatullin et al., 2000,2001), the results are consistent with a pMMO model containing two type 2 (Cu^{2+}) copper centers in the active site. Some support for the arrangement of copper in both models have been obtained from studies on the pMMO from *M. trichosporium* OB3b (Takeguchi et al., 1998, 1999a,b, 2000).

Although controversial, non-heme iron also appears to be either a component of the active site of the pMMO or as an electron shuttle to the copper centers. The evidence of the involvement of iron in methane catalysis is as follows: (a) preparations of pMMO from *M. capsulatus* Bath and *M. trichosporium* OB3b contain 1–2 Fe atoms (Zahn and DiSpirito, 1996; Takeguchi et al., 1998, 1999b; Basu et al., 2002; Choi et al., 2003; Lieberman et al., 2003) that are associated with the αβ subunits of the pMMO (Zahn and DiSpirito, 1996); (b) a copper-dependent iron uptake is observed in cells expressing the pMMO (Nguyen et al., 1994; Zahn and DiSpirito, 1996); (c) nitric oxide derivative EPR spectra of membrane fractions containing pMMO, but not in the membrane fraction of cells expressing the sMMO, as well as in purified preparations of pMMO from *M. capsulatus* Bath and *M. trichosporium* OB3b indicate the presence of a nitrosyl-iron signal (Zahn and DiSpirito, 1996; Takeguchi et al., 1998,1999); (e) Nitapyrin, an inhibitor of the ammonia monooxygenase (AMO) and pMMO, prevented the formation of the nitrosyl-iron complex (Zahn et al., 1996c); (f) presence of a weak high spin $g = 6$ signal in purified preparations of pMMO from *M. capsulatus* Bath and *M. trichosporium* OB3b (Zahn and DiSpirito, 1996; Takeguchi et al., 1998,1999b); and (g) propargylamine, an inhibitor of the pMMO, decreases the intensity of the $g = 6$ signal (Takeguchi et al., 1999b).

B. Methanol Dehydrogenase (MDH)

In methanotrophs, methanol is oxidized to formaldehyde by a periplasmic methanol dehydrogenase. In all methanotrophs tested, the enzyme is a two subunit tetrameric protein with molecular masses of 60,000–67,000 and 8,500 to 12,000 Da with an $\alpha_2\beta_2$ subunit configuration (Anthony 1992a,b,1996; Waechter-Brulla et al., 1993) (see also Chapter 10 by C. Anthony, Vol. 1, this book). The holoenzyme contains 2 mol of pyrroloquinoline quinone (PQQ) associated with the larger subunit and 1 mol of Ca (Anthony, 1992b, 1996). The reader is referred to the Chapter 10 on quinoproteins by C. Anthony, Vol. 1 of this book, for a description of the molecular properties of this enzyme. With respect to this chapter, the physiological electron acceptors for PQQ containing MDH are acidic monoheme *c*-type cytochromes, originally classified as cytochrome c_L(Anthony 1982,1992a, 2002).

C. Formaldehyde Dehydrogenase (FalDH)

Of the oxidative enzymes involved in methane oxidation, the formaldehyde oxidation step is by far the most complex. The formaldehyde formed during the oxidation of methanol is either assimilated into cell carbon via the serine or ribulose monophosphate cycle or oxidized for energy to formate by the formaldehyde dehydrogenase (FalDH)(Anthony, 1982; Hanson and Hanson, 1996). The central role of formaldehyde in methanotrophic metabolism is reflected in the variety of different formaldehyde oxidizing enzymes identified in methylotrophic bacteria. Based on the nature of the electron acceptor, formaldehyde oxidizing enzymes are divided into two groups, NAD(P)$^+$-dependent or dye(cytochrome)-linked. The NAD(P)$^+$-linked enzymes are further subdivided based on the need for the secondary cofactors such as thiol compounds, tetrahydrofolate, methylene tetrahydromethanopterin, or modifier proteins (Johnson and Quayle, 1964; Stirling and Dalton, 1978; Marison and Attwood, 1980, 1982; Attwood, 1990; Speer et al., 1994; Chistoserdova et al., 1998; Vorholt et al., 1998b, 1999; Tate and Dalton, 1999). The NAD$^+$-linked formaldehyde dehydrogenase has been isolated from *Methylococcus capsulatus* Bath (Stirling and Dalton, 1978; Tate and Dalton, 1999). This glutathione-independent, FalDH was originally reported as a homodimer (Stirling and Dalton, 1978), but subsequently shown to be a homotetramer, with a subunit molecular mass of 63,615 Da (Tate and Dalton, 1999). The substrate specificity and kinetics of the purified enzyme is regulated by a small (8,600 Da) heat stable protein called modifin (Tate and Dalton, 1999). In the presence of modifin, the enzyme is specific for formaldehyde, in the absence of modifin, the enzyme is a general class III alcohol/aldehyde dehydrogenase. The structural gene and enzymatic properties of a class III alcohol dehydrogenase has been reported in the marine methanotroph, *Methylobacter marinus* A45. The NAD(P)$^+$-dependent, glutathione-independent enzyme has a predicted subunit molecular mass of 46,000 Da and an isoelectric point of 4.8 (Speer et al., 1994). As in the methanol utilizing bacterium, *Methylobacterium extorquens* AM1, many methane oxidizing bacteria also appear to utilize a tetrahydrofolate (H_4F) and methylene tetrahydromethopterin (H_4MPT)-dependent formaldehyde oxidizing pathway(s) (Large and Quayle, 1963; Marison and Attwood, 1982; Chistoserdova et al., 1998; Vorholt et al., 1998b; Pomper, 1999; Vorholt et al., 1999). The activities and structural genes for a methenyl H_4MPT cyclohydrase, a methenyl H_4MPT, and NADP$^+$-dependent methylene H_4MPT have been identified in *M. capsulatus* Bath (Vorholt et al., 1999).

Dye linked formaldehyde oxidation has been observed in a number of methylotrophic bacteria. In general, the activity in methylotrophic bacteria is associated with non-specific aldehyde dehydrogenases, most of which are not induced, or induced at low levels during growth on C_1 compounds (Johnson and Quayle, 1964; Anthony 1975,1982,1996; Patel et al., 1979,1980; Marison and Attwood, 1980,1982; Attwood and Quayle, 1984,1990; Duine et al., 1986; Groen et al., 1986; Klein, 1994). Thus, most DL-FalDHs are not believed to be physiologically significant in formaldehyde metabolism. The exceptions to this rule are found in the dye-linked aldehyde dehydrogenases from *Pseudomonas* sp. RJ1 (Mehta, 1975) and *Hyphomicrobium zavarzinii* ZV 580 (Klein et al., 1994). Both enzymes are induced in cells cultured on C_1 compounds and show optimal activity with formaldehyde. The DL-FalDH has been isolated from *Hyphomicrobium zavarzinii* ZV 580 (Klein, 1994). In methanotrophs, the DL-FalDH has only been isolated from *M. capsulatus* Bath (Zahn et al., 2001). The enzyme is the major formaldehyde oxidizing enzyme in cells cultured in high copper medium and expressing the pMMO. The soluble NAD(P)$^+$-linked formaldehyde oxidation was observed in cells expressing the sMMO (Stirling and Dalton, 1978; Tate and Dalton, 1999). The DL-FalDH from *M. capsulatus* Bath is a membrane-associated homotetramer with a subunit molecular mass of 49,500 Da. PQQ was the only cofactor identified in the enzyme, with a PQQ-to-subunit stochiometry of approximately 1:1. The enzyme is specific for formaldehyde and utilizes the cytochrome $b_{559/569}$ complex as the physiological electron acceptor.

D. Formate Dehydrogenase

In the methanotrophs tested, formate is oxidized to CO_2 by a NAD^+-dependent formate dehydrogenase (FDH) (Anthony, 1982; Hanson and Hanson, 1996). The enzyme has been isolated from *Methylosinus trichosorium* OB3b (Jollie and Lipscomb, 1990, 1991; Yoch et al., 1990). As isolated by Jollie and Lipscomb (1990, 1991), the enzyme is composed of 4 subunits with molecular masses of 98,000 (α), 56,000 (β), 20,000 (γ), and 11,500 (δ) with an $(\alpha\beta\gamma\delta)_2$ subunit structure. The holonezyme contains approximately 2 flavin, 2 molybdenum, 47 iron, and 48 inorganic sulfide. The iron and sulfide are arranged in 5 to 6 different $[Fe\text{-}S]_x$ clusters. In contrast, the FDH isolated by Yoch et al. (Yoch et al., 1990) was shown to be composed of two subunits with molecular masses of 53,800 (α) and 102,600 (β) with a subunit structure of $(\alpha\beta)_2$. As in the case of the two MMOs, and multiple FalDHs, the difference in the subunit structure may reflect the different culture conditions used by each laboratory, Yoch et al. (1990) isolated the FDH from cells expressing the pMMO whereas the enzyme isolated by Jollie and Lipscomb (1990, 1991) (Kiene, 1991) were from cells expressing the sMMO. Unlike the multiple MMO and FalDH examples, the FDH isolated from both laboratories appear to contain similar subunits. The difference in molecular masses between the two reporting laboratories is difficult to explain, but the absence of the two lower molecular mass subunits observed in the enzyme preparations from Yoch et al. (1990) may explain the three-fold lower activity levels.

The cofactor composition of the FDH is complex, with a flavin, a molybdenum and five to six active $[Fe\text{-}S]_x$ centers for an enzyme catalyzing a two electron oxidation (Jollie and Lipscomb, 1991). The cofactor complexity suggests the enzyme may serve multiple functions within the cell. For example in addition to NAD^+, the enzyme has been shown to reduce a number of redox dyes as well as ferricytochrome c_{554} from *M. trichosporium* OB3b (Jollie and Lipscomb, 1991). The probable cytoplasmic locations of NAD^+ and FDH, and periplasmic location for ferricytochrome c_{554}, indicates the reduction of ferricytochrome c_{554} is not physiologically relevant. However, the variety of electron acceptors observed with the purified enzyme does indicate that the enzyme may be a more general source of reducing potential for the cell (Jollie and Lipscomb, 1991).

E. Oxidation of Non-C_1 Substrates

In addition to the known growth substrates, methane and methanol, methanotrophs will oxidize ammonia (Dalton, 1977; Zahn et al., 1994), molecular hydrogen (De Bont, 1976; Kawamura et al., 1983; Chen and Yoch, 1987; Shah et al., 1995; Csaki et al., 2001; Hanczar et al., 2002), and a variety of organic compounds (Colby et al., 1977; Stirling et al., 1979; Anthony 1982; Stanley et al., 1983; Dalton et al., 1984; DiSpirito et al., 1992; Hanson and Hanson, 1996; Lontoh et al., 1999; Han and Semrau, 2000, 2001). With the exception of methane, and in some cases methanol, the oxidation of other substrates does not support growth and has been termed co-oxidation when the initial oxidation is catalyzed by one of the MMOs. Implicit in the use of the term co-oxidation is that the oxidation provides no metabolic energy. However, some co-substrates do generate metabolic energy and may prove physiologically significant (Han and Semrau, 2000; Hanczar et al., 2002).

1. Heterotrophic Substrates

In addition to methane, both MMOs will oxidize or co-oxidize a number of alkanes, alkenes, and in the case of the sMMO a variety of aromatics compounds (Colby et al., 1977; Stirling et al., 1979; Anthony, 1982; Dalton et al., 1984; DiSpirito et al., 1992; Hanson and Hanson,1996; Lontoh et al., 1999, 2000; Han and Semrau 2000, 2001). In general, substrates initially catalyzed by the MMOs are believed to provide little benefit to the cell. In fact, the initial oxidation by the MMOs consumes reducing equivalents, which would prove detrimental to the cell. However, if the product(s) of the MMO catalyzed reaction(s) produce metabolic intermediates that can be utilized as carbon or energy sources these transformations may provide benefits to the cell. For example, chloromethane has been shown to stimulate growth of *Methylomicrobium album* BG8 on methanol (Han and Semrau, 2000; Hanczar et al., 2002).

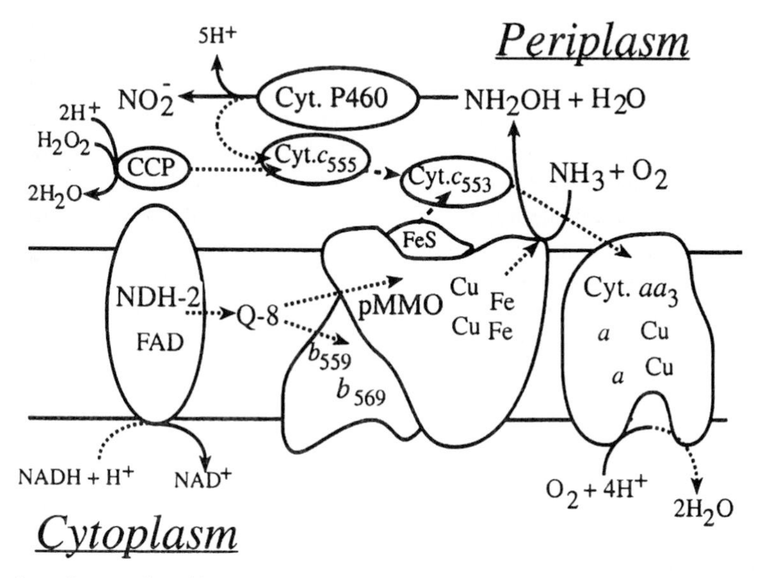

Fig. 2. Proposed mechanism of ammonia oxidation in cells of *M. capsulatus* Bath expressing the pMMO. Substrate oxidation steps are depicted as solid lines, dashed lines represent the path of electron flow. Abbreviations: CCP, cytochrome *c* peroxidase; Cyt., cytochrome; NDH-2; type 2 NADH:quinone oxidoreductase; pMMO, particulate methane monooxygenase.

2. Ammonia

As with aerobic chemoautotrophic ammonia oxidizing bacteria (see Chapter 6 by A. Hooper et al., Vol. 2, this book), methanotrophs oxidize ammonia to nitrite in a two-step process (Fig. 2). The MMO catalyzes the first energy dependent oxidation of ammonia to hydroxylamine (Colby et al., 1977; Dalton, 1977). Both forms of the MMO have been shown to catalyze this reaction (Colby et al., 1977; Dalton 1977; Prior and Dalton 1985; Stirling et al., 1979). The second step involves the four electron oxidation of hydroxylamine to nitrite (Dalton, 1977; Sokolov et al., 1980). Originally, hydroxylamine was believed to be oxidized by a hydroxylamine oxidoreductase-like enzyme (Sokolov et al., 1980) but was later shown to be catalyzed by cytochrome P460 (Zahn et al., 1994) (see cytochrome P460). Based on the K_m and V_{max}, cytochrome c_{555} appears to be the physiological electron acceptor although the enzyme will also reduce cytochrome c_{553} and cytochrome c'.

3. Hydrogen

H_2 uptake and evolutions has been observed in a number of methanotrphs (De Bont, 1976; Kawamura et al., 1983; Chen and Yoch, 1987; Shah et al., 1995; Csaki et al., 2001; Hanczar et al., 2002). In general, the hydrogenase activities in methanotrophs are constitutive, although inducible hydrogenase activity has been observed in *M. trichosporium* OB3b (Chen and Yoch, 1987). Of the strains examined, hydrogen oxidation activity in *M. capsulatus* Bath is the best characterized system. *M. capsulatus* Bath contains two constitutive hydrogenases, one soluble and the other membrane-bound (Csaki et al., 2001; Hanczar et al., 2002). The soluble hydrogenase is an NAD^+-linked enzyme that was shown to provide reducing equivalents to both MMOs (Hanczar et al., 2002).

Based on the structural gene sequence (*hupE*) (Csaki et al., 2001), the membrane-associated hydrogenase in *M. capsulatus* Bath is a class I [NiFe]-hydrogenase (Cauvin et al., 1991; Dross et al., 1992; Wu and Mandrand 1993; Vignais and Toussaint 1994; Vignais et al., 2001). Consistent with this class of bacterial hydrogenases, *hupE* is found on an operon containing the structural gene for a *b*-type cytochrome (*hupC*), which has been shown to mediate electron flow from the hydrogenase to a quinone in class I hydrogenases (Cauvin et al., 1991; Dross et al., 1992; Vignais et al., 2001).

III. Respiratory Components

A. Soluble Cytochromes

In contrast to the limited role of *c*-type cytochromes

in the oxidation of growth substrates, methanotrophs show complex cytochrome *c* patterns similar to that observed in the facultative methylotrophs (Anthony, 1975, 1991, 1992a; DiSpirito, 1990; DiSpirito et al., 1990; Long and Anthony, 1991; Zahn et al., 1994, 1996a, 1997; Bergmann et al., 1999). The monoheme *c*-type cytochromes are similar in both methylotrophs and methanotrophs, but the two groups differ in the nature of multiheme cytochromes and non-*c*-type cytochromes. The following discussion will focus on the soluble cytochromes in *M. marinus* A45 and *M. capsulatus* Bath.

1. c_L and c_H-type Cytochromes

Like other methylotrophic bacteria, methanotrophs have two major, on a concentration basis, *c*-type monoheme cytochromes (O'Keeffe and Anthony, 1980; Ohata and Tobari, 1981; Ambler et al., 1986; DiSpirito 1990; DiSpirito et al., 1990; Anthony, 1992a; Zahn et al., 1994, 1997). Originally, the two major periplasmic *c*-type cytochromes in methylotrophic bacteria were labeled cytochrome c_H and cytochrome c_L in accordance with their isoelectric points (Anthony, 1992a). In methylotrophic bacteria the low isoelectric point cytochrome, cytochrome c_L, is the electron acceptor during methanol oxidation by PQQ-containing MDHs. The amino acid sequence of cytochrome c_L indicates it constitutes a novel class of *c*-type cytochromes (Nunn and Anthony, 1988; Anthony 1992a). Cytochrome c_{555} from *M. capsulatus* Bath and possibly cytochrome c_{554} from *M. marinus* A45 (Table 1) fall into the cytochrome c_L group (Ambler et al., 1986; DiSpirito et al., 1990; Zahn et al., 1996a).

The second major soluble cytochrome in methylotrophic bacteria, cytochrome c_H, fall into Ambler's the class I *c*-type cytochromes (Ambler, 1982,1991). This class of cytochromes usually serves as an electron shuttle between cytochrome bc_1 and the terminal oxidase. In *M. capsulatus* Bath cytochrome c_{553} (Table 1) functions as the electron shuttle between the *bc* complex and cytochrome aa_3, as well as between ferrocytochrome c_{555} and cytochrome aa_3 (DiSpirito et al., 1994; Zahn et al., 1996b). Consistent with its physiological role, cytochrome c_{553} is a typical class I *c*-type cytochrome in size, with a heme site near the N- terminus: N-SEDLAKALNCVMCHSVDK-KILGPAFKDVAQK- (Table 1) (Ambler, 1982,1991; Zahn et al., 1996b,1997).

2. Cytochrome P460

In *M. capsulatus* Bath, cytochrome P460 is responsible for the four electron oxidation of hydroxylamine to nitrite (Table 1) (Zahn et al., 1994). The enzyme is similar but not identical to cytochrome P460 from *Nitrosomonas europaea* (Bergmann and Hooper, 1994; Bergmann et al., 1998). In *N. europaea* two different enzymes have been shown to oxidize hydroxylamine to nitrite, hydroxylamine oxidoreductase (HAO) and cytochrome P460 (Hooper and Nason, 1965; Erickson and Hooper, 1972; Hooper et al., 1984; Numata et al., 1990; Hooper et al., 1997) (see also Chapter 6 by A. Hooper et al., Vol. 2, this book). HAO is the periplasmic enzyme responsible for hydroxylamine oxidation in chemoautotrophic ammonia oxidizing bacteria (Hooper and Nason, 1965; Hooper et al., 1997; Hooper, 2002). The physiological role of cytochrome P460 in chemolithoautotrophic nitrifiers is still in question, it may serve as a secondary enzyme to protect the organism from this toxic and mutagenic intermediate. In *M. capsulatus* Bath, cytochrome P460 is the main, if not sole, enzyme responsible for the oxidation of hydroxylamine (Zahn et al., 1994).

3. Cytochrome c'

Cytochromes c' represent a class of type II periplasmic cytochromes with highly conserved physiochemical and structural properties over a wide range of species (Ren et al., 1993). The crystal structures of *Rhodospirillum molischianum* and *Chromatium vinosum* cytochromes c' have shown the cytochromes to have a dimeric (α_2) quarternary structure with each subunit consisting of a 4 helix antiparallel bundle (Weber et al., 1981; Finzel et al., 1985; McRee et al., 1990; Renz et al., 1993). Analysis of the amino acid sequences from 11 other eubacterial cytochromes c' by secondary structure prediction models has confirmed that the 4 helix antiparallel bundle is also present in all other eubacterial cytochromes c' with available sequence data (Ambler et al., 1981). Despite the conserved physiochemical and structural properties that exist for this class of cytochromes, it is surprising to note that only about 6 to 11% of the amino acids which compose the primary structure of cytochromes c' are found to be conserved throughout the class (Ambler, 1982,1991).

Previous studies have suggested that several properties of cytochrome c' from *M. capsulatus* Bath are atypical of the conserved physiochemical and struc-

Table 1. Properties of the soluble cytochromes from *M. capsulatus* Bath and *M. marinus* A45

Property	Value for cytochrome:									
	Methylobacter marinus A45				*Methylococcus capsulatus* Bath					
	c_{554}	c_{552}	c_{553}	c_{551}	c_{553}	c_{555}	CCP	*c'*	P460	c_{553O}
Mass (kDa)										
Subunit	8.5	14	34	16.5	9.4	11.1	38.8	16.2	16.4	124.4
Subunit composition			α_2	α_2			α_2	α_2	α_2	α_2
Isoelectric point	5.6	4.7	4.9	4.8	8.4	5.9	4.5	7.0	7.0	6.0
Concentration (pmol/mg cell protein)	974	449	128	150	130	610	185	257	420	38
Heme (mol/subunit)	1	1	2	2	1	1	2	1	1	8
Redox Potential (mV)	nd	nd	nd	nd	nd	+175 to +195	–254 and +432	–205	–300 to –380	nd
Absorption maxima (nm)										
Oxidized	413	408	406	410	410	411	413	401	419	411
Dithionite-reduced										
γ-band	418	417	419	416	418	419	421	420	463	421
β-band	524	522	524	523	524	523	526	–	–	530
α-band	554	552	553	551	553	555	556	–	–	553
EPR (signals)										
LS1										
g_z	3.15	–	3.32	–	nd	2.90	3.23	–	–	3.66
g_y	2.19	–	–	–		2.31	2.02	–	–	1.8
g_x	1.32	–	2.0	–		2.03	1.80	–	–	< 0.7
LS 2										
g_z	–	–	2.88	–		–	2.79	–	–	2.97
g_y	–	–	2.35	–		–	2.40	–	–	2.26
g_x	–	–	1.55	–		–	1.58	–	–	1.49
HS1										
g_z	–	6.20	5.39	6.05		–	6.05	6.29	6.19	6.0
g_y	–	3.42	4.3	2.98		–	4.28	5.35	5.70	4.3
g_x	–	2.1	2.0	2.28		–	2.00	2.0	2.0	2.0
HS2										
g_z	–	–	–	–		–	–	6.06	6.0	–
g_y	–	–	–	–		–	–	5.34	5.30	–
g_x	–	–	–	–		–	–	2.00	2.00	–
References	1,2	1,2	1,2	1,2	3	3,4,7	6, 7	6, 7	4, 8, 10	9

[1]DiSpirito 1990; [2]DiSpirito et al., 1990; [3]Zahn and DiSpirito, unpublished; [4]Ambler et al. 1986; [4]Zahn et al. 1994, [5]1996a, [6]1996b; [7]1997; [8]Bergman et al. 1998; [9]1999; [10]2000

tural properties exhibited by other c′-cytochromes (Zahn et al., 1996a; Bergmann et al., 2000). Most notable are the differences observed for the native and subunit molecular mass and the redox potential, while other properties, including the electron paramagnetic spectra, UV-visible spectra, ligand-binding behavior, and amino acid composition are virtually identical to other characterized c′-cytochromes. The greater molecular mass of the c′ cytochrome from *M. capsulatus* and lack of conserved amino acid residues makes alignment of this sequence to others extremely difficult. However, several residues of the cytochrome c′ of *M. capsulatus* Bath appear to be found in most other c′ cytochromes, the Arg 12, which forms a hydrogen bond to one of the heme proprionate groups, Tyr 53, Gly 56, Gly 116, the three residues of the c-heme binding site motif (Cys 119, Cys 122, and His 123), and Lys 127 (Ambler, 1991; Yasui et al., 1992). The rather low sequence similarity between cytochrome c′ of *M. capsulatus* Bath and other c′ cytochromes does not necessarily mean that the cytochrome c′ of *M. capsulatus* does not share important features of secondary and tertiary structure with other c′ cytochromes. Even c′-type cytochromes with low similarity at the level of amino acid sequence, such as those from *Chromatium vinosum* and *Rhodospirillium molischianum*, can have very similar three dimensional structures (Ren et al., 1993).

The physiological function of cytochrome c′ from *M. capsulatus* Bath is still in question. Initial observations indicated the cytochrome c′ was involved in electron flow from cytochrome P460 to cytochrome c_{555}. However, kinetic experiments have shown that the rate of electron flow in *M. capsulatus* Bath from cytochrome P460 to cytochrome c_{555} is identical in the presence or absence of cytochrome c′ (Zahn et al., 1994,1996b). In addition, the genes encoding the cytochrome c′ (*ccp*) and cytochrome P460 (*cyp*) appear to be regulated by different factors. The transcription of *ccp* appears to be induced by ammonium, while transcription of *cyp* is not (Bergmann et al., 2000). This is consistent with the observation that while levels of cytochrome P460 remain constant following addition of ammonium to growth media (Bergmann et al., 1998), whereas levels of cytochrome c′ increase three-fold (Zahn et al., 1994). Recent observations indicate the cytochrome may play a protective or detoxification role similar to cytochrome P460 in *N. europaea* and not as the initial electron acceptor from cytochrome P460. For example, cytochrome c′ rapidly and irreversibly binds hydroxylamine and NO.

4. Phylogenetic Relationship Between Cytochrome c′ and Cytochrome P460

Cytochrome c′ from *M. capsulatus* Bath shows considerable sequence similarity to cytochromes P460 from both *M. capsulatus* Bath and the autotrophic nitrifying bacterium *N. europaea* (Bergmann and Hooper, 1994; Bergmann et al., 1998, 2000). Cytochrome c′ and cytochrome P460 of *M. capsulatus* Bath have 31% identical amino acid residues, and cytochrome c′ of *M. capsulatus* Bath and cytochrome P460 of *N. europaea* have 18% identical amino acid residues. Consistent with a phylogenetic relationship to cytochromes c′, both sequenced cytochromes P460 have the basic traits of this class of type II cytochromes. For example, both cytochromes P460 have the C-terminal heme binding regions, the alignment with glycine 56 and lysine 127 of cytochrome c′. It is interesting to note that the residue at position 19 is arginine in *N. europaea* cytochrome P460 aligns with arginine 12 in *M. capsulatus* Bath cytochrome c′ and is replaced by a similar lysine in *M. capsulatus* Bath cytochrome P460. In all three cases the long, positively charged side chain may interact with a heme proprionate group, as observed in all known c′-type cytochromes (Ren et al., 1993). These results suggest an overall similarity in the secondary and tertiary structures of all three cytochromes, which may consist of four antiparallel α-helices, with the heme attached to the carboxy-terminal helix by two sulfhydryl linkages, as in other c′. cytochromes (Yasui et al., 1992; Ren et al., 1993; Tahirov et al., 1996). One important structural difference between the two cytochromes P460 and cytochrome c′ of *M. capsulatus* Bath is the lysine residue at position 74 in cytochrome P460 from both *N. europaea* and *M. capsulatus* Bath which is absent in *M. capsulatus* Bath cytochrome c′. This lysine residue is believed to be the site of a novel covalent crosslink between the polypeptide of cytochrome P460 of *N. europaea* and the heme (Arciero and Hooper, 1997).

The sequence similarity between cytochrome c′ of *M. capsulatus* Bath and the cytochromes P460 indicates that the cytochromes P460 may have evolved from a c′-type cytochrome, possibly as a result of a gene duplication event in a methanotroph. One of the duplicated genes for the cytochrome c′ developed a covalent crosslink between a lysine residue of the polypeptide and the heme, enabling the new cytochrome, cytochrome P460, to catalyze the oxidation of hydroxylamine.

5. Cytochrome c Peroxidase

The di-heme cytochrome *c* peroxidase (CCP), initially called cytochrome c_{557}, was first identified during the purification of cytochrome aa_3 from *M. capsulatus* Bath (DiSpirito et al., 1994). Subsequent characterization (Zahn et al., 1997) demonstrated the enzyme is a classic cytochrome *c* peroxidase (Table 1) similar to the CCPs observed in other methylotrophic (Bosma et al., 1987; Goodhew et al., 1990; Long and Anthony 1991) and the ammonia oxidizing bacterium *N. europaea* (Arciero and Hooper, 1994). The peroxidases from all three bacterial groups have similar molecular masses (33,700–44,000 Da) with one low-spin, high potential heme, and one high-spin, low-potential heme. The physiological function of bacterial cytochrome *c* peroxidases has not been determined. In nitrifiers and methanotrophs, where substrate oxidation requires both activation of dioxygen via the AMO or MMO as well as the reduction of dioxygenase by there respective terminal oxygenases, the presence of a cytochrome *c* peroxidases may reflect the need for a periplasmic H_2O_2 detoxification enzyme.

Di-heme *c*-type cytochromes with similar molecular masses and spectral properties have been isolated from *M. marinus* A45 and identified in *Methylomonas* sp. MN (DiSpirito, 1990; DiSpirito et al., 1990). However, peroxidase activity was not tested in either cytochrome.

6. Cytochrome c_{553}

Studies on ammonia regulation of cytochrome *c*′ and cytochrome P460 in *M. capsulatus* Bath lead to the identification a novel group of high molecular mass mutiheme cytochromes (Bergmann et al., 1998). Of the cytochromes identified in this group, only cytochrome c_{5530} has been characterized (Bergman et al., 1999). Cytochrome c_{553} is a homodimer with subunit molecular mass of 124,350 Da and an isoelectric point of 6.0. The heme *c* concentration was estimated to be 8.2 ± 0.4 mol heme *c* per subunit consistent with the 8 heme binding motif (Bergmann et al., 1999). The electron paramagnetic resonance spectrum showed the presence of multiple low spin, S = 1/2, hemes (Table 1). The structural gene for cytochrome c_{553} is part of the Occ gene cluster which contains three other open reading frames (ORFs). ORF 1 encodes a putative periplasmic *c*-type cytochrome with a molecular mass of 118,620 Da, that shows approximately 40% amino acid sequence identity with *occ*, and contains nine *c*-heme binding motifs. ORF3 encodes a putative periplasmic *c*-type cytochrome with a molecular mass of 94,000 Da, contains seven *c*-heme binding motifs, but shows no sequence homology to *occ* or ORF1. ORF4 encodes a putative 11,100 Da protein. The four ORFs have no apparent similarity to any proteins in the GenBank database. The subunit molecular masses, arrangement and number of hemes, and amino acid sequences demonstrate that cytochrome c_{553}, as well as the gene products of ORF1 and ORF3, constitute a new class of *c*-type cytochrome.

The physiological role for these proteins is still unknown. However, one or more of these high molecular mass cytochrome appear to be induced by ammonia and may function like cytochrome c_{554} in *N. europaea* (Bergmann et al., 1998)(see also Chapter 6, Hooper, this volume, for discussions on cytochrome c_{554}).

B. Membrane-Associated Respiratory Components

Compared to the soluble components, the membrane-associated components of the respiratory chains in methanotrphs are not well characterized. With the exception of the quinines (Urakami and Komagata, 1986), the membrane-associated respiratory components of the electron transport chain has only been examined in *M. capsulatus* Bath.

1. NADH:Quinone oxidoreductase (NDH)

Cook and Shiemke (2002) have recently described the isolation of a type-2 NADH:quinone oxidoreductase (NDH-2) from *M. capsulatus* Bath. The enzyme has a subunit molecular mass of 36,000 Da and contained flavin adenine dinucleotide (FAD) as the sole cofactor. The purified enzyme oxidized NADH, but not NADPH, and reduced a variety of quinones including ubiquinone-0, duroquinone, and menaquinone. Co-purification of pMMO-NDH-2 and stimulation of methane oxidation activity by the addition of NDH-2 and NADH to pMMO preparations also suggests the electron flow from NADH to pMMO is as follows: NADH → NDH-2 → ubiquinol → pMMO (Choi et al., 2003).

2. Cytochrome bc_1 Complex

Little is know on the molecular properties of the cytochrome bc_1 complex in methanotrophs. *M. capsulatus* Bath has been shown to contain ubiquinone-8 (Urakami and Komagata, 1986) and high concentrations of heme B (0.35 to 0.4 nmol/mg cell protein) compared to heme A (0.07 to 0.09 nmol/mg cell protein) (Fig. 3) (Zahn and DiSpirito, 1996). Inhibitor data indicates methanotrophs contain a bc_1-type complex (see Table 2, Fig. 4), however, the complex has only been partially purified, called cytochrome $b_{559/569}$, and little information outside of the spectral characterization of the two heme *b* groups on the molecular properties was presented (Zahn and DiSpirito, 1996; Zahn et al., 2001).

3. Terminal Oxidases

A cytochrome aa_3-type oxidase has been isolated from *M. capsulatus* Bath (DiSpirito et al., 1994). The oxidase was a typical bacterial oxidase composed of three subunits with molecular masses of 46,000, 28,000, and 20,000 Da. The optical and electron paramagnetic resonance also suggested the enzyme was an aa_3-type oxidase. However, in contrast to other aa_3-type oxidiases, which normally contain a 0.5 to 1.5 copper to heme *a* ratio, the oxidase from *M. capsulatus* Bath showed a copper to heme *a* ratio of 3 to 1. In addition, dialysis against buffers containing 5 mM Na-ethylenediaminetetracetate failed to alter the copper to heme ratios. Subsequent studies have shown the excess copper associated with cytochrome aa_3 is due to the presence of the copper binding compound

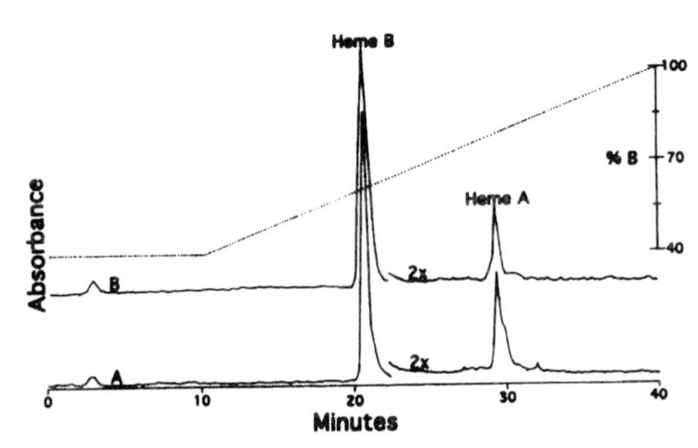

Fig. 3. Reverse-phase high performance liquid chromatographic (HPLC) profile of non-covalently bound heme in membrane fractions. Solvent extracted heme from KCl-washed membranes of *M. capsulatus* Bath expressing the pMMO (A) and sMMO (B) (Zahn and DiSpirito, 1996). HPLC instrument conditions were as follows: flow rate, 4 ml/minute; detector wavelength, 400 nm; column, VyDac semi-preparative C_{18} column (10 x 250 nm).

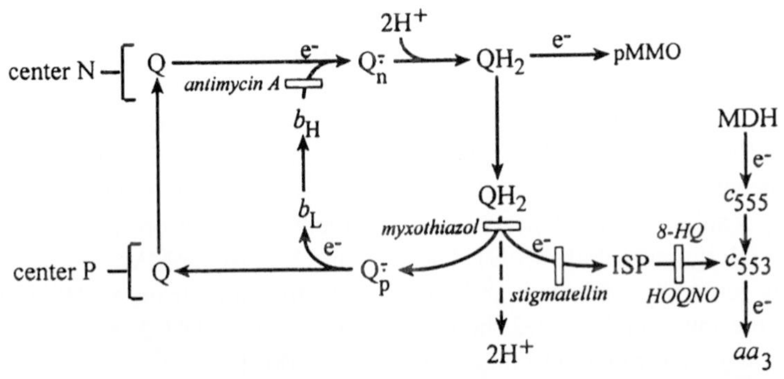

Fig. 4. Model of the electron flow in the Q-cycle, and to the pMMO *M. capsulatus* Bath. Open rectangles show the reactions that are blocked (figure based after Brandt and Trumpower (Brand and Trumpower, 1994).

Table 2. Effect of inhibitors of the cytochrome bc_1-complex on methane and methanol oxidation by cell free fractions and on methane oxidation by purified pMMO preparations. The 100% values for propylene oxidation by purified preparation of pMMO was 12 to 38 nmol•min^{-1}•mg protein^{-1} using duroquinol as the reductant and by cell free fractions 48 nmol•min^{-1}•mg protein^{-1} using NADH as the reductant. Rate of methanol dependent oxygen uptake was 143 nmol•min^{-1}•mg protein^{-1}.

Inhititor/Uncoupler	Concn (μM)	% Inhibition or Stimulation of Propylene Oxidation		% Inhibition of CH_3OH oxidation	Ref.*
		Cell Extract	pMMO	Cell Extracts	
Antimycin A	5	47	5	102	1
	10	17	0	96	
	50	0	0	93	
8-Hydroxyquinone	25	106	nd	100	1,2,3
	50	120	nd	100	
	100	0	nd	86	
2-*M*-heptyl-4-Hydoxyquinone-	10	122	-	100	1,2
N-Oxide	25	184	156	100	
	50	210	0	100	
	100	0	0	117	
Myxothiazol	20	100	95	104	1,2
	150	138	167	100	
	350	150	128	109	
Stigmatellin	2	97	nd	100	1
	5	111	nd	104	
	10	132	nd	102	

*[1]Kunz, Choi and DiSpirito unpublished; [2]Zahn and DiSpirito, 1996; Dalton et al., 1984

which co-purifies with the oxidase, as it does with the pMMO (Zahn and DiSpirito, unpublished).

The presence of ba_3-type oxidase has also been reported in *M. capsulatus* Bath (Nguyen et al., 1998). Unfortunately, no information on the properties of the oxidase was provided in the report.

IV. Respiratory Chains

A. *Respiratory Components in Cells Expressing Different MMOs*

In methanotrophs which express both forms of the MMO, a major difference in the copper and iron concentration is observed in cells expressing the different MMOs (Nguyen et al., 1994; Zahn and DiSpirito, 1996; Choi et al., 2003). The majority of the copper and iron in cells expressing the pMMO appear to be associated with the pMMO and Cu-cbc (Zahn and DiSpirito, 1996; DiSpirito et al., 1998). Comparison of the EPR spectrum of the washed membrane fraction from cells expressing the pMMO verses the membrane fraction from cells expressing the sMMO show a very intense copper signal with a $g_{\perp} = 2.057$ and hyperfine splitting constants of $g_{\parallel} = 2.27$ and $|A_{\parallel}| = 174$ G (Fig. 5) similar to the type II copper center of the pMMO (Zahn and DiSpirito, 1996; Yuan et al., 1997,1998a,b,1999; Tagekuchi et al., 1999a; Basu et al., 2002; Choi et al., 2003; Liebermann et al., 2003). Subtractive EPR spectra of the membrane samples shows a minor increase for the g = 4.05 and g = 4.61 signals from the membranes of cells expressing the sMMO and a significant increase in the population of a g = 6.0 signal from the membrane samples from cells expressing the pMMO (Fig. 5). Preliminary studies of *b*-type cytochrome from *M. capsulatus* Bath have indicated the signals at g = 4.05 and g = 4.61 arise from

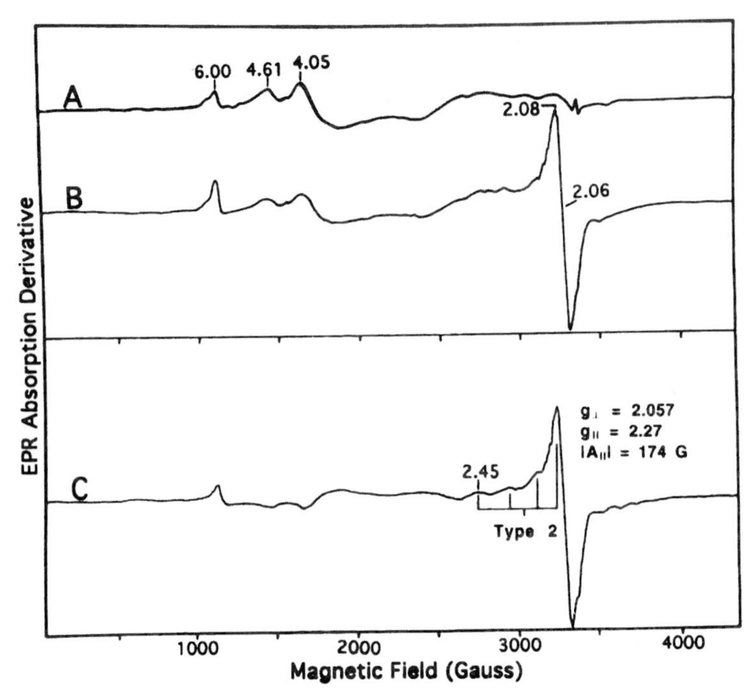

Fig. 5. EPR spectra for the washed membrane fractions from *M. capsulatus* Bath expressing the sMMO (A) and pMMO (B) and the difference spectra of trance B minus trace A. Protein concentration was 57.5 mg/ml, operating parameters were; modulation frequency, 100 kHz; modulation amplitude, 10 G; time constant 100 ms. The microwave frequency was 9.421 GHz and the microwave power was 2.02 mW.

cytochrome $b_{559/569}$ (Zahn and DiSpirito, unpublished results). The results are consistent with the small, 2–4%, increase in the heme B concentration per mg membrane protein in the membrane fraction of cells expressing the sMMO (Fig. 3) (Zahn and DiSpirito, 1996). The signal at g = 6 appears to be associated with the non-heme iron center of the pMMO (Zahn and DiSpirito, 1996b; Takeguchi et al., 1998, 1999b; Takeguchi, 2000).

With respect to respiratory components, the only major change was a 45% increase in the concentration of heme C in the soluble fraction of cells expressing the sMMO although the increase has not been associated with any soluble *c*-type cytochrome (Zahn and DiSpirito, 1996). Thus, with the possible exception of the soluble cytochrome *c* composition, the respiratory components in cells expressing the sMMO are probably similar in both type and composition to cells expressing the pMMO. Electron flow during methanol and formate oxidation steps, as well electron flow from NAD-2 and cytochrome bc_1 complex to the terminal oxidase appears similar in cells expressing the pMMO and sMMO. However, the change in MMO composition, as well as FalDH composition, does alter the overall nature of electron flow in methanotrophs.

B. Respiration in Methanotrophs Expressing the sMMO

A working model for electron flow during methane oxidation in cells expressing the sMMO is shown in Fig. 6. With respect to respiratory chains the major difference between cell expressing the sMMO and pMMO lies in the nature of electron donor to the MMO and electron acceptor of the FalDH. In cells expressing the sMMO, NADH is the physiological electron donor to the sMMO and NAD^+ is the major physiological electron acceptor during formaldehyde oxidation (Colby et al., 1977; Colby and Dalton, 1978; Dalton et al., 1984; Tate and Grisshammer, 1996; Wallar and Lipscomb, 1996; Murrell et al., 2000; Zahn et al., 2001).

C. Respiration in Methanotrophs Expressing the pMMO

A number of studies have indicated the pMMO is coupled to the electron transport chain probably at the quinone/ semiquinone or cytochrome bc_1 complex level (Colby et al., 1977; Dalton et al., 1984; Zahn and DiSpirito, 1996). The effects of several cytochrome bc_1 inhibitors on the oxidation of methane

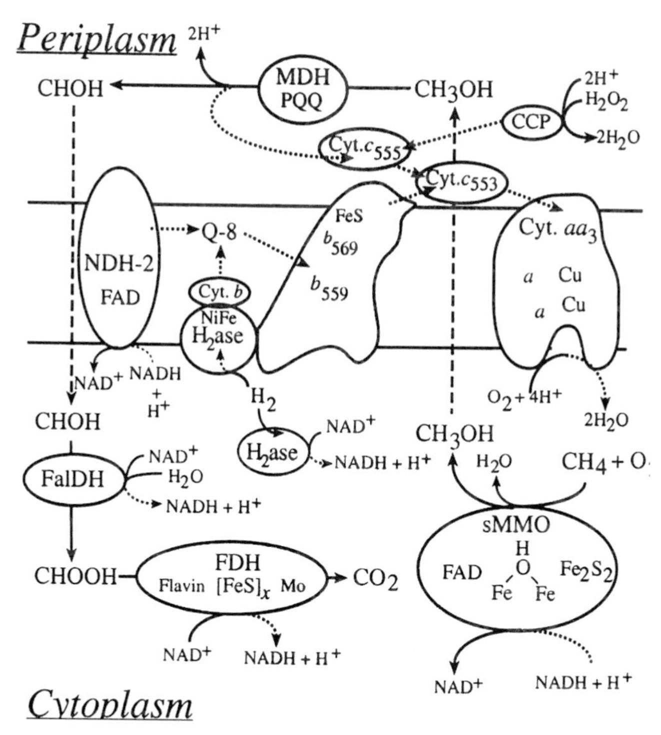

Fig. 6. Model for the oxidation of methane to carbon dioxide by cells of *M. capsulatus* Bath expressing the sMMO. Substrate oxidation steps are depicted as solid lines, dashed lines represent the path of electron flow. Abbreviations: Cyt, cytochrome; DL-FalDH, dye-linked formaldehyde dehydrogenase; FDH, formate dehydrogenase; H_2ase, hydrogenase; MDH, methanol dehydrogenase; NDH-2, type 2 NADH: quinone oxidoreductase; NAD(P)-FalDH, NAD(P)-linked formaldehyde dehydrogenase; pMMO, particulate methane monooxygenase; Q-8, ubiquinone-8; sMMO, soluble methane monooxygenase.

in cell free extracts expressing the pMMO, and on isolated pMMO preparations were consistent with these earlier observations (Fig. 4, Table 2). Assuming the cytochrome *bc*-complex in *M. capsulatus* Bath is similar to studied mitochondrial and bacterial complexes (Jagow and Link, 1986; Brand and Trumpower, 1994; Matsumo-Yagi and Hatefi 1996, 1997, 1999; Junemann et al., 1998; Zhang et al., 1998; Snyder et al., 2000), the results point to an electron flow from the high potential heme *b* center to pMMO possibly via ubiquinone 8. The three best characterized inhibitors of the bc_1-complexs, antimycin A, mixothiazol and stigmatellin, all provide an indication that electrons flow to the pMMO via the high potential heme *b* probably via ubiquinone 8. Antimycin, a N center inhibitor, binds near the high potential *b*-heme, and blocks reduction of cytochrome *b* through center N, but does not inhibit reduction of cytochrome *b* through center P (Brand and Trumpower, 1994; Zhang et al., 1998; Matsumo-Yagi and Hatefi, 1999). Stigmatellin and mixothiazol (P center inhibitors) bind at similar but different sites near cytochrome b_L (Zhang et al., 1998). Both inhibitors block reduction of cytochrome c_1 (cytochrome c_{553} in *M. capsulatus* Bath), but do not block reduction of the *b* cytochromes through center N. If antimycin blocks cytochrome *b* reduction and the P center inhibitors, mixothiazol and stigmatellin, blocks the reduction of cytochrome c_1 (cytochrome c_{553} in *M. capsulatus* Bath), one would predict pMMO activity would be inhibited by antimycin, but not by mixothiazol or stigmatellin if the physiological reductant to the pMMO was via cytochrome *b*. When tested antimycin was found to inhibit pMMO and both mixothiazol and stigmatellin stimulated pMMO activity. In cell free extracts, the stimulation of pMMO activity by mixothiazol and stigmatellin is interpreted as the preferential shuttling of electrons to pMMO. Less is known of the mechanism of inhibition of 8-hydroxyquinone (8-HQ) and HQNO (Jagow and Link, 1986), but if both inhibit electron from the FeS center in the cytochrome bc_1 complex the results are consistent with the electron flow proposed in Figs. 3 and 7. Both inhibitors stimulated pMMO activity in cell free extracts at low concentrations, but completely

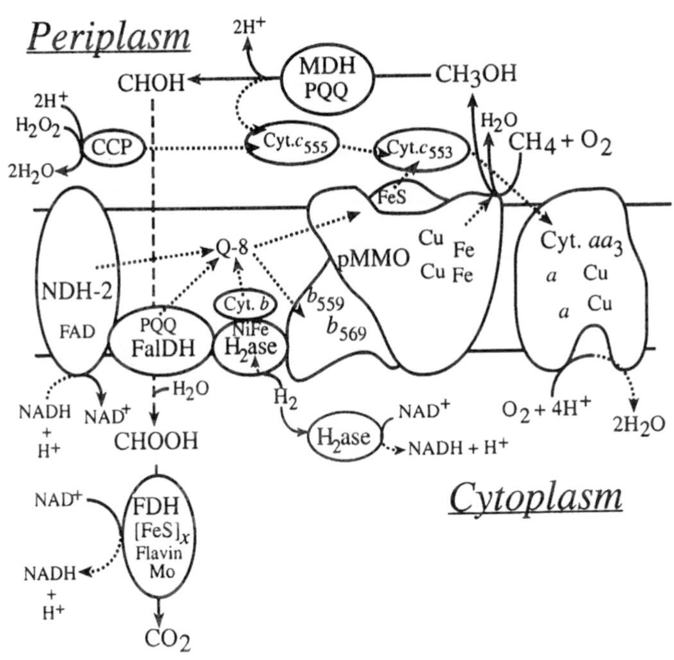

Fig. 7. Model for the oxidation of methane to carbon dioxide by cells of *M. capsulatus* Bath expressing the pMMO. Substrate oxidation steps are depicted as solid lines, dashed lines represent the path of electron flow. Abbreviations: Cyt, cytochrome; DL-FalDH, dye-linked formaldehyde dehydrogenase; FDH, formate dehydrogenase; H_2ase, hydrogenase; MDH, methanol dehydrogenase; NDH-2, type 2 NADH: quinone oxidoreductase; NAD(P)-FalDH, NAD(P)-linked formaldehyde dehydrogenase; pMMO, particulate methane monooxygenase; Q-8, ubiquinone-8; sMMO, soluble methane monooxygenase.

inhibited pMMO activity at higher concentrations.

An alternative interpretation of these results is that the pMMO has a quinone or semiquinone binding site that the above inhibitors bind to and either stimulate or inhibit electron transfer to the pMMO. The effect of these inhibitors on the purified enzyme is consistent with this interpretation. Overall, results of the inhibitor studies indicate that the cytochrome bc_1 complex directs electrons the pMMO probably via ubiquinone 8.

Additional evidence that the pMMO is coupled to the electron transport chain at the cytochrome bc_1 complex is provided by the results from studies on the membrane-associated DL- FalDH (Zahn et al., 2001). This enzyme is the major formaldehyde oxidizing enzyme in cells cultured in high copper media and expressing the pMMO. The soluble $NAD(P)^+$-linked formaldehyde oxidation was the major activity in cells cultured in low copper medium and expressing the soluble methane (Vorholt et al., 1998b; Tate and Dalton, 1999). The DL-FalDH is specific for formaldehyde, oxidizing formaldehyde to formate and utilized the ubiquinone or diheme *b*-type cytochrome, cytochrome $b_{559/569}$, as the physiological electron acceptor. Our working model for electron flow during methane oxidation in cells expressing the pMMO is shown in Fig. 7.

Acknowledgments

This work was supported by a grant from the Department of Energy (02-96ER20237).

References

Ambler RP (1982) The structure and classification of cytochromes *c*. In: Robinson AB, Kaplan NO (eds) From Cyclotrons to Cytochromes, pp 263–280. Academic Press, New York

Ambler RP (1991) Sequence variability in bacterial cytochromes *c*. Biochim Biophys Acta 1058: 263–280

Ambler RP, Bartsch RG, Daniel M, Kamen MD, McLellan L, Meyer TE and van Beeumen J (1981) Amino acid sequences of bacterial cytochrome *c*, and cytochrome c_{556}. Proc Nat Acad Sci USA 7: 6854–6857

Ambler RP, Dalton H, Meyer TE, Bartsch RG and Kamen MD (1986) The amino acid sequence of cytochrome c_{555} from *Methylococcus capsulatus* Bath. Biochem J 233: 333–337

Anthony C (1975) The microbial metabolism of C_1 compounds.

The cytochromes of *Pseudomonas* AM1. Biochem J 146: 289–298

Anthony C (1982) The Biochemistry of Methylotrophs. Academic Press, London

Anthony C (1992a) The *c*-type cytochromes of methylotrophic bacteria. Biochim Biophys Acta 1099: 1–15

Anthony C (1992b) The structure of bacterial quinoprotein dehydrogenase. Int J Biochem 24: 29–38

Anthony C (1996) Quinoprotein-catalysed reactions. Biochem J 320: 697–711

Arciero DM and Hooper AB (1994) A di-heme cytochrome *c* peroxidase from *Nitrosomonas europaea* catalytically active in both the oxidized and half-reduced states. J Biol Chem 269: 11878–11886

Arciero DM and Hooper AB (1997) Evidence for a crosslink between *c*-heme and a lysine residue in cytochrome P460. FEBS Lett 410: 457–460

Ashcroft SJ and Ashcroft FM (1990) Properties and functions of ATP-sensitive K-channels. Cell Signal. 2:1 97–214

Attwood MM (1990) Formaldehyde dehydrogenases from methylotrophs. Meth Enzymol 188: 314–324

Attwood M and Quayle JR (1984) Formaldehyde as a central intermediary metabolite in methylotrophic metabolism. In: Crawford RL and Hanson RS (eds), Microbial Growth on C_1 Compounds, pp 315–323. American Society for Microbiology Press, Washington DC

Basu P, Katterle B, Anderson KA and Dalton H (2002) The membrane-associated form of methane monooxigenase from *Methylococcus capsulatus* Bath is a copper/iron protein. Biochem J 369: 417–427

Bergmann DJ and Hooper AB (1994) The primary structure of cytochrome P460 of *Nitrosomonas europaea*: the presence of a heme-*c*-binding motif. FEBS Lett 353: 324–326

Bergmann DJ, Zahn JA, Hooper AB and DiSpirito AA (1998) Cytochrome P460 genes from the methanotroph, *Methylococcus capsulatus* Bath. J Bacteriol 180: 6440–6445

Bergmann DJ, Zahn JA and DiSpirito AA (1999) High-molecular-mass multi-*c*-heme cytochromes from *Methylococcus capsulatus* Bath. J Bacteriol 181:9 91–997

Bergmann D, Zahn JA and DiSpirito AA (2000) Primary structure of cytochrome *c*′ from *Methylococcus andapsulatus* Bath: evidence of a phylogenetic link between P460 and *c*′-type cytochromes. Arch Microbiol 173: 29–34

Bosma G, Braster M, Southamer A and Vereveld WH (1987) Subfractionation and characterization of soluble *c*-type cytochromes from *Paracoccus denitrificans* cultured under various limiting conditions in the chemostat. Eur J Biochem 164: 665–670

Brand U and Trumpower B (1994) The protonmotive Q cycle in mitochondria and bacteria. Biochem. Mol Biol 29: 165–197

Cauvin B, Colbeau A and Vignais PM (1991) The hydrogenase structural operon in *Rhodobacter capsulatus* contains a third gene, *humM*, necessary for the formation of physiologically competent hydrogenase. Mol Microbiol 5: 2519–2527

Chang S-L, Wallar BJ, Lipscomb JD and Mayo KH (1999) Solution structure of component B from methane monooxygenase derived through heteronuclear NMR and molecular modeling. Biochemistry 38: 5799–5812

Chen Y-P and Yoch DC (1987) Regulation of two nickel-requiring (inducible and constitutive) hydrogenases and their coupling to nitrogenase in *Methylosinus trichosporium* OB3b. J Bacteriol 169: 4778–4783

Chistoserdova L, Vorholt JA, Thauer RK and Lidstrom ME (1998) C_1 transfer enzymes and coenzymes linking methylotrophic bacteria and methanotrophic Archaea. Science 281:99–102

Choi D-W, Kunz RC, Boyd ES, Semrau JD, Antholine WE, Han J-I, Zahn JA, Boyd JM, de la Mora AM and DiSpirito AA (2003) The membrane-associated methane monooxygenase (pMMO) and pMMO-NADH:quinone oxidoreductase complex from *Methylococcus capsulatus* Bath. J Bacteriol 185: 5755–5764

Colby J and Dalton H (1978) Resolution of the methane monooxygenase of *Methylococcus capsulatus* (Bath) into three components. Purification and properties of component C, a flavoprotein. Biochem J 171: 461–468

Colby J and Dalton H (1979) Characterization of the second prosthetic group of the flavoprotein NADH-acceptor reductase (component C) of the membrane monooxygenase from *Methylococcus capsulatus* (Bath). Biochem J 177: 903–908

Colby J, Stirling DI and Dalton H (1977) The soluble methane monooxygenase from *Methylococcus capsulatus* Bath: Its ability to oxygenate *n*-alkanes, *n*-alkenes, ethers, and alicyclic, aromatic and heterocyclic compounds. Biochem J 165: 395–402

Cook SA and Shiemke AK (2002) Evidence that a type-2 MADH: quinone oxidoreductase mediates electron transfer to particulate methane monooxygenase in *Methylococcus capsulatus*. Arch Biochem Biophys 398: 32–40

Csaki R, Hanczar T, Bodrossy L, Murrell JC and Kovacs KL (2001) Molecular characterization of structural genes coding for a membrane bound hydrogenase in *Methylococcus capsulatus* (Bath). FEMS Microbiol Lett 205: 203–207

Dalton H (1977) Ammonia oxidation by the methane oxidizing bacterium *Methylococcus capsulatus* Bath. Arch Microbiol 144: 273–279

Dalton H, Prior SD, Leak DJ and Stanley SH (1984) Regulation and control of methane monooxygenase. In: Crawford RL and Hanson RS (eds) Microbial Growth on C_1 Compounds, pp 75–82. American Society for Microbiology, Washington, D.C.

De Bont JA (1976) Hydrogenase activity in nitrogen-fixing methane-oxidizing bacteria. Antoine van Leeuwenhoek 42: 255–259

DiSpirito AA (1990) Soluble cytochromes *c* from *Methylomonas* sp. A4. Methods Enzymol 188: 299–304.

DiSpirito AA, Lipscomb JD and Lidstrom ME (1990) Soluble cytochromes from the marine methanotrophs, *Methylomonas* sp. A4. J Bacteriol 172: 5360–5367

DiSpirito AA, Gulledge J, Shiemke AK, Murrell JC and Lidstrom ME (1992) Trichloroethylene oxidation by the membrane-associated methane monooxygenase in type I, type II, and type X methanotrophs. Biodegradation 2: 151–164

DiSpirito AA, Shiemke AK, Jordan SW, Zahn JA and Krema CL (1994) Cytochrome aa_3 from *Methylococcus capsulatus* Bath. Arch Microbiol 161: 258–265

DiSpirito AA, Zahn JA, Graham DM, Kim HJ, Larive CK, Derrick TS, Cox CD and Taylor A (1998) Copper-binding compounds from *Methylosinus trichosporium* OB3b. J Bacteriol 180: 3606–3613

Dross F, Geisler V, Lenger R, Theis F, Krafft T, Fahrenholz F, Kojro E, Duchene A, Tripier D, Joveal K and Kröger A (1992) The quinone-reactive Ni/Fe-hydrogenase of *Wolinella succinogenes*. Eur J Biochem 206: 93–102

Duine JA, Frank J and Jongejan AJ (1986) PQQ and quinoprotein enzymes in microbial oxidations. FEMS Lett 32: 165–178

Erickson RH and Hooper AB (1972) Preliminary characterization of a variant co-binding heme protein from *Nitrosomonas*. Biochim Biophys Acta 275: 231–244

Finzel B, Weber PC, Hardman KD and Salemme FR (1985) Structure of ferricytochrome *c′* from *Rhodospirillum molischianun* at 1.6 A resolution. J Mol Biol 186: 627–643

Fox BG, Surerus KK, Münck E and Lipscomb JD (1988) Evidence for a μ-oxo-bridged binuclear iron cluster in the hydroxylase component of methane monooxygenase. J Biol Chem 264: 10553–10556

Fox BG, Froland WA, Dege JE and Lipscomb JD (1989) Membrane monooxygenase from *Methylosinus trichosporium* OB3b. J Biol Chem 251: 1105–1115

Gilbert B, McDonald IA, Finch R, Stafford GA, Neilsen AK and Murrell KC (2000) Molecular analysis of the *pmo* (particulate methane monooxygenase) operons from two type II methanotrophs. J Bacteriol 66: 966–975

Goodhew CF, Wilson IBH, Hunter DJB and Pettigrew GW (1990) The cellular location and specificity of bacterial cytochrome *c* peroxidases. Biochem J 271: 707–712

Green J and Dalton H (1985) Protein B of soluble methane monooxygenase from *Methylococcus capsulatus* (Bath). A novel regulatory protein of enzyme activity. J Biol Chem 260: 15795–15802

Groen BW, van Kleef AJ and Duine JA (1986) Quinohaemoprotein alcohol dehydrogenase apoenzyme from *Pseudomonas testosteroni*. Biochem J 234: 611–615

Han J-I and Semrau JD (2000) Chloromethane stimulates growth of *Methylomicrobium album* BG8 on methanol. FEMS Microbiol Lett. 187: 77–81

Han J-I and Semrau JD (2001) Quantification of the expression of *pmoA* in methanotrophs using RT-PCR. Proceed Am Chem Soc. 221

Hanczar T, Csaki R, Bodrossy L, Murrell JC and Kovacs K (2002) Detection and localization of two hydrogenases in *Methylococcus capsulatus* Bath and their potential role in methane metabolism. Arch Microbiol 177: 167–172

Hanson RS and Hanson TE (1996) Methanotrophic bacteria. Microbiol Rev 60: 439–471

Hooper AB and Nason A (1965) Characterization of hydroxylamine-cytochrome *c* reductase from the chemoautotrophs *Nitrosomonas europaea* and *Nitrosocytic oceanus*. J Biol Chem 240: 4044–4057

Hooper AB, DiSpirito AA, Olsen TC, Andersson KK, Cunningham W and Taaffe LR (1984) Generation of the proton gradient by a periplasmic dehydrogenase. In: Crawford RL, Hanson RS (ed) Microbial Growth on C-1 Compounds, pp 53–58. American Society for Microbiology, Washington, D.C.

Hooper AB, Vannelli T, Bergmann BJ and Arciero DM (1997) Enzymology of the oxidation of ammonia to nitrite by bacteria. Antoine van Leeuwenhoek 71: 59–69

Jagow C von and Link TA (1986) Use of specific inhibitors on the mitochondrial bc_1 complex. Meth Enzymol 126: 252–271

Johnson PA and Quayle JR (1964) Microbial growth on C-1 compounds. 6. Oxidation of methanol, formaldehyde and formate by methanol grown *Pseudomonas* AM1. Biochem J 93: 281–290

Jollie DR and Lipscomb JD (1990) Formate dehydrogenase from *Methylosinus trichosporium* OB3b. Meth Enzymol 188: 331–334

Jollie DR and Lipscomb JD (1991) Formate dehydrogenase from *Methylosinus trichosporium* OB3b. J Biol Chem 266: 21853–21863

Junemann S, Heathcote P and Rich P (1998) On the mechanism of quinol oxidation in the bc_1 complex. J Biol Chem 273: 21603–21607

Kawamura S, O'Neil JG and Wilkinson JF (1983) Hydrogen production by methylotrophs under anaerobic conditions. J Ferment Technol 61: 151–156

Kiene RP (1991) Production and Consumption of Methane in Aquatic Systems. In: Rogers JE and Whitman WB (eds) Microbial Production and Consumption of Greenhouse Gases: Methane, Nitrogen Oxides, and Halomethanes, pp 111–146. American Society for Microbiology, Washington, DC.

Kim HJ (2003) Isolation, structural elucidation, and characterization of a novel copper-binding compound from *Methylosinus trichosporium* OB3b. A 1.1 Å crystal structure. PhD Thesis, pp 162. University of Kansas, Lawrence

Klein CR, Kesseler FP, Perreli C, Frank J, Duine JA and Schwartz AC (1994) A novel dye-linked formaldehyde dehydrogenase with some properties indicating the presence of a protein-bound redox-active quinone cofactor. Biochem J 301: 289–295

Large E and Quayle JR (1963) Enzyme activities in extracts of *Pseudomonas* AM1. Biochem J 87: 386–396

Lieberman RL, Shrestha DB, Doan PE, Hoffman BM, Stemmler TL and Rosenzweig AC (2003) Purified particulate methane monooxigenase from *Methylococcus capsulatus* Bath is a dimer with both mononuclear copper and a copper-containing cluster. Proceed Nat Acad Sci USA 100: 3820–3825

Lemos SS, Collins MLP, Eaton SS, Eaton GR and Antholine WE (2000) Comparison of EPR-visible Cu^{2+} Sites in pMMO from *Methylococcus capsulatus* (Bath) and *Methylomicrobium album* BG8. Biophys J 79: 1085–1094

Long AR and Anthony C (1991) Characterization of the periplasmic cytochromes *c* of *Paracoccus denitrificans*: Identification of the electron acceptor for methanol dehydrogenase, and description of a novel cytochrome *c* heterodimer. J Gen Microbiol 137: 415–425

Lontoh S, DiSpirito AA and Semrau JD (1999) Dichloromethane and trichloroethylene inhibition of methane oxidation by the membrane-associated methane monooxygenase of *Methylosinus trichosporium*OB3b. Arch Microbiol 173: 29–34

Lontoh S, Zahn JA, DiSpirito AA and Semarau JD (2000) Identification of intermediates of in vivo trichoroethylene oxidation by the membrane-associated methane monooxygenase of *Methylosinus trichosporium* OB3b. FEMS Microbiol Lett 186: 109–113

Marison IW and Attwood MM (1980) Partial purification and characterization of a dye-linked formaldehyde dehydrogenase from *Hyphomicrobium* X. J Gen Microbiol 177: 305–313

Marison IW and Attwood MM (1982) Possible alternative mechanism of the oxidation of formaldehyde to formate. J Gen Microbiol 128: 1441–1446

Matsuno-Yagi A and Hategi Y (1996) Ubiquinol-cytochrome *c* oxidoreductase. The redox reactions of the bis-heme cytochrome *b* in ubiquinone-sufficient and ubiquinone-deficient systems. J Biol Chem 271: 6164–6171

Matsuno-Yagi A and Hatefi Y (1977) Ubiquinol:cytochrome *c* oxidoreductase. The redox reactions of the bis-heme cytochrome *b* in unergized submitochondrial particles. J Biol Chem 272: 16928–16933

McRee DA, Redford SM, Meyer TE and Cusanovich MA (1990) Crystallization and characterization of *Chromatium vinosum* cytochrome *c'*. J Biol Chem 265: 5364–5365

Mehta RJ (1975) A novel inducible formaldehyde dehydrogenase of *Pseudomonas* sp; (RJ1). Antonie van Leeuwenhoek J Microbiol Serol. 41: 89–95

Murrell JC and Holmes A (1996) Molecular biology of particulate methane monooxygenase. In: Lidstrom ME, Tabita FR (eds) Proceedings of the 8th International Symposium on Microbial Growth on C-1 Compounds, pp 133–140, Kluwer Academic Publishers, Boston

Murrell JC, McDonald IR and Gilbert B (2000) Regulation of expression of methane monooxygenases by copper ions. Trends Microbiol 8: 221–225

Nguyen A-N, Schiemke AK, Jacobs SJ, Hales BJ, Lidstrom ME and Chan SI (1994) The nature of the copper ions in the membranes containing the particulate methane monooxygenase from *Methylococcus capsulatus* (Bath). J Biol Chem 269: 14995–15005

Nguyen H-N, Nakagawa KH, Gednab B, Elliott S, Lidstrom ME, Hodgson KO and Chan SI (1996) X-ray absorption and EPR studies on the copper ions associated with particulate methane monooxygenase from *Methyococcus capsulatus* Bath. Cu(I) ions and their implications. J Am Chem Soc 118: 12766–12776

Nguyen H-H, Elliott SJ, Yip JH-K and Chan SI (1998) The particulate methane monooxygenase from *Methylococcus capsulatus* (Bath) is a novel copper-containing three-subunit enzyme. J Biol Chem 273: 7957–7966

Nielsen AK, Gerders K and Murrell JC (1996) Regulation of bacterial methane oxidation: transcription of the soluble methane monooxygenase operon of *Methylococcus capsulatus* (Bath) is repressed by copper ions. Microbiology 142: 1289–1296

Nielsen AK, Gerders K and Murrell JC (1997) Copper-dependent reciprocal transcriptional regulation of methane monooxygenase genes in *Methylococcus capsulatus* and *Methylosinus trichosporium*. Mol Microbiol 25: 399–409

Numata M, Saito T, Yamazaki T, Fukumori Y and Yamanaka T (1990) Cytochrome P460 of *Nitrosomonas europaea*: Further purification and further characterization. J Biochem 108: 1016–1021

Nunn DN and Anthony C (1988) The nucleotide sequence and deduced amino acid sequence of the genes for cytochrome c_L of *Methylobacterium* AM1. Nucleic Acids Res 16: 7722–7723

O'Keeffe DT and Anthony C (1980) The two cytochromes *c* in the facultative methylotroph *Pseudomonas* AM1. J Biochem 192: 411–419

Ohata S and Tobari J (1981) Two cytochromes *c* from *Methylomonas* J. Biochem J 90: 215–224

Oremland RS, Culbertson CW (1992) Importance of methane-oxidizing bacteria in the methane budget as revealed by the use of a specific inhibitor. Science 356:421–423

Patel RN, Hou CT and Felix A (1979) Microbial oxidation of methane and methanol: Purification and properties of a heme-containing aldehyde dehydrogenase from *Methylomonas methylovora*. Arch Microbiol 122: 241–247

Patel RN, Hou CT, Derelanko P and Felix A (1980) Purification and properties of a heme-containing aldehyde dehydrogenase from *Methylosinus trichosporium*. Arch Biochem Biophys 203: 654–662

Pilkington SJ and Halton H (1990) Soluble methane monooxygenase from *Methylococcus capsulatus* Bath. Meth Enzymol 188: 181–190

Pomper BK, Vorholt A, Chistoserdova L, Lidstrom ME and Thauer RK (1999) A methyenyl tetrahydromethanopterin cyclohydrolase and an methyl tetrahydrofolate cyclohydrolase in *Methylobacterium extorquens* AM1. Eur J Biochem 261: 476–480

Prior SD and Dalton H (1985) Copper stress underlies the fundamental change in intracellular localization of methane monooxygenase in methane oxidizing organisms: Studies in batch and continuous culture. J Gen Microbiol 131: 155–163

Ren Z, Meyer T and McRee D (1993) Atomic structure of a cytochrome *c'* with an unusual ligand-controlled dimer dissociation at 1.8Å resolution. J Mol Biol 234: 433–445

Semrau JD, Chistoserdov A, Lebron J, Costello A, Davagnino J, Kenna E, Holmes AJ, Finch R, Murrell JC and Lidstrom ME (1995) Particulate methane monoxygenase genes in methanotrophs. J Bacteriol 177: 3071–3079

Shah NN, Hanna ML, Jackson KK and Taylor RT (1995) Batch cultivation of *Methylosinus trichosporium* OB3b. 4. Production of hydrogen-driven soluble or particulate methane monooxygenase activity. Biotechnol Bioeng 45: 229–238

Snyder SH, Gutierrez-Cirlos EB and Trumpower BL (2000) Evidence for a concerted mechanism of ubiquinol oxidation by the cytochrome bc_1 complex. J Biol Chem 275: 13535–13541

Sokolov IG, Romanovskaya VA, Shkurko YB and Malashenko YR (1980) Comparative characterization of the enzyme systems of methane-utilizing bacteria that oxidize NH_2OH and CH_3OH. Microbiology 49: 202–209

Speer B, Chistoserdova L and Lidstrom ME (1994) Sequence of the gene for a NAD(P)-dependent formaldehyde dehydrogenase (class III alcohol dehydrogenase) from a marine methanotroph *Methylobacter marinus* A45. FEMS Lett 121: 349–356

Stanley SH, Prior SD, Leak DJ and Dalton H (1983) Copper stress underlies the fundamental change in intracellular location of methane monooxygenase in methane-oxidizing organisms. Biotechnol Lett 5: 487–492

Stirling DI and Dalton H (1978) Purification and properties of an $NAD(P)^+$-linked formaldehyde dehydrogenase from *Methylococcus capsulatus* (Bath). J Gen Microbiol 107: 19–29

Stirling DI, Colby J and Dalton H (1979) A comparison of the substrate and electron-donor specificity of the methane monooxygenase from three strains of methane oxidizing bacteria. Biochem J 177: 362–364

Stolyar SM, Franke M and Lidstrom ME (2001) Expression of individual copies of *Methylococcus capsulatus* Bath particulate methane monooxygenase genes. J Bacteriol 183: 1810–1812

Tahirov T, Misaki A, Myer T, Cusanovitch M, Higuchi. Y and Yasuoka N (1996) High resolution crystal structure of two polymorphs of cytochrome *c'* from the purple phototrophic bacterium *Rhodobacter capsulatus*. J Mol Biol 259: 467–479

Takeguchi IO (2000) Role of iron and copper in particulate methane monooxygenase of *Methylosunus trichosporium* OB3b. Catal Sur-Japan 4: 51–63

Takeguchi M, Miyakawa K and Okura I (1998) Purification and properties of particulate methane monooxygenase from *Methylosinus trichosporium* OB3b. J Mol Catal 132: 145–153

Takeguchi M, Miyakawa K and Okura I (1999a) The role of copper in particulate methane monooxygenase from *Methylosinus trichosporium* OB3b. J. Mol. Catal. 137: 161–168

Takeguchi M, Miyakawa K and Okura I (1999b) Role of iron in particulate methane monooxygenase from *Methylosinus trichosporium* OB3b. BioMetals 12: 123–129

Tate CG and Grisshammer RR (1996) Heterologous expression of G-protein coupled receptors. Trends Biotechnol 14: 426–430

Tate S and Dalton H (1999) A low-molecular-mass protein from *Methylococcus capsulatus* (Bath) is responsible for the regulation of formaldehyde dehydrogenase activity in vitro. Microbiology 145: 159–167

Topp E and Hanson RS (1991) Metabolism of radiatively important trace gases by methane-oxidizing bacteria. In: Rogers JE, Whitman WB (eds) Microbial Production and Consumption of Greenhouse Gases: Methane, Nitrogen Oxides, and Halomethanes, pp 71–90. American Society for Microbiology, Washington, DC

Tukhvatullin IA, Gvozdev RI and Andersson KA (2000) The structure of the active center of b-peptide membrane-bound methane monooxygenase (pMMO) from Methylococcus capsulatus Bath. Biochem Biophys Mol Biol 374: 177–182

Tukhvatullin IA, Gvozdev RI and Andersson KA (2001) Structural and functional model of methane hydroxylase of membrane-bound methane monoxygenase from *Methylococcus capsulatus* (Bath). Russian Chem Bull 50: 1783–1792

Urakami T and Komagata K (1986) Occurrence of isoprenoid compounds in gram negative bacteria. J Gen Appl Microbiol 32: 317–341

Vignais PM and Toussaint B (1994) Molecular biology of membrane-bound H_2 uptake hydrogenases. Arch Microbiol. 161: 1–10

Vignais PM, Billoud B and Meyer J (2001) Classification and phylogeny of hydrogenases. FEMS Microbiol Rev 25: 455–501

Vorholt JA (2002) Cofactor-dependent pathways of formaldehyde oxidation in methylotrophic bacteria. Arch Microbiol 178: 239–249

Vorholt J, Chistoserdova L, Lidstrom KE and Thauer RK (1998a) Distribution of tetrahydromethanopterin-dependent enzyme in methylotrophic bacteria and phylogeny of methenyl tetrahydromethanopterin cyclohydrolases. J Bacteriol 181: 5750–5757

Vorholt JA, Chistoserdova L, Lidstrom ME and Thauer RK (1998b) The NADP-dependent methylene tetrahydromethanopterin dehydrogenase in *Methylobacterium extorquens* AM1. J Bacteriol 180: 5351–5356

Vorholt JA, Chistoserdova L, Stolyar SM, Thauer RK and Lidstrom ME (1999) Distribution of tetrahydromethanopterin-dependent enzymes in methylotrophic bacteria and phylogeny of methenyl tetrahydromethanopterin cyclohydrolases. J Bacteriol 181: 5750–5757

Waechter-Brulla D, DiSpirito AA, Chistoserdova LV and Lidstrom ME (1993) Methanol oxidation genes in the marine methanotroph, *Methylomonas* sp. A4. J Bacteriol 161: 258–265

Wallar BJ and Lipscomb JD (1996) Dioxygen activation by enzymes containing binuclear non-heme iron clusters. Chem Rev 96: 2625–2657

Walters KJ, Gassner GT, Lippard SJ and Wagner G (1999) Structure of the soluble methane monooxygenase regulatory protein B. Proceed Nat Acad Sci U.S.A. 96: 7877–7882

Weber PC, Howard A, Xuong N and Salemme F (1981) Crystallographic structure of *Rhodospirillum molischianum* ferricytochrome *c′* at 2.5 A resolution. J Biochem 153: 399–424

Whittenbury R and Dalton H (eds) (1981) The methylotrophic bacteria. Springer-Verlag, Berlin

Woodland MP and Dalton H (1984) Purification and characterization of component A of the methane monooxygenase from *Methylococcus capsulatus* Bath. J Biol Chem 259: 53–60

Wu L-F and Mandrand MA (1993) Microbial hydrogenases: Primary structure, classification, signatures, and phylogeny. FEMS Microbiol Rev 104: 243–270

Yasui M, Harada S, Kai Y, Kasai N, Kusanki M and Matsuura Y (1992) Three dimensional structure of ferricytochrome *c′* from *Rhodospirillum rubrum* at 2.8 Å resolution. J Biochem 111: 317–324

Yoch DC, Chen Y-P and Hardin MG (1990) Formate dehydrogenase from the methane oxidizer *Methylosinus trichosporium* OB3b. J Bacteriol 172: 4456–4463.

Yuan H, Collins MLP and Antholine WA (1997) Low frequency EPR of the copper in particulate methane monooxygenase from *Methylomicrobium albus* BG8. J Am Chem Soc 119:5073–5074

Yuan H, Collins MLP and Antholine WA (1998a) Analysis of type 2 Cu^{2+} in pMMO from *Methylomicrobium albus* BG8. Biophys J 74: A300

Yuan H, Collins MLP and Antholine WA (1998b) Concentration of Cu, EPR detectable Cu, and formation of cupric-ferrocyanide in membrane with pMMO. J Inorg Biochem 72: 179–185

Yuan H, Collins MLP and Antholine WA (1999) Type 2 Cu^{2+} in pMMO from *Methylomicrobium albus* BG8. Biophys J 76: 2223–2229

Zahn JA and DiSpirito AA (1996) Membrane associated methane monooxygenase from *Methylococcus capsulatus* (Bath). J Bacteriol 178:1018–1029

Zahn JA, Duncan C and DiSpirito AA (1994) Oxidation of hydroxylamine by cytochrome P460 of the obligate methanotroph *Methylococcus capsulatus* Bath. J Bacteriol 176: 5879–5887

Zahn JA, Arciero DM, Hooper AB and DiSpirito AA (1996a) Cytochrome *c′* of *Methylococcus capsulatus* Bath. Eur J Biochem 240: 684–691

Zahn JA, Arciero DM, Hooper AB and DiSpirito AA (1996b) Evidence for an iron center in the ammonia monooxygenase from *Nitrosomonas europaea.* FEBS Lett 397: 35–38

Zahn JA, Arciero DM, Hooper AB and DiSpirito AA (1997) Cytochrome *c* peroxidase from *Methylococcus capsulatus* Bath. Arch Microbiol 168: 362–372

Zahn JA, Bergmann DJ, Boyd JM, Kunz RC and DiSpirito AA (2001) Membrane-associated Quinoprotein formaldehyde dehydrogenase from *Methylococcus capsulatus* Bath J Bacteriol 183: 6832–6840

Zhang Z, Huang L, Shuimeister VM, Chi Y-I, Kim KK, Hung L-W, Croft AC, Berry EA and Kim S-H (1998) Electron transfer by domain movement in cytochrome bc_1. Nature 392: 677–687

Chapter 8

The Enzymes and Bioenergetics of Bacterial Nitrate, Nitrite, Nitric Oxide and Nitrous Oxide Respiration

Stuart J. Ferguson
Department of Biochemistry, University of Oxford, South Parks Road, Oxford OX13QU U.K.

David J. Richardson*
School of Biological Sciences, University of East Anglia, Norwich NR4 7TJ U.K.

Author for correspondence, email: d.richardson@uea.ac.uk

Davide Zannoni (ed): Respiration in Archaea and Bacteria. Vol 2: Diversity of Prokaryotic Respiratory Systems, pp. 169–206.

Summary

Respiratory reactions involving inorganic nitrogen species provide a rich variety of systems with which to study bacterial bioenergetics and biological chemistry. These respiratory reactions encompass the reduction of nitrate, nitrite, nitric oxide and nitrous oxide; in all cases catalysis involves redox chemistry that takes place at metal centers. These catalytic metal centers include mono-nuclear type II copper, *c* heme or d_1 heme in nitrite reductases, the Mo-*bis*-molybdopterin guanine dinucleotide cofactor in nitrate reductases, a non heme Fe-heme Fe dincuclear center in nitric oxide reductases and a tetra-nuclear copper sulfide center in nitrous oxide reductase. Recent structures and new spectroscopic data for these enzymes have rapidly advanced our knowledge of the molecular and dynamic properties of the catalytic centers, as well as how electrons are transferred to them; models of catalysis can thus be proposed. Bioenergetic studies have also provided an understanding of the way in which these enzymes are coupled to energy-conserving electron transport pathways.

I. Introduction

The reduction of nitrate to nitrite (E_o' = +420 mV), of nitrite to nitric oxide (E_o' = +375 mV) or ammonium (E_o' = +320 mV), of nitric oxide to nitrous oxide (E_o' = +1175 mV) and of the latter to nitrogen gas (E_o' = +1355 mV) are each reactions with E_0' values that are sufficiently positive for the reactions to be linked to bacterial energy-conserving respiratory chains. This then allows oxidative phosphorylation to occur under anaerobic conditions, although it should be noted that many of the organisms that use these reactions do so as a second choice when oxygen is not available as the terminal electron acceptor. The reactions can occur either individually within a bacterial species or in various combinations. Thus, in a denitrifying bacterium, nitrate can be sequentially reduced through to nitrogen gas, in which case the product of nitrite reduction will be nitric oxide. Typical examples of organisms equipped in this way include *Paracoccus denitrificans, Paracoccus pantotrophus, Pseudomonas stutzeri* and *Pseudomonas aeruginosa*. On the other hand, some organisms, for example *Escherichia coli*, have enzymes that sequentially reduce nitrate to ammonium via nitrite; hence this organism is not a denitrifier but is sometimes called an ammonifier. Other organisms may have other combinations of these enzyme activities; for example the pathogen *Neisseria gonorrhoeae* has a nitrite reductase and nitric oxide reductase only (Householder et al., 2000; Lissenden et al., 2000). The photosynthetic bacteria *Rhodobacter sphaeroides* and *Rhodobacter capsulatus* always appear to have the nitric oxide reductase but the presence of one or more of the reductases for nitrate, nitrite and nitrous oxide varies between strains (Bell et al., 1992). Historically the organisms recognized as involved in these reactions have been from a limited number of eubacteria. Perspective on the occurrence of these enzymes is rapidly changing owing to the acquisitions of many genome sequences. The latter are establishing that these enzymes, especially the nitric oxide reductase, are more widespread than hitherto suspected and some at least are now known to occur in Archaea (Hendriks et al., 2000; Richardson et al., 2001). The roles of these enzymes within the nitrogen cycle are presented in Fig. 1.

The relatively positive redox potentials of all the reactions involved in reduction of nitrate, and species formed therefrom, means that the enzymes for each of the reactions can be linked to bacterial electron transport systems at either the quinone or *c*-type cytochrome/cupredoxin region. Reduction of nitrate to nitrite is always linked to the quinone level (E_o' in region of –80 to +80 mV), whereas reduction of nitrous oxide is only known to be associated with the cytochrome *c*/cupredoxin level (E_o' in region of 250 mV) of an electron transfer chain. This makes sense in terms of the relative redox potentials of the reactions. The linkage of the nitrite and nitric oxide reduction reactions shows more variation. When nitrite is reduced to ammonia the connection is at the level of quinone. However, cytochromes *c* or cupredoxins usually provide electrons to the reductases that have nitric oxide as reaction product, although an exception may be in some *Pseudomonas* species where the connection may be at the level of quinol.

Abbreviations: Mo-bis-MGD – molybdenum bis molybdopterin guanine dinucleotide; Nap – periplasmic nitrate reductase; Nar – membrane-bound nitrate reductase; Nir – nitrite reductase; Nos – nitrous oxide reductase; Nor – nitric oxide reductase; $q^+/2e$ – number of positive charges moved across the membrane per two electrons transferred

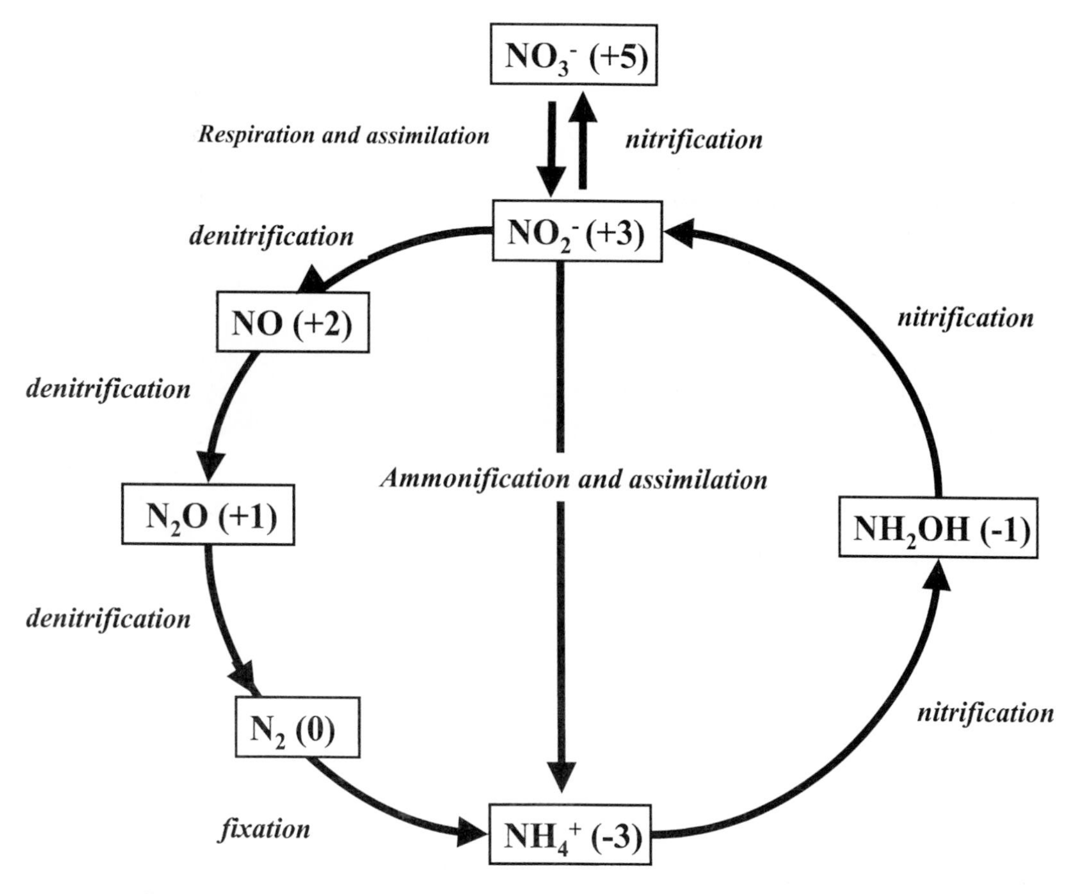

Fig. 1. The Nitrogen Cycle. The numbers in brackets refer to the oxidation state of nitrogen.

Nitric oxide reductases have a more complex pattern with some having quinols as electron donors and others *c*-type cytochromes. Although at first sight this seems unexpected, it should be recalled that the same variation is found amongst cytochrome oxidases, a point of comparison that is reinforced by the structural similarity between oxidases and nitric oxide reductases that will be discussed below. Although these enzymes all function in conjunction with the membrane-bound respiratory electron transport systems, not all the enzymes are themselves membrane-bound. While one type of nitrate reductase (Fig. 3) and all types of nitric oxide reductase (Fig. 6) have subunits that are integral membrane proteins, in Gram negative bacteria a second type of nitrate reductase, all three types of nitrite reductases and the nitrous oxide reductase are water-soluble periplasmic proteins (Figs. 2, 4 and 5). Gram positive organisms do not have a periplasm and thus they tend to have variant forms of nitrite and nitrous oxide reductases that are anchored to the cytoplasmic membrane, while the periplasmic type of nitrate reductase is absent from Gram positive organisms. The ultrastructure of the Archaea means that they do not have a formal periplasm, hence all their enzymes have some form of membrane anchor on the outer surface of the cytoplasmic membrane.

An important bioenergetic point concerning the respiration of nitrate, nitrite, nitric oxide and nitrous oxide is to establish the stoichiometry of charge translocation that accompanies each reaction. Electrons passing from ubiquinol (or menaquinol) to nitrate will be coupled to charge translocation only if the nitrate reductase itself is catalyzing net proton movement across the membrane. As we shall see, one kind of nitrate reductase, that in the periplasm, cannot do this, and thus the $q^+/2e = 0$, whereas the second type of nitrate reductase, which is membrane-bound, achieves $q^+/2e = 2$. The passage of electrons via the cytochrome bc_1 complex from quinol to the *c*-type cytochrome level in the respiratory chain will result in a $q^+/2e = 2$, but no additional proton translocation

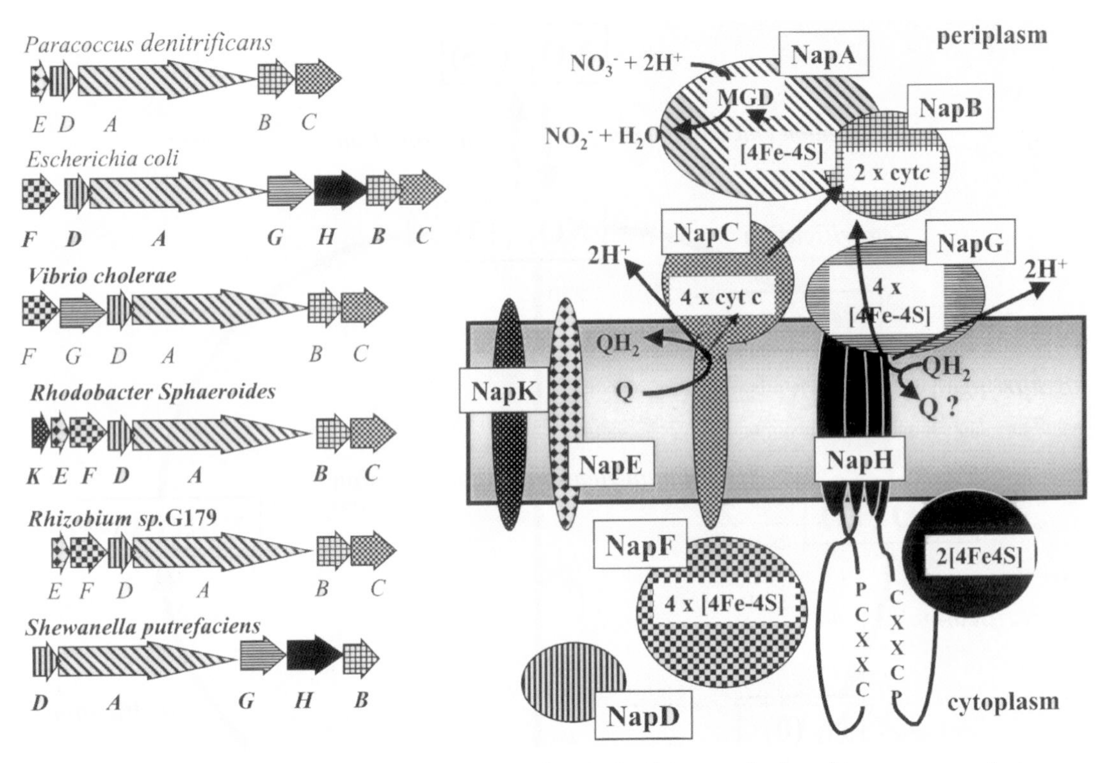

Fig. 2. The organization of *nap* gene clusters and predicted subcellular organization of the *nap* gene products.

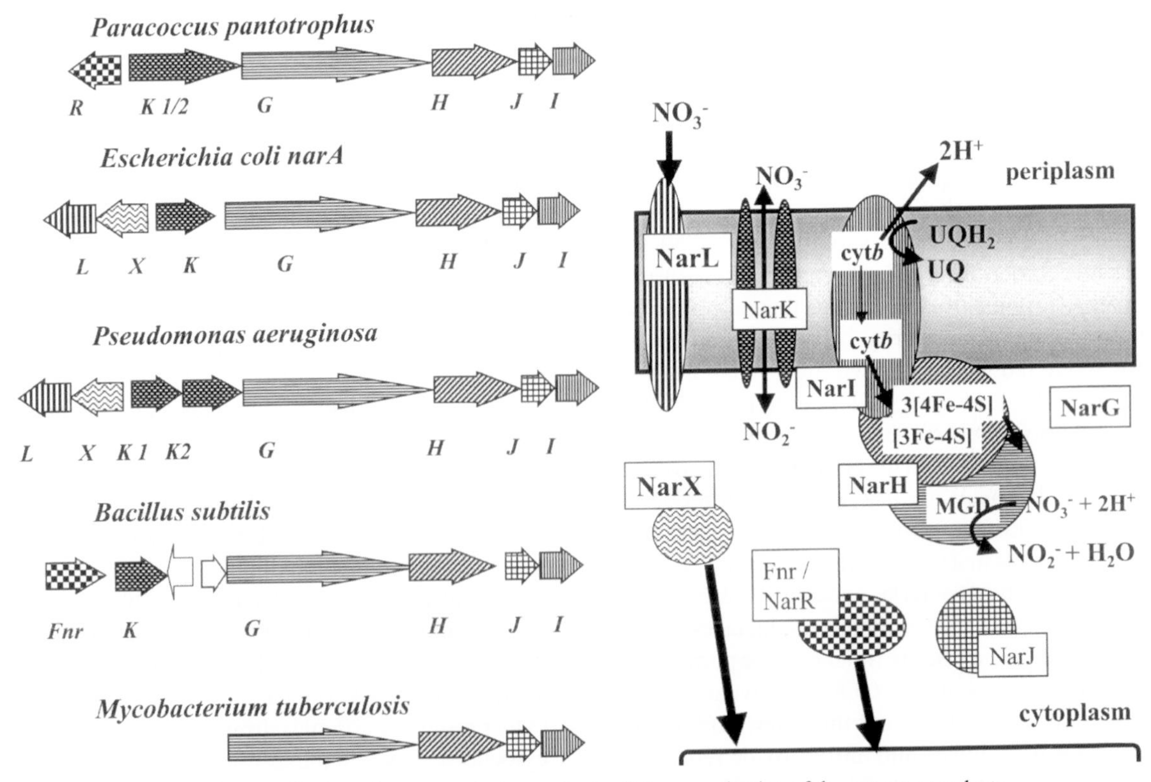

Fig. 3. The organization of *nar* gene clusters and subcellular organization of the *nar* gene products.

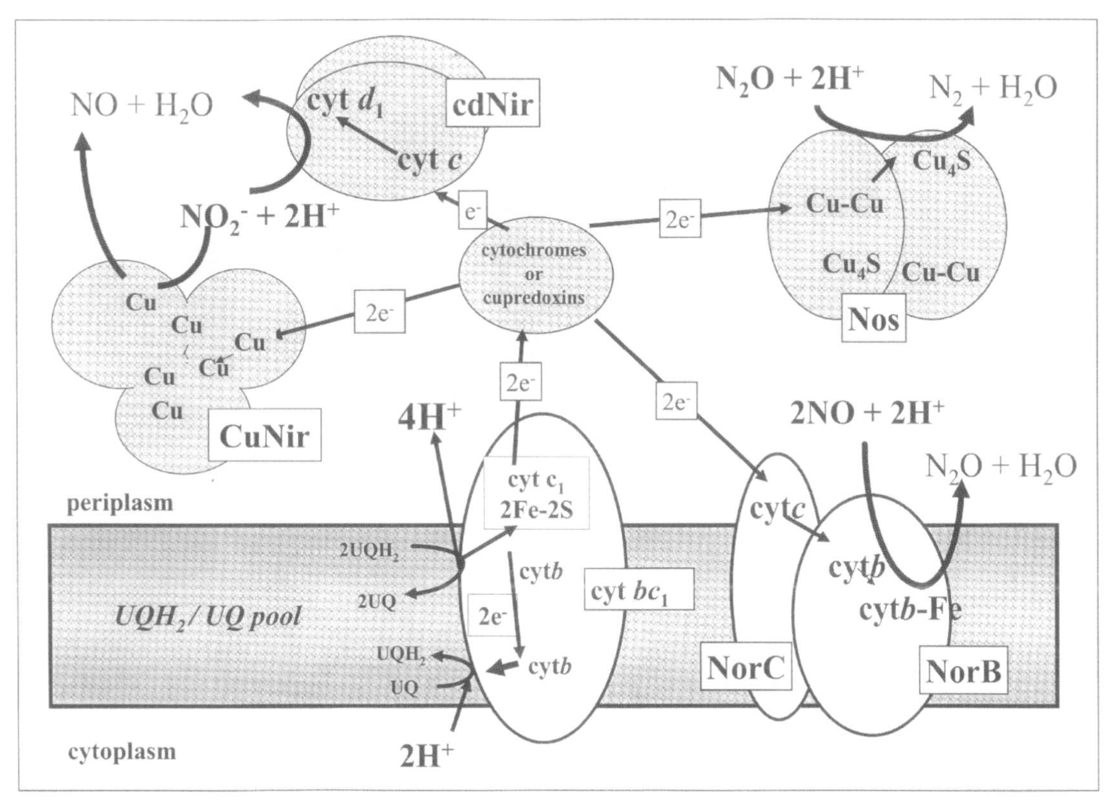

Fig. 4. The organization of the cytochrome bc_1 complex dependent nitrite reductases, nitrous oxide reductase and nitrous oxide reductase. (note that the cdNir and CuNir have never conclusively been found in the same organism)

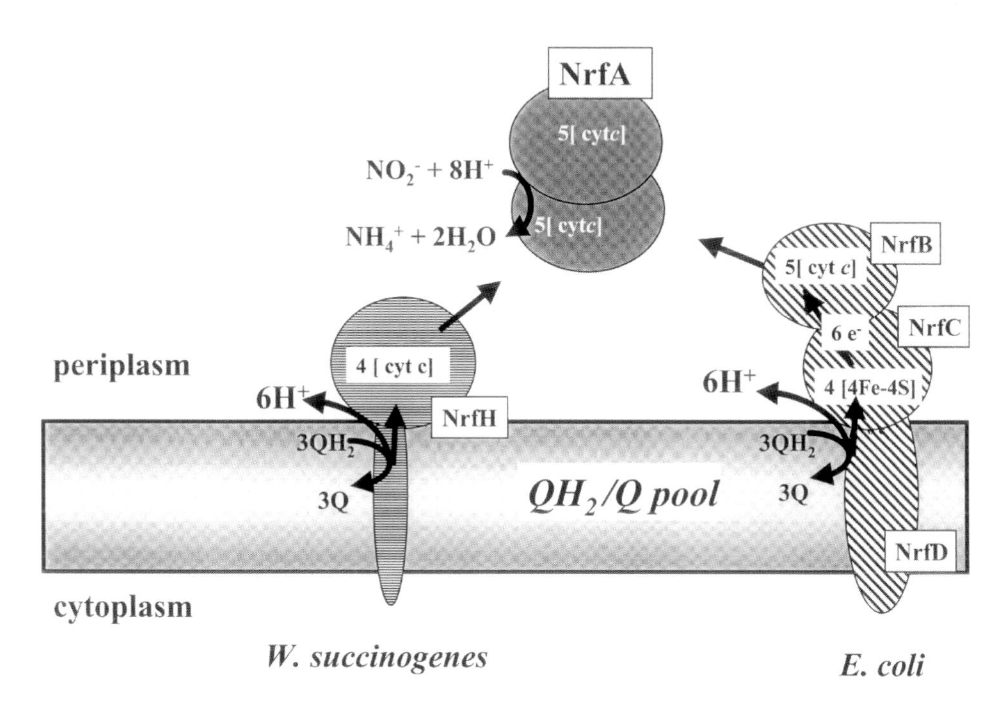

Fig. 5. The organization of the NrfH and NrfB dependent cytochrome *c* nitrite reductase systems.

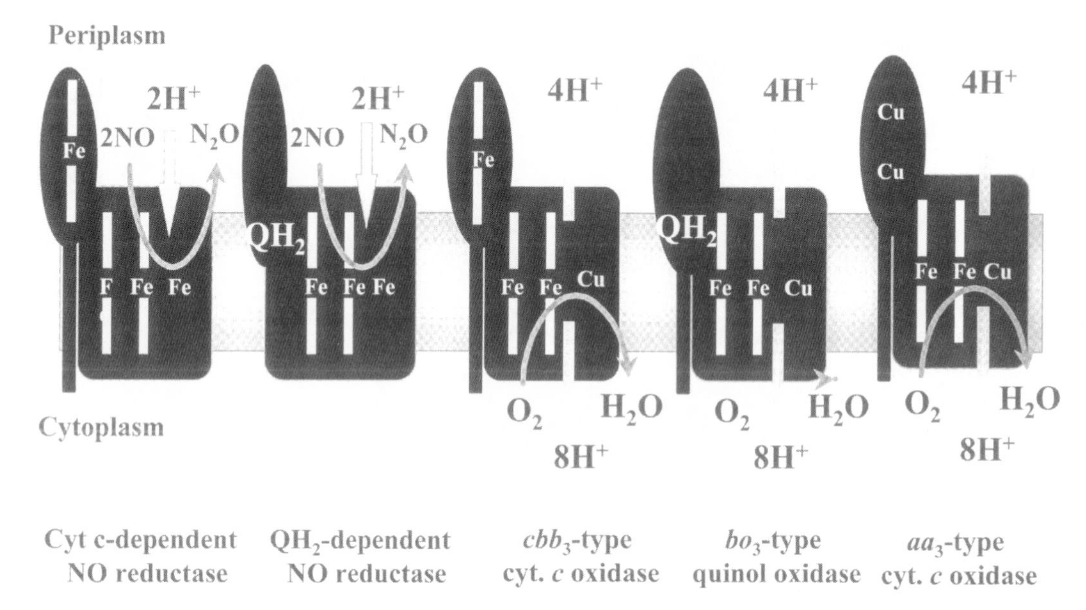

Fig. 6. The different subgroups of nitric oxide reductases and oxidases.

occurs as the electrons pass on to the nitrite, nitric and nitrous oxide reductases. When direct oxidation of quinol by a nitric oxide reductase occurs then $q^+/2e = 0$. However, in general the situation is that the $q^+/2e$ is never greater than 2 for any of the electron acceptors we are considering here when quinol is the electron donor. This contrasts with the use of oxygen by electron transfer chains because $q^+/2e$ values of between 4 and 6 are found for most pathways of electron transfer leading to a cytochrome oxidase. These higher stoichiometries correlate with the general observation that cells make most ATP per pair of electrons when they reduce oxygen and thus the latter is the preferred electron acceptor. We note that in many circumstances the electrons will reach ubiquinone via a proton-translocating enzyme. This can be an NADH dehydrogenase, for which usually $q^+/2e = 4$, or another type of enzyme, for example a formate dehydrogenase, for which $q^+/2e = 2$. Thus even when electrons are ultimately delivered from quinol to the acceptor by the periplasmic nitrate reductase or a quinol-dependent nitric oxide reductase, they are still often accompanied by some proton translocation. Finally, it is important to note that there is not a simple correlation between the redox potential of the acceptor couple and the $q^+/2e$ ratio; if there were then the ratio should be higher for nitrous oxide (E_o' *ca.* 1100 mV) reduction than that for oxygen (E_o' *ca.* 800 mV). For some reason it seems that a proton-translocating, and thus membrane-bound, nitrous oxide reductase has not evolved.

The aim of this chapter is to summarize current knowledge and uncertainties concerning (i) the delivery of electrons to reductases that catalyze reduction of nitrate, nitrite, nitric oxide and nitrous oxide and (ii) the structure/function relationships for the enzymes. For the most part we deal only with recent developments and refer the reader to several previous reviews (e.g. (Berks et al., 1995a) (Ferguson, 1998a; Richardson and Watmough, 1999) (Moura and Moura, 2001)) for background information. Particular emphasis is placed on the cytochrome cd_1 nitrite reductase as this system provides an excellent example of how X-ray crystallography is being combined with a range of spectroscopic and biophysical techniques to inform on the enzyme structure and function. It also demonstrates why it is important to study an individual enzyme system in more than one species of bacteria as rather unexpected differences can emerge.

II. Respiratory Nitrate Reductases

There are two types of respiratory nitrate reductase in bacteria (Fig. 2), each of which should be distinguished from an assimilatory, and cytoplasmic, nitrate reductase that is found in many organisms. All three classes of bacterial nitrate reductase contain a Mo-*bis*-guanine dinucleotide cofactor (Mo-bis-MGD) at the active site and catalyze the reaction:

$$NO_3^- + 2H^+ + 2e^- \rightarrow NO_2^- + H_2O \quad (1)$$

Discussion of the assimilatory enzyme is beyond the scope of this chapter, but details have recently been described (Lin and Stewart, 1998; Richardson et al., 2001).

A. The Periplasmic Nitrate Reductase (Nap)

The genes for this enzyme can been found in the genomes of a number of α,β,δ,ε and γ proteobacteria (Richardson, 2000; Richardson et al., 2001). The enzyme can serve a range of physiological functions that include excess energy dissipation under aerobic conditions in *P. denitrificans*, redox poising of the photosynthetic electron transport system in *Rhodobacter capsulatus*, anaerobic high-affinity nitrate respiration in *E. coli* and catalyzing the first step in anaerobic denitrification in *Pseudomonas* sp. G179 (Richardson, 2000; Potter et al., 2001; Richardson et al., 2001). The regulation of *nap* gene expression reflects these different physiological functions. Thus in *P. pantotrophus nap* expression is maximum during aerobic growth on highly reduced carbon substrates (Sears et al., 2000; Ellington et al., 2002), while in *E. coli nap* expression is highest under anaerobic nitrate-limited growth conditions (Wang et al., 1999).

1. Structure and Spectroscopy of the Periplasmic Nitrate Reductase

In most cases the periplasmic nitrate reductase purifies as a two subunit enzyme complex containing a 16 kDa di-heme, NapB, and a 90 kDa catalytic subunit, NapA, that binds an [4Fe-4S] cluster and the *bis*-molybdopterin guanine dinucleotide (Mo-bisMGD) cofactor (Berks et al., 1994, 1995c). However, the enzyme from *D. desulfuricans* purifies as a single subunit enzyme and the X-ray crystal structure of this has recently been solved (Dias et al., 1999). The protein folds into four domains. One of these is formed entirely from an N-terminal segment of the polypeptide and binds the [4Fe-4S] cluster. The remaining three domains have αβ structure and fold around the Mo-bisMGD cofactor providing numerous H-bonds to the MGD moieties. The MGD cofactor lies at the bottom of a substrate funnel between domains II and III which have a similar fold that exhibits some resemblance to the typical NAD-binding fold of dehydrogenases. Residues lining this cleft and at its base will play roles in defining substrate specificity and orientating the substrate onto the Mo. Of particular significance is a conserved arginine at the base of the pocket that may interact with nitrato oxygens of the substrate. This arginine is conserved in the homologous Fdh-H subunit of the formate-hydrogen lyase system of *E. coli* for which the crystal structure of a nitrite-inhibited form shows the arginine to bind a nitrite oxygen (Boyington et al., 1997).

Spectroscopic studies of *P. pantotrophus* NapAB complex and *Desulfovibrio desulfuricans* NapA, have revealed EPR signals in the dithionite reduced enzyme that are characteristic of a s=1/2 $[4Fe4S]^{1+}$ center (Breton et al., 1994; Bursakov et al., 1997; Butler et al., 1999). The $[4Fe4S]^{1+}$ center is of catalytic relevance as it is oxidized to an EPR-silent $[4Fe\,4S]^{2+}$ state following addition of nitrate to initiate enzyme turnover. The [4Fe-4S] center is 12 Å away from the Mo ion of the Mo-bis-MGD center. One of the MGD moieties lies between the FeS center and the Mo, raising the possibility that it mediates electron transfer. The midpoint redox potential of this FeS center has been determined from EPR redox titrations to be –160 mV (Breton et al., 1994). This is much lower than expected for a redox center involved in electron transfer from ubiquinol (E_m = +60 mV) to nitrate (E_m = 420 mV) and provides an example of an endergonic electron transfer step in an electron transfer process that is exergonic overall. Although this may seem odd, it is more important for kinetic reasons to position a redox center on a pathway so as to offer the electron a route that involves tunneling up to 14Å between centers than it is to set the redox potential to match those of other components on the pathway (Ferguson, 1998b; Page et al., 1999). We will return to this point in the context of the membrane-bound nitrate reductase.

The crystal structure of NapA from *D. desulfuricans* (Dias et al., 1999) has confirmed the earlier predictions based on amino acid sequence comparisons (Berks et al., 1995c), that the molybdenum amino acid ligand in periplasmic nitrate reductases is a cysteine residue. The Mo ion is additionally coordinated by four sulfur ligands provided by the two MGD moieties and a water/hydroxo ligand. The structure has been interpreted as being that of Mo(VI) redox state. Based upon this oxidized structure it was proposed that the enzyme cycles between des-oxo-Mo(IV) and mono-oxo-Mo(VI), which is then protonated to form the water/hydroxo ligand (Dias et al., 1999). This scheme is different from that derived from X-ray absorption studies on the *P. pantotrophus* NapAB enzyme. Mo K-edge EXAFS of ferricyanide-oxidized enzyme suggested a di-oxo Mo(VI) species with five sulfur ligands which changed to mono-oxo Mo (IV) species with three sulfur ligands on reduction with dithionite (Fig. 7). The addition of nitrate to the reduced enzyme resulted in re-oxidation to a di-oxo Mo(VI) species similar to the ferricyanide-oxidized enzyme (Butler et al., 1999). Thus the *P. pantotrophus* NapAB enzyme may cycle between mono-oxo and di-oxo states during the catalytic cycle, rather than the mono-oxo/des-oxo states predicted for *D. desulfuricans* NapA. Of course, this difference may lie in the different experimental methods used to provide the data on which the catalytic cycle is proposed. However, it should also be noted that, based on primary sequence similarity, NapA from *D. desulfuricans* is more closely related to Fdh from *E. coli* than is NapA from *P. pantotrophus* (Richardson et al., 2001). Furthermore, unlike NapA from *P. pantotrophus*, NapA from *D. desulfuricans* has been demonstrated to have low levels of formate dehydrogenase activity (Bursakov et al., 1997). Thus the possibility exists that the subtle differences between the two enzymes may account for different oxygen co-ordination at Mo during the catalytic cycle.

The EXAFS data on *P. pantotrophus* NapAB also suggest that upon reduction by dithionite two sulfur atoms, presumably from one pterin, are lost from the co-ordination sphere of Mo(IV), but re-ligate following reoxidation by nitrate. The catalytic relevance of changes in the co-ordination of the pterins to the Mo ion in Mo-bisMGD enzymes is debated. For example, the lability of one of the pterin dithiolene ligands has been observed for the crystal structures of the DMSO reductase from *R. sphaeroides* and *Rba. capsulatus* (Schindelin et al., 1996; Schneider et al., 1996). However, at least in the case of non-physiological dithionite-driven turnover of Nap, it would appear that redox state-dependent pterin dissociation can be part of a catalytic cycle (Butler et al., 1999). The possibility that the Mo coordination sphere of Nap can adopt a range of states is reflected in the EPR spectra of this enzyme which show that the Mo(V)-*bis*-MGD of *P. pantotrophus* NapA can exhibit a number of different EPR signals (Butler et al., 1999) (reviewed in (Potter et al., 2001)). Distinct signals thought to arise from Mo(V) coordinated by five sulfurs and by three sulfurs have been described as well as signals arising from enzyme with nitrate, azide, cyanide or thiocyanate in the catalytic pocket (Butler et al., 1999, 2000). *P. pantotrophus* NapA has also been studied by X-band proton ENDOR which, together with EPR analysis, suggested that nitrate does not bind directly to Mo(V) during turnover, but most likely binds to the Mo(IV) state (Butler et al., 2002). A model for the catalytic cycle of *P.*

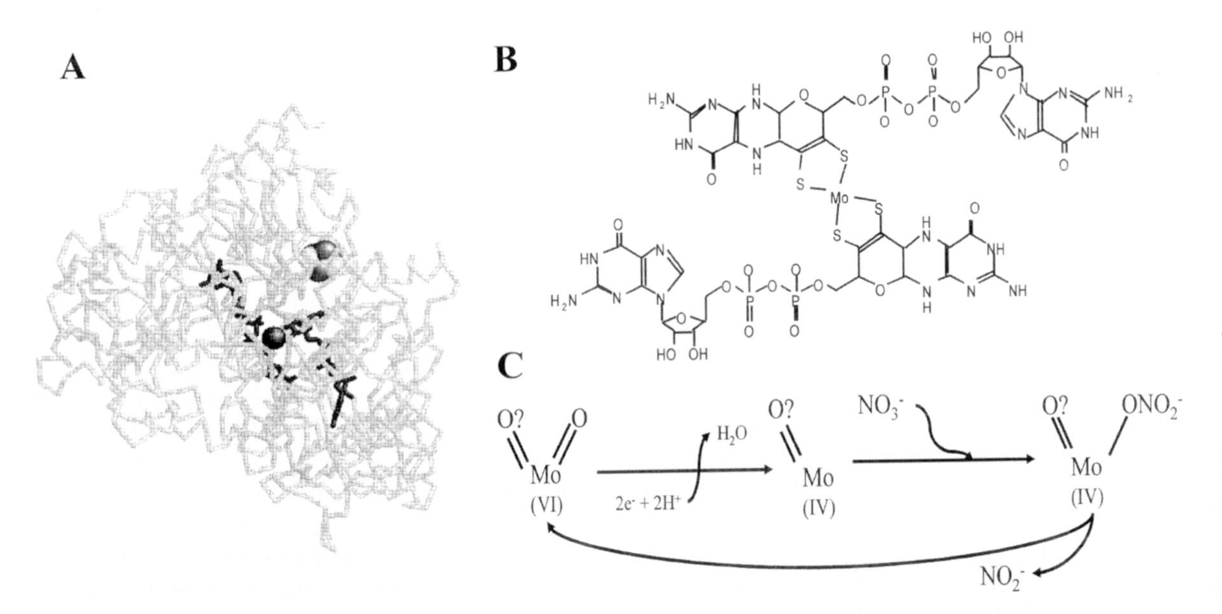

Fig. 7. The structural organization of the NapA subunit of the periplasmic nitrate reductase (A), the Mo-bis-MGD factor (B) and an oxo-transferase mechanism for nitrate reduction (C). Note, the question of whether Nop Mo (VI) is a di-oxo or mono-oxo species is debatable, hence the question mark on the second oxo group. The bis-MGD and cysteine S-ligands are not shown in C.

pantotrophus NapA Mo-bis-MGD based on EPR, EXAFS and ENDOR data is given in Fig. 7.

2. Electron Transport to NapA.

Electron transport to NapA (Fig. 2) illustrates a more general question in bacterial respiration of how electrons are transferred from the quinol pool to some periplasmic oxido-reductases without the participation of an energy-conserving cytochrome bc_1 complex. The *nap* operon contains a gene *(napC)* which is predicted to encode a membrane-anchored tetra-heme *c-type* cytochrome, homologues of which are found in a number of gene clusters associated with periplasmic oxidoreductases (Berks et al.,1995c). Primary structure analysis of this protein suggests that it is a member of a large family of bacterial tetra-heme and penta-heme cytochromes that is proposed to participate in electron transfer between the quinol/quinone pool and periplasmic redox enzymes (Roldan et al., 1998). These enzymes include the trimethylamine N-oxide reductase, dimethylsulfoxide reductase, soluble fumarate reductase, some nitrite reductases and hydroxylamine oxidoreductase. The quinol oxidation activity of some of these family members has been demonstrated (Roldan et al., 1998; Shaw et al., 1999; Simon et al., 2000; Simon et al., 2001). Topological analysis suggests that, in their native state, these cytochromes include a single N-terminal transmembrane α-helix that serves to anchor a globular multi-heme domain(s) onto the periplasmic surface of the cytoplasmic membrane. To facilitate biochemical and spectroscopic analyses, NapC has been expressed as a water-soluble periplasmic protein, NapC^{sol} (Roldan et al., 1998). Studies using magnetic circular dichroism (MCD) and electron paramagnetic resonance (EPR) suggested that all four hemes are low spin and feature *bis*-histidine axial ligation (Roldan et al., 1998). Redox potentiometric analysis indicated that the hemes had low midpoint redox potentials ($E_{m\,8.0}$) of –56mV, –181mV, –207mV –235mV. Primary structure and mutational analysis of NapC, and multiple sequence alignments of other members of the NapC family, led to the identification of four histidine residues as candidates for the distal heme ligands of the four covalently-bound bound hemes. Mutation of each of these results in a folded protein that binds a high spin His-H_2O/OH^- ligated heme (Cartron et al., 2002). This primary structure analysis and the proteolysis pattern of NapC has led to the proposal that the globular region comprises two structurally similar di-heme binding domains (Roldan et al., 1998; Cartron et al., 2002). Around 50 amino acids of the N-terminal and C-terminal segments, each containing two CXXCH motifs, can be aligned. The two CXXCH motifs in each segment are spaced around 28 amino acids apart and a conserved His residue lies between the two motifs in each segment. These are candidates for the distal ligand of two of the *bis*-His co-ordinated heme irons predicted from the EPR/MCD. A second conserved His residue lies 3 to 5 amino acids after the second CXXCH motif in each segment and is a candidate for the ligand of the second *bis*-His ligated heme in each domain. The parallel di-heme pairs envisaged would be similar to that identified in the recent X-ray crystal structure NapB (see below). Other members of the NapC family which have been purified include NrfH from *W. succinogenes*, CymA from *Shewanella putrefaciens* and the penta-heme DorC from *Rhodobacter capsulatus*. Spectroscopic analysis of CymA has also demonstrated that the four hemes are bis-histidine coordinated and the four candidates for the NapC distal heme ligands are conserved in CymA (Field et al., 2000). However, this is not the case for NrfH and DorC. Sequence analysis suggests that these cytochromes lie in different sub-groups of the family to NapC and CymA and so the possibility arises that not all members of the NapC family have equivalent heme ligation patterns.

An important unanswered question concerning NapC and related proteins is how it interacts with quinones and quinols, which are widely believed to be confined to the hydrophobic environment of the bilayer. The single transmembrane helix predicted for this class of protein is unlikely to provide a quinone binding site analogous to those that have been established in the transmembrane helical regions of photosynthetic reaction centers and the cytochrome bc_1 complex. It is more likely that NapC contains some amphiphilic helices that can dip into the membrane bilayer, analogously for example to the peripheral membrane protein D-lactate dehydrogenase in *E. coli* (Dym et al., 2000). It is notable that 'linker' region between the two putative domains of NapC contains a conserved EW sequence. Such a motif has been proposed to contribute a general quinone binding site in some proteins (Fisher and Rich, 2000).

The di-heme NapB cytochrome subunit of the NapAB complex is the electron-accepting redox partner for NapC. Studies using magnetic circular dichroism (MCD) and electron paramagnetic resonance (EPR) suggested that both hemes are low spin and feature *bis*-histidine axial ligation (Butler et al.,

2001). Primary structure analysis of NapB led to the identification of two conserved histidine residues as candidates for the distal heme ligands. Redox potentiometric analysis *of P. pantotrophus* NapAB indicated that the hemes had midpoint redox potentials of +80mV and –10 mV (Berks et al., 1994). The redox cycling of the hemes has been observed by both EPR and visible spectroscopies during turnover of the NapAB complex with nitrate, demonstrating the competence of these cofactors in mediating electron transfer to the catalytic site of the enzyme (Berks et al., 1994; Butler et al., 1999). *Hemophilus influenzae* NapB has been expressed independently from NapA in *E. coli* (Brige et al., 2001). The midpoint redox potentials of the two hemes were determined to be –25 mV and –175 mV. These are lower than those determined for the *P. pantotrophus* NapB and could reflect the fact that in the Nap system is dependent on menaquinol (E_0' = –80 mV) in *H. influenzae*, rather than ubiquinone (E_0' = +60 mV) as in *P. pantotrophus*. The crystal structure of a proteolyzed form of the recombinant NapB from *H. influenzae*, has been solved and revealed a new protein fold for the NapB class of cytochromes (Brige et al., 2002). The two heme groups have nearly parallel heme planes and are stacked at van der Waals distances with an iron-to-iron distance of only 9.9 Å. These structural features have now been found in a number of multi-heme enzymes (Barker and Ferguson 1999; Richardson, 2000), including the cytochrome *c* nitrite reductase and hydroxylamine cytochrome *c* oxidoreductase that will be discussed in section III.A. On the basis of the *H. influenzae* NapB and *D. desulfuricans* NapA structures a model of the NapAB complex has been proposed in which the four redox centers are positioned in a virtually linear configuration which spans a distance of nearly 40 Å (Brige et al., 2002). However, a recent crystal structure of a NapAB complex has not confirmed this model (Arnoux et al., 2003). The cofactors are arranged as a slightly bent molecular wire, with heme II of NapB being in the vicinity of the iron sulfur cluster in NapA. Heme I of NapB is exposed to solvent and can be proposed to be the electron acceptor of NapC, which is the physiological electron donor.

3. The Possible Roles of Accessory nap Gene Products

In addition to the *napABC* genes, six other *nap* genes, *napDEKFGH*, are found in different combinations in different bacteria (Fig. 2). NapK and NapE are monotopic integral membrane proteins of no currently known function and NapD may be a private chaperone for NapA that is involved in folding and cofactor insertion prior to export via the Tat system. However Nap F,G and H are predicted to bind iron-sulfur clusters raising the possibility of a role in electron transfer to NapA. NapF and NapG are predicted it to be 20 kDa proteins that bind four [4Fe-4S] clusters and to be located in the cytoplasm and periplasm, respectively. NapH (32kDa) is predicted to be an integral membrane protein with four transmembrane helices, a cytoplasmic C-terminal domain that contains two [4Fe-4S] binding motifs and two cytoplasmic Cys-(Xaa)3-Cys-Pro motifs that may be redox active or bind a metal center (Berks et al., 1995a). Introduction of in-frame deletions to yield *napC*, *napF*, *napGH* and *napFGH* mutants in ubiquinol-only and menaquinol-only backgrounds in *E. coli* has demonstrated that NapGH, but not NapF, are essential for electron transfer from ubiquinol to NapA and that NapC is essential to electron transfer from both ubiquinol and menaquinol (Brondijk et al., 2002). This raises the possibility that NapGH form a quinol dehydrogenase complex. Although NapF was not required for electron transfer to NapAB there was evidence from growth rate and growth yield analysis that NapGHF might form a protonmotive complex in which coupling was regulated by NapF. The possibility that NapGH can form a quinol-oxidizing complex could rationalize the absence of NapC, but presence of NapGH, in the *nap* gene cluster of *Campylobacter jejuni* (Parkhill et al., 2000). It also raises the possibility of parallel non-protonmotive NapC-only and proton-motive NapGH routes of electron transfer from QH_2 to NapAB in bacteria that have *napC* and *napGH* genes (Fig. 2). The growth conditions under which the different routes operate should now be studied in more detail to assess their physiological significance.

B. The Membrane-Bound Nitrate Reductase (Nar)

The membrane-bound nitrate reductase is a complex, three-subunit quinol dehydrogenase. It contains a *ca.* 140 kDa bis-MGD catalytic subunit (NarG, α) *ca.* 60kDa electron transfer subunit (NarH, β) which binds four iron-sulfur clusters and a *ca.* 25 kDa di-*b*-haem integral membrane quinol dehydrogenase subunit (NarI, γ). The *narGHI* genes encoding these subunits are present in all of the *nar* gene clusters so far identified (Fig. 3), as is *narJ* which encodes a private chaperone (Richardson et al., 2001). However, upstream and downstream of *narGHJI* the gene content can vary. For example in *E. coli narGHJI* is an operon and is preceded by the *narK* gene encoding a nitrogen oxyanion transporter. Further upstream of *narK* there are genes encoding the two component nitrate responsive sensor-regulator system NarXL. In some gene clusters, e.g. *Pseudomonas aeruginosa* there are two *narK* genes upstream of NarG but in *P. pantotrophus*, and probably also *P. denitrificans*, the *nar* transcriptional unit is *narKGHJI*, with the *narK* being a gene fusion of two smaller *narK* units (Wood et al., 2001,2002). In *Mycobacterium tuberculosis* the *narK* gene does not cluster with the *narGHJI* genes, but there are four *narK*-like genes elsewhere on the chromosome. In *B. subtilis* and *P. pantotrophus* genes encoding transcription factors of the Fnr family are also present in the *nar* gene cluster (Wood et al., 2001, 2002). In *E. coli* Fnr binds an oxygen-sensitive iron sulfur cluster that serves to mediate the anaerobic induction of the *narGHJI* operon, acting in conjunction with the nitrate-responsive NarXL sensor-regulator and integration host factor to maximize nitrate reductase expression (Potter et al., 2001). In *P. pantotrophus* the situation is different. *narKGHJI* is again under the control of an oxygen-responsive iron-sulfur cluster containing Fnr protein, FnrP. However no NarXL system has been identified. Instead a Fnr homologue, NarR, that lies upstream of *narKGHJI* serves a nitrate or nitrite responsive transcription factor that co-regulates the *narK* operon (Wood et al., 2001, 2002). Another recently discovered variation is the presence of *narC*, coding for a *c*-type cytochrome as the first gene in the operon coding for Nar in *Thermus thermophilus* (Zafra et al., 2002). NarC has an essential role in the synthesis of an active enzyme and its integration into the membrane (Zafra et al., 2002). Why this extra component is needed in this thermophile is unclear. Notable differences between Nap and Nar are that only the latter uses chlorate or bromate as alternative substrates and is strongly inhibited by azide.

1. *The Structure and Spectroscopy of NarGHI*

This enzyme has been the subject of a great deal of biochemical and spectroscopic analysis (reviewed in Blasco et al., 2001), which can now be mapped on to recent structural resolution (Jormakka et al., 2004). In contrast to Nap in which a Cys-S serves as Mo-ligand, the two structures of the *E. coli* NarG reveals that an Asp provides either one (Jormakka et al., 2004) or two (Bertero et al., 2003) oxygen ligands to Mo. A Ser-derived oxygen ligand is found in the *Rhodobacter capsulatus* DMSO reductase (see Chapter 9 by McEwan et al., Vol 1 of this book), which has Mo(V) EPR signals that are more similar to Nar than to Nap (Bennett et al., 1994, 1996; Sears et al., 1995; Butler et al., 1999). Considering the Nar structures in conjunction with early EXAFS and EPR studies it seems most likely that at pH <7, when Nar is most active, the Mo(VI) is coordinated by four thiols provided by the two MGD moieties, one Asp-, and one oxo group. The oxo group would convert to hydroxide on reduction to Mo(V) and be lost as water on further reduction to Mo(IV). The significance of the monodentate (1-O ligand from aspartate) versus the bidentate (2-O ligands for Asp) Mo coordination remains uncertain, but they could reflect different forms of the enzyme entered in different redox states or at different pH and ionic strength, both of which can cause significant changes in the Mo(V) EPR spectra. This will require further study. There is evidence that, in contrast to Nap, the Mo(V) state will bind nitrate and that this substrate-bound state may be relevant to the catalytic cycle and a bidentate coordination may sterically hinder such a binding process (George et al., 1985, 1989; Anderson et al., 2001).

The NarG and NarH subunits of Nar are peripheral to the cytoplasmic side of the membrane and readily released as water-soluble proteins. These two subunits show no activity with quinols as electron donors; only when the transmembrane NarI subunit is present is the ability to oxidize quinols conferred (Ballard and Ferguson, 1988). Sequence analysis of NarI showed that there are four conserved histidines, two on each of two helices out of the five predicted for this subunit (Berks et al., 1995b). These histidines bind two hemes stacked across the membrane as other work had shown for the enzyme from *P. denitrificans* that the two hemes had distinct redox potentials (+95 mV and +210 mV) (Ballard and Ferguson, 1988), it could be safely proposed that the function of the NarI subunit was to oxidize quinol at the outer (P side of the membrane) and transfer electrons via the two hemes to the catalytic subunit at the cytoplasmic (N) side of the membrane (Berks et al., 1995b). This organization of the Nar enzyme means that it would act as the electron carrying arm of a redox loop system for generating protonmotive force as originally proposed as a general mechanism for respiratory chain action by Mitchell. Substantial support for this model came from a range of site-directed mutagenesis studies, coupled with analysis of redox potentials and the effects of quinol binding to NarI (see Blasco et al., 2001; Rothery et al., 2001, for reviews).

Recently, structural biology has provided the first molecular description of a full protonmotive redox loop, comprising the *E. coli* nitrate-induced formate dehydrogenase (Fdh-N) and the membrane-bound nitrate reductase (Jormakka et al., 2002, 2004; Bertero etal., 2003). The structure of Fdh-N was the first to be resolved. Fdh-N is closely related to Nar, especially in respect of the equivalent FeS subunit (β) and bis-Mo-MGD catalytic subunit (α) and initially provided a good model for Nar (Jormakka et al, 2002; Richardson and Sawers, 2002). This type of formate dehydrogenase had also been predicted to have two hemes within an integral membrane subunit but which differ from the counterpart in nitrate reductase in having only four helices, three of which provide the four histidine ligands (Berks et al., 1995b). A critical difference between the nitrate reductase and the formate dehydrogenase is topology, because in the formate dehydrogenase the alpha and beta subunits are in the periplasm. Thus this dehydrogenase catalyses the inward movement of electrons from the P side to the N-side where menaquinone is reduced. Therefore it resembles the nitrate reductase in constituting the electron-carrying arm of a redox loop. In Fdh-N the electrons extracted from formate at the periplasmic bis-MGD active site pass down a 90Å 'ladder' of 8 redox centers ultimately to reduce menaquinone at the N-face of the cytoplasmic membrane (Fig. 8). This redox ladder comprises the bis-MGD, five iron sulfur clusters and two hemes. Each are within 12Å of its nearest neighbor ensuring rapid electron transfer. A large ~ 340 mV potential 'drop' (–420 mV to –75 mV at the 'N' face) allows efficient electron transfer against the membrane potential to occur. Much of the redox 'ladder' is membrane extrinsic, and it is the two membrane intrinsic hemes that are critical to charge separation across the membrane. Of note is the interaction of the 'N' face heme to the bound menasemiquinone analogue HQNO, which accepts a hydrogen bond from one of the histidine ligands. This is the first time that a heme-ligand has been shown to be directly involved in quinone binding and it may be that this will also be the case in the nitrate reductase γ subunit.

The basic Fdh-N framework of an integral membrane di-heme subunit, a peripheral membrane ferredoxin with four associated iron-sulfur clusters, and a peripheral membrane subunit with an iron-sulfur cluster and an active site Mo-*bis*-MGD, is also found in the NarGHI. In the FdhN beta subunit the four iron sulfur centers (all [4Fe-4S]) are arranged in two pairs in each of two domains. This arrangement was already predicted for Nar from detailed spectroscopic studies on *E. coli* NarH, the only difference being that one of the Nar FeS centers is a [3Fe-4S] cluster (Blasco et al., 2001). In the case of NarH, the low redox potentials of two of the centers had raised the possibility that they are not directly involved in electron transfer between quinol and nitrate. However, consideration of the structure of the Nar and Fdh-N β subunits leaves no doubt that all four centers will be directly involved in mediating electron transfer between formate and nitrate. Thus, it seems that the extended electron transfer chains of NarGHI and Fdh-N add examples to a growing number of cases of electron transfer complexes in bacteria that contain a mixture of endergonic and exergonic electron transfer steps and which include NapA (see section II.B.ii), the Ni-Fe hydrogenase and fumarate reductase (see (Ohnishi et al., 2000) for review). The arrangement of the redox centers in the FdhN-Nar electron transfer system of *E.coli* is shown in Fig. 8. One question that has been resolved is whether, like the alpha subunit of FdhN, NarG binds an iron sulfur center in addition to the *bis*-MGD. The possibility that a His-xx-CysxxCysx$_n$Cys motif towards the N-terminal of NarG binds an iron sulfur center had been raised (Berks et al., 1995a), but such a center had not been detected spectroscopically, although mutational analysis has suggested that these Cys residues are important for electron transfer (see (Blasco et al., 2001)). However, the NarG structure confirms that this motif does bind a [4Fe-4S] cluster for which elucidation of the redox properties must now await spectroscopic indentification.

The Nar and Fdh-N structures provide a framework up on which studies on electron transfer through the enzyme complexes can be built. One of the techniques being applied is protein film electrochemistry. Early studies by Craske showed that NarGH can be purified separately from NarI, retaining enzymatic activity (Craske and Ferguson, 1986). Recently soluble NarGH has been shown to form films on graphite and gold electrodes within which direct and facile exchange of electrons between the electrode and the enzyme occurs. Protein film voltammetry has been used to define the catalytic behavior of NarGH in the potential domain and a complex pattern of reversible, nitrate concentration dependent, modulation of

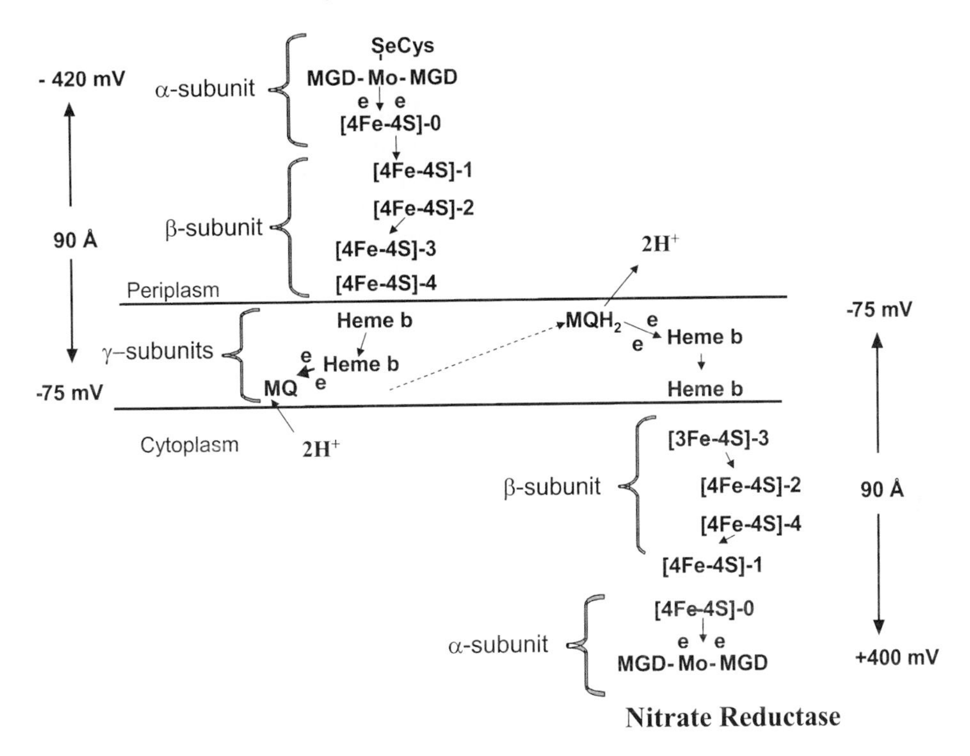

Fig. 8. The organization of the FdhN-Nar electron transfer system of *Escherichia coli*.

activity has been resolved (Anderson et al., 2001). The study shows that NarGH can catalyze nitrate reduction via two pathways, each having distinct k_{cat}/K_M. Catalysis is directed to occur via one of the pathways by an electrochemical event within NarGH which may lie at the level of the Mo-bisMGD or one of the iron-sulfur centers.

2. Nitrate Transport to Nar

One important question that the orientation of Nar poses is how is the substrate delivered to the active site? One or more transport proteins are implicated by the finding that while (and in contrast to Nap) the purified Nar enzyme can use chlorate as an alternative substrate, in the cell Nar is inaccessible to chlorate unless the permeability barrier of the cytoplasmic membrane is breached through addition of a detergent. The nitrate anion is encountered in the external environment of the bacterium and so must be transported into the cell (against the 'negative-inside' membrane potential) to serve as the substrate for NarGH. This could, in principle, be a *pmf*-consuming process that would then affect the net energetics of the QH_2-NR electron transfer system (Fig. 3). In the anaerobic denitrification pathway there is a need to transport the nitrate anion substrate for Nar into the cell and the cytoplasmically generated nitrite out of the cell to the periplasm where it can serve as substrate for nitrite reductase. Failure to do this will result in the cell being unable to denitrify. Thus the nitrate/nitrite transport process provides an excellent level at which to control denitrification. This was demonstrated in a series of studies in the 1980s in which it was demonstrated that nitrate reduction was inhibited by oxygen in intact cells, but not in membrane vesicles or triton X-100 permeabilized cells (Alefounder and Ferguson, 1980). This oxygen-inhibition could also be mimicked by nitrous oxide or ferricyanide, indicating that it was not an effect of molecular oxygen itself but a redox effect that could possibly be mediated by the Q-pool which will be in contact with an integral membrane transporter (Alefounder et al., 1981, 1983). The basis of oxygen/redox-regulation of nitrate transport has not yet been resolved but progress is now being made through the identification of the gene, *narK* that encodes the N-oxyanion transport system and is regulated by FnrP and NarR (Wood et al., 2002). In *Paracoccus pantotrophus narK* encodes a 24 transmembrane helix protein that is a fusion of two transporters, NarK1 and NarK2. Recent work in which the contribution that each domain makes to nitrate uptake in the cell has been assessed separately has suggested that NarK1 is a nitrate-proton symporter and NarK2 is a nitrate-nitrite antiporter (Wood et al., 2002). This confirms earlier work on *P. denitrificans* (Boogerd et al., 1981), that suggested that a nitrate-2H^+ symporter initiates nitrate respiration and then, when sufficient intracellular nitrite has built up, a nitrate-nitrite antiporter operates. The later system is bioenergetically more efficient as it does not consume PMF and it may be that the fused nature of the two transporters allows co-operative regulation so that 'switch on' of the nitrate-nitrite antiporter switches off the nitrate-proton symporter. NarK1 and NarK1 cluster separately in phylogenetic trees and it is notable that a number of bacteria (e.g. *Ps. aeruginosa*) have two separate genes encoding NarK1 and NarK2 homologues (Wood et al., 2002); while *E. coli*, by contrast, has genes encoding two NarK2 type proteins (NarK and NarU). Although it has been argued that *E. coli narK*, associated with the *narGHJI* gene cluster, encodes a nitrite antiporter (Rowe et al., 1994), recent work has suggested that both NarK and NarU do in fact contribute to nitrate uptake (Clegg et al., 2002), but that NarU, the gene for which is associated with a second *nar* operon (*narZ*) in *E. coli*, may contribute more than NarK to nitrite excretion. Clearly there is still much to be learnt about the interaction and physiological roles of different NarK-type proteins in different bacteria.

III. Respiratory Nitrite Reductases

As explained in the Introduction, there are two types of reaction catalysed by nitrite reductase. One is the reduction of nitrite to ammonium.

$$NO_2^- + 6e^- + 8H^+ \rightarrow NH_4^+ + 2H_2O \quad (2)$$

This is catalysed by a dimeric enzyme in which each monomer has five heme groups. Each of the hemes is attached to the protein by two thioether bonds and so this type of nitrite reductase can be termed a pentaheme c-type cytochrome, although it is increasingly being called NrfA after its gene. The characteristic of c-type cytochrome is that the heme is attached to the two cysteines of a CXXCH motif. As discussed below, one of the hemes in this type of nitrite reductase is exceptional in being attached to a CXXCK motif and is the catalytic heme. One exception to this is found in the NrfA of *C. jejuni*, in which all five hemes are attached to CXXCH motifs. The functional significance of this somewhat surprising latter finding remains to be determined.

The second reaction catalysed by nitrite reductases is the reduction of nitrite to nitric oxide.

$$NO_2^- + e^- + 2H^+ \rightarrow NO + H_2O \quad (3)$$

This is catalysed by two very different enzymes, one called cytochrome cd_1, the other copper nitrite reductase. They are less commonly known by their gene names, i.e. NirS and NirK respectively. As the name suggests, cytochrome cd_1 contains a *c*-type cytochrome center and a second specialized heme, which is non-covalently bound and provides the active site. One of heme centers of each type is associated with each polypeptide chain in the dimeric molecule. The Cu-type of nitrite reductase is a trimeric molecule with six copper ions. Three of these are Type 1 coppers and serve to collect electrons from donor proteins and pass them on to the Type 2 coppers at the three active sites. The recruitment by biology of both iron and copper to catalyze the same reaction is common in biology. As we shall describe below, these two types of nitrite reductase have very different reaction mechanisms. Finally, we note that only one of these three types of nitrite reductase has ever been found conclusively in any one organism.

A. Cytochrome c Nitrite Reductase (NrfA)

The reduction of nitrate to ammonium in the periplasm of a number of bacteria, including *Escherichia coli*, involves two enzymes, periplasmic nitrate reductase (NapA) and periplasmic nitrite reductase (NrfA). The process is found in many enteric bacteria and is important for anaerobic nitrate and nitrite respiration at low nitrate concentrations (Potter et al., 2001). The NrfA protein catalyses the six electron reduction of nitrite to ammonium:

$$NO_2^- + 6e^- + 8H^+ \rightarrow NH_4^+ + 2H_2O \quad (4)$$

NrfA can also catalyze the 5 electron reduction of NO and 2 electron reduction of hydroxylamine, both of which may be bound intermediates in the catalytic cycle for nitrite reduction (Berks et al., 1995a). Interestingly, it has recently been shown that the NO reductase activity of NrfA may also play a physiological role in detoxification of exogenous NO (Poock et al., 2002). Primary structure and biochemical analysis revealed that the *E. coli* NrfA protein is a 52 kDa pentaheme cytochrome in which four hemes are covalently-bound to the conventional motif, CXXCH. The fifth heme is attached to the novel CXXCK motif (Darwin et al., 1993; Hussain et al., 1994). This attachment requires specialized heme cytochrome c maturation proteins (Eaves et al., 1998).

1. The Structure and Spectroscopy of NrfA

The crystal structures of cytochrome *c* nitrite reductase from *E. coli*, *Sulfurospirillum deleyianum* and *Wolinella succinogenes* have been determined (Einsle et al., 1999, 2000; Bamford et al., 2002). In all three systems NrfA crystallized as a homo-dimer (Fig. 9), with the hemes within each monomer closely packed to form arrangements of near-parallel and near-perpendicular heme pairs that have also been observed in the octa-heme hydroxylamine oxidoreductase (Igarashi et al., 1997). Indeed there is also structural similarity in protein folding to HAO that reveals a clear evolutionary relatedness between the two enzymes which is intriguing given that Nrf can reduce nitrite to a bound hydroxylamine intermediate and HAO can oxidize hydroxylamine to nitrite. However, both enzymes appear to be essentially unidirectional. In *E. coli* NrfA the active site heme displays a distal lysine ligand and proximal oxygen ligand that is likely to arise from water (Bamford et al., 2002). In the case of the *W. succinogenes* enzyme NrfA structures with water, sulfate, nitrite, hydroxylamine and azide bound in the active site have been reported (Einsle et al., 2000).

A number of common features of the cytochrome *c* nitrite reductases are apparent in the active sites of all of the *W. succinogenes*, *S. deleyianum* and *E. coli* structures. These include the catalytic heme with the lysine primary amine nitrogen as the proximal ligand, a putative substrate inlet channel with positively charged electrostatic potential and a putative product efflux channel that exhibits a more negative electrostatic potential. In addition, an active site calcium ion is conserved across all three structures, both in terms of the identity of ligating residues and in the presence of two conserved water ligands (Fig. 9). The crystal structure of the *W. succinogenes* enzyme with bound nitrite shows that the nitrogen is coordinated to the heme iron with the two oxygen atoms being H-bonded by His264 and Arg114 (*E. coli* numbering). In the hydroxylamine coordinated derivative the single oxygen is H-bonded by Arg114. Proton delivery to the active site may involve conserved tyrosine residues. A possible scheme for reduction of nitrite at the active site has been proposed (Einsle et al., 2002).

The structure of the *E. coli* enzyme has recently been analyzed in combination with an MCD and spectro-potentiometric EPR (Bamford et al., 2002). This has allowed the assignment of spectroscopic signals and equilibrium redox potentials to individual hemes in the crystal structure. MCD analysis confirmed the presence of four bis-His coordinated Fe(III) hemes in a NrfA solution at pH7. The MCD analysis also confirmed the presence of a high spin heme Fe(III) with N and O ligation in which the oxygen ligand arose predominantly from water. EPR spectra were collected from NrfA samples poised electrochemically at number of different potentials. Broad perpendicular mode X-band EPR signals with positive features at g = 13.4 and 3.5, that are characteristic of weakly spin-coupled $S=5/2$, $S=1/2$ paramagnets, titrated with an apparent $E_m = -107$ mV and are likely to arise from the active site Lys-OH_2 coordinated heme (heme 1) and the nearby *bis*-His coordinated heme (heme 3). These hemes have a closest ring distance of 4.3Å and inter plane angle of 25°. A rhombic heme Fe(III) EPR signal at $g_z = 2.91$, $g_y = 2.3$, $g_x = 1.5$ titrates as a n=1 center with $E_m = -37$ mV and is likely to arise from *bis*-His coordinated heme (heme 2) in which the interplanar angle of the imidazole rings is 21.2°. The final two *bis*-His coordinated hemes (hemes 4 and 5) have imidazole interplanar angles of 64.4° and 71.8°. Either or both of these hemes could give rise to a 'Large g max' EPR signal at $g_z = 3.17$ that titrates at potentials between –250 mV and –400 mV.

The catalytic cycle of NrfA may proceed via bound NO and NH_2OH derivatives, although the high K_M for NH_2OH does raise the question of whether it can be a legitimate intermediate (Fig 10). Catalysis has also been probed with protein film electrochemistry (Angove et al., 2002). The catalytic current-potential profiles observed on progression from substrate-limited to enzyme-limited nitrite reduction revealed a fingerprint of catalytic behavior distinct from that observed during hydroxylamine reduction. Suggesting that the cytochrome c nitrite reductase interacts

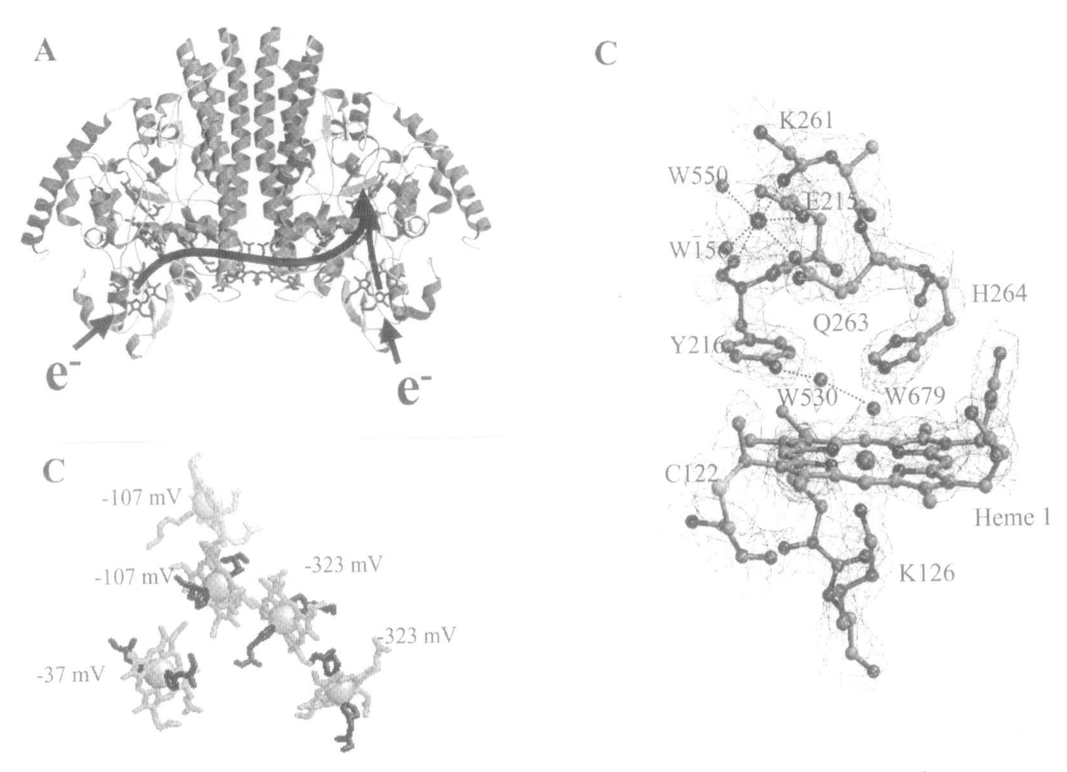

Fig. 9. The structure of *E. coli* cytochrome *c* nitrite reductase (NrfA) and its active site pocket.

differently with these two substrates, a notion supported by the crystal structures. However, a sigmoidal catalytic wave with an n=2 slope with Em = *ca.* −103 mV was a common feature of the development of the voltammetric response with increasing nitrite or hydroxylamine concentration. The midpoint of this catalytic wave is similar to that determined for the active site di-heme pair and thus suggests that a rate-limiting coordinated two-electron reduction of the active site for both substrates and suggest that the mechanisms for reduction of both substrates are underpinned by common rate defining processes. One question that now needs to be addressed is the significance of the dimeric organization of NrfA. Hemes 5 at the dimer interface are close enough for inter-dimer electron transfer and this could prove to be a mechanism by which dimers of the penta-heme monomer store sufficient electrons to catalyze the rapid 6-electron reduction of nitrite (Fig. 9). It is then possible that a different route of electron transfer operates for the 2-electron reduction of hydroxylamine.

2. Electron Transfer to NrfA

Analysis of the organization of *nrfA* gene clusters from a range of bacteria reveals that they can be divided into two groups. In one group, that includes *W. succinogenes* and *S. deleyianum* (Fig. 5), the *nrfA* clusters with an adjacent gene *nrfH* which encodes a membrane-anchored tetra-heme quinol dehydrogenase of the NapC family discussed in Section II.B.2 (Simon et al., 2000, 2001). Analysis of NrfHA reconstituted into proteoliposomes has clearly shown the electron transport from quinol-nitrite is not proton-motive. In the second group, which includes *E. coli* (Fig. 5), the *nrfA* clusters with genes encoding a periplasmic penta-heme cytochrome (*nrfB*), a periplasmic (4 × [4Fe4S]) ferredoxin (*nrfC*) and an integral membrane putative quinol-dehydrogenase (*nrfD*) containing seven transmembrane helices (Hussain et al., 1994; Potter et al., 2001). Clearly electron transfer from quinol to the NrfA in the different groups is distinct and it can not be excluded that electron transfer from quinol to nitrite via NrfDCAB is protonmotive. The different protein-protein and cofactor-cofactor interactions between NrfH-A and NrfB-A may be reflected by insertions and deletions in loop regions of the polypeptide chain in the two NrfA subgroups that can be identified in primary structure analysis. This leads to differences in the solvent exposure of heme

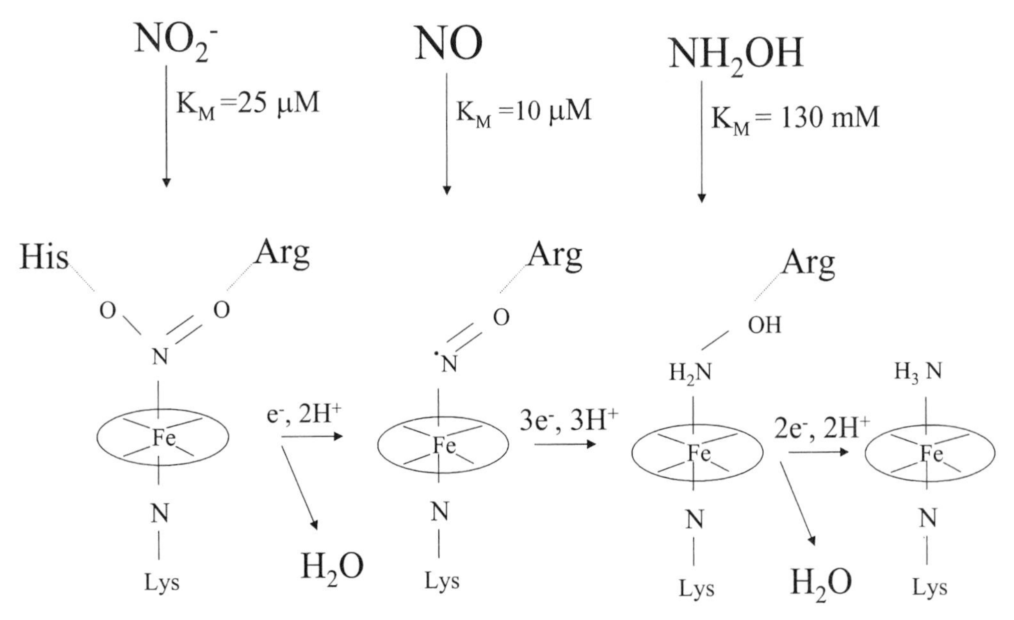

Fig. 10. A catalytic scheme for the cytochrome *c* nitrite reductase.

2 and in electrostatic surface of the protein around this heme (Bamford et al., 2002). Consideration of the crystal structure makes heme 2 the most likely electron input site (Fig. 9).

B. Cytochrome cd_1 Nitrite Reductase (cdNir)

The name cytochrome cd_1 recognizes the content of one *c*-type cytochrome center and one d_1 heme per polypeptide chain. The former center is defined by the covalent attachment of Fe protoporphyrin IX to the polypeptide chain as a consequence of thioether bond formation between the vinyl groups of the porphyrin and two thiols of cysteine residues in a CXXCH motif. The d_1 heme is unique to this type of enzyme and is characterized by partial saturation of two of the rings and the presence of carbonyl groups. Structures of cytochrome cd_1 have been solved for the proteins from the bacteria *P. pantotrophus* and *P. aeruginosa*, with the surprising outcome that there are significant differences between the two, at least in the oxidized states. Thus to avoid potential confusion in the reader's mind, we discuss here sequentially some of the findings concerning the two enzymes before returning to compare the two enzymes and seeking to synthesize an overall picture of current understanding of this type of enzyme (see also Allen et al., 2001 for another recent review).

1. Paracoccus pantotrophus Cytochrome cd_1

The crystal structure of the dimeric oxidized cytochrome cd_1 from *P. pantotrophus* showed that the d_1 heme was enclosed in an eight bladed beta propeller structure with the iron having one expected axial ligand, a proximal histidine, and one unexpected ligand, a distal tyrosine (Fulop et al., 1995) (Fig. 11). Even more unexpected was that this tyrosine, residue 25, was part of the alpha helical domain of the protein that formed the c-type cytochrome sector of the enzyme. There was also a surprise in this part of the protein because the heme iron had two axial histidine ligands, one of which, residue 17, was close in sequence to the tyrosine ligand to the d_1 heme iron. The other histidine ligand was provided from the CXXCH motif. Solution spectroscopic measurements were in accord with these heme ligations (Cheesman et al., 1997). Reduction of this crystal form resulted in loss of the tyrosine from the d_1 heme and thus the provision of a binding site on the iron for the substrate (Fig. 11). Histidine 17 had also dissociated from the Fe of the *c*-type cytochrome domain and been replaced by the sulfur of methionine 106 which had undergone a considerable spatial displacement following reduction of the enzyme (Williams et al., 1997). The first mechanistic proposal for the enzyme envisaged that these ligand switches would occur during the catalytic cycle. This was an attractive idea as the reaction product, nitric oxide, can have a high affinity for hemes and a tyrosine rebinding mechanism could be envisaged to contribute to the displacement of the product. However, subsequent experiments in which fully reduced enzyme was oxidized by either oxygen (an alternative substrate) or nitrite under rapid reaction conditions showed that the *c* type cytochrome heme had spectroscopic properties consistent with a ferric state having histidine plus methionine, rather than bis-histidinyl, coordination (Allen et al., 2000a,b; George et al., 2000; Koppenhofer et al., 2000a). Use of either oxygen or nitrite as oxidant in such experiments has complications because products remain bound to the enzyme, and neither of these substrates, requiring four and one electrons respectively, can remove the electron from each of the two heme centers on one polypeptide concomitantly with turnover of a single acceptor ion/molecule at the active site. In contrast, a third substrate, hydroxylamine, requires two electrons for its reduction to ammonium. When fully reduced enzyme, in the absence of any excess reductant, was oxidized by hydroxylamine, it was noted that the visible absorption spectrum of the resulting fully oxidized enzyme was distinct from the initial as prepared oxidized enzyme (Allen et al., 2000a,b). Examination of this form of the enzyme by both EPR and NIR MCD showed that the heme of the *c* type cytochrome had histidine/methionine (Allen et al., 2000a,b). Over a period of approximately 20 minutes this form of the oxidized enzyme reverted to the as prepared state , with the two histidine ligands to the *c*-type cytochrome domain that have already been described. This reversion process also correlated with a loss of the ability of the cytochrome cd_1 to catalyze

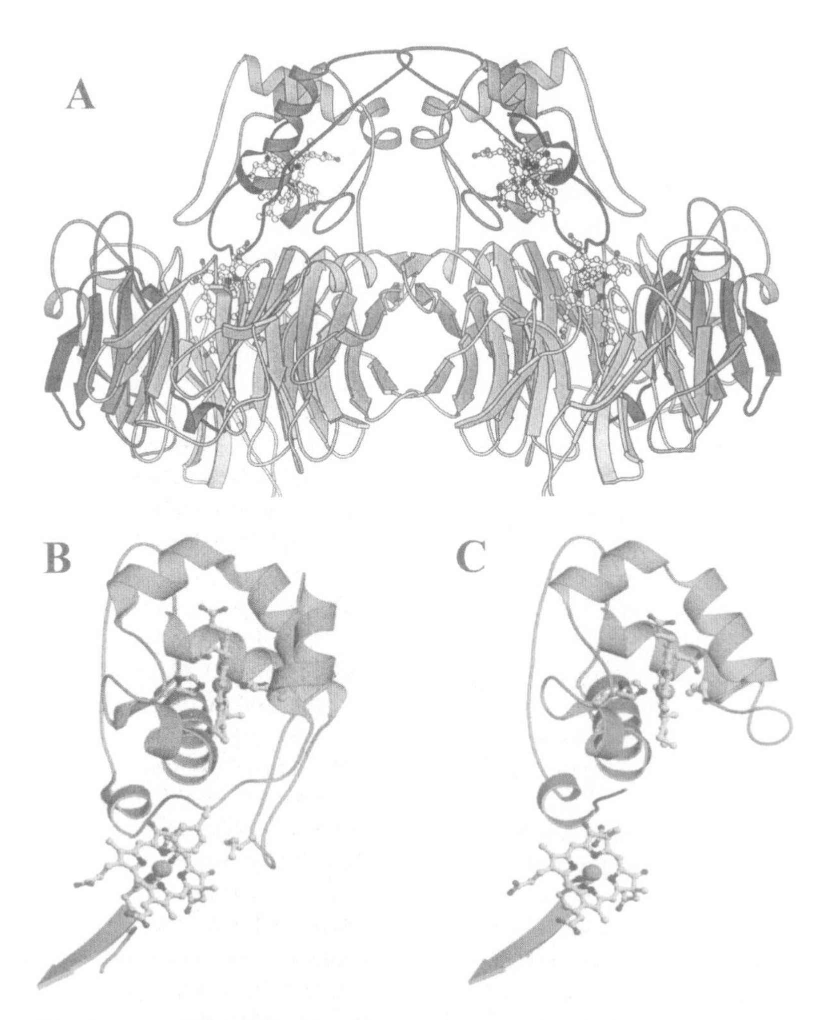

Fig. 11. The structure of the *P. pantotrophus* cytochrome cd_1 nitrite reductase. (A) overall structure of oxidized enzyme. (B) detail of heme coordination in oxidized enzyme. (C) detail of heme coordination in reduced enzyme.

the oxidation of reduced pseudoazurin, a physiological type I Cu electron donor protein, by nitrite. Thus the form of oxidized enzyme obtained immediately following a cycle of oxidation by hydroxylamine was considerably more catalytically competent than the as prepared oxidized enzyme (Allen et al., 2000a,b). Subsequently it has been shown that the latter form of the enzyme requires activation by a cycle of reduction and reoxidation in order to elicit high rates of activity towards the oxidation of either pseudoazurin, cytochrome c_{550} from *P. denitrificans* or horse heart mitochondrial cytochrome *c* with any of the three electron acceptors, nitrite, oxygen or hydroxylamine (Richter et al., 2002). Thus a current hypothesis is that the as prepared form of the oxidized enzyme represents a resting state of the enzyme. The purpose of this form of the enzyme and its mechanism of activation in vivo are not known.

The conclusion from spectroscopic and activity measurements that the oxidized cytochrome cd_1 molecule can adopt a conformation in which it is activated might reasonably be correlated with the structure seen originally for the reduced enzyme, which, recall, has histidine/methionine coordination at the c-type heme. However, a new crystallographic study on this enzyme has added an unexpected twist of complexity to the story. Crystallization of reduced protein under anaerobic conditions has demonstrated the existence of a third conformation for cytochrome cd_1 from *P. pantotrophus* (Sjogren and Hajdu, 2001). This differs from that originally described for reduced enzyme, not in respect of the ligands to the iron atoms, but rather in terms of a rotation by approximately 60° of the *c* heme domain relative to the d_1 heme domain. Treatment of this crystal form with oxygen gave an oxidized crystal in which the structure of the protein was unchanged. Thus it is tempting to conclude that this new crystal structure represents the protein in the activated conformation that is observed in solution after a cycle of oxidation and reduction. In this context, therefore, it could be considered that both the original oxidized and reduced crystal structures are not catalytically relevant forms. However, the 'new' reduced crystal could not be reoxidized by nitrite which therefore injects a note of caution about accepting this conclusion too precipitately (Sjogren and Hajdu, 2001). Inspection of the 'old' and 'new' crystal forms shows that the two forms are related by a rotation of the cytochrome domain around what appears to be a hinge region. Future work will need to address the possibility that such rotation is part of the catalytic mechanism, especially since a similar movement has been seen under certain circumstances with the *P. aeruginosa* enzyme (see Section III.B.2). The various conformational changes now seen for cytochrome cd_1 may well explain the remarkable hysteretic redox titration that is seen for the enzyme from *P. pantotrophus* (Koppenhofer et al., 2000b).

One of the enduring issues concerning cytochrome cd_1 is the purpose of the specialized d_1 heme. It differs from the usual heme in several respects, amongst which is the partial saturation and presence of electron-withdrawing carbonyl groups. A plausible mechanism for the enzyme involves the binding of the anionic nitrite to this heme in its reduced state. However, this would be at variance with the usual expectation that it is the ferric state of a heme that binds an anion. Evidence that the ferrous d_1 heme is tailored to bind an anion has come from a crystallographic study of cyanide binding to the cytochrome cd_1 from *P. pantotrophus*. A structure of cyanide bound to the ferrous enzyme was obtained but when the crystal was oxidized the cyanide was lost and the tyrosine ligand returned to the d_1 heme iron coordination sphere (Jafferji et al., 2000). The dissociation constant for cyanide from the ferrous d_1 heme in solution could be shown to be in the micromolar range, a low value for a ferrous heme. These observations show that the driving force for tyrosine ligation to the d_1 heme iron, whatever its purpose, is substantial and that the d_1 heme is tuned in its ferrous state to bind anions. It is notable that another variation of the standard heme structure, siroheme, also binds cyanide in its ferrous state and participates in anion reduction reactions. Study of the binding of cyanide to the reduced *P. aeruginosa* enzyme and a mutant lacking one of the active site histidines has emphasized the importance of this side chain (residue 369) in controlling the binding of anionic ligands (Sun et al., 2002).

The binding of nitrite to the d_1 heme has also been investigated recently. For the enzyme from *P. pantotrophus* it was found that addition of nitrite to the oxidized enzyme prevented, unlike cyanide, the return of tyrosine 25 to the heme iron. Nitrite was the only ion which was significantly effect in this respect (Allen et al., 2002). An approximate dissociation constant for the oxidized enzyme nitrite complex of 2 mM could be obtained. This observation suggests that the active site of cytochrome cd_1 is tailored to bind nitrite rather than any other anion. One can expect that the reduced form of the enzyme will have even higher affinity for nitrite but this is difficult to measure owing to its binding being followed by formation of nitric oxide. A current estimate for the dissociation constant for nitrite from the reduced enzyme is of the order of 1 μM (Cutruzzola et al., 2001). However, this may be an overestimate since cyanide binds with this affinity to the reduced enzyme and nitrite appears to bind more strongly than cyanide to the active site. Finally, we note that nitrite is a β-accepting ligand (Nasri et al., 2000), a feature that could promote its binding to an Fe(II) porphyrin rather than the less electron rich Fe(III) species. Such backbonding can strengthen the Fe-N bond while contributing to the weakening of an N-O bond in nitrite. This is what is required for formation of nitric oxide from nitrite but whether the special features of the d_1 heme contribute to this effect is not known. It may in fact be the case that the d_1 heme is less effective at such back bonding than the normal porphyrin owing to an inversion of energy levels (see below).

As we have already discussed, cytochrome cd_1 is able to reduce not only nitrite to nitric oxide but also hydroxylamine to ammonia, a reaction that it shares with the NrfA type of nitrite reductase. However, unlike NrfA, it is not able to reduce nitric oxide to hydroxylamine. Reduction of nitric oxide to hydroxylamine requires an active site that can operate at a potential of around –100 mV, which is the case for NrfA (see section III.A), whereas the nitrite to nitric oxide and hydroxylamine to ammonia reactions are compatible with 'operating' potentials in the range of +300–400 mV. Although determination of the potential of the d_1 heme in cytochrome cd_1 has proved problematic, it can safely be taken to be approximately 300 mV, consistent with the enzyme receiving electrons from the cytochrome *c*/cupredoxin level of the electron transfer system. This in turn means that electrons destined for cytochrome cd_1 nitrite reductase can pass through the proton translocating cytochrome bc_1 complex en route from ubiquinol (Fig. 4). In contrast, the operating potentials of the NrfA hemes have to have less positive values than that for d_1, with the result that electrons must reach NrfA ubiquinol or menaquinol without passing through a cytochrome bc_1 complex. Thus the stoichiometry of proton translocation per electron, and thus of ATP synthesis, is higher when nitrite is reduced by cytochrome cd_1 (or the copper-type nitrite reductase) rather than by NrfA. Avoidance of reduction of NO to ammonium via hydroxylamine by cytochrome cd_1 also avoids short circuiting of the nitrogen cycle (Fig. 1).

Detailed spectroscopic analysis of the d_1 heme center suggests that the relative energy levels of the d orbitals associated with the d_1 heme iron differ from the usual pattern in that for the low-spin ferric state the 4-electron reduced heme d_1 macrocycle stabilizes $(d_{xz},d_{yz})^4(d_{xy})^1$ relative to the more usual $(d_{xz},d_{yz})^3(d_{xy})^2$ electronic configuration (Cheesman et al., 1997). This 'inverted' ground state represents a relocation of unpaired spin density from orbitals extending above and below the heme plane to an orbital lying in that plane and this will have important consequences for the binding of π-unsaturated ligands such as nitric oxide, in this case the product molecule. Specifically, NO would be less strongly bound to the d_1 heme then to the usual *b*-type heme. Recently spectroscopic work with the enzyme from *P. aeruginosa* has also addressed the unusual electronic properties of the heme d_1 ring and how these influence the binding of nitric oxide (Das et al., 2001). Thus these properties may be a significant reason for the development of the specialized d_1 heme

Another uncertainty surrounding d_1 heme is its route of biosynthesis and how it is incorporated into the apo enzyme in the periplasm. The polypeptide for cytochrome cd_1 is translocated as an unfolded apo protein by the Sec system and acquires its c-type cytochrome center in the periplasm as a result of the action of the Ccm biogeneisis apparatus (Heikkila et al., 2001 and see article by Thony-Meyer in this volume). The NirD protein, of unknown function but believed to be involved in the assembly of the d_1 part of the enzyme, has a signal sequence suggesting it is carried by the Tat system (Heikkila et al., 2001). As loss of the Tat system prevents loss of an active cytochrome cd_1 it may be tentatively concluded that NirD translocates some key component needed for enzyme assembly in the periplasm.

2 Pseudomonas aeruginosa Cytochrome cd_1

The cytochrome cd_1 that has received most attention over the years is that from *P. aeruginosa*. Despite the fact that cytochrome cd_1 is a specialized enzyme, it is remarkable that the structure of the *P. aeruginosa* enzyme is not identical in several critical respects to the enzyme from *P. pantotrophus*, most notably in terms of the coordination of the *c* heme and d_1 heme iron centers. The difference is reflected in the visible absorption spectra of the oxidized enzymes. In the oxidized structure of the *P. aeruginosa* enzyme one sees that the overall structure of the dimer is the same in the sense that the *c*-type cytochrome domain as for *P. pantotrophus* is helical and separated from the 8-bladed β-propeller domain that binds the d_1 heme center (Nurizzo et al., 1997). However, there are intriguing differences. First, the fold of the cytochrome *c* domain of the oxidized *P. aeruginosa* enzyme is essentially the same as in the reduced *P. pantotrophus* (Fig. 11C) enzyme and the heme iron coordination is His/Met. The iron of the d_1 heme does not have a tyrosine ligand; in contrast to the enzyme from *P. pantotrophus* the sixth ligand is hydroxide, but this in turn is coordinated to Tyr 10. However, the latter residue is in no sense equivalent to the Tyr 25 of the *P. pantotrophus* enzyme. The Tyr 10, which is not an essential residue (Cutruzzola et al., 1997), is provided by the other subunit to that in which it is positioned close to the d_1 heme iron. In other words there is a crossing over of the domains. A reduced state structure of the *P. aeruginosa* enzyme has only been obtained with nitric oxide bound to the d_1 heme iron (Nurizzo et al., 1998). As expected, the heme *c* domain is unaltered by the reduction but the Tyr 10 has moved away from the heme d_1 iron and clearly the hydroxide ligand to the d_1 heme has dissociated so as to allow the binding of the nitric oxide. This form of the enzyme was prepared by first reducing with ascorbate and then adding nitrite.

Although the observed conformational changes in the wild type *P. aeruginosa* enzyme are clearly less pronounced than those in the *P. pantotrophus* enzyme, the driving forces for the changes still require to be understood. It had been postulated (Nurizzo et al., 1998) that reduction of the *c*-type cytochrome domain might lead to conformational changes leading to the release of the hydroxide from the d_1 heme Fe. However, a recent study contradicts this view. It has been possible to obtain crystals of the *P. aeruginosa* enzyme in which the *c* heme is reduced but d_1 heme is oxidized and this condition persists for sufficient time for the structure to be obtained. This shows that hydroxide is still bound to the d_1 heme and that therefore reduction of the d_1 heme, either alone or in combination of reduction of the *c* type heme, is responsible for the conformational change at the d_1 heme iron (Nurizzo et al., 1999). Reduction of the d_1 heme iron alone seems likely to be the essential factor, not least because no change in the *c* heme domain occurs upon reduction and therefore it is difficult to see how there could be any effect of reduction of the *c* heme center relayed to the d_1 heme. Recently pulse radiolysis work with the *P. aeruginosa* enzyme has shown that in the absence of nitrite the electron transfer from the *c* heme to the d_1 heme is very slow (order of seconds) (Kobayashi et al., 2001). This would be consistent with the idea that it is arrival of an electron at the d_1 heme that triggers the conformational change. This change could easily limit the electron transfer rate. There is clearly a parallel here with the *P. pantotrophus* enzyme where the reduction of the d_1 heme has been argued to trigger the conformational change.

The preparation of site-directed mutants of cytochrome cd_1 is not easy as the usual expression systems do not make the d_1 heme. In the case of the cytochrome cd_1 from *P. aeruginosa*, this problem has been overcome by expression of a semi-apo form of the enzyme, i.e. possessing the *c* but not the d_1 heme, in *Pseudomonas putida*, and then adding back d_1 heme that had been extracted from the wild type enzyme. The first mutant to be studies was Y10F which showed that loss of the hydroxyl group had no effect on the activity of the enzyme (Cutruzzola et al., 1997). The first structure of the enzyme from *P. pantotrophus* showed that the active site contained two histidines that appeared to be well placed to act as proton donors to the oxygen of the nitrite destined to become water. Mutagenesis of the corresponding residues in *P. aeruginosa* has shown that indeed that these two residues are important for nitrate reduction (although curiously not for the oxidase reaction also catalysed by the enzyme. However, the replacement of one of these histidines by alanine resulted in a large conformational change (Cutruzzola et al., 1997, 2001; Brown et al., 2001) analogous to that seen recently in a new structural form of *the P. pantotrophus* enzyme (Sjogren and Hajdu, 2001). These findings reinforce the need for further study of the conformational changes within the cytochrome cd_1 molecule. As noted earlier, studies by X-ray crystallography of the binding of cyanide to the enzyme also implicated one of the active site histidines (369 in the *P. aeruginosa* sequence) as critical for the binding of anionic ligands (Sun et al., 2002).

3. Intramolecular Electron Transfer and the Mechanism of Nitrite Reduction

It has long been known that, under some conditions at least, electron transfer between the *c* and d_1 hemes of the *P. aeruginosa* enzyme is slow, in the order of seconds (Cutruzzola, 1999; Kobayashi et al., 2001). What does this mean? It is not necessarily related to the loss of the hydroxide ligand from the d_1 heme iron, because, under some experimental conditions used, the enzyme was reduced at the outset, with azurin present, and it is the movement of an electron from the c to the d_1 heme that is slow despite the presence of a ligand, nitric oxide, on the d_1 heme. This raises a problem because the nitric oxide gets trapped and thus what seems to be a dead end ferrous d_1 heme nitric oxide complex is formed. An expected gating mechanism that might be anticipated to prevent the formation of an Fe(II)-d_1 heme nitric oxide complex appears not to operate. However, it is clear that under some conditions the rate of interheme electron transfer can be much faster. A recent example is provided by a study in which electron transfer was triggered by photodissociation of carbon monoxide from a mixed valence form of the enzyme (Wilson et al., 1999). Electrons moved from the d_1 to the c heme at a rate of thousands per second. The dissociation of carbon monoxide and study of the kinetics of its rebinding also gave insight into the dynamics of the d_1 heme pocket. Again the meaning of these observations for the functioning of the enzyme is not clear.

Information about the kinetics of electron transfer between the hemes and of chemical events at the hemes is now increasing for the enzyme from *P. pantotrophus*. The first reported kinetic study on the *P. pantotrophus* enzyme utilized pulse radiolysis (Kobayashi et al., 1997). In this work exceedingly rapid reduction of the c-type cytochrome center in the enzyme was followed by electron transfer to the d_1 heme on the millisecond timescale. This method involves one electron processes only, and so the c heme, once reoxidized by the d_1 heme, remained oxidized. Interpretation of this millisecond rate of electron transfer is not straightforward because we do not know the driving force, i.e. the redox potentials of the c and d_1-type hemes. However, under the conditions of the pulse radiolysis experiment the electron transfer from the c-type center to the d_1 heme occurs essentially to completion. This implies that, at under these conditions at least, the redox potential difference is of the order of at least 100 mV. A difference of 100 mV and an edge to edge heme distance of 11 Å would suggest that electron transfer might be faster than is observed, as judged by current theories (Page et al., 1999). However, these theories suppose that no chemical bond rearrangements accompany the electron transfer event. In the case of cytochrome cd_1 from *P. pantotrophus*, at least two chemical bond rearrangements might accompany the oxidation/reduction processes. These are the ligand switching at the c heme and the dissociation of Tyr 25 from the d_1 heme iron. For the following reasons it is likely in the pulse radiolysis experiment that the ligands did not change at the c-type center but did so at the d_1 heme:

(1) The spectrum in the Soret region, indicative of the c-type center of the enzyme immediately following the reduction is not identical to that obtained when the enzyme is fully reduced under steady state/equilibrium conditions. This suggests that the reduced c-type cytochrome center formed under the conditions of the pulse radiolysis experiment retained the His/His coordination.

(2) The speed of the reduction by the solution radical generated in the pulse radiolysis experiment is also consistent with this proposal; conformational rearrangements are highly unlikely on the microsecond timescale.

(3) The midpoint redox potential of a His/His coordinated heme is likely to be less than +100 mV. Given that the d_1 heme is catalyzing a reaction with a mid point potential of approximately +350 mV, it can be expected to have a potential considerably more positive than +100 mV. This would account for the stoichiometric transfer of electrons from the c heme to the d_1 heme under the conditions of the pulse radiolysis experiment.

(4) The d_1 heme probably loses its Tyr 25 ligand under these conditions, because if nitrite is present during the pulse radiolysis experiment then, although the rate of electron transfer between the hemes is essentially unaltered, there is evidence for a chemical process taking place at the d_1 heme center (Kobayashi et al., 1997, 2001). This is suggestive that, at least in the presence of nitrite, the arrival of an electron at the d_1 heme triggers the dissociation of the Tyr 25 ligand.

All these observations prompt the question of what values are obtained for the redox potentials of the c and d_1 hemes under equilibrium conditions. As explained earlier, this cannot be answered owing to the hysteretic redox titration. Electron transport from the c to the d_1 heme on the msecs timescale has also now been reported when the reduced *P. pantotrophus* enzyme is reoxidized by oxygen, nitrite or hydroxylamine (Allen et al., 2000a; George et al., 2000; Koppenhofer et al., 2000a).

4. The Mechanism of Nitrite Reduction

It is generally accepted that the reaction catalysed by cytochrome cd_1 begins by nitrite binding through its nitrogen atom to the ferrous d_1 heme iron (Fig. 12). For *P. pantotrophus*, this is supported by crystallographic data (Williams et al., 1997) and is consistent with some chemical considerations (Nasri et al., 2000). However strictly speaking it cannot be definitely excluded that during steady-state turnover nitrite might bind to the ferric state of the cd_1 heme. When, in rapid reaction studies, nitrite binds to the reduced cd_1 heme there is evidence for very rapid dehydration of nitrite, probably involving donation of two protons from active site residues to the oxygen destined to form water. Heterolysis of the N-O bond, and oxidation of the cd_1 heme iron, would result in a species that can formally be regarded as Fe^{3+}-NO or Fe^{2+}-NO^+. Under some conditions the latter species must be relatively long lived because it can undergo rehydration in ^{18}O containing water to give ^{18}O labeled nitrite which eventually yields ^{18}O labeled nitric oxide. The Fe^{3+}-NO species can also be trapped by nucleophiles such as azide, hydroxylamine or aniline. This species has not so far been directly detected by a spectroscopic method during steady state catalysis. However, it has been detected by adding nitric oxide to the oxidized form of *P. stutzeri* cytochrome cd_1. and more recently in a stopped-flow kinetic study of the *P. pantotrophus* enzyme (George et al., 2000). The method used was FTIR. Quantum mechanical calculations also support the importance of proton donation to nitrite bound to a ferrous d_1 heme moiety (Ranghino et al., 2000).

It is expected that the Fe^{3+}-NO species will decompose to release product nitric oxide and generate a ferric form of the d_1 heme iron center. The iron atom may then be five or six coordinated. Six coordination would arise in the mechanism proposed by Fülöp et al. (1995) in which the Tyr 25 ligand was predicted to become a d_1 heme iron ligand following the departure of nitric oxide. In principle an alternative mechanism would be that the ferric d_1 heme-NO complex might become reduced before releasing nitric oxide. Such reduction might be relatively facile because an electron can be available on the c heme which is only approximately 11 Å away, and ferrous heme-NO complexes are usually much more stabile then their ferric counterparts which would provide a driving force for the electron transfer from the c heme center. However, the great stability of ferrous heme-NO complexes suggests that formation of such a species during catalysis would lead to an inhibited enzyme from which NO release would be very slow. Thus we might expect cytochrome cd_1 to have an important design feature for avoiding the formation of ferrous d_1 heme-NO complex. The slow rate of inter-heme electron transfer has suggested such a mechanism. But the often observed formation of the ferrous d_1 heme-NO complex suggest that experimental conditions which would promote the release of NO have not yet been identified. A role for tyrosine 25 in displacing nitric oxide now seems less likely as the Y255 variant enzyme is fully active (Gordon et al., 2003).

In original stopped-flow rapid reaction studies on cytochrome cd_1 from *P. aeruginosa* the ferrous d_1 heme-NO species was formed despite the electron transfer from the c to the d_1 heme being relatively slow, rate constant approximately 1 s^{-1}, given the relatively short distance between the two hemes. Such a distance would normally predict much faster electron transfer rates. This relatively slow inter-heme electron transfer rate has been frequently observed and before the structure of the protein was known was thought to reflect relatively large inter-heme separation distance and/or relative orientation of the two hemes that was not conducive to rapid electron transfer (Makinen et al., 1983). The crystal structures provide no evidence for either of these proposals; there is nothing unusual about the relative orientation of the c and d_1 heme groups. A slow electron transfer rate must, therefore, be regarded as diagnostic of rate-limiting chemical steps at the d_1 heme center.

Further rapid reaction analysis of the *P. aeruginosa* enzyme has recently been reported (Cutruzzola et al., 2002). Mixing of the fully reduced enzyme with between 10 and 150 μM nitrite was followed, judged by visible absorption spectra , by nitrite binding to the active site on the 6 msec timescale. This was followed on the 0.1 to 10 sec timescale by oxidation of the c-heme to give enzyme that had nitric oxide bound to the d_1 heme. This is rather slow if the enzyme in steady state is to turnover many times per second. Subsequently, at times between 10 s and 4 min the enzyme was fully reduced by the ascorbate present to give essentially a dead end complex with nitric oxide bound to the ferrous heme. The changes in the rates of these steps were also reported for mutant forms of the enzyme that each lacked one of the two active site histidine ligands.

The first rapid reaction study of the reduction

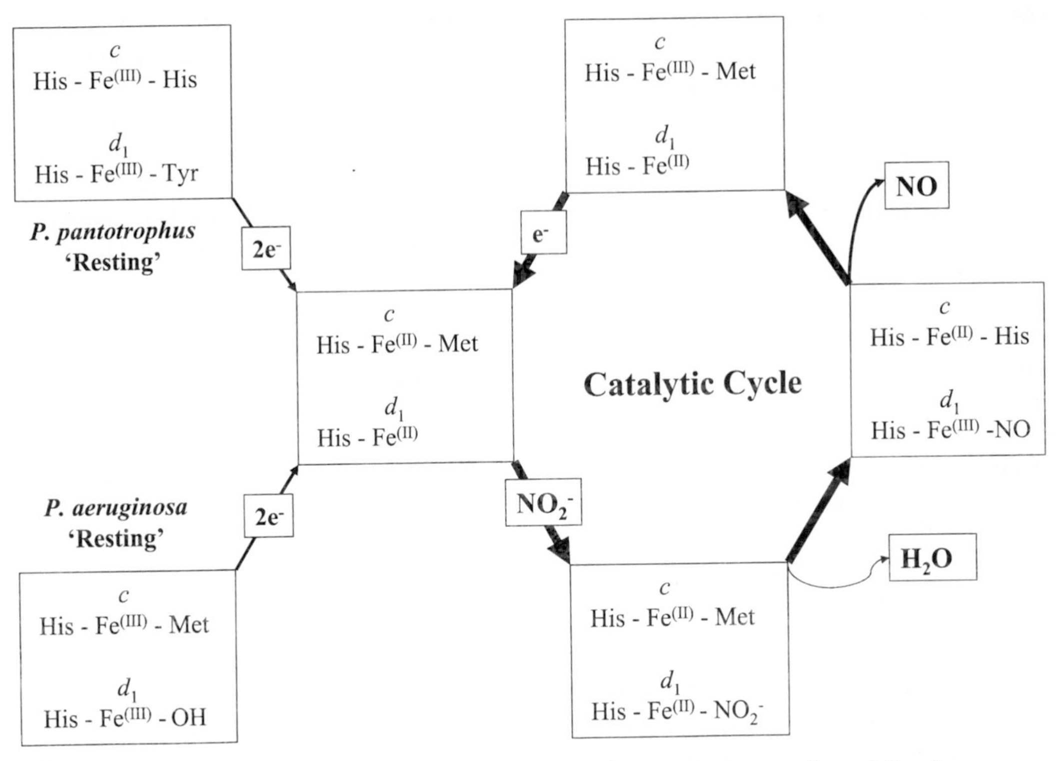

Fig. 12. Possible catalytic cycles for the cytochrome cd_1 nitrite reductases of *Paracoccus pantotrophus* and *Pseudomonas aeruginosa*

of nitrite by the enzyme from *P. pantotrophus* was recently reported (George et al., 2000). The fully reduced enzyme, notably in the absence of any excess reductant, was mixed with nitrite and the timecourse of the subsequent reaction followed by both stopped flow FTIR and absorbance spectroscopy. On the tens to hundreds of millisecond timescale only one of the two electrons on each polypeptide chain was used to reduce nitrite to nitric oxide, but the latter remained bound to the d_1 heme. A fraction of the enzyme molecules had a ferric d_1 heme with NO bound (readily detectable by FTIR) with the c heme in the reduced state while the remainder had an oxidized c-type cytochrome center and a ferrous d_1 heme-NO complex at the active site. EPR analysis clearly showed that the oxidized c heme iron had methionine and histidine coordination, thus reinforcing the evidence that the bishistindinyl coordination seen in the as prepared enzyme is not necessarily significant during turnover (see above). The puzzling outcome from this rapid reaction study is why the enzyme did not go into turnover and catalyze synthesis of two molecules of nitric oxide with regeneration of the fully oxidized enzyme.

In summary, there is still much to understand about the nitrite reduction reaction. The crystal structures have shown how nitrite can bind to the d_1 heme iron and protons can be provided to one of its oxygen atoms from two histidine residues. However, as yet no rapid reaction study has detected the release of product nitric oxide rather that the formation of the inhibitory dead-end ferrous d_1 heme-NO complex. It is also not clear why the rate of inter-heme electron transfer is so slow over 11 Å when nitrite or nitric oxide is the ligand to the d_1 heme. Understanding these issues and also the role of the flexibility of the c heme domain remain major challenges for the future.

5 Electron Donors to Cytochrome cd_1 Nitrite Reductase

As mentioned in the Introduction, nitrite reductases for which nitric oxide is the reaction product are linked to the underlying respiratory chain at the level of *c*-type cytochromes and/or cupredoxins. Thus the electrons destined for cytochrome cd_1 pass through the cytochrome bc_1 complex. How they pass on from this complex to cytochrome cd_1 is still a matter of uncertainty. It had long been assumed that azurin is an in vivo electron donor to cytochrome cd_1 of *P. aeruginosa*. The construction of mutants of *P. aeruginosa* in which one or both of the genes for azurin and cytochrome c_{551} have been deleted has led to the conclusion that in vivo cytochrome c_{551} is essential for the donation of electrons to the nitrite reductase and that azurin was ineffective (Vijgenboom et al., 1997). The discrepancy between in vivo and in vitro observations could be reconciled if it is the failure of azurin to except electrons from the cytochrome bc_1 complex, or other donor, that is responsible for its ineffectiveness in vivo.

In the case of the enzyme from *P. pantotrophus* it has been proposed that either pseudoazurin or cytochrome c_{550} are the electron carriers between the cytochrome bc_1 complex and cytochrome cd_1. The original indications for this view were that nitrite reduction was unaffected in a cytochrome c_{550} deficient mutant of the related organism *P. denitrificans* and that a copper chelator inhibited nitrite reduction in the same mutant (Moir and Ferguson , 1994). The latter observation naturally implicated a cupredoxin. Subsequently this view has been substantiated by the observation that a mutant of *P. denitrificans* constructed to be deficient in both pseudoazurin and cytochrome c_{550} is unable to reduce nitrite (Pearson et al., 2003)

In anticipation of demonstrating that cytochrome c_{550} and pseudoazurin were the two alternate electron donors to cytochrome cd_1, it has been discussed how two proteins of such different structures could each interact with cytochrome cd_1 (Williams et al., 1995). It was proposed that hydrophobic patches studded with positive charges on the surfaces of either pseudoazurin or cytochrome c_{550} recognize a similar patch, but of complementary charge, on the cytochrome cd_1. This was termed pseudospecificity.

The supply of electrons to the cytochrome cd_1 of *P. stutzeri* raises an interesting question. Several years ago Zumft and colleagues demonstrated that disruption of the *nirT* gene, which occurred in a cluster with other genes coding for components of the cytochrome cd_1 system prevented electron transfer to cytochrome cd_1 (Jungst et al., 1991). NirT is now recognized as a member of the NapC family of proteins (Section II.A.2) that are involved in cytochrome bc_1-independent electron transfer to or from periplasmic reductases or dehydrogenases. Thus the requirement of NirT for nitrite reduction in *P. stutzeri* would imply that the pathway of electron flow is cytochrome bc_1 complex independent. However, it has been shown that electron transfer to cytochrome cd_1 in *P. stutzeri* was as fully sensitive to the bc_1 inhibitors antimycin and mucidin as in *P. denitrificans* (Kucera et al., 1988). On this basis another role for NirT in nitrite reductase of *P. stutzeri* needs to be sought.

C. Copper-containing Nitrite Reductases (CuNir)

The copper-containing nitrite reductase (CuNir) is a product of the *nirK* gene and is widely distributed amongst Bacteria (Braker et al., 1998) and has also been found in Archaea and the fungus *Fusarium oxysporum*. However, most biochemical studies have focused on the enzymes from *Achromobacter cycloclastes* and *Alcaligenes xylosoxidans*. The CuNIRs are homotrimers which bind three type I and three type II copper centers (Fig. 13). The type I copper is coordinated by two Cys, one Met and one His residue (Fig. 13) and has intense visible absorption bands at *ca.* 450 nm and 600 nm which arise from Cys(S)-Cu(II) charge transfer transitions. CuNirs can be classified into two spectrally distinct classes, 'Blue' and 'Green', depending on the relative intensities of the 450 nm and 600 nm bands which may be influenced by the angle of the His-Cu-Met bonds (Dodd et al., 1998). The type II coppers lie at the subunit-subunit interfaces and are ligated by three histidine residues; two being provided by one subunit, the third from the adjacent subunit (Godden et al., 1991; Kukimoto et al., 1994) (Fig. 12). The tetrahedral coordination sphere of the Cu(II) is completed by a water (or hydroxide) molecule, but this is absent in the reduced Cu(I) species. Spectroscopic and structural studies of the oxidized enzyme in the presence and absence of nitrite suggest that the type II copper centers are the sites of substrate binding. The type I copper then plays a role in electron transfer from soluble electron donors to the active site (Kukimoto et al., 1995). It is generally assumed that the type I copper center of a copper nitrite reductase receives electrons from a cupredoxin, pseudoazurin or azurin depending upon the organism. A His→Ala mutation of the type 1 copper His ligand of the *A. xylosoxidans* CuNir leads to a *ca.* 500 mV increase in the E_0' which prevents electron transfer via this route (Prudencio et al., 2002). In vitro studies combined with site specific mutagenesis have implicated broadly the same positively charged docking patch on pseudoazurin as discussed above for the docking of this class of protein on cytochrome cd_1 (Murphy et al., 2002). At present it is not clear whether any *c*-type cytochrome can substitute for pseudoazurin in this respect, although it is notable that the *Pseudomonas aureofaciens* CuNir is active in vivo in mutants of *Ps. stutzeri* deficient in the cytochrome cd_1 and in this case electron transfer may proceed via *c*-type cytochromes (Glockner et al., 1993). Electron flow to the copper nitrite reductase can be deduced to be via the cytochrome bc_1 complex, at least in *A. xylosoxidans,* because it is fully inhibited in intact cells by antimycin or mucidin (Kucera et al., 1988).

Studies on the mechanism of CuNir have been greatly aided by crystal structures of nitrite bound forms. These reveal that, unlike the cd_1NIRs which bind substrate at the d_1 heme via the nitrogen, the oxidized type II copper centers of CuNIR coordinate nitrite in a bidentate fashion via the oxygens (Adman et al., 1995; Murphy et al., 1997). One of these oxygens is further stabilized by a hydrogen bond to a conserved aspartate residue that, together with a conserved histidine, also H-bonds a water molecule. EPR, ENDOR and X-ray absorption analysis has also confirmed that nitrite binds to the oxidized cupric type II center (Howes et al., 1994; Strange et al., 1995). By contrast, studies on the reduced cuprous type II center reveal that nitrite and other ligands (e.g. azide) bind poorly (Strange et al., 1999). Thus a mechanism in which nitrite binds to an oxidized type II copper center displacing a hydroxide ion has been proposed (Murphy et al., 1997). This center is in turn reduced by transfer of an electron from the type I center. Nitrite decomposes after protonation at the now reduced type II center to give a transient intermediate which has NO and hydroxide ion bound. The catalytic cycle is completed when the bound NO is released. Unlike the mechanisms proposed for cd_1NIR neither substrate, nor product is bound through the nitrogen during the catalytic cycle. However, based both on the structure of the nitrite bound form of the *A xylosoxidans* enzyme and an earlier EXAFS study Eady and co-workers (Dodd et al., 1998) have proposed an alternative mechanism which involves reorientation of the substrate at the copper after binding and the participation of a bound nitrosyl (Cu(I)-NO+) intermediate. It is notable that in the ligand free-state the redox potentials of the type I ($E_0' \approx 250$ mV) and type II ($E_0' < 200$ mV) centers do not favor rapid electron transfer. However, ENDOR and EXAFS analysis of nitrite bound type 2 centers suggest that the E_0' will increase, so making intramolecular electron transfer more favorable (Strange et al., 1995; Veselov et al., 1998). Hence in the presence of an appropriate electron donor (Fig. 2B) the CuNIR will be in a mixed valence state, Type I-Cu(I), Type II -Cu(II), and substrate binding may trigger electron transfer.

The importance of the conserved active site aspartate and histidine has been investigated though site-directed mutagenesis. Mutation of Asp92, in *A. xylosoxidans*, to either Glu or Asn led enzymes that retained type I Cu center with wild-type properties, but which had altered activities and spectroscopic properties of the type II centers. The D92E mutant was no longer capable of binding nitrite as monitored by EPR spectroscopy and had very low enzyme activity with the artificial electron donors reduced Methyl-

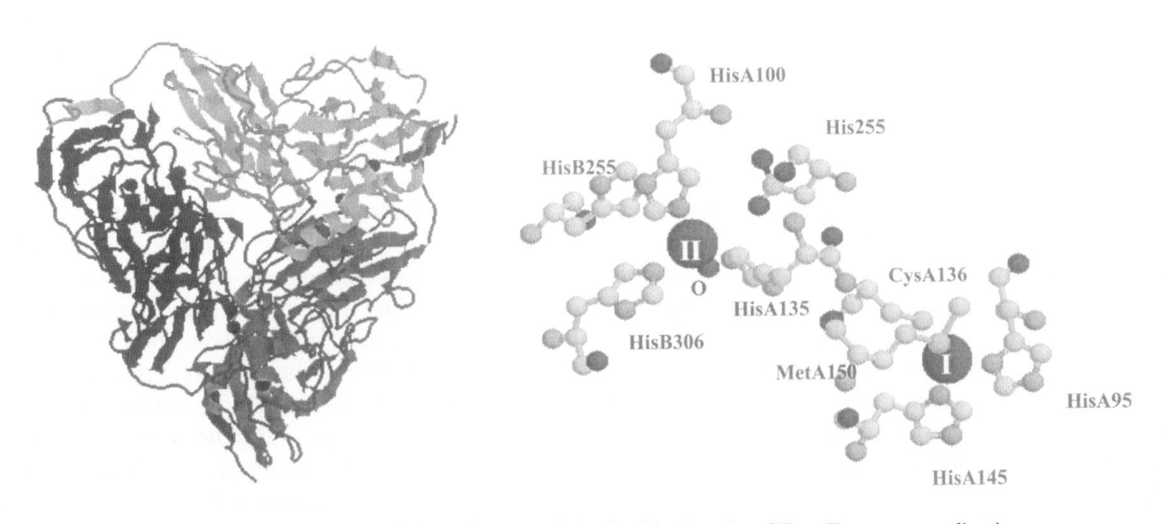

Fig. 13. The structure copper nitrite reductase and detail of the Type I and Type II copper coordination.

viologen and sodium dithionite. The D92N also had very low enzyme activities with these donors. However, when the physiological electron donor reduced azurin I was used, both mutant proteins exhibited some restoration of enzyme activity, which suggested that on formation of an electron transfer complex with azurin, a conformational change in NiR occurs that returns the catalytic Cu center to a functionally active state capable of binding and reducing nitrite. It is intriguing to note that enzyme carrying a His254F mutation has Zn, rather than copper at the active site and is inactive (Ellis et al., 2002). This then makes the active site very similar to that of Zn-dependent carbonic anhydrase (Ellis et al., 2002).

Another site directed mutagenesis study on the same residues but with enzyme from *A. faecalis* S-6 has been reported recently. In the oxidized D98N nitrite-soaked structures, nitrite is coordinated to the type II copper via its oxygen atoms in an asymmetric bidentate manner; however, elevated B-factors and weak electron density indicated that both nitrite and Asn98 are less ordered than in the native enzyme. This disorder likely results from the inability of the N delta 2 atom of Asn98 to form a hydrogen bond with the bound protonated nitrite, indicating that the hydrogen bond between Asp98 and nitrite in the native NIR structure is essential for anchoring nitrite in the active site for catalysis. In the oxidized nitrite soaked H255N crystal structure, nitrite did not displace the ligand water and is instead coordinated in an alternative mode via a single oxygen to the type II copper. His255 is clearly essential in defining the nitrite binding site despite the lack of direct interaction with the substrate in the native enzyme. The resulting pentacoordinate copper site in the H255N structure may also serve as a model for a proposed transient intermediate in the catalytic mechanism consisting of a hydroxyl and nitric oxide molecule coordinated to the copper. Interestingly, an unusual dinuclear type I copper site formed in the reduced nitrite soaked D98N and H255N crystal structures. This had similar structural properties to the dinuclear Cu_A sites of nitrous oxide reductase (see below) and cytochrome aa_3 oxidase and may represent an evolutionary link between this center and the mononuclear type I copper centers (Boulanger and Murphy, 2001).

IV. Nitric Oxide Reductases (Nor)

Bacterial Nitric oxide reductases (Nor) catalyze the two electron reduction of NO to N_2O which involves formation of an N-N bond.

$$2NO + 2H^+ + 2e \rightarrow N_2O + H_2O \qquad (5)$$

Three sub-classes of bacterial Nor can be identified: the cytochrome *c*-dependent NorCB enzymes (Watmough et al., 1999; Hendriks et al., 2000); the quinol-oxidizing NorZ enzymes (Busch et al., 2002) and the quinol oxidizing Cu_A-containing Nor (Suharti et al., 2001) (Fig. 6). Most work to date on bacterial Nors has been carried out on the NorCB class of enzyme and this will be discussed first.

A. The Cytochrome c-Dependent Nor

NorCB enzymes have been purified from *P. denitrificans*, *Pseudomonas stutzeri*, and *Paracoccus halodenitrificans* (Girsch and de Vries, 1997; Sakurai and Sakurai, 1997; Cheesman et al., 1998; Hendriks et al., 1998; Gronberg et al., 1999), but sequence information is available from genome analysis of a large number of bacteria, including *Rhodobacter sphaeroides* and *Pseudomonas aeruginosa* (Hendriks et al., 2000). The NorC subunit is a mono-heme *c*-type cytochrome that possesses an N-terminal transmembrane helix that anchors the heme domain to the periplasmic face of the cytoplasmic membrane. The NorB subunit is a close relative of the catalytic subunit of respiratory heme-copper oxidases. Indeed, *P. denitrificans* NorCB possesses oxidase, as well as NO reductase, activity. The catalytic subunit of heme-copper oxidases (e.g. cytochrome aa_3 oxidase) characteristically folds into 12 transmembrane helices that binds two metal centers: (i) a *bis*-histidine coordinated magnetically isolated low spin heme and (ii) a magnetically coupled dinuclear center formed by a high spin heme and a copper ion (Cu_B) which will bind and reduce dioxygen. Secondary structure modeling of NorB suggests a similar 12 helical arrangement and cofactor binding property (Saraste and Castresana, 1994; van der Oost et al.,1994). Seven conserved histidine residues, responsible for ligating the three redox active metal centers, can be identified in helices II, VI, VII and X. The key difference between NorB and other heme-copper oxidases is the composition of the dinuclear center, which contains non-heme iron (Fe_B) rather than copper (Cu_B).

The three histidine residues responsible for ligating Cu_B in oxidases are completely conserved in NorB where they are likely to be Fe_B ligands. It is well established that Fe(III) prefers different coordination geometry (e.g. octahedral and penta- or hexa-dentate) to Cu(II) (e.g. distorted tetrahedral and tetra-dentate). Primary sequence analysis of NorB reveals three conserved glutamic acid residues (E198, E202, E267; *P. denitrificans* numbering), located in potential transmembrane helices, that are absent from heme-copper oxidases (Butland et al., 2001). Given that glutamate is frequently found in biological systems as a non-heme Fe ligand, it is plausible that one or more of these conserved glutamates is involved in coordinating Fe_B, along with the three conserved 'Cu_B' histidine ligands In addition they may serve to modulate the properties of the catalytic center leading to differences in the redox potentials of the heme that may in turn influence the catalytic cycle. A recent study of the interaction of the substrate analogue carbon monoxide (CO) with reduced Nor used resonance Raman spectroscopy to study the resultant Fe(II)-CO heme. The results suggested that in Nor Fe(II) $d_\pi \rightarrow$ CO π^* backbonding is minimized, possibly as a result of a negatively charged heme distal pocket. This is pertinent because primary sequence analysis shows that a region of helix VI of Nor that is close to the dinuclear center contains two conserved glutamate residues that are not found in heme copper oxidases, one of which E198, is essential for activity (Butland et al., 2001).

Electron transfer between the redox centers in Nor has been studied and supports the view that the arrangement of the redox centers is similar to heme copper oxidases. Intramolecular electron redistribution after photolysis of the partially reduced CO-bound enzyme showed that electron transfer in Nor proceeds from heme c, via the low-spin heme b to the heme b_3-Fe_B active site (Hendriks et al., 2002). The electron-transfer rate between hemes c and b was *ca.* $3 \times 10^4\ s^{-1}$ while the rate of electron transfer between hemes b and b_3 was $>10^6\ s^{-1}$. Electron transfer between heme *c* and heme *b* was coupled to the generation of an electric potential, consistent with an electron moving from the outer face of them membrane into the membrane interior.

1. Spectroscopic Analysis of NorCB

In the absence of a crystal structure, much of what we have recently learned about the actice site of Nor has come from spectro-potentiometric studies. *P. pantotrophus* NorCB is normally purified in a form that exhits an absorption shoulder at 595 nm in the fully oxidized species. Resonance Raman studies on NorCB and model compounds has suggested that this '595 nm' band is a ligand to metal charge transfer (LMCT) band of active site high spin ferric heme b_3 which has no proximal iron ligand and forms a μ-oxo bridge with the non-heme iron (Pinakoulaki et al., 2002) (Fig. 14). It is pertinent to note that this μ-oxo bridged form of NorB does not readily bind exognous ligands such as CN^- or protons and so may not be a relevant state for NO binding.

Redox potentiometry has revealed that the non-heme iron of the dinuclear center has an E_m(pH 7.6) of +320 mV, while the high-spin heme b_3 has a surprisingly low midpoint redox potential (E_m, pH 7.6 = +60 mV) (Gronberg et al., 1999). In the absence of bound substrate this imposes a large thermodynamic barrier to reduction by the low spin electron transferring heme *c* (E_m, pH 7.6 = +310 mV) and heme *b* (E_m, pH 7.6 = +345 mV). This raises the possibility that one-electron reduced (mixed valence) active site (Fe^{2+}-hemeFe^{3+}) is the relevant substrate-binding state for the catalytic cycle. Reduction of the Fe_B in the dinuclear center results in a shift in the absorption maximum of the ferric heme b_3 LMCT band to longer wavelengths; pH 6.0 the LMCT band λ_{max} is 605 nm, while at pH 8.5 it is 635 nm (Field et al., 2002). Magnetic Circular Dichroism spectroscopy suggests that at all pHs examined the proximal ligand to the ferric heme b_3 in the three electron reduced form is histidine but that at pH 8.5 the distal ligand is hydroxide while, at pH 6.0, when the enzyme is most active, it is water (Fig. 14). It is also notable that the recominant NorCB expressed in *E. coli* purifies with a 605 nm band in the fully oxidized state suggesting that in this species the heme b_3 of di-nuclear center is not μ-oxo bridged but binds a hydroxide ligand (Butland et al., 2001). This enzyme is still fully active, adding to the view that the μ-oxo bridged form of NorB may not be not relevant to the catalytic cycle but is a resting state.

At present there is little agreement on the catalytic cycle of Nor. An early study on NO reduction in the cytochrome aa_3 oxidase suggested the involvement of a [heme-NO NO-Cu_B] intermediate. An analogous [heme-NO NO-Fe_B] intermediate has been forwarded for Nor (the *trans* mechanism) which then leads to the formation and loss of N_2O, leaving an oxo-bridged species (Girsch and de Vries, 1997; Hendriks et al., 1998). An alternative mechanism is the binding of two NO molecules to a coordinately unsaturated Fe_B (*cis*-Fe_B mechanism). This has been observed on the Cu_B of the cytochrome *bo* oxidase (Butler et al., 1997; Watmough et al., 1998) and has the advantage of preventing the formation of a ferrous heme-nitrosyl species which in myoglobins bind NO very tightly (Watmough et al., 1999). One means by which this may be avoided is the low redox potential of the heme of the dinculear center which is around 200 mV lower than that of cytochrome aa_3 oxidase (Gronberg et al., 1999).

A final mechanism could be one in which all of the chemistry takes place on the high spin heme b_3 (*cis*-heme mechanism) in a modification of the mechanisms proposed for the fungal P450 nitric oxide reductase or flavohemoglobin (Park et al., 1997; Hausladen et al., 2001; Pinakoulaki et al., 2002). Here one NO molecule would bind to ferric heme b_3 and intramolecular electron transfer would yield the two electron reduced Fe^{2+}-NO^-. A second NO molecule would then bind in the active site to form hyponitrite ($HONNO^-$) and achieve N-N bond formation. Cleavage of the N-O bond would then release N_2O and a ferric heme b_3. This mechanism has the advantage of accomodating a mixed valence form as the starting point. Indeed it also allows for a catalytic mechanism that does not include the 'locked' oxo-bridged fully oxidized active site.

Recent electrometric studies on *P. denitrificans* NorCB have begun to provide some insight into possible catalytic mechanisms (Hendriks et al., 2002). During turnover of the fully reduced enzyme with NO, a number of phases could be resolved. The first phase ($k = 5 \times 10^5\ s^{-1}$) was electrically silent and characterized by the disappearance of absorbance at 433 nm and the appearance of a broad peak at 410 nm and was assinged to the formation of a ferrous NO adduct of heme b_3. Such an intermediate would be expected in a *trans* mechanism involving both heme b_3 and Fe_B or with *cis*-heme mechanism, but not with a *cis*-Fe_B mechanism. However, it should be noted that these studies begin with the fully reduced enzyme rather than a mixed valance state.

2. Nor and Proton Movement

Proton uptake is integral to any of the models for the catalytic cycle of Nor. In the case of cytochrome aa_3 oxidase, in addition to proton uptake for catalysis, the enzyme also pumps four protons across the membrane for every O_2 that is reduced to water. Recent high resolution X-ray crystal structures of cytochrome aa_3 oxidase, together with site-specific mutagenesis, have lead to the identification of amino acid residues that form the so-called D- and K-channels which are important in the delivery of protons from the cytoplasm to the dinuclear center (Vygodina et al., 1997). However, these residues are absent from Nor and, furthermore, it has been demonstrated that in chromatophores of *Rb. capsulatus* there is no generation of membrane potential when electrons were fed into the electron transfer system at the level of periplasmic *c*-type cytochrome (Bell et al., 1992). In addition in liposomes, *P. denitrificans* Nor does not show any generation of an membrane potential during steady-state turnover (Hendriks et al., 2002). Together these results rule out both a cytoplasmic site of nitric oxide reduction (i.e. with protons taken from the cytoplasm) and a proton pumping activity of the nitric oxide reductase, and imply that protons are taken up from the same side (the periplasmic side) of the membrane as electrons during catalysis. It seems probable that the some of conserved glutamic acid residues that are present in periplasmic loop regions and putative transmembrane helices in Nor, but absent in other C*c*O's, play a role in proton movements, forming an 'E' channel. Investigations into this possibility have shown that E125 which lies at periplasmic side of helix IV and E198, which lies in the middle of helix VI, are essential for activity, but not assembly, of *P. denitrificans* NorCB (Butland et al., 2001). Further support for a route of proton uptake form the periplasm has recently come from electrometric studies on NorCB (Hendriks et al., 2002). Little is currently known about the mechanism of proton output from cytochrome aa_3 and cytochrome *bo* oxidases and it is possible that the proton input channel used to move protons from the periplasm to the di-nuclear center in NorB evolved into the proton output channel in heme copper oxidases. Comparitive studies on proton input in Nor and proton output in cytochrome aa_3 oxidase may prove informative in this respect.

B. The Three Classes of Nitric Oxide Reductase

Analysis of the amino acid sequences of Nor's in the current databases reveals that there three Nor subgroups. In the NorCB subgroup, the *norC* and *norB* genes are found adjacent to each other. The NorC mediates electron transfer between the protonmotive cytochrome bc_1 complex, periplasmic cytochrome or cupredoxins and NorB. In the second class of

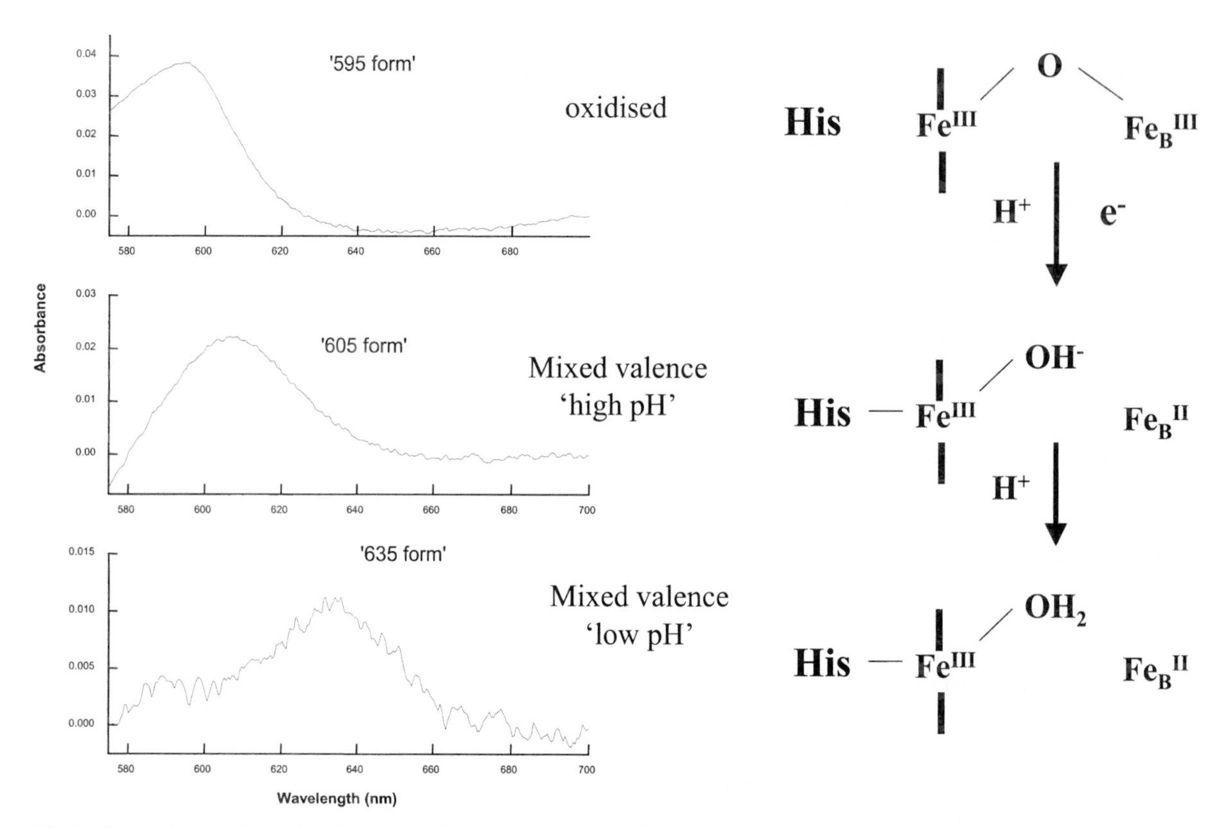

Fig. 14. Spectral properties and models for the dinuclear active site of the nitric oxide reductase in different redox and protonated states. Amino acid ligends to Fe_B are ommitted (see text for details). Spectra are derived from Field et al., 2001.

Nor's, for example in the denitrifyer *Ralstonia eutropha* (Cramm et al., 1997, 1999), the *norB* gene is always found in isolation, *norC* is not present. In these cases the NorB protein is predicted to have an C-terminal extension which folds to give two extra transmembrane helices linked by a globular region in the periplasm. It has been suggested that this extension serves as a functional substitute for NorC, with a possibility being that it serves as a direct quinol dehydrogenase (Cramm et al., 1999) so that by analogy to oxidases there are cytochrome *c* dependent and quinol-dependent systems. This would lead to a cytochrome bc_1 complex independent route for electron transport and so the NorC-dependent and quinol-dependent Nor's would have different coupling efficiencies. The q^+/e^- for the NorC-dependent system would be 6 and 2 with NADH and succinate as electron donor respectively, but it would be 4 and 0 for the quinol-dependent Nor. These differences in energy coupling then resemble the situation described earlier for the membrane-bound and periplasmic nitrate reductases.

An unusual quinol oxidizing Nor has been purified from the Gram-positive bacterium *Bacillus azotoformans* (Suharti et al., 2001). The enzyme consists of two subunits with Mw's of 16 and 40 kDa. As for *P. denitrificans* Nor, the *B. azotoformans* enzyme complex binds one low-spin *b*-type heme and one high spin b heme that is magnetically coupled to a non-heme iron presumably forming a binuclear center where reduction of NO occurs. Heme *c* is absent but, uniquely, the enzyme binds two copper atoms in the form of a CuA center that is characteristic of cytochrome aa_3 oxidases and nitrous oxide reductases. The enzyme uses menaquinol as electron donor, whereas cytochrome c, which is the substrate of other NO reductases, is not used. Copper A and both hemes are reducible by menaquinol. Thus the *B. azotoformans* NO reductase is a hybrid between copper A containing cytochrome oxidases and NO reductases present in Gram-negative bacteria. In this sense it may represent an ancient progenitor of the family of heme-copper oxidases.

Comparison of the gene clusters for cytochrome *c*-dependent Nor and q-Nor reveal a number of extra genes in the former that are absent in the latter.

Thus in *P. denitrificans* the *norCB* genes are part of a *norCBQDEF* operon, where as in *Synechocystis* and *R. eutrophus* the only *norB* gene is present in a single-gene operon (Cramm et al., 1997, 1999). NorQ is predicted to be a cytosolic protein that binds ATP, NorD is predicted also predicted to be a cytosolic protein but has not features that give a clue as to its function, NorE is predicted to be an integral membrane protein with five transmembrane helices that has homology to the subunit III of cytochrome aa_3 oxidase, and NorF is predicted to encode a small integral membrane protein with 2 transmembrane helices (de Boer et al., 1996). The absence of *norQDEF* in the gene clusters of encoding the single subunit q-Nor suggest that they play a role in regulation, assembly or stability of the NorCB complex. Although NorCB is normally purified as a two subunit complex it cannot be excluded that additional subunits, particularly NorE or NorF are lost during purification. Indeed, early purification protocols of Nor did result in preparations that had additional polypeptides present (Carr and Ferguson, 1990).

V. Nitrous Oxide Reductase

Nitrous oxide reductase (Nos) catalyses the reduction of nitrous oxide to dinitrogen:

$$N_2O + 2e^- + 2H^+ \rightarrow N_2O + H_2O \quad (6)$$

In environmental terms this is a very important reaction as N_2O is a greenhouse gas, the emission of which from biological sources makes a significant contribution to global warming. Nos is a homo-dimer of a 65 kDa copper-containing subunit that binds a di-nuclear Cu_A electron entry site, similar to that found in the cytochrome aa_3 oxidase, and a tetra-nuclear Cu_Z catalytic center. Each monomer is made up of two domains, the 'Cu_A domain' that has a cupredoxin fold and the 'Cu_z domain' which is a seven-bladed propeller of β-sheets. The dimer organization is such that inter-dimer electron transfer must take place between the Cu_A center of one monomer and the Cu_Z center of the second monomer (Brown et al., 2000a,b). Electron input into Cu_A is usually via mono-haem *c*-type cytochromes or cupredoxins (E_m values around 250 mV).

The molecular nature of the Cu_Z center has only emerged recently through structural studies on the enzymes *Pseudomonas nautica* and *P. denitrificans* (Brown et al., 2000a,b). These structures revealed that the Cu_Z center belongs to a new type of metal cluster, in which four copper ions are liganded by seven histidine residues, two hydroxide molecules and, most notably, a bridging inorganic sulfide (Fig. 15). The presence of a $[Cu_4S]$ center is supported by elemental analyses and resonance Raman spectroscopy (Rasmussen et al., 2000). It has been suggested that N_2O binds to the Cu_Z center via a single copper ion, with the remaining copper ions acting as an electron reservoir (Brown et al., 2000a,b). This would then allow for fast electron transfer avoiding the formation of dead-end products. However, it is also possible that the inorganic sulfide is the binding site for N_2O with the oxygen atom then being transferred to form sulfoxide and N_2 being released. The bound oxygen would then be protonated and released as water.

The $[Cu_4S]$ center can, in principle, exist in five different oxidation states (Fig. 15) with all but the fully reduced all-Cu(I) state being expected to give rise to strong optical bands in the visible absorption spectrum as a consequence of S^{2-} to Cu(II) charge transfer and intervalence transitions between Cu(I) and Cu(II) (see Rasmussen et al., 2002). Only two states, $[Cu_4S]^{3+}$ and $[Cu_4S]^{5+}$ are expected to give rise to strong S=½ EPR signals (Fig. 15). The range of redox states clearly could allow for gating of electron transfer from the Cu_A (which is one-electron donor) to facilitate coordinated 2-electron reduction of N_2O. A recent spectroscopic study on a dithionite-reduced form of nitrous oxide reductase from *P. nautica* has shown that although the Cu_A center is reduced the Cu_Z center has a S=½ state (Chen et al., 2002). This is attributed mainly to one of the four coppers in the Cuz center, known as Cu_I, with significant contribution from Cu_{II} and the bridging sulfide. The presumed substrate binding edge of the tetranuclear center would thus bind nitrous oxide via the oxidized Cu_I and reduced Cu_{IV}. Binding might be followed by electron transfer from Cu_{IV} directly and from Cu_{II} via the bridging sulfur on a superexchange pathway (Chen et al., 2002).

It has long been recognized that Nos can be purified in different spectroscopic forms, which are characterized by different colours (see Rasmussen et al., 2002). These colour changes arise from differences in the Cu_z center. Detailed studies on the *P. pantotrophus* Nos have recently established that exposure of the $[Cu_4S]$ to air leads to a change in its redox properties (Rasmussen et al., 2002). Thus in anaerobically purified Nos the Cu_z center exhibits n=1

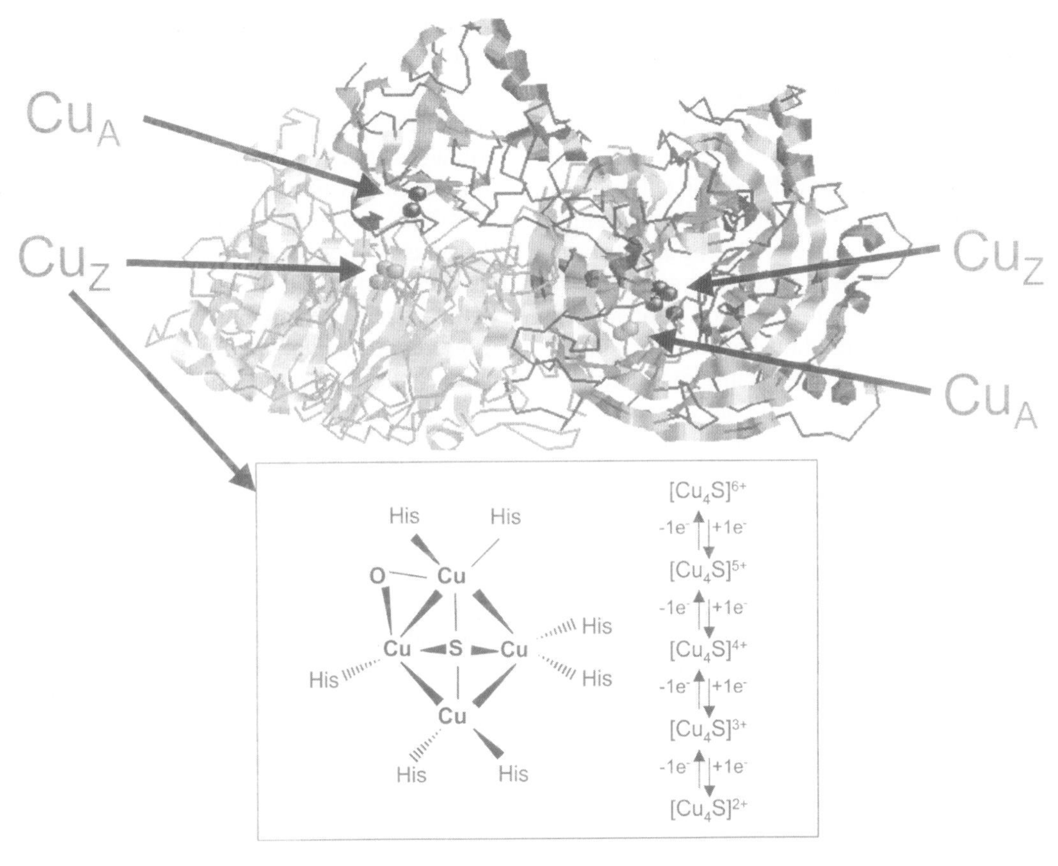

Fig. 15. The structure of the nitrous oxide reductase and detail of active site.

redox character with E°′ = +60 mV, but the air-exposed center exhibits no detectable redox change under similar experimental conditions suggesting a strong positive shift in the redox potential. Spectroscopic properties of this 'redox fixed' center (termed Cu_Z^*) are similar, but not identical, to spectra of the reduced 'redox active' form of Cu_Z. The structural relationship between these two forms of the catalytic site is unclear, as is their catalytic relevance. However, EPR and magnetic circular dichroism spectra suggest that the basic [Cu_4S] structure is common to both and it is notable that the reduced methyl-viologen linked steady-state activity of the enzymes with each form of the center is comparable. Activity was not assessed with physiological electron donors and it should be noted that electron transport to the Cu_Z site from electron donors such as cytochrome c_{550} would be thermodynamically unfavourable. However, it has been shown that changes in the coordination environment of the Cu_A site can effect the spectral properties of the Cu_Z site (Charnock et al., 2000) and it is conceivable that structural changes in the Cu_A domain may occur on binding of a redox partner that modulate the properties of the Cu_Z domain and create a driving force for electron transfer.

The assembly of the novel [Cu_4S] is intriguing as Nos is a periplasmic enzyme. Nos is a substrate for the Tat export system that translocates folded proteins and this suggests that the tetra-nuclear copper center is assembled in the cytoplasm. However, a recent study showed that a mutant of *P. stutzeri* with a defective Tat system accumulated an apo protein in the cytoplasm, suggesting that the copper centers are incorporated in the periplasm (Heikkila et al., 2001). On the other hand, disturbance by mutation of the Cu_A site in the C terminal region of the protein was deleterious for protein export to the periplasm (Heikkila et al., 2001), suggesting that the Tat system may accommodate a substantially folded apoprotein in this case rather than a holo protein containing a redox cofactor as is generally thought to be the case for this protein transport system. The nitrous oxide reductase subunit is encoded by the *nosZ* gene. This usually clusters with a number of other Nos genes, which include *nosDFYL* that are required for assembly of the active site. NosDFY may form an ATP dependent copper transporter and NosL may be a copper-binding chaperone. NosL has recently been characterized and is a monomeric 18.5 kDa protein that specifically and stoichiometrically binds Cu(I) which is ligated by a Cys residue, and, probably, one Met and one His are thought to serve as the other ligands (McGuirl et al., 2001).

VI. Concluding Remarks

In recent years the impact of structural biology on the enzymology of the N-cycle has been immense. Not only are structures of many key enzymes now available, but high resolution structures of systems closely related to the membrane-bound nitrate reductase and nitric oxide reductase provide models which can form the basis of mechanistic studies. Coupling this structural information to spectroscopic and electrochemical analyses is now providing great insight into the enzymatic mechanisms that underlie the interconversion of nitrogen species and the way in which these reactions are coupled to energy-conserving electron transport pathways.

Acknowledgments

We would like to thank Gary Sawers, Ben Berks, Nick Watmough and Tim Rasmussen for helpful contributions to this manuscript and some of the figures and the BBSRC for supporting work arising from our own laboratories

References

Adman ET, Godden JW and Turley S (1995) The structure of copper-nitrite reductase from *Achromobacter cycloclastes* at five pH values, with NO_2^- bound and with type II copper depleted. J Biol Chem 270: 27458–27474

Alefounder PR and Ferguson SJ (1980) The location of dissimilatory nitrite reductase and the control of dissimilatory nitrate reductase by oxygen in *Paracoccus denitrificans*. Biochem J 192: 231–240

Alefounder PR, McCarthy JEG and Ferguson SJ (1981) The basis for the control of nitrate reduction by oxygen in *Paracoccus denitrificans*. FEMS Microbiol Lett 12: 321–226

Alefounder PR, Greenfield AJ, McCarthy JEG and Ferguson SJ (1983) Selection and organization of denitrifying electron-transfer pathways in *Paracoccus denitrificans* Biochim Biophys Acta. 724, 20–39

Allen JWA, Cheesman MR, Higham CW, Ferguson SJ and Watmough NJ (2000a) A novel conformer of oxidized *Paracoccus pantotrophus* cytochrome cd_1 observed by freeze-quench NIR-MCD spectroscopy. Biochem Biophys Res Commun 279: 674–677

Allen JWA, Watmough NJ and Ferguson SJ (2000b) A switch in heme axial ligation prepares *Paracoccus pantotrophus* cytochrome cd_1 for catalysis. Nat Struct Biol 7: 885–888

Allen JWA, Ferguson SJ and Fulop V (2001) Cytochrome cd_1 nitrite reductase. In: Messerschmidt A, Huber R, Poulos T and Wieghardt K (eds) Handbook of Metalloproteinsk pp 424–439. John Wiley & Sons, New York

Allen JWA, Higham, CW, Zajicek, RS, Watmough NJ and Ferguson SJ (2002) A novel kinetically stable, catalytically active, all ferric nitrite-bound complex of *Paracoccus pantotrophus* cytochrome cd_1. Biochem J 366: 883–888

Anderson LJ, Richardson DJ and Butt JN (2001) Catalytic protein film voltammetry from a respiratory nitrate reductase provides evidence for complex electrochemical modulation of enzyme activity. Biochemistry 40 11294–11307

Angove HA, Cole JA, Richardson DJ and Butt JN (2002) Protein film voltammetry reveals distinctive fingerprints of nitrite and hydroxylamine reduction by a cytochrome *c* nitrite reductase. J Biol Chem 22: 22

Arnoux P, Sabaty M, Alric J, Frangioni B, Guigliarelli B, Adriano JM and Pignol D (2003) Structural and redox plasticity in the heterodimeric periplasmic nitrate reductase. Nat Struct Biol 10: 928–934

Ballard AL and Ferguson SJ (1988) Respiratory nitrate reductase from *Paracoccus denitrificans*. Evidence for two *b*-type haems in the gamma subunit and properties of a water-soluble active enzyme containing alpha and beta subunits. Eur J Biochem 174: 207–212

Bamford VA, Angove HC, Seward HE, Thomson AJ, Cole JA, Butt JN, Hemmings AM and Richardson DJ (2002). Structure and spectroscopy of the periplasmic cytochrome *c* nitrite reductase from *Escherichia coli*. Biochemistry 41: 2921–2931

Barker PD and Ferguson SJ (1999) Still a puzzle: Why is haem covalently attached in *c*-type cytochromes? Struct Fold Des 7(12):R281–290

Bell LC, Richardson DJ and Ferguson SJ (1992) Identification of nitric oxide reductase activity in *Rhodobacter capsulatus*: the electron transport pathway can either use or bypass both cytochrome c_2 and the cytochrome bc_1 complex. J Gen Microbiol 138: 437–443

Bennett B, Berks BC, Ferguson SJ, Thomson AJ and Richardson DJ (1994) Mo(V) electron paramagnetic resonance signals from the periplasmic nitrate reductase of *Thiosphaera pantotropha*. Eur J Biochem 226 789–798

Bennett B, Charnock JM, Sears HJ, Berks BC, Thomson AJ, Ferguson SJ, Garner CD and Richardson DJ (1996) Structural investigation of the molybdenum site of the periplasmic nitrate reductase from *Thiosphaera pantotropha* by X-ray absorption spectroscopy. Biochem J 317: 557–563

Berks BC, Richardson DJ, Robinson C, Reilly A, Aplin RT and Ferguson SJ (1994) Purification and characterization of the periplasmic nitrate reductase from *Thiosphaera pantotropha*. Eur J Biochem 220: 117–124

Berks BC, Ferguson SJ, Moir JW and Richardson DJ (1995a) Enzymes and associated electron transport systems that catalyze the respiratory reduction of nitrogen oxides and oxyanions. Biochim Biophys Acta 1232: 97–173

Berks BC, Page MD, Richardson DJ, Reilly A, Cavill A, Outen F and Ferguson SJ (1995b) Sequence analysis of subunits of the membrane-bound nitrate reductase from a denitrifying bacterium: the integral membrane subunit provides a prototype for the dihaem electron-carrying arm of a redox loop. Mol Microbiol 15: 319–331

Berks BC, Richardson DJ, Reilly A, Willis AC and Ferguson SJ (1995c) The napEDABC gene cluster encoding the periplasmic nitrate reductase system of *Thiosphaera pantotropha*. Biochem J 309: 983–992

Bertero MG, Rothery RA, Palak M, Hou C, Lim D, Blasco F, Weiner JH and Strynadka NC (2003) Insights into the respiratory electron transfer pathway from the structure of nitrate reductase A. Nat Struct Biol 10:681–687

Blasco F, Guigliarelli B, Magalon A, Asso M, Giordano G and Rothery RA (2001) The coordination and function of the redox centers of the membrane-bound nitrate reductases. Cell Mol Life Sci 58: 179–193

Boogerd FC, Van Verseveld HW and Stouthamer AH (1981) Respiration-driven proton translocation with nitrite and nitrous oxide in *Paracoccus denitrificans*. Biochim Biophys Acta 638: 181–191

Boulanger MJ and Murphy ME (2001) Alternate substrate binding modes to two mutant (D98N and H255N) forms of nitrite reductase from *Alcaligenes faecalis* S-6: Structural model of a transient catalytic intermediate. Biochemistry 40: 9132–9141

Boyington JC, Gladyshev VN, Khangulov SV, Stadtman TC and Sun PD (1997) Crystal structure of formate dehydrogenase H: Catalysis involving Mo, molybdopterin, selenocysteine, and an Fe_4S_4 cluster. Science 275: 1305–1308

Braker G, Fesefeldt A and Witzel KP (1998) Development of PCR primer systems for amplification of nitrite reductase genes (*nirK* and *nirS*) to detect denitrifying bacteria in environmental samples. Appl Environ Microbiol 64: 3769–3775

Breton J, Berks BC, Reilly A, Thomson AJ, Ferguson SJ and Richardson DJ (1994) Characterization of the paramagnetic iron-containing redox centers of *Thiosphaera pantotropha* periplasmic nitrate reductase. FEBS Lett 345: 76–80

Brige A, Cole JA, Hagen WR, Guisez Y and Van Beeumen JJ (2001) Overproduction, purification and novel redox properties of the dihaem cytochrome *c*, NapB, from *Haemophilus influenzae*. Biochem J 356: 851–858

Brige A, Leys D, Meyer TE, Cusanovich MA and Van Beeumen JJ (2002) The 1.25 Å resolution structure of the diheme NapB subunit of soluble nitrate reductase reveals a novel cytochrome *c* fold with a stacked heme arrangement. Biochemistry 41: 4827–4836

Brondijk TH, Fiegen D, Richardson DJ and Cole JA (2002) Roles of NapF, NapG and NapH, subunits of the *Escherichia coli* periplasmic nitrate reductase, in ubiquinol oxidation. Mol Microbiol 44: 245–255

Brown K, Djinovic-Carugo K, Haltia T, Cabrito I, Saraste M, Moura JJ, Moura I, Tegoni M and Cambillau C (2000a) Revisiting the catalytic Cu_Z cluster of nitrous oxide (N_2O) reductase. Evidence of a bridging inorganic sulfur. J Biol Chem 275: 41133–41136

Brown K, Tegoni M, Prudencio M, Pereira AS, Besson S, Moura JJ, Moura I and Cambillau C (2000b) A novel type of catalytic copper cluster in nitrous oxide reductase. Nat Struct Biol 7: 191–195

Brown K, Roig-Zamboni V, Cutruzzola F, Arese M, Sun W, Brunori M, Cambillau C and Tegoni M (2001) Domain swing upon His to Ala mutation in nitrite reductase of *Pseudomonas aeruginosa*. J Mol Biol 312: 541–554

Bursakov SA, Carneiro C, Almendra MJ, Duarte RO, Caldeira J, Moura I and Moura JJ (1997) Enzymatic properties and effect of ionic strength on periplasmic nitrate reductase (NAP) from *Desulfovibrio desulfuricans* ATCC 27774. Biochem Biophys Res Commun 239: 816–822

Busch A, Friedrich B and Cramm R (2002) Characterization of the *norB* gene, encoding nitric oxide reductase, in the nondenitrifying cyanobacterium *Synechocystis sp.* strain PCC6803. Appl Environ Microbiol 68: 668–672

Butland G, Spiro S, Watmough NJ and Richardson DJ (2001) Two conserved glutamates in the bacterial nitric oxide reductase are essential for activity but not assembly of the enzyme. J Bacteriol 183: 189–199

Butler CS, Seward HE, Greenwood C and Thomson AJ (1997) Fast cytochrome *bo* from *Escherichia coli* binds two molecules of nitric oxide at Cu_B. Biochemistry 36: 16259–16266

Butler CS, Charnock JM, Bennett B, Sears HJ, Reilly AJ, Ferguson SJ, Garner CD, Lowe DJ, Thomson AJ, Berks BC and Richardson DJ (1999) Models for molybdenum coordination during the catalytic cycle of periplasmic nitrate reductase from *Paracoccus denitrificans* derived from EPR and EXAFS spectroscopy. Biochemistry 38: 9000–9012

Butler CS, Charnock JM, Garner CD, Thomson AJ, Ferguson SJ, Berks BC and Richardson DJ (2000) Thiocyanate binding to the molybdenum centre of the periplasmic nitrate reductase from *Paracoccus pantotrophus*. Biochem J 352: 859–864

Butler CS, Ferguson SJ, Berks BC, Thomson AJ, Cheesman MR and Richardson DJ (2001) Assignment of haem ligands and detection of electronic absorption bands of molybdenum in the di-haem periplasmic nitrate reductase of *Paracoccus pantotrophus*. FEBS Lett 50: 71–74

Butler, C. S., Fairhurst, S. A., Ferguson, S. J., Thomson, A. J., Berks, B. C., Richardson, D. J., and Lowe, D. J. (2002). Mo(V) co-ordination in the periplasmic nitrate reductase from *Paracoccus pantotrophus* probed by electron nuclear double resonance (ENDOR) spectroscopy. Biochem J 363: 817–823

Carr GJ and Ferguson SJ (1990) The nitric oxide reductase of *Paracoccus denitrificans*. Biochem J 269: 423–429

Cartron M, Roldan MD, Berks BC, Richardson DJ and Ferguson SJ (2002) Identification of two domains and distal histidine ligands to the four hemes in the bacterial *c*-type cytochrome NapC; the prototype connector between quinol/quinone and periplasmic oxido-reductases. Biochem. J 368:425–432

Charnock JM, Dreusch A, Korner H, Neese F, Nelson J, Kannt A, Michel H, Garner CD, Kroneck PMH and Zumft WG (2000) Structural investigation of the Cu_A centre of nitrous oxide reductase from *Pseudomonas stutzeri* by site directed mutagenesis and X-ray spectroscopy. Eur J Biochem 267: 1368–1381

Cheesman MR, Ferguson SJ, Moir JW, Richardson DJ, Zumft WG and Thomson AJ (1997) Two enzymes with a common function but different heme ligands in the forms as isolated. Optical and magnetic properties of the heme groups in the oxidized forms of nitrite reductase, cytochrome cd_1, from *Pseu-*

domonas stutzeri and *Thiosphaera pantotropha*. Biochemistry 36: 16267–16276

Cheesman MR, Zumft WG and Thomson AJ (1998) The MCD and EPR of the heme centers of nitric oxide reductase from *Pseudomonas stutzeri*: Evidence that the enzyme is structurally related to the heme-copper oxidases. Biochemistry 37: 3994–4000

Chen P, DeBeer George S, Cabrito I, Antholine WE, Moura JJG, Moura I, Hedman B, Hodgson KO and Solomon EI (2002) Electronic structure description of the sulfide bridged tetranuclear Cu_Z center in N_2O reductase. J Am Chem Soc 124: 744–745

Clegg S, Yu F, Griffiths L and Cole JA (2002) The roles of the polytopic membrane proteins NarK, NarU and NirC in *Escherichia coli* K-12: Two nitrate and three nitrite transporters. Mol Microbiol 44: 143–155

Cramm R, Siddiqui RA and Friedrich B (1997) Two isofunctional nitric oxide reductases in *Alcaligenes eutrophus* H16. J Bacteriol 179: 6769–6777

Cramm R, Pohlmann A and Friedrich B (1999) Purification and characterization of the single-component nitric oxide reductase from *Ralstonia eutropha* H16. FEBS Lett 460: 6–10

Craske A and Ferguson SJ (1986) The respiratory nitrate reductase from *Paracoccus denitrificans*. Molecular characterisation and kinetic properties. Eur J Biochem 158: 429–436

Cutruzzola F (1999) Bacterial nitric oxide synthesis. Biochim Biophys Acta 1411: 231–249

Cutruzzola F, Arese M, Grasso S, Bellelli A and Brunori M (1997) Mutagenesis of nitrite reductase from *Pseudomonas aeruginosa*: Tyrosine-10 in the *c* heme domain is not involved in catalysis. FEBS Lett 412: 365–369

Cutruzzola F, Brown K, Wilson EK, Bellelli A, Arese M, Tegoni M, Cambillau C and Brunori M (2001) The nitrite reductase from *Pseudomonas aeruginosa:* Essential role of two active-site histidines in the catalytic and structural properties. Proc Natl Acad Sci USA 98: 2232–2237

Cutruzzola F, Arese M, Ranghino G, von Pouderoyen G, Canters G and Brunori M (2002) *Pseudomonas aeruginosa* c(551): Probing the role of the hydrophobic patch in electron transfer. J Inorg Chem 88:353–361

Darwin A, Hussain H, Griffiths L, Grove J, Sambongi Y, Busby S and Cole J (1993) Regulation and sequence of the structural gene for cytochrome *c*552 from *Escherichia coli*: Not a hexahaem but a 50 kDa tetrahaem nitrite reductase. Mol Microbiol 9: 1255–1265

Das TK, Wilson EK, Cutruzzola F, Brunori M and Rousseau DL (2001) Binding of NO and CO to the d_1 heme of cd_1 nitrite reductase from *Pseudomonas aeruginosa.* Biochemistry 40: 10774–10781

de Boer AP, van der Oost J, Reijnders WN, Westerhoff HV, Stouthamer AH and van Spanning RJ (1996) Mutational analysis of the *nor* gene cluster which encodes nitric-oxide reductase from *Paracoccus denitrificans*. Eur J Biochem 242: 592–600

Dias JM, Than ME, Humm A, Huber R, Bourenkov GP, Bartunik HD, Bursakov S, Calvete J, Caldeira J, Carneiro C, Moura JJ, Moura I and Romao MJ (1999) Crystal structure of the first dissimilatory nitrate reductase at 1.9 Å solved by MAD methods. Structure Fold Des 7: 65–79

Dodd FE, Van Beeumen J, Eady RR and Hasnain SS (1998) X-ray structure of a blue-copper nitrite reductase in two crystal forms. The nature of the copper sites, mode of substrate binding and recognition by redox partner. J Mol Biol 282: 369–382

Dym O, Pratt EA, Ho C and Eisenberg D (2000) The crystal structure of D-lactate dehydrogenase, a peripheral membrane respiratory enzyme. Proc Natl Acad Sci USA 97: 9413–9418

Eaves DJ, Grove J, Staudenmann W, James P, Poole RK, White SA, Griffiths I and Cole JA (1998) Involvement of products of the *nrfEFG* genes in the covalent attachment of haem c to a novel cysteine-lysine motif in the cytochrome *c*552 nitrite reductase from *Escherichia coli.* Mol Microbiol 28: 205–216

Einsle O, Messerschmidt A, Stach P, Bourenkov GP, Bartunik HD, Huber R and Kroneck PM (1999) Structure of cytochrome *c* nitrite reductase. Nature 400: 476–480

Einsle O, Stach P, Messerschmidt A, Simon J, Kroger A, Huber R and Kroneck PM (2000) Cytochrome c nitrite reductase from *Wolinella succinogenes.* Structure at 1.6 Å resolution, inhibitor binding, and heme-packing motifs. J Biol Chem 275: 39608–39616

Einsle O, Messerschmidt A, Huber R, Kroneck PM and Neese F (2002) Mechanism of the six-electron reduction of nitrite to ammonia by cytochrome *c* nitrite reductase. J Am Chem Soc 124: 11737–11745

Ellington MJK, Bhakoo KK, Sawers G, Richardson DJ and Ferguson SJ (2002) Hierarchy of carbon source selection in *Paracoccus pantotrophus*: Strict correlation between reduction state of the carbon substrate and aerobic expression of the *nap* operon. J. Bacteriol 184:4767–4774

Ellis MJ, Prudencio M, Dodd FE, Strange RW, Sawers G, Eady RR and Hasnain SS (2002) Biochemical and crystallographic studies of the Met144Ala, Asp92Asn and His254Phe mutants of the nitrite reductase from *Alcaligenes xylosoxidans* provide insight into the enzyme mechanism. J Mol Biol 316: 51–64

Ferguson SJ (1998a) Nitrogen cycle enzymology. Curr Opin Chem Biol 2: 182–193.

Ferguson SJ (1998b) The *Paracoccus denitrificans* electron transport system: Aspects of the organization, structures and biogenesis. In: Canters GW and Vigenboom E (eds) Biological Electron Transfer Chains: Genetics, Composition and Mode of Operation, pp 77–88. Kluwer Academic Publishers, Dordrecht

Field SJ, Dobbin PS, Cheesman MR, Watmough NJ, Thomson AJ and Richardson DJ (2000) Purification and magneto-optical spectroscopic characterization of cytoplasmic membrane and outer membrane multiheme *c*-type cytochromes from *Shewanella frigidimarina* NCIMB400. J Biol Chem 275: 8515–8522

Field SJ, Prior L, Roldan MD, Cheesman MR, Thomson AJ, Spiro S, Butt JN, Watmough NJ and Richardson DJ (2002) Spectral properties of bacterial nitric-oxide reductase: Resolution of pH-dependent forms of the active site heme b_3. J Biol Chem 23: 20146–20150

Fisher N and Rich PR (2000) A motif for quinone binding sites in respiratory and photosynthetic systems. J Mol Biol 296: 1153–1162

Fulop V, Moir JW, Ferguson SJ and Hajdu J (1995) The anatomy of a bifunctional enzyme: structural basis for reduction of oxygen to water and synthesis of nitric oxide by cytochrome cd_1. Cell 81: 369–377

George GN, Bray RC, Morpeth FF and Boxer DH (1985) Complexes with halide and other anions of the molybdenum centre of nitrate reductase from *Escherichia coli.* Biochem J 227: 925–931

George GN, Turner NA, Bray RC, Morpeth FF, Boxer DH and Cramer S (1989) X-ray-absorption and electron-paramagnetic-resonance spectroscopic studies of the environment of molybdenum in high-pH and low-pH forms of *Escherichia coli* nitrate reductase. Biochem J 259: 693–700

George SJ, Allen JWA, Ferguson SJ and Thorneley RNF (2000) Time-resolved infrared spectroscopy reveals a stable ferric heme-NO intermediate in the reaction of *Paracoccus pantotrophus* cytochrome cd_1 nitrite reductase with nitrite. J Biol Chem 275: 33231–33237 (and correction in Vol 276: 47742)

Girsch P and de Vries S (1997) Purification and initial kinetic and spectroscopic characterization of NO reductase from *Paracoccus denitrificans*. Biochim Biophys Acta 1318: 202–216

Glockner AB, Jungst A and Zumft WG (1993) Copper-containing nitrite reductase from Pseudomonas aureofaciens is functional in a mutationally cytochrome cd_1-free background (NirS$^-$) of *Pseudomonas stutzeri*. Arch Microbiol 160: 18–26

Godden JW, Turley S, Teller DC, Adman ET, Liu MY, Payne WJ and LeGall J (1991) The 2.3 Ångstrom X-ray structure of nitrite reductase from *Achromobacter cycloclastes*. Science 253: 438–442

Gordon EH, Sjogren T, Lofqvist M, Richter CD, Allen JW, Higham CW, Hajdu J, Fulop V and Ferguson SF (2003) Structure and kinetic properties of *Paracoccus pantotrophus* cytochrome cd_1 nitrite reductase with the d_1 heme active site ligand tyrosine 25 replaced by serine. J Biol Chem 278: 11773–11781

Gronberg KL, Roldan MD, Prior L, Butland G, Cheesman MR, Richardson DJ, Spiro S, Thomson AJ and Watmough NJ (1999) A low-redox potential heme in the dinuclear center of bacterial nitric oxide reductase: Implications for the evolution of energy-conserving heme-copper oxidases. Biochemistry 38: 13780–13786

Hausladen A, Gow A and Stamler JS (2001) Flavohemoglobin denitrosylase catalyzes the reaction of a nitroxyl equivalent with molecular oxygen. Proc Natl Acad Sci USA 98: 10108–10112

Hendriks J, Warne A, Gohlke U, Haltia T, Ludovici C, Lubben M and Saraste M (1998) The active site of the bacterial nitric oxide reductase is a dinuclear iron center. Biochemistry 37: 13102–13109

Hendriks J, Oubrie A, Castresana J, Urbani A, Gemeinhardt S and Saraste M (2000) Nitric oxide reductases in bacteria. Biochim Biophys Acta 1459: 266–273

Hendriks JH, Jasaitis A, Saraste M and Verkhovsky MI (2002) Proton and electron pathways in the bacterial nitric oxide reductase. Biochemistry 41: 2331–2340

Heikkila MP, Honisch U, Wunsch P and Zumft WG (2001) Role of the Tat transport system in nitrous oxide reductase translocation and cytochrome cd_1 biosynthesis in *Pseudomonas stutzeri* J. Bacteriol. 183, 1663–1671

Householder TC, Fozo EM, Cardinale JA and Clark VL (2000) Gonococcal nitric oxide reductase is encoded by a single gene, *norB*, which is required for anaerobic growth and is induced by nitric oxide. Infect Immun 68: 5241–5246

Howes BD, Abraham ZH, Lowe DJ, Bruser, T, Eady RR and Smith BE (1994) EPR and electron nuclear double resonance (ENDOR) studies show nitrite binding to the type 2 copper centers of the dissimilatory nitrite reductase of *Alcaligenes xylosoxidans* (NCIMB 11015). Biochemistry 33: 3171–3177

Hussain H, Grove J, Griffiths L, Busby S and Cole J (1994) A seven-gene operon essential for formate-dependent nitrite reduction to ammonia by enteric bacteria. Mol Microbiol 12: 153–163

Igarashi N, Moriyama H, Fujiwara T, Fukumori Y and Tanaka N (1997) The 2.8 Å structure of hydroxylamine oxidoreductase from a nitrifying chemoautotrophic bacterium, *Nitrosomonas europaea*. Nat Struct Biol 4: 276–284

Jafferji A, Allen JW, Ferguson SJ and Fulop V (2000) X-ray crystallographic study of cyanide binding provides insights into the structure-function relationship for cytochrome cd_1 nitrite reductase from *Paracoccus pantotrophus*. J Biol Chem 275: 25089–25094

Jormakka M, Tornroth S, Byrne B and Iwata S (2002) Molecular basis of proton motive force generation: Structure of formate dehydrogenase-N. Science 295: 1863–1868

Jormakka M, Richardson D, Byrne B and Iwata S (2004) Architecture of NarGH reveals a structural classification of Mo-bisMGD enzymes. Structure 12: 95–104

Jungst A, Wakabayashi S, Matsubara H and Zumft WG (1991) The *nirSTBM* region coding for cytochrome cd_1-dependent nitrite respiration of *Pseudomonas stutzeri* consists of a cluster of mono-, di- , and tetraheme proteins. FEBS Lett 279: 205–209

Kobayashi K, Koppenhofer A, Ferguson SJ and Tagawa S (1997) Pulse radiolysis studies on cytochrome cd_1 nitrite reductase from *Thiosphaera pantotropha*: Evidence for a fast intramolecular electron transfer from *c*-heme to d_1-heme. Biochemistry 36: 13611–13616

Kobayashi K, Koppenhofer A, Ferguson SJ, Watmough NJ and Tagawa S (2001) Intramolecular electron transfer from *c* heme to d_1 heme in bacterial cytochrome cd_1 nitrite reductase occurs over the same distances at very different rates depending on the source of the enzyme. Biochemistry 40: 8542–8547

Koppenhofer A, Little RH, Lowe DJ, Ferguson SJ and Watmough NJ (2000a) Oxidase reaction of cytochrome cd_1 from *Paracoccus pantotrophus*. Biochemistry 39: 4028–4036

Koppenhofer A, Turner KL, Allen JW, Chapman SK and Ferguson SJ (2000b) Cytochrome cd_1 from *Paracoccus pantotrophus* exhibits kinetically gated, conformationally dependent, highly cooperative two-electron redox behavior. Biochemistry 39: 4243–4249

Kucera I, Hedbavny R and Dadak V (1988) Separate binding sites for antimycin and mucidin in the respiratory chain of the bacterium *Paracoccus denitrificans* and their occurrence in other denitrificans bacteria. Biochem J 252: 905–908

Kukimoto M, Nishiyama M, Murphy ME, Turley S, Adman ET, Horinouchi S and Beppu T (1994) X-ray structure and site-directed mutagenesis of a nitrite reductase from *Alcaligenes faecalis* S-6: Roles of two copper atoms in nitrite reduction. Biochemistry 33: 5246–5252

Kukimoto M, Nishiyama M, Ohnuki T, Turley S, Adman ET, Horinouchi S and Beppu T (1995) Identification of interaction site of pseudoazurin with its redox partner, copper-containing nitrite reductase from *Alcaligenes faecalis* S-6. Protein Eng 8: 153–158

Lin JT and Stewart V (1998) Nitrate assimilation by bacteria. Adv Microb Physiol 39: 1–30

Lissenden S, Mohan, S, Overton T, Regan T, Crooke H, Cardinale JA, Householder TC, Adams P, O'Conner CD, Clark VL, Smith H and Cole JA (2000) Identification of transcription activators that regulate gonococcal adaptation from aerobic to anaerobic or oxygen-limited growth. Mol Microbiol 37: 839–855

Makinen MW, Schichman SA, Hill SC and Gray HB (1983) Heme-heme orientation and electron transfer kinetic behavior of multisite oxidation-reduction enzymes. Science 222: 929–931

McGuirl MA, Bollinger JA, Cosper N, Scott RA and Dooley DM (2001) Expression, purification, and characterization of NosL, a novel Cu(I) protein of the nitrous oxide reductase (*nos*) gene cluster. J Biol Inorg Chem 6: 189–195

Moir JWB and Ferguson SJ (1994) Properties of a *Paracoccus denitrificans* mutant deleted in cytochrome *c*550 indicate that a copper protein can substitute for this cytochrome in electron transport to nitrite, nitric oxide and nitrous oxide. Microbiology 140: 389–397

Moura I. and Moura JJ (2001) Structural aspects of denitrifying enzymes. Curr Opin Chem Biol 5: 168–175

Murphy ME, Turley S and Adman ET (1997) Structure of nitrite bound to copper-containing nitrite reductase from *Alcaligenes faecalis*. Mechanistic implications. J Biol Chem 272: 28455–28460

Murphy LM, Dodd FE, Yousafzai FK, Eady RR and Hasnain SS (2002) Electron donation between copper containing nitrite reductases and cupredoxins: The nature of protein-protein interaction in complex formation. J Mol Biol 315: 859–871

Nasri H, Ellison MK, Krebs C, Huynh BH and Scheidt WR (2000) Highly variable β-bonding in the interaction of iron(II) porphyrinates with nitrite. J Am Chem Soc 122: 10795–10804

Nurizzo D, Cutruzzola F, Arese M, Bourgeois D, Brunori M, Cambillau C and Tegoni M (1998) Conformational changes occurring upon reduction and NO binding in nitrite reductase from *Pseudomonas aeruginosa*. Biochemistry 37: 13987–13996

Nurizzo D, Silvestrini MC, Mathieu M, Cutruzzola F, Bourgeois D, Fulop V, Hajdu J, Brunori M, Tegoni M and Cambillau C (1997) N-terminal arm exchange is observed in the 2.15 Å crystal structure of oxidized nitrite reductase from *Pseudomonas aeruginosa*. Structure 5: 1157–1171

Nurizzo D, Cutruzzola F, Arese M, Bourgeois D, Brunori M, Cambillau C and Tegoni M (1999) Does the reduction of *c* heme trigger the conformational change of crystalline nitrite reductase? J Biol Chem 274: 14997–15004

Ohnishi T, Moser CC, Page CC, Dutton PL and Yano T (2000) Simple redox-linked proton-transfer design: New insights from structures of quinol-fumarate reductase. Structure Fold Des 8: 23–32

Page CC, Moser CC, Chen X and Dutton PL (1999) Natural engineering principles of electron tunneling in biological oxidation-reduction. Nature 402: 47–52

Park SY, Shimizu H, Adachi S, Nakagawa A, Tanaka I, Nakahara K, Shoun H, Obayashi E, Nakamura H, Iizuka T and Shiro Y (1997) Crystal structure of nitric oxide reductase from denitrifying fungus *Fusarium oxysporum*. Nat Struct Biol 4: 827–832

Parkhill J, Wren BW, Mungall K, Ketley JM, Churcher C, Basham D, Chillingworth T, Davies RM, Feltwell T, Holroyd S, Jagels K, Karlyshev AV, Moule S, Pallen MJ, Penn CW, Quail MA, Rajandream MA, Rutherford KM, van Vliet AH, Whitehead S and Barrell BG (2000) The genome sequence of the food-borne pathogen *Campylobacter jejuni* reveals hypervariable sequences. Nature 403: 665–668

Pearson IV, Page MD, van Spanning RJ and Ferguson SJ (2003) A mutant of *Paracoccus denitrificans* with disrupted genes coding for cytochrome c_{550} and pseudoazurin establishes these two proteins as the in vivo electgron donors to cytochrome cd_1 nitrite rreductase. J Bacteriol 185: 6308–6315

Pinakoulaki E, Gemeinhardt S, Saraste M and Varotsis C (2002) Nitric oxide reductase: Structure and properties of the catalytic site from resonance Raman scattering. J Biol Chem 23: 23–33

Poock S, Moir JWB, Cole JAC, Richardson DJ (2002) Respiratory detoxification of nitric oxide by the cytochrome *c* nitrite reductase of *Escherichia coli*. J Biol Chem 26: 23664–23669

Potter L, Angove H, Richardson D and Cole J (2001) Nitrate reduction in the periplasm of Gram-negative bacteria. Adv Microb Physiol 45: 51–112

Prudencio M, Sawers G, Fairhurst SA, Yousafzai FK, Eady RR (2002) *Alcaligenes xylosoxidans* dissimilatory nitrite reductase: alanine substitution of the surface exposed histidine 139 ligand of the type 1 copper centre prevents electron transfer to the catalytic centre. Biochem 41: 3430–3438

Ranghino G, Scorza E, Sjogren T, Williams PA, Ricci M and Hajdu J (2000) Quantum mechanical interpretation of nitrite reduction by cytochrome *cd*1 nitrite reductase from *Paracoccus pantotrophus*. Biochemistry 39, 10958–10966

Rasmussen T, Berks BC, Sanders-Loehr J, Dooley DM, Zumft WG and Thomson AJ (2000) The catalytic center in nitrous oxide reductase, Cu_Z, is a copper-sulfide cluster. Biochemistry 39: 12753–12756

Rasmussen T, Berks BC, Butt JN and Thomson AJ (2002) Multiple forms of the catalytic centre, Cu_Z, in the enzyme nitrous oxide reductase from *Paracoccus pantotrophus*. Biochem J 364: 807–815

Richardson DJ (2000) Bacterial respiration: A flexible process for a changing environment. Microbiology 146: 551–571

Richardson DJ and Watmough NJ (1999) Inorganic nitrogen metabolism in bacteria. Curr Opin Chem Biol 3: 207–219

Richardson D and Sawers G (2002) Structural biology. PMF through the redox loop. Science 295: 1842–1843

Richardson DJ, Berks BC, Russell DA, Spiro S and Taylor CJ (2001) Functional, biochemical and genetic diversity of prokaryotic nitrate reductases. Cell Mol Life Sci 58: 165–178

Richter CD, Allen JW, Higham CW, Koppenhofer A, Zajicek RS, Watmough NJ and Ferguson SJ (2002) Cytochrome cd_1, reductive activation and kinetic analysis of a multifunctional respiratory enzyme. J Biol Chem 277: 3093–3100

Roldan MD, Sears HJ, Cheesman MR, Ferguson SJ, Thomson AJ, Berks BC and Richardson DJ (1998) Spectroscopic characterization of a novel multiheme *c*-type cytochrome widely implicated in bacterial electron transport. J Biol Chem 273: 28785–28790

Rothery RA, Blasco F, Magalon A and Weiner JH (2001) The diheme cytochrome b subunit (NarI) of *Escherichia coli* nitrate reductase A (NarGHI): Structure, function and interaction with quinols. J. Mol Microbiol and Biotechnol. 3, 273–283

Rowe JJ, Ubbink-Kok T, Molenaar D, Konings WN and Driessen AJ (1994) NarK is a nitrite-extrusion system involved in anaerobic nitrate respiration by *Escherichia coli*. Mol Microbiol 12: 579–586

Sakurai N and Sakurai T (1997) Isolation and characterization of nitric oxide reductase from *Paracoccus halodenitrificans*. Biochemistry 36: 13809–13815

Saraste M and Castresana J (1994) Cytochrome oxidase evolved by tinkering with denitrification enzymes. FEBS Lett 341: 1–4

Schindelin H, Kisker C, Hilton J, Rajagopalan KV and Rees DC (1996) Crystal structure of DMSO reductase: redox-linked changes in molybdopterin coordination. Science 272: 1615–1621

Schneider F, Lowe J, Huber R, Schindelin H, Kisker C and Knablein J (1996) Crystal structure of dimethyl sulfoxide reductase from *Rhodobacter capsulatus* at 1.88 Å resolution. J Mol Biol 263: 53–69

Sears HJ, Bennett B, Spiro S, Thomson AJ and Richardson DJ (1995) Identification of periplasmic nitrate reductase Mo(V) EPR signals in intact cells of *Paracoccus denitrificans*. Biochem J 310: 311–314

Sears HJ, Sawers G, Berks BC, Ferguson SJ and Richardson DJ (2000) Control of periplasmic nitrate reductase gene expression (*napEDABC*) from *Paracoccus pantotrophus* in response to oxygen and carbon substrates. Microbiology 146: 2977–2985

Shaw AL, Hochkoeppler A, Bonora P, Zannoni D, Hanson GR and McEwan AG (1999) Characterization of DorC from *Rhodobacter capsulatus*, a *c*-type cytochrome involved in electron transfer to dimethyl sulfoxide reductase. J Biol Chem 274: 9911–9914

Simon J, Gross R, Einsle O, Kroneck PM, Kroger A and Klimmek O (2000) A NapC/NirT-type cytochrome *c* (NrfH) is the mediator between the quinone pool and the cytochrome *c* nitrite reductase of *Wolinella succinogenes*. Mol Microbiol 35: 686–696

Simon J, Pisa R, Stein T, Eichler R, Klimmek O and Gross R (2001) The tetraheme cytochrome *c* NrfH is required to anchor the cytochrome *c* nitrite reductase (NrfA) in the membrane of *Wolinella succinogenes*. Eur J Biochem 268: 5776–5782

Sjogren T and Hajdu J (2001) The Structure of an alternative form of *Paracoccus pantotrophus* cytochrome cd_1 nitrite reductase. J Biol Chem 276: 29450–29455

Strange RW, Dodd FE, Abraham ZH, Grossmann JG, Bruser T, Eady RR, Smith BE and Hasnain SS (1995) The substrate-binding site in Cu nitrite reductase and its similarity to Zn carbonic anhydrase. Nat Struct Biol 2: 287–292

Strange RW, Murphy LM, Dodd FE, Abraham ZH, Eady RR, Smith BE and Hasnain SS (1999) Structural and kinetic evidence for an ordered mechanism of copper nitrite reductase. J Mol Biol 287: 1001–1009

Suharti, Strampraad MJ, Schroder I and de Vries S (2001) A novel copper A containing menaquinol NO reductase from *Bacillus azotoformans*. Biochemistry 40: 2632–2639

Sun W, Arese M, Brunori M, Nurizzo D, Brown K, Cambillau C, Tegoni M and Cutruzzola F (2002) Cyanide binding to cd_1 nitrite reductase from *Pseudomonas aeruginosa*: Role of the active-site His369 in ligand stabilization. Biochem Biophys Res Commun 291: 1–7

van der Oost J, de Boer AP, de Gier JW, Zumft WG, Stouthamer AH and van Spanning RJ (1994) The heme-copper oxidase family consists of three distinct types of terminal oxidases and is related to nitric oxide reductase. FEMS Microbiol Lett 121: 1–9

Veselov A, Olesen K, Sienkiewicz A Shapleigh JP, Scholes CP (1998) Electronic structural information from Q-band ENDOR on the type 1 and type 2 copper liganding environment in wild-type and mutant forms of copper containing nitrite reductase. Biochem 37: 6095–6105

Vijgenboom E, Busch JE and Canters GW (1997) In vivo studies disprove an obligatory role of azurin in denitrification in *Pseudomonas aeruginosa* and show that *azu* expression is under control of *rpoS* and ANR. Microbiology 143: 2853–2863

Vygodina TV, Capitanio N, Papa S and Konstantinov AA (1997) Proton pumping by cytochrome *c* oxidase is coupled to peroxidase half of its catalytic cycle. FEBS Lett 412: 405–409

Wang H, Tseng CP and Gunsalus RP (1999) The *napF* and *narG* nitrate reductase operons in *Escherichia coli* are differentially expressed in response to submicromolar concentrations of nitrate but not nitrite. J Bacteriol 181: 5303–5308

Watmough NJ, Cheesman MR, Butler CS, Little RH, Greenwood C and Thomson AJ (1998) The dinuclear center of cytochrome bo_3 from *Escherichia coli*. J Bioenerg Biomembr 30: 55–62

Watmough NJ, Butland G, Cheesman MR, Moir JW, Richardson DJ and Spiro S (1999) Nitric oxide in bacteria: Synthesis and consumption. Biochim Biophys Acta 1411: 456–474

Williams PA, Fulop V, Leung YC, Chan C, Moir JW, Howlett G, Ferguson SJ, Radford SE and Hajdu J (1995) Pseudospecific docking surfaces on electron transfer proteins as illustrated by pseudoazurin, cytochrome *c*550 and cytochrome cd_1 nitrite reductase. Nat Struct Biol 2: 975–982

Williams PA, Fulop V, Garman EF, Saunders NF, Ferguson SJ and Hajdu J (1997) Haem-ligand switching during catalysis in crystals of a nitrogen-cycle enzyme. Nature 389: 406–412

Wilson EK, Bellelli A, Liberti S, Arese M, Grasso S, Cutruzzola F, Brunori M and Brzezinski P (1999) Internal electron transfer and structural dynamics of cd_1 nitrite reductase revealed by laser CO photodissociation. Biochemistry 38: 7556–7564

Wood NJ, Alizadeh T, Bennett S, Pearce J, Ferguson SJ, Richardson DJ and Moir JW (2001) Maximal expression of membrane-bound nitrate reductase in *Paracoccus* is induced by nitrate via a third FNR-like regulator named NarR. J Bacteriol 183: 3606–3613

Wood NJ, Alizadeh T, Richardson DJ, Ferguson SJ and Moir JW (2002) Two domains of a dual-function NarK protein are required for nitrate uptake, the first step of denitrification in *Paracoccus pantotrophus*. Mol Microbiol 44: 157–170.

Zafra O, Ramirez, S., Castan, P., Moreno, R., Cava, F., Valles C., Caro, E and Berenguer J (2002) A cytochrome *c* encoded by the nar operon is required for the synthesis of active respiratory nitrate reductase in *Thermus thermophilus*. FEBS Lett 523: 99–102

Chapter 9

Fe(II) Oxidation by *Thiobacillus ferrooxidans*: The Role of the Cytochrome *c* Oxidase in Energy Coupling

W. John Ingledew*
School of Biology, University of St. Andrews, St. Andrews KY16 9JF, U.K.

Summary

Possible mechanisms by which energy maybe conserved from the aerobic oxidation of ferrous iron by *Thiobacillus ferrooxidans* are discussed. A rationale based on a consideration of the thermodynamic constraints of the system and an analysis of sequence information of the respiratory cytochrome oxidase is applied. Limitations are imposed on possible models and it is suggested that *Tb. ferrooxidans* cytochrome oxidase may be uniquely partially decoupled from proton-pumping.

I. Introduction

The thiobacilli are placed in the *Proteobacteria* division close to the junction between the β and α sub-divisions, on the basis of the sequence of their 16SRNA (Rawlings, 2001). *Thiobacillus* (*Tb.*) *ferrooxidans* can grow solely on the aerobic oxidation of ferrous iron [Fe(II)]. This occurs in a medium of dilute sulfuric acid (down to pH 1.0) (for reviews see Ingledew 1982, 1990; Ehlrich et al., 1990; Blake et al., 1993; Rawlings, 2001). The redox potentials of the respiratory substrate and oxidant are both high and the redox potential difference between them rather small. The bacterium can grow in the chemostat at a Fe(II)/(III) electrode potential (E_h), as high as +770 mV at pH 2.0. At pH 2.0 the mid-point potential (E_m) for the O_2/H_2O couple is +1120 mV, giving a redox potential difference (ΔE) of approximately 350 mV (electrode potential values are given with respect to the hydrogen half cell). The bacterium has to conserve energy available from this small ΔE to live, including fixing its own CO_2 via a Calvin Cycle and, if required, its own N_2 (to fix one CO_2 via a Calvin Cycle requires 3ATP and 2NADPH). This thus represents one of the narrowest thermodynamic margins capable of sustaining life. These processes are schematically outlined in Fig. 1, where the downhill flow of reducing equivalents from Fe(II) to O_2 is shown coupled

Email: wji@st-andrews.ac.uk

Davide Zannoni (ed): Respiration in Archaea and Bacteria. Vol 2: Diversity of Prokaryotic Respiratory Systems, pp. 207–215.

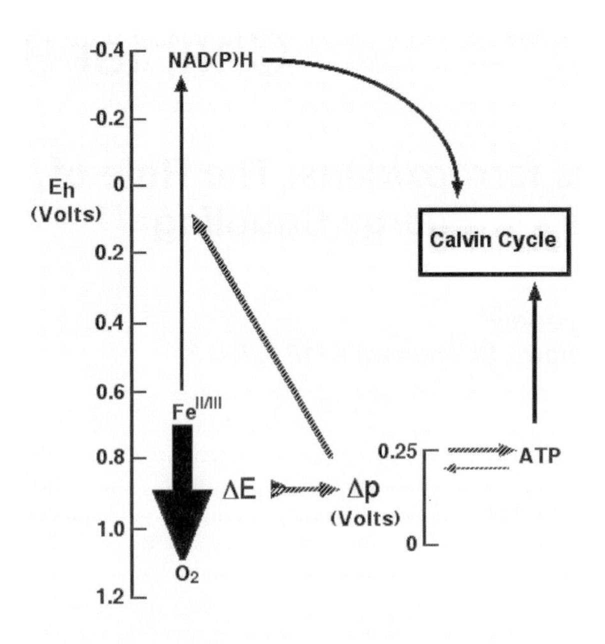

Fig. 1. Redox and metabolic profile of energy conservation during aerobic respiration on ferrous iron. The dominant arrow indicates the flow of reducing equivalents from Fe(II) to oxygen and their approximate respective redox potentials. This process generates a transmembrane Δp, which is given its own voltage scale. The Δp is used either to make ATP or to drive reversed electron transfer—uphill from Fe(II) to $NADP^+$. Both ATP and NADPH are used to fix CO_2 via a Calvin Cycle.

to the generation of a transmembrane Δp, this Δp is used to drive ATP synthesis and uphill electron transfer from Fe(II) to $NADP^+$. A redox potential scale is shown on the left, a scale is also shown for the Δp. The latter scale is also in volts, the maximal value shown represents the maximum measured Δp in *Tb. ferrooxidans* during Fe(II) oxidation (Cox et al., 1978). These scales serve to illustrate the thermodynamic constraints under which *Tb. ferrooxidans* grows on Fe(II) oxidation. The ΔE is not much greater than Δp, this has implications for the stoichiometry which links the processes.

Abbreviations: Δψ – the membrane potential; $\Delta E_{(bulk)}$ – the electrode potential difference between the donor and acceptor couples in the bulk phase; Δp – the proton motive force; ΔpH – the pH difference between the inside and the outside of the cell; COX – the cytochrome *c* oxidase; Cyt – cytochrome; Eh_{Fe} – the electrode potential of the Fe(II)/(III) couple (pH 2.0) relative to the hydrogen half cell; $Eh_{O_2}^{pHx}$ – the electrode potential of the O_2/H_2O couple at the specified pH relative to the hydrogen half cell; E_m – the midpoint potential of the specified couple at the specified pH relative to the hydrogen half cell; F – the Faraday; R – the Gas constant; SWISS-PROT – protein sequence database; T – the absolute temperature; TrEMBL – gene sequence database

II. Energy Conservation in Fe(II) Oxidizing *Thiobacillus ferrooxidans*

As there is only a small ΔE available from aerobic ferrous iron oxidation the mechanism of energy conservation can appear enigmatic. Any proposal must account for the generation of a proton motive force (Δp). This Δp is subsequently used to drive uphill electron transfer (to $NADP^+$), ATP synthesis and all the other energy requiring processes of the cell. Twenty-five years ago a simple scheme was proposed to explain this phenomenon (Ingledew et al., 1977). This proposal is outlined in Fig. 2A, and is revisited in this paper. In the model Fe(II) is oxidized on the external surface of the cell (there is compelling logic and strong evidence for this location; Ingledew, 1982), the reducing equivalents pass to the cytochrome *c* oxidase (COX) for oxygen reduction and the protons for oxygen reduction come from the cytoplasm. The side of access and egress of the O_2 and H_2O does not need to be defined as it does not contribute to the transmembrane charge separation. It is the use of cytoplasmic protons for O_2 reduction that was controversial, as is discussed below. The separation of the two electrochemical half-reactions (defined by the sidedness of Fe(II) oxidation and proton consumption) causes the generation of a transmembrane Δp. Misunderstanding arose because the ΔE between the E_h of the Fe(II)/(III) couple in the chemostat (+770 mV) and the E_h of the O_2/H_2O couple in the cytosol (nominally the E_m at pH 7.0; +820 mV) is only 50 mV, whereas if the O_2 reduction was located in equilibrium with the external phase (E_m at pH 2.0 of the O_2/H_2O couple; 1.120 mV) the ΔE is much larger, 350 mV and therefore able to pump protons against the Δp. So why would the organism reduce O_2 in the cytoplasm where the ΔE is less? This argument overlooks the fact that in the former case the generation of a Δp is included in the thermodynamics, i.e. the small ΔE been the two couples is the residual ΔE after the generation of a Δp. In fact if you write the scheme with the scalar protons for O_2 reduction coming from the periplasm and the process linked to the pumping of vectorial protons across the membrane, as in diagram Fig. 2B the net result is thermodynamically identical: in scheme 2A $4H^+$ are lost from inside and $4e^-$ are lost from outside, in scheme 2B precisely the same changes occur. Therefore the two schemes (Fig. 2 A and B) must be thermodynamically equivalent. The governing thermodynamics are the thermodynamics

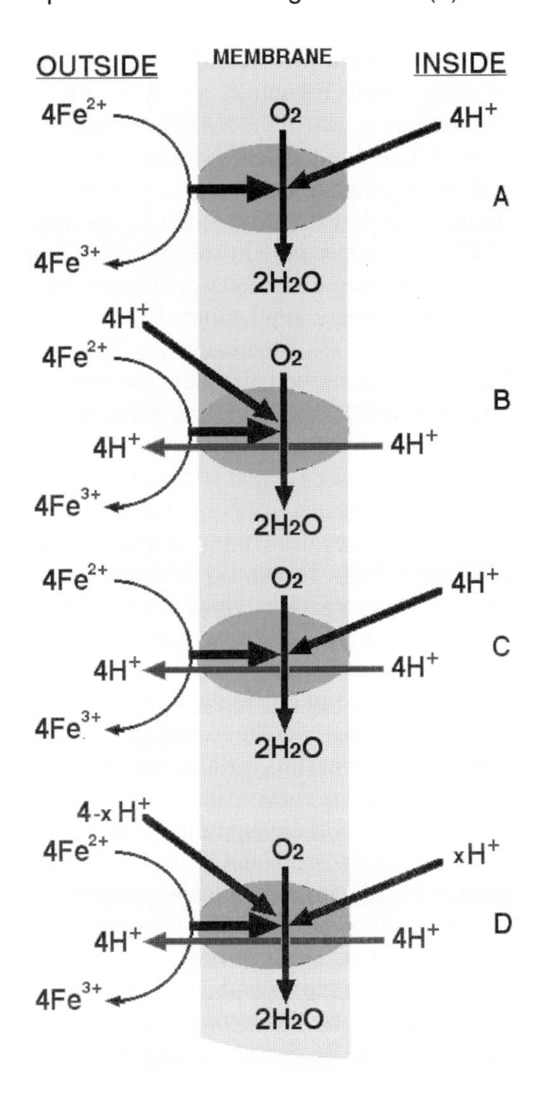

Fig. 2. Schemes for energy coupling. A. The coupling scheme proposed by Ingledew et al. (1977). The external pH is 2.0 and the internal pH is close to neutrality. Four electrons (negative charges) are moved from outside to inside, this results in the generation of four positive charges outside and the loss of four positive charges inside. The latter are due to the removal of the four protons for O_2 reduction. This generates a Δp. B. An alternative to scheme A, with the same coupling stoichiometry but with the protons for O_2 reduction entering from the outside and vectorial protons being pumped across the membrane. C. The normal coupling scheme for cytochrome oxidases, whereby, in addition to the separation of the electrochemical half reactions (Scheme A), additional protons are translocated. If this were a mitochondrial system the Fe(II) could be representing ferrocyanide oxidation or ferro-cytochrome *c* oxidation. The two proton entry points in the diagram are not meant to indicate necessarily separate channels. D. A hybrid partial coupling scheme in which with the loss of the K-channel some protons may access the O_2 reduction site from the outside surface. This gives a stoichiometry between that of scheme A (or B) and C, which may be accommodated by the thermodynamics.

of the bulk phase, and there are a number of ways of presenting this more formally. One way is to define ΔE_{ET} as the redox potential difference between the Fe(II)/(III) couple and the O_2/H_2O couple (ET stands for electron transfer) (Ingledew, 1986). If the protons for O_2 reduction come from the inside $\Delta E_{ET} = E_{Fe} - E_{O_2}{}^{pH7}$ (for the sake of argument putting pH_{in} at 7.0) and $E_{(bulk)} = E_{Fe} - E_{O_2}{}^{pH2.0}$.

$$\Delta E_{bulk} = \Delta E_{ET} - (Eh_{O_2}{}^{pHout} - Eh_{O_2}{}^{pHin}) \quad (1)$$

i.e. the overall redox potential difference is equivalent to the redox potential difference of the electron transfer reaction plus the redox potential difference 'lost' through locating the O_2 reaction in the cytosol ('lost' due to the pH dependence of that couple). This gives:

$$\Delta E_{(bulk)} = Eh_{ET} - (Eh_{O_2}{}^{pH0} - (2.3RT/F)pH_{in}) - (Eh_{O_2}{}^{pH0} - (2.3RT/F)pH_{out}) \quad (2)$$

$$\text{as } \Delta E_{ET} = Eh_{Fe} - Eh_{O_2}{}^{pHin}$$

Eq. (2) reduces to:

$$\Delta E_{(bulk)} = Eh_{Fe} - Eh_{O_2}{}^{pH7} - (2.3RT/F)\Delta pH \quad (3)$$

Under normal respiring conditions the external pH is 2.0 and the internal pH is close to 7.0 (nominal) therefore the ΔpH is close to –5 (Cox et al., 1978). The potential of the O_2/H_2O couple in each phase is related to the pH and the electrode potential at pH0:

$$Eh_{O_2}{}^{pH2} = Eh_{O_2}{}^{pH0} - (2.3RT/F) \times 2.0 \text{ for pH 2.0}$$

$$\text{and } Eh_{O_2}{}^{pH7} = Eh_{O_2}{}^{pH0} - (2.3RT/F) \times 7.0 \text{ for pH 7.0.}$$

Substituting for $Eh_{O_2}{}^{pH7}$ and ΔpH into Eq. (3):

$$\Delta E_{(bulk)} = Eh_{Fe} - (Eh_{O_2}{}^{pH0} - (2.3RT/F)\,7.0) - (2.3RT/F)(-5) \quad (4)$$

Which rearranges to

$$\Delta E_{(bulk)} = Eh_{Fe} - (Eh_{O_2}{}^{pH0} - (2.3RT/F)\,2.0) \quad (5)$$

From the relationship between pH and electrode potential for the O_2/H_2O couple the last term can become $Eh_{O_2}{}^{pH2}$, therefore

$$\Delta E_{(bulk)} = Eh_{Fe} - Eh_{O_2}{}^{pH2} \tag{6}$$

not surprisingly, the thermodynamics of the bulk phase.

Fe(II) oxidation has been shown to generate a Δp (Ingledew et al., 1977). The Δp was estimated to be at 256 mV in coupled respiring cells, it was mainly a ΔpH (4.5 units, the internal pH measured at pH 6.5), the $\Delta\psi$ being small (–10 mV). The ΔE under which this Δp was measured is not precisely defined, the Eh of the O_2/H_2O couple is 1.12 V, but the potential of the Fe(II)/Fe(III) couple during the experiment can only be crudely assessed.

III. The Respiratory Chain Components

The components of the respiratory chain of Fe(II) grown *Tb. ferrooxidans* have been quite well characterized over the years, although some of the fine details of Fe(II) oxidation and electron transfer pathways in the periplasm are not fully resolved (Ingledew and Cobley, 1980; Barr et al., 1990; Rawlings, 2001; Yarzabal et al., 2002). Redox titrations monitored by optical and EPR spectroscopy revealed multiple *c*-type cytochromes (Cyt), two *a*-type cytochromes (one high-spin, one low-spin), two *b*-type hemes and copper centers (Ingledew and Cobley, 1980). A number of the respiratory chain components now appear in the protein and gene sequence databases (Hall et al., 1996; Appia-Ayme et al., 1999). These include the H^+-translocating ATPase, an iron oxidase, and rusticyanin in the protein database [Swiss-Prot, 2002] and in the gene sequence database [TrEMBL, 2002] a putative four subunit COX, all the components of a cytochrome bc_1-Reiske complex, a high potential iron protein (HiPIP), plus cytochromes *c*. A recent report on sequences mined in the *Tb. ferrooxidans* ATCC 23270 genome finds 11 genes encoding for putative cytochromes *c*. At least 8 cytochromes *c* were differentiated on gels, of these one was specific for cells grown on sulfur and three specific for cells grown on Fe(II) (Appia-Ayme et al., 1999; Yarzabal et al., 2002). A diheme *c* (4)-type cytochrome has also been isolated (Giudici-Orticoni et al., 2000).

This plethora of redox components is rationalized when they are sorted into respiratory chain complexes; the cytochrome oxidase contains two *a*-type cytochromes and copper centers (Ingledew and Cobley, 1980; Appia-Ayme et al., 1999; [TrEmbl, 1998]), the bc_1 complex contains the Reiske iron sulfur center and cytochromes *b*(2) and c_1 (Ingledew and Cobley, 1980; Elbehti et al., 2000, [TrEmbl, 2002]). Little is known of the NADH dehydrogenase complex. Located in the periplasm are a number of individual enzymes; an Fe(II) Cyt *c*552 oxidoreductase (Kusano et al., 1992), Cyt *c*552 (Appia-Ayme et al., 1998), the copper protein rusticyanin (Cox and Boxer, 1978) and some additional *c*-type cytochromes (Ingledew and Cobley, 1980; Giudici-Orticoni et al., 2000; [TrEmbl, 2002]; Yarzabal et al., 2002). Some of these components will facilitate the transfer of reducing equivalents from Fe(II) to the cytochrome oxidase. The cytochrome oxidase uses the reducing equivalents to reduce oxygen to water, conserving the energy available in a Δp. The gene sequence of a putative COX is known and it is clearly a member of the COX superfamily. This family contains a binuclear (copper-Fe(II) heme) site at which oxygen reduction occurs and which also functions to generate Δp during turnover. The oxidases of this superfamily pump protons from inside to outside and remove protons from the internal phase for the reduction of oxygen to water during turnover (this is illustrated in Fig. 2 C). The *Tb. ferrooxidans* enzyme has some differences from other members of the superfamily, the λ-max of its reduced α-band is at 597 nm instead of the normal 603 nm (this is why it was originally referred to as a Cyt a_1, a class of heme which has been shown to be either an anomalous aa_3-type or a high-spin *b*-type heme). Also, given the thermodynamic constraints of Fe(II) oxidation by *Tb. ferrooxidans* I will pose the question, 'is the *Tb. ferrooxidans* enzyme different from the rest of the superfamily when it comes to H^+-pumping?' Because there may not be enough redox potential energy available to drive the additional protons across the membrane against the Δp.

A. *Thiobacillus ferrooxidans* Cytochrome Oxidase (Cox)

The *Tb. ferrooxidans* cytochrome oxidase is the only membrane-bound component of the putative Fe(II) oxidation pathway and the only section that can therefore be directly coupled to energy conservation. The enzyme is a member of the well characterized COX superfamily but the *Tb. ferrooxidans* enzyme is unusual in a number of respects; its λ-max in the α-band of the reduced enzyme is at 597 nm instead of at 603 nm, it copes with an environment where its external surface is in an acidic medium and its internal surface is in a neutral medium; it is uniquely thermodynamically constrained; and it has some interesting amino acid sequence differences from the rest of its superfamily. The thermodynamic considerations raise the question, is the *Tb. ferrooxidans* cytochrome oxidase an 'orthodox' member of its superfamily? This question is important because if the enzyme functions the same as other members of its group the mechanism of energy conservation must be as shown in Fig. 2 C, i.e. translocating twice the number of protons against the Δp than in models 2 A and 2 B. Is this thermodynamically feasible? The Δp in *Tb. ferrooxidans* cells respiring on Fe(II) at pH 2.0 has been measured at 256 mV (Cox et al., 1978), thus to move the 'equivalent' of $2H^+$ across the membrane per electron transferred requires a ΔE in excess of 2 × 256 mV (512 mV). The thermodynamic requirements of a pathway involving a normal cytochrome oxidase may be greater than the available energy (Fig. 2 C). To move the equivalent of $1H^+$ per electron requires a ΔE in excess of 256 mV, as shown in Fig.2A and B and the energy available in this case is sufficient.

Can the requirements of the model be more accurately defined? Unfortunately the thermodynamic parameters are not all accurately known. To fix thermodynamic constraints on the models we need three values; the E_h for the O_2/H_2O couple in the bulk phase, the E_h for the Fe(II)/(III) couple in the bulk phase and the Δp under conditions for which the preceding two parameters are known. The Δp has been estimated as high as 256 mV during Fe(II) oxidation at pH 2.0 (Cox et al., 1978; Alexander et al., 1987). The E_m for the O_2/H_2O couple at pH 2.0 is +1120 mV, the E_h for this couple under the conditions used for the measurement of Δp is probably close to this value. The operational value for the Fe(II)/(III) couple is more complex, in the usual sulfate media at pH 2.0 the E_m is approximately +650 mV although in the chemostat in which the cells are growing the value of E_h can rise as high as +770 mV. The conditions in which the Δp was measured in respiring cells is probably close to the +650 mV value. If we use that value, this gives a ΔE of 1120–650 mV, i.e. 470 mV. Probably not enough to support the mechanism in Fig. 2 C, but there is enough uncertainty in the measurements and estimates not to rule it out completely.

In the scheme of Fig. 2C, the Fe(II) could be replaced by ferrocyanide or ferrocytochrome *c* and represent the situation in mammalian mitochondria and other systems (Jasaitis et al., 1999). However, the preceding arguments indicate that this mechanism may be beyond the thermodynamic capacity of the Fe(II) oxidizing system. Is it possible that the *Tb. ferrooxidans* COX differs in its proton pumping from other members of the family? The structures of cytochrome oxidases from *Paracoccus denitrificans* (Iwata et al., 1995) and from bovine heart mitochondria (Tsukihara et al., 1995) have been determined by X-ray crystallography and hundreds of established and putative protein and gene sequences are also available in databases. A comparison of these with the putative oxidase sequence of *Tb. ferrooxidans* could be informative.

The pumping of protons by orthodox COX is well established although the mechanism by which the protons are pumped remains hotly debated. What is known is that the process involves turnover at the catalytic core of the enzyme and two proton-conducting channels lead from the cytosol/matrix to the catalytic site at which O_2 is reduced. The structures of these two proton-conducting channels are known, they are called the 'D-channel' and the 'K-channel' (for details on this topic see Chapters 6 and 7 by Abramson et al. and Brzezinski et al., Vol. 15 of this series, respectively). The respective roles of these channels in delivering either the 'vectorial' protons for translocation or the 'scalar' protons for the chemical reaction are not resolved. In summary it is thought that the D-channel delivers the vectorial protons and perhaps some of the protons for O_2 reduction and the K-channel delivers protons for O_2 reduction. The gene sequence of a putative *Tb. ferrooxidans* cytochrome oxidase is available in databases [TrEmbl]. This allows us to explore structure of the *Tb. ferrooxidans* enzyme by comparison with the known structures, of particular interest are the structures of these proton-conducting channels. Comparing the sequence to other sequences and to the known structures shows some interesting differences in the K-channel and in the catalytic core regions of the enzyme. Throughout the large family (nearly 400 sequences, from difference organisms, were compared), some residues are conserved throughout the family, others were conserved with the notable exception of *Tb. ferrooxidans.*

1. Is the K-Channel Absent?

The K-channel takes its name from the conserved lysine (K354 in *Paracoccus (P.) denitrificans* numbering). In the *Paracoccus* enzyme, a proton-channel leads from serine S291, which is in contact with external water molecules, via lysine K354 to tyrosine

Y280. Two water molecules are placed in this channel. The lysine (K354) is close to S291 and most likely non protonated. The channel consequently has a gap between K354 and T351 that is not bridged by water molecules (Iwata et al., 1995, Tsukihara et al., 1995; Hofacker and Schulten, 1998). Simulations of the Paracoccus enzyme show that K354 is flexible enough to bridge the gap by movement of its Cβ-Nε chain. Once protonated, the lysine head group moved within 50ps by 3.1 Å to a position close to T351 (Hofacker and Schulten, 1998). Figure 3 shows a summary of the results of a simple alignment carried out on sequences selected by a Blast on a fragment of the *Tb. ferrooxidans* sequence (Altschul et al., 1997). This focuses on the region of the principle residues of the K-channel and also picks out some ligands to the prosthetic groups of the catalytic core. The principle residue of the K-channel, the conserved lysine K354 (*P. denitrificans*), is uniquely replaced in *Tb. ferrooxidans*, which has an isoleucine in this position. I have looked at nearly 400 sequences and the conservation of the lysine is remarkable, only one other case of its absence is found in the literature, this is with glutamine in this position (there can be as little one base different between a Q and a K, and a previous report of Q in this position was subsequently revised to a K). This remaining example is the putative oxidase from *Rhodobacter* (*Rb.*) *sphaeroides* (Q9Z605).

```
347  IAVPTGIKIFSWIATMWGG 365   the consensus motif
            |   |
     VSIPTGFIYLSAIGTIWGG       Tb. ferrooxidans
```

Thus *Tb. ferrooxidans* appears to be the only example with a hydrophobic residue in this position. The K-channel leading up to the crucial linking lysine incorporates other highly conserved residues; a tryptophan W358 and a serine S291, both of these are absent. Proximal to the lysine and close to the reaction center the threonine T351 and tyrosine Y280 are still conserved. In the *Tb. ferrooxidans* enzyme the W358 is replaced by an alanine. Out of nearly 400 sequences probed the W is replaced in 14 cases, in 12 of these it is replaced by an arginine. In only two cases (*Deinococcus radiodurans* and *Thermus thermophilus*) is it replaced with a hydrophobic residue. In a quick alignment of the 50 closest sequences to the *Tb. ferrooxidans* sequence S291 is only replaced in the *Tb. ferrooxidans* enzyme, its replacement is the hydrophobic leucine (Fig. 3). The adjacent histidine H292, which is at the surface opening of the putative channel, is replaced by a glutamate. However this histidine is not so highly conserved and there are other examples of this substitution. There are no apparent alternative potential proton transporting residues in this location. The *Tb. ferrooxidans* enzyme is thus unique in lacking most of the essential residues of the outer K-channel, including the mechanistically crucial lysine K354, these are highlighted in bold in Fig. 3. Preliminary molecular modeling studies show that an alternative H-bonding proton channel network cannot be generated, so it maybe that this channel is uniquely inoperative in this enzyme.

2. The D-Channel is Present

The D-channel is formed by a chain of water molecules leading from aspartate D124 to glutamate E278 (*Paracoccus* numbering), the polar residues lining the channel serve to stabilize the water molecules. The D-channel is so called because of the aspartate (D124) at its entrance but this residue is not highly conserved. A key residue of the channel is the glutamate E278 near the catalytic center of the enzyme, labeled D in Fig. 3 (Iwata et al., 1995; Hofacker et al., 1998). The equivalent of E278 is present in the *Tb. ferrooxidans* enzyme. Other residues forming the channel are; aspartate D124, threonine T203, asparagine N199, and serines S134 and S193. In the *Tb. ferrooxidans* oxidase the D124 is replaced by an asparagine, but this is not a unique replacement. All of the other residues are found and preliminary structural modeling investigation shows the channel to be intact.

3. Other Modifications in Thiobacillus ferrooxidans Cytochrome Oxidase

As stated in a preceding section, the *Tb. ferrooxidans* cytochrome oxidase normally has one exposed surface in a medium at pH 2.0 and the other exposed surface in a medium close to neutrality, in between it is buried in the membrane. The consequences of this in terms of its amino acid sequence may be that some modifications would be expected on the acid surface to cope with the acidity, these modifications would not be unique to *Tb. ferrooxidans* but would also be predicted in other acidophiles. However, the loss of the K-channel may require further adaptations of the

```
Agrobacterium tume.         QHLFWFFGHPEVYILILPGFGIISHIVATFSKKPVFGY VWAHHMFT YFTAATMIIAVPTGIKIFSWIAT
Synechocystis sp            QHLFWFFGHPEVYILILPGFGIISHIVATFSKKPVFGY VWAHHMFT YFTAATMIIAVPTGIKIFSWIAT
Chattonella antiqua         QHLFWFFGHPEVYILILPGYGIISHIVSTFSKKPVFGY VWAHHMFT YFTAATMIIAVPTGIKIFSWIAT
Anabaena variabilis         QHLFWFFGHPEVYIMILPSFGVISHVISRFANKPVFGS VWSHHMYV YFTAATMVIAVPTGIKLFSWLAT
Rhizobium loti              QHLFWFFGHPEVYILIMPAFGVVSHIIPSLAHKPIFGK VWSHHLFT YFSAATMVIAIPTGIKIFSWLAT
Cafeteria roenbergens       QHLFWFFGHPEVYILILPAFGIISQVAAAFAKKNVFGY VWAHHMFT YFSAATMIIAVPTGIKIFSWLAT
Ralstonia solanacear.       QHLFWFFGHPEVYVLILPAFGIISQVSASFAKKNVFGY VWAHHMFT YFSAATMIIAVPTGIKIFSWLAT
Prototheca wickerh.         QHLFWFFGHPEVYILILPAFGIISQVVATFSRKPVFGY VWAHHMFT YFTTASIIIAVPTGIKVFSWIAT
Plantago media              QHLFWFFGHPEVYILILPAFGIISQIIATYAKKPVFGY VWAHHMFT YFTAATMIIAVPTGIKIFSWIAT
Rhizosolenia setigera       QHLFWFFGHPEVYILVLPAFGIISQILSRNARKPIFGY VWAHHMYT YFTAATMIIGVPTGIKIFSWIAT
Undaria pinnatifida         QHLFWFFGHPEVYILIIPGFGIISHILCKYANKPIFGY VWAHHMFT YFTAATMIIAVPTGIKMFSWIAT
Cyanidioschyzon me.         QHLFWFFGHPEVYILIIPGFGIISHILSKYSNKPIFGY VWAHHMFT YFTAATMIIAVPTGIKIFSWIAT
Plantago rigida             QHLFWFFGHPEVYILIIPGFGIISQILCKFSKKPVFGY VWAHHMYT YFTAATMIIAVPTGIKIFSWIAT
Chlorella vulgaris          QHLFWFFGHPEVYILILPGFGIISHILSTFARKPVFGY VWAHHMFT YFTAATMIIAVPTGIKIFSWIAT
Botrydiopsis alpina         QHLFWFFGHPEVYILILPGFGIVSHILSTFARKPVFGY VWAHHMFT YFTAATMIIAVPTGIKIFSWVAT
Nannochloropsis ocu.        QHLFWFFGHPEVYILILPGFGIVSHILSTFARKPVFGY VWAHHMFT YFTAATMIIAVPTGIKIFSWIAT
Agrobacterium tumef.        QHLFWFFGHPEVYILILPGFGIVSHILSTFARKPVFGY VWAHHMFT YFTAATMIIAVPTGIKIFSWIAT
Melosira ambiqua            QHLFWFFGHPEVYILILPGFGIVSHVLATFARKPVFGY VWAHHMFT YFTAATMIIAVPTGIKIFSWIAT
Corynebacterium glu.        QHLFWFFGHPEVYILILPGFGIVSHILATFARKPVFGY VWAHHMFT YFTAATMIIAVPTGIKIFSWVAT
Mycobacterium leprae        QHLFWFFGHPEVYILILPGFGIVSHILSTLSRKPVFGY VWAHHMFT YFTAATMIIAVPTGIKIFSWVAT
Caulobacter crescentus      QHLFWFFGHPEVYILILPAFGIISHIVSSFANKPVFGY VWAHHMYT YFTAATMIIAVPTGIKIFSWVAT
Monodus sp.                 QHLFWFFGHPEVYILILPAFGIISHIVSSFSNKPVFGY VWAHHMYT YFTAATMIIAVPTGIKIFSWVAT
Bacillus subtilis           QHLFWFFGHPEVYILILPGFGIVSHIVSTFSRKPVLGY VWAHHMYT HFTAATTIIAVPTGIKIFSWIAT
Vaucheria sessilis          QHLFWFFGHPEVYILILPAFGIISHIIVSASRKPIFGY VWAHHMFT YFTAATMIIAVPTGIKVFSWIAT
Xanthomonas camp.           QHLFWFFGHPEVYILILPGFGIISHIIVSAARKPIFGY VWAHHMYT YFTAATMIIAIPTGIKIFSWLAT
Xanthomonas axono.          QHLFWFFGHPEVYILILPGFGIVSHVISTFSKKPIFGY VWAHHMYT YFTAATMIIAVPTGIKIFSWIAT
Ochromonas danica           QHLFWFFGHPEVYILIIPGFGIISHVIATFSKKPIFGY VWAHHMYV YFTAATMIIAVPTGIKIFSWVAT
Phytophthora infes.         QHLFWFFGHPEVYILILPGFGIISHIVSTFSKKPVFGY VWAHHMYT YFVFATMVIAVPTGIKIFSWIAT
Phytophthora megasp.        QHLFWFFGHPEVYILILPGFGIISHIVSTFSKKPVFGY VWAHHMYT YFVFATMVIAVPTGIKIFSWIAT
Mycobacterium tuber.        QHLFWFFGHPEVYILILPGFGIISHIVSTFSKKPVFGY VWAHHMYT YFVFATMVIAVPTGVKIFSWIAT
Pseudochorda nagaii         QHLFWFFGHPEVYILILPGFGMISHVISTFSKKPIFGY VWAHHMYT YFVAATMVIAVPTGVKVFSWIAT
Colpomenia bullosa          QHLFWFFGHPEVYILILPGFGIISHIVSTFSKKPVFGY VWAHHMYT YFVAATMIIAVPTGVKIFSWIAT
Tribonemaaequale            QHILWFFGHPEVYIIILPGFGIISHVISTFAKKPIFGY VWAHHMYT YFMLATMTIAVPTGIKVFSWIAT
Pylaiella littoralis        QHILWFFGHPEVYMLILPGFGIISHVISTFARKPIFGY VWAHHMYT YFQMATMTIAVPTGIKVFSWIAT
Tribonema marinum           QHILWFFGHPEVYIIVLPAFGIVSHVIATFAKKPIFGY VWAHHMYT YFMMATMVIAVPTGIKIFSWIAT
Anabaena variabilis         QHIFWFFGHPEVYIMILPAFGVVSEIIPTFSRKPLFGY VWAHHMFT YFMFATMLISIPTGVKVFNWVST
Laminaria digitata          QHIFWFFGHPEVYIMILPAFGVVSEIIPTFSRKPLFGY VWAHHMFT YFMFATMLISIPTGVKVFNWVST
Paracoccus denitrif (a)     QHIFWFFGHPEVYIMILPAFGIVSQIVPAFARKPLFGY VWAHHMFT FFMYATMLIAVPTGVKIFNWIAT
Schizosaccharo. pom.        QHLFWFFGHPEVYIIALPFFGIVSEIFPVFSRKPIFGY VWAHHMFA FFSFMTYLIAVPTGIKFFNWIGT
Ectocarpus sp.              QHLFWFFGHPEVYIIALPFFGIVSEIFPVFSRKPIFGY VWAHHMFA FFSFMTYLIAVPTGIKFFNWVGT
Botrydium granulatum        QHLFWFFGHPEVYVLALPFFGIVSEIIPVFSRKPMFGY VWAHHMFV FFSFMTFLISVPTGVKFFNWVGT
Rhodobacter sphaer.         EHLFWIFGHPEVYILILPAFGIFSEVIPVFARKRLFGY VWVHHMFT IFAVATMAIAIPTGIKIFNWLLT
Hyaloraphidium curv.        QHLFWFYSHPAVYIMILPFFGVISEVIPVHARKPIFGY VWAHHMFT FFMATTMLIAVPTGIKIFSWCGT
Paracoccus denitrif. (b)    QHMFWFYSHPAVYIMILPFFGAISEIIPIHSRKPIFGY VWAHHMFT FFMITTMIIAVPTGIKIFSWLAT
Synechocystis sp.           QHLFWFYSHPAVYLMILPIFGIMSEVIPVHARKPIFGY VWVHHMFT FFTISTLIVAVPTGVKIFGWVAT
Anabaena sp.                QHYFWFYSHPAVYVIILPIFGIFSEIFPVYSRKPLFGY VWVHHLYV FFMLTTMLVSVPTGIKVFAWVAT
Nitrobacter agilis          QHLFWFYSHPAVYVMALPAFGVFSEILPVFARKPLFGY VWVHHMFT FFMASTMLIAVPTGIKVLAWTAT
Thiobacillus ferrooxi.      LDQFWFLFHPEVYVFILPAFAIWLEILPAAAKRPLFAR SGVHHYFT IFMTITETVSIPTGFIYLSAIGT
                             . :*:  ** **:: :* :.   .:    : : ::.    .** :.  *   :  :.:***.  :    *

                         267⇑⇑        ⇑ ⇑          ⇑⇑              ⇑⇑   ⇑               ⇑  ⇑   ⇑    361
                            33        D K          KK              11   2               K  K   K
```

Fig. 3. Alignment of selected sections of subunit I sequence. The three principal outer K-channel residues are highlighted in bold, additional ones are labeled K. The 49 most homologous sequences to the putative *Tb. ferrooxidans* cytochrome oxidase sequence where selected by BLAST [blastp, blosum62, EMBnet-CH/SIB (Altschul et al., 1997)], tailored for range and length and aligned using CLUSTALW (EMBnet-CH). Gaps were allowed although none remain in the ranges shown, the ranges shown are 267–304, 322–329 and 339–361. K denotes residues involved in the 'K-channel', D denotes a residue of the 'D-channel', numbers 1–3 indicate residues discussed in the text; *, indicates a fully conserved residue; **:** and **.** indicate strength of conservation of class of residue.

Tb. ferrooxidans cytochrome oxidase. In COX with this channel blocked by mutation of essential residues the enzyme turnover is inhibited (Konstantinov, et al., 1997). A further modification of the enzyme would be required; allowing protons for O_2 reduction to access the reaction site from the external phase (either Fig.2 B or Fig.2 D). Apart from the principal prosthetic group-ligating residues and the proton-channel residues there are a number of other highly conserved residues in the COX superfamily which are not retained in the *Tb. ferrooxidans* enzyme. In the consensus Cu_b binding sequence:

```
322 VWAHHMFT   329   consensus motif
     ||  **|
    SGVHH YFT        Tb. ferrooxidans
```

The VWxHHM is highly conserved except in *Tb. ferrooxidans*. The valine-trytophan (V322 W323) is labeled 1 in Fig.3, the methionine M327 is labeled 2. The conserved HH forms the bidentate histidine ligand to the copper center Cu_b, in all sequences. A quick Blast probing the 50 most homologous sequences shows are no other cases the valine-trytophan (VW) being replaced, the M327 is less highly conserved. These replacements could impact on the functioning of the catalytic site. Another unique replacement is of the conserved glutamine–histidine Q267-H268 for leucine and aspartate (labeled 3 in Fig.3). Investigation of these and other unique substitutions in the *Tb. ferrooxidans* cytochrome oxidase may prove informative to mechanistic studies on the enzyme (see also Chapters 6 and by Abramson et al. and by Brzezinski et al., Vol. 15 of this series).

4. The Consequences of a Modified Cytochrome Oxidase

The possible absence of one of the two proton-conducting channels which allow protons to pass from the cytoplasm to the catalytic core may result in a modification of the proton-translocating stoichiometry of the enzyme. However, it is not clear what the consequences of an absence of the K-channel might be. The K-channel is thought to provide access for the 'scalar' protons for O_2 reduction, and perhaps not even all of these. Wikström and co-workers (Wikström et al., 2000) have gone as far as to suggest that only one of the eight protons taken up by the enzyme during its catalytic cycle is transferred via the K-pathway and that the unique K-pathway proton may be specifically required to aid O-O bond scission by the enzyme. So the outcomes of the loss of the K-channel are not clear, although in line with the thermodynamic imperatives it is probably related to the lowering of the proton-translocating stoichiometry. The implication is that at least some of the protons for O_2 reduction must be allowed to access the catalytic site from the external surface. These do not need to be only the specific K-channel protons, some protons normally carried by the D-channel may access from the outer surface in this modified oxidase and a hybrid scheme, such as that shown in Fig. 2D may be possible and within the thermodynamic restrictions (for further discussion see Chapters 6 and 7 by Abramson et al. and by Brzezinski et al., Vol. 15 of this series).

IV. Concluding Remarks

Tb. ferrooxidans is a chemolithotrophic autotroph, it can grow on Fe(II) oxidation alone, utilizing the energy available from the oxidation for all its growth requirements. The energy available is small so the turnover is high. The tight thermodynamic constraints may limit the proton translocating stoichiometry of the terminal respiratory oxidase. This lead to an analysis of the thermodynamics of the system, a discussion of potential coupling models and an investigation into some unique facets of the *Tb. ferrooxidans* cytochrome oxidase. This investigation raises the possibility that the *Tb. ferrooxidans* is uniquely partially decoupled from proton-pumping. More extensive investigation of this unusual cytochrome oxidase may help resolve this question and prove helpful in resolving outstanding questions to mechanistic questions on the enzyme.

References

Alexander B, Leach S and Ingledew WJ (1987) The relationship between chemiosmotic parameters and sensitivity to anions and organic acids in the acidophile *Thiobacillus ferrooxidans*. J Gen Microbiol 133: 1171–1179.

Altschul S F, Madden TL, Alejandro A, Schaffer AA, Zhang J, Zhang Z, Miller W and Lipman DJ (1997) Gapped BLAST and PSI-BLAST: A new generation of protein database search programs. Nucleic Acids Res 25: 3389–3402

Appia-Ayme C, Benrine A, Cavazza C, Guidici-Orticoni M-T, Bruschi M, Chippaux M and Bonnefoy V (1998) Characterization and expression of the co-transcribed *cyc*1 and *cyc*2 genes encoding the cytochrome *c*(4) (*c*552) and a high-molecular-mass cytochrome *c* from *Thiobacillus ferrooxidans*

ATCC33020. FEMS Microbiol Lett 167: 171–177

Appia-Ayme C, Guiliani N, Ratouchniak J and Bonnefoy V (1999) Characterisation of an operon encoding two *c*-type cytochromes, an aa_3-type cytochrome oxidase, and rusticyanin in *Thiobacillus ferrooxidans* ATCC 33020. Appl Env Microbiol 65: 4781–4787

Barr, DW, Ingledew WJ and Norris PR (1990) Respiratory chain components of iron-oxidizing acidophilic bacteria. FEMS Microbiological Lett. 70: 85–90.

Blake RC, Shute EA, Greenwood MM, Spencer GH and Ingledew WJ (1993) Enzymes of aerobic respiration on iron. FEMS Microbiol Rev 11: 9-18

Cox JC and Boxer DH (1978) The purification and some properties of rusticyanin, a blue copper protein involved in iron(II) oxidation from *Thiobacillus ferrooxidans.* Biochem J 174: 497–502

Cox JC, Nicholls DJ and Ingledew WJ (1978) Trans-membrane electrical potential and pH gradient in the acidophile *Thiobacillus ferrooxidans*. Biochem J 178: 195–200.

Elbehti A, Brasseur G and Lemesle-Meunier D (2000) First evidence for existence of an uphill electron transfer through the bc1 and NADH-Q oxidoreductase complexes of the acidophilic obligate chemolithotrophic ferrous ion-oxidising bacterium *Thiobacillus ferrooxidans*. J Bact 182: 3602–3606

Elhrich H, Salerno JC and Ingledew WJ (1990) Iron- and Manganese-oxidising Bacteria. In: Shively JM and Barton LL (eds) Variations in Autotrophic Life, pp 147–170. Academic Press, London

Giudici-Orticoni MT, Leroy G, Nitschke W and Bruschi M (2000) Characterization of a new diheme *c*(4)-type cytochrome isolated from *Thiobacillus ferrooxidans*. Biochemistry 39: 7205–7211

Hall JF, Hasnain SS and Ingledew WJ (1996) The structural gene for rusticyanin from *Thiobacillus ferrooxidans*: Cloning and sequencing of the rusticyanin gene. FEMS Microbiol Lett 137: 85–89

Hofacker I, and Schulten, K (1998) Oxygen and proton pathways in cytochrome *c* oxidase. Proteins: Structure, Function, and Genetics. 30: 100–107

Ingledew WJ (1982) The bioenergetics of an acidophilic chemolithotroph. Biochim Biophys Acta 683: 89–117

Ingledew WJ (1986) Ferrous iron oxidation by *Thiobacillus ferrooxidans*. Biotechnol Bioeng 16: 23–33

Ingledew WJ (1990) The physiology and biochemistry of acidophiles. In: Edwards C (ed) The Microbiology of Extreme Environment, pp 33–54. Open University Press, Milton Keynes and Philadelphia

Ingledew WJ and Cobley JG (1980) A potentiometric and kinetic study on the respiratory chain of ferrous-iron grown *Thiobacillus ferrooxidans*. Biochim Biophys Acta 590: 141–158

Ingledew WJ, Cox JC and Halling PJ (1977) A proposed mechanism for energy conservation during Fe(II) oxidation by *Thiobacillus ferrooxidans*: Chemiosmotic coupling to net proton influx. FEMS Microbiol Lett 2: 193–197

Iwata S, Ostermeier C, Ludwig B and Michel H (1995) Structure at 2.8-Å resolution of cytochrome *c* oxidase from *Paracoccus-denitrificans*. Nature 376: 660–669

Jasaitis A, Verkhovskaya M, Morgan JE, Verkhovsky M and Wikström M (1999) Assignment and charge translocation stoichiometries of the major electrogenic phases in the reaction of cytochrome *c* oxidase with dioxygen. Biochemistry 38: 2697–2706

Konstantinov AA, Siletsky S, Mitchell D, Kaulen, A and Gennis RB (1997) The roles of the two proton input channels in cytochrome *c* oxidase from *Rhodobacter sphaeroides* probed by the effects of site-directed mutations on time-resolved electrogenic intraprotein proton transfer. Proc Natl Acad Sci USA 94: 9085–9090

Kusano T, Takeshima T, Sugawara K, Inoue C. Shiratori, T Yano T, Fuhumori Y and Yamanaka T (1992) Molecular cloning of the gene encoding *Thiobacillus ferrooxidans* Fe(II) oxidase—high homology of the gene product with HiPIP. J Biol Chem 267: 11242–11247

Rawlings DE (2001) The molecular genetics of *Thiobacillus ferrooxidans* and other mesophilic, acidophilic, chemolithotrophic, iron- or sulfur-oxidising bacteria. Hydrometallurgy 59: 187–201

Tsukihara T, Aoyama H, Yamashita E, Tomizaki T, Yamaguchi H, Shinnzawha-Itoh K, Nakashima R, Yaono R and Yoshikawa S (1995) The whole structure of the 13-subunit oxidized cytochrome c oxidase at 2.8 Å. Science 269: 1069–1074

Wikström M, Jasaitis A, Backgren C, Puustinen A, and Verkhovsky MI (2000) The role of the D- and K-pathways of proton transfer in the function of the haem-copper oxidases. Biochim Biophys Acta 1459:514–520

Yarzabal A, Brasseur G, and Bonnefoy V (2002) Cytochromes of *Acidithiobacillus ferrooxidans*. FEMS Microbiol Lett 209:189–195

Chapter 10

Sulfur Respiration

Oliver Klimmek*[1], Wiebke Dietrich[1], Felician Dancea[2], Yi-Jan Lin[2], Stefania Pfeiffer[2], Frank Löhr[2], Heinz Rüterjans[2], Roland Gross[1], Jörg Simon[1] and Achim Kröger[1†]

[1]Institut für Mikrobiologie, Johann Wolfgang Goethe-Universität, Marie-Curie-Str. 9, D-60439 Frankfurt am Main, Germany; [2]Institut für Biophysikalische Chemie, Johann Wolfgang Goethe-Universität, Marie-Curie-Str. 9, D-60439 Frankfurt am Main, Germany

Summary

Certain anaerobic prokaryotes grow by oxidative phosphorylation with elemental sulfur as the terminal electron acceptor instead of O_2 (sulfur respiration or anaerobic respiration with sulfur). The mechanism of sulfur respiration has been investigated in great detail only in *Wolinella* (*W.*) *succinogenes*. This ε-proteobacterium uses, as terminal electron acceptor, polysulfide which is formed abiotically from elemental sulfur and sulfide. Polysulfide reduction is facilitated by a periplasmic sulfur-binding protein whose structure was determined. Polysulfide respiration with H_2 or formate as electron donor is catalyzed by the membrane of *W. succinogenes* and is coupled to apparent proton translocation across the membrane. The electrochemical proton potential (Δp) thus generated and the corresponding H^+/e^- ratio were determined to be 170 mV (negative inside) and 0.5, respectively. The electron transport chain catalyzing polysulfide respiration by H_2 or formate consists of polysulfide reductase, methyl-menaquinone and hydrogenase or formate dehydrogenase. Each enzyme consists of two different hydrophilic and a hydrophobic subunit which anchors the enzyme in the membrane. The membrane anchors of hydrogenase and formate dehydrogenase are similar di-heme cytochromes *b* which carry the site of quinone reduction. Methyl-menaquinone is thought to be bound to the membrane anchor of polysulfide reductase and to accept electrons from the cytochromes *b* of the dehydrogenases. Proteoliposomes containing polysulfide reductase, methyl-menaquinone, and formate dehydrogenase isolated from *W. succinogenes* catalyze polysulfide reduction by formate which is coupled to Δp generation. The coupling mechanism is discussed.

---

*Author for correspondence, email: klimmek@em.uni-frankfurt.de. † See Obituary at the beginning of this volume

Davide Zannoni (ed): Respiration in Archaea and Bacteria. Vol 2: Diversity of Prokaryotic Respiratory Systems, pp. 217–232.

I. Introduction

It is well established that most of the known anaerobic prokaryotes perform 'oxidative phosphorylation' without O_2. Depending on the species and the metabolic conditions, these bacteria use a large variety of inorganic (e. g. nitrate, nitrite, sulfate, thiosulfate, elemental sulfur) or organic compounds (e. g. fumarate, dimethylsulfoxide, trimethylamine-N-oxide, vinyl- or arylchlorides) as terminal electron acceptors instead of oxygen. The redox reactions with these acceptors are catalyzed by membrane-integrated electron transport chains and are coupled to the generation of an electrochemical proton potential (Δp) across the membrane. Oxidative phosphorylation in the absence of O_2 is also termed 'anaerobic respiration.' Oxidative phosphorylation with elemental sulfur is called 'sulfur respiration.'

Certain anaerobic Archaea (e.g. members of the genera *Acidianus*, *Thermoproteus*, *Pyrodictium*) and Bacteria (e. g. *Wolinella succinogenes*, members of the genera *Desulfuromonas* and *Sulfurospirillum*) use H_2 as the electron donor in sulfur respiration (see below, reaction *a*). The organisms grow with H_2 and elemental sulfur as the sole metabolic substrates (Schauder and Kröger, 1993; Hedderich et al., 1999). Therefore it is likely that reaction (*a*) is coupled to the consumption of protons from the inside (H_i^+) and to proton release on the outside (H_o^+) of the membrane, so that the Δp thus generated drives ATP synthesis.

$$H_2 + S^\circ \longrightarrow HS^- + H^+ \quad (1H_i^+ \rightarrow 1H_o^+) \qquad (a)$$

$$H_2 + \tfrac{1}{2}O_2 \longrightarrow H_2O \quad (8H_i^+ \rightarrow 8H_o^+) \qquad (b)$$

Aerobic bacteria (e.g. *Paracoccus denitrificans* and *Bacillus subtilis*) grow by oxidative phosphorylation driven by reaction (*b*) where the sulfur in reaction (*a*) is replaced by O_2 (Richardson, 2000). The two reactions differ in the number of protons apparently translocated per electron transported from H_2 to the acceptor substrate (H^+/e^- ratio, n_{H^+}/n_e). The H^+/e^- ratio is estimated to be 0.5 for reaction (*a*) and 4 for reaction (*b*). These numbers were calculated from the redox potentials of the substrates (Table 1) according to Eq. (1) with $\Delta p = 0.17$ Volt and assuming that 50% of

$$n_H^+ / n_e = q \cdot \Delta E / \Delta p \qquad (1)$$

the free energy of the redox reactions is conserved ($q = 0.5$). The relatively low H^+/e^- ratio of reaction (*a*) is due to the much lower redox potential of the couple S/HS^- as compared to that of O_2/H_2O. Assuming that three protons have to be translocated for the synthesis of a molecule of ATP, the H^+/e^- ratio of reaction (*a*) suggests that six electrons have to be transported from H_2 to sulfur for the synthesis of one molecule of ATP. This is a striking example of the advantage of oxidative phosphorylation as compared to substrate phosphorylation. In contrast to substrate phosphorylation, oxidative phosphorylation allows ATP synthesis to be driven by reactions which do not allow the synthesis of one mol of ATP per mol of substrate. In oxidative phosphorylation, the number of protons translocated can be adjusted to the free energy of the driving redox reaction. This would possibly explain the prevalence of oxidative phosphorylation especially in anaerobic organisms, where the free energy change of the catabolic reactions is usually much lower than in aerobic organisms.

In this chapter, sulfur respiration of *W. succinogenes* will be described and the coupling mechanism of apparent proton translocation to sulfur reduction will be discussed. Sulfur respiration and anaerobic respiration with other acceptors have been reviewed previously (Schauder and Kröger, 1993; Unden and Bongaerts, 1997; Hedderich et al., 1999; Richardson, 2000; Kröger et al., 2002).

II. Polysulfide as Intermediate in Sulfur Respiration

As elemental sulfur is nearly insoluble in water (5 µg / l at 25 °C, Boulégue, 1978), it is unlikely that elemental sulfur is the substrate of sulfur respiration for mesophilic bacteria. Elemental sulfur readily dissolves in aqueous solutions containing sulfide

Abbreviations: Δp – electrochemical proton potential across a membrane; $\Delta\psi$ – electrical proton potential across a membrane; DMN – 2,3-dimethyl-1,4-naphthoquinone; FdhA/B/C – formate dehydrogenase; HQNO – 2-(n-heptyl)-4-hydroxyquinoline-N-oxide; HydA/B/C – hydrogenase; MM – methyl-menaquinone; MMH_2 – hydroquinone of MM; MM_b – methyl-menaquinone bound to PsrC; MM_bH^- – hydroquinone anion of MM_b; PsrA/B/C – polysulfide reductase; [S] – polysulfide sulfur; TPP^+ – tetraphenylphosphonium; TTFB – 4,5,6,7-tetrachloro-2-trifluoromethylbenzimidazol

Table 1. Redox potentials

	E_o' (mV)
$H_2 \rightarrow 2H^+ + 2e^-$	–420
$HCO_2^- + H_2O \rightarrow HCO_3^- + 2H^+ + 2e^-$	–413 [1]
$MMH_2 \rightarrow MM + 2H^+ + 2e^-$	–90 [2]
$HS^- \rightarrow 1/8\ S_8 + H^+ + 2e^-$	–275 [1]
$4\ HS^- \rightarrow S_4^{2-} + 4H^+ + 6e^-$	–260 [3]
Succinate → Fumarate + $2H^+ + 2e^-$	+30 [1]
$H_2O \rightarrow 1/2 O_2 + 2H^+ + 2e^-$	+815 [1]

The numbers were taken from Thauer et al., 1977 [1], Clark, 1960 [2], Dietrich and Klimmek, 2002 [3]. MM, methyl-menaquinone.

(HS^-) which is the product of sulfur respiration. Thus polysulfide is formed abiotically according to reaction (*c*). The amount of sulfur dissolved increases with the

$$n/8\ S_8 + HS^- \rightarrow S_{n+1}^{2-} + H^+ \qquad (c)$$

concentration of HS^-, with pH, and with the temperature of the solution (Schauder and Müller, 1993). In a solution containing 10 mM HS^-, the equilibrium concentration of polysulfide sulfur is 0.1 mM at pH 6.7 and 30 °C. At 90 °C 0.1 mM sulfur is already dissolved at pH 5.5.

Polysulfide may be the intermediate of sulfur reduction in organisms growing under conditions which allow the dissolution of at least 0.1 mM sulfur (Schauder and Kröger, 1993). All eubacterial sulfur reducers and most of the archaeal strains grow maximally under conditions where polysulfide is stable up to at least 0.1 mM dissolved polysulfide sulfur. Therefore, it is likely that these organisms use polysulfide as the actual substrate of sulfur respiration. Polysulfide is not available to the extremely acidophilic Archaea growing at pH 3 or below (e. g. members of the genera *Acidianus*, *Stygioglobus* and *Thermoplasma*) (Hedderich et al., 1999). Under the growth conditions of these sulfur reducers elemental sulfur has to be mobilized in a different, and as yet unknown, way.

Tetrasulfide (S_4^{2-}) and pentasulfide (S_5^{2-}) are the predominant species of polysulfide at pH values above 6 (Schwarzenbach and Fischer, 1960; Giggenbach, 1972). The pK of acid dissociation of HS_4^- and HS_5^- were measured to be 6.3 and 5.7, respectively. The two species dismutate rapidly according to reaction (*d*).

$$3\ S_5^{2-} + HS^- \leftrightarrows 4\ S_4^{2-} + H^+ \qquad (d)$$

The equilibrium constant of reaction (*d*) was determined to be $4 \cdot 10^{-9}$ M at 20 °C (Giggenbach 1972). From the equilibrium constant it can be calculated that the concentration of S_4^{2-} is twice that of S_5^{2-} in a solution (pH 8.5 and 37 °C) containing 1 mM polysulfide sulfur and 2 mM HS^- (Klimmek et al. 1991). The redox potential of polysulfide can be calculated from that of elemental sulfur/HS^- (Table 1) and the equilibrium constant of reaction (*c*) ($3.6 \cdot 10^{-9}$ M at 37 °C) with the simplification that only S_4^{2-} is formed upon dissolution of sulfur. Thus, the redox potential of the polysulfide obtained is only slightly more positive than that of S_8/HS^- (Table 1).

III. Polysulfide Respiration in *W. succinogenes*

The mechanism of sulfur respiration in *W. succinogenes* has been investigated thoroughly. *W. succinogenes* belongs to the ε-group of Proteobacteria and has been isolated from bovine rumen (Simon et al., 2000). It does not ferment sugars and grows solely by anaerobic respiration with sulfur, fumarate, dimethylsulfoxide, nitrate, nitrite or N_2O as electron acceptor. H_2 or formate serve as electron donors of anaerobic respiration. Polysulfide was identified as the intermediate of sulfur respiration. Disulfides (R–S–S–R), thiosulfate ($S_2O_3^{2-}$) or tetrathionate ($S_4O_6^{2-}$) do not serve as electron acceptors.

W. succinogenes grows by polysulfide respiration with either H_2 (reaction *e*) or formate (reaction *f*) as electron donor (Macy et al., 1986; Hedderich et al., 1999).

$$H_2 + S_n^{2-} \rightarrow HS^- + S_{n-1}^{2-} + H^+ \qquad (e)$$

$$HCO_2^- + S_n^{2-} + H_2O \rightarrow HCO_3^- + HS^- + S_{n-1}^{2-} + H^+ \qquad (f)$$

Acetate and glutamate serve as carbon sources. The cell density in the culture increased proportionally to the amounts of polysulfide reduced, and equimolar amounts of formate and polysulfide sulfur were consumed. The specific activity of polysulfide reduction by formate of cells of *W. succinogenes* grown at the expense of reaction (*f*) was 20 % higher than the activity calculated from the specific growth rate and the growth yield of the growing bacteria (Klimmek et al., 1991; Schauder and Kröger, 1993). This confirms the view that reaction (*f*) is responsible for growth.

Table 2. Bioenergetic data of polysulfide respiration with formate of *W. succinogenes* (Hedderich et al., 1999). The data are compared to those of fumarate respiration. The values of pH, Y (growth yield) and ΔE refer to the middle of the exponential growth phase at 37°C. ΔE was calculated from the E_0' given in Table 1 with the given values of pH and equal concentrations of HCO_2^- and HCO_3^-, polysulfide sulfur and HS^-, and fumarate and succinate. The numbers in parentheses were estimated as described in the text.

Electron acceptor	pH	Y (g cells/mol formate)	-ΔE (Volt)	Δp (Volt)	H^+/e^-	ATP/e^-	$\frac{\Delta E \cdot F}{ATP/e^-}$ (kJ/mol ATP)
Polysulfide	8.4	3.2	0.20	0.17	(1/2)	(1/6)	116
Fumarate	7.9	7.0	0.44	0.18	1	1/3	127

A. Bioenergetics of Polysulfide Respiration

Cells of *W. succinogenes* catalyzing reaction (*e*) or (*f*) were found to take up tetraphenylphosphonium (TPP^+) from the external medium (Wloczyk et al., 1989; Klimmek, 1996). TPP^+ uptake was prevented by the presence of a protonophore. From the amount of TPP^+ taken up, the $\Delta\psi$ (0.17 Volt) was calculated to be approximately the same as that generated by fumarate respiration with formate (reaction *g*) (Table 2). The values of $\Delta\psi$ and Δp were nearly the same, since the ΔpH

$$HCO_2^- + \text{Fumarate} + H_2O \rightarrow HCO_3^- + \text{Succinate} \qquad (g)$$

across the membrane was negligible. The H^+/e^- and the ATP/e^- ratios were determined for fumarate respiration but not for polysulfide respiration. However, the latter values can be estimated from the growth yields (Y) of *W. succinogenes* growing at the expense of reactions (*f*) and (*g*). The growth yield of polysulfide respiration was measured to be approximately half that of fumarate respiration, suggesting that the H^+/e^- and ATP/e^- ratio of polysulfide respiration was also half that of fumarate respiration (Schauder and Kröger, 1993). This view is confirmed by the redox potential differences (ΔE) between formate and each of the two electron acceptors under the growth conditions of the bacteria. The ΔE of polysulfide respiration is approximately half that of fumarate respiration. The free energy required for ATP synthesis calculated from ΔE and the ATP/e^- ratios are 116 and 127 kJ / mol in anaerobic respiration with polysulfide and fumarate, respectively. Both values are consistent with the general observation that in most instances phosphorylation requires about 100 kJ per mol of ATP in growing bacteria.

B. The Electron Transport Chain

The membrane fraction isolated from *W. succinogenes* cells grown with formate and either polysulfide or fumarate catalyzes polysulfide reduction by H_2 or formate (Schauder and Kröger, 1993; Fauque et al., 1994). The corresponding electron transport chain consists of polysulfide reductase, methyl-menaquinone (Fig. 1), and either hydrogenase or formate dehydrogenase (Fig. 2) (Hedderich et al., 1999). This was shown by reconstituting a functional electron transport chain from the isolated enzymes (Table 3). Proteoliposomes containing polysulfide reductase, methyl-menaquinone and either hydrogenase or formate dehydrogenase catalyzed polysulfide reduction by H_2 (reaction *e*) or formate (reaction *f*). The turnover number of polysulfide reductase in electron transport was commensurate with that measured in the membrane fraction of *W. succinogenes*. When prepared according to the method designed by Rigaud et al. (1995), proteoliposomes containing polysulfide reductase, methyl-menaquinone and formate dehydrogenase catalyzed polysulfide reduction by formate. This reaction was coupled to the uptake of TPP^+ from the external medium (Fig. 3). TPP^+ uptake was prevented by a protonophore. The $\Delta\psi$ (0.14 Volt) calculated from the amount of TPP^+ taken up under steady state electron transport was not significantly lower than that generated across the membrane of cells by the same reaction (Table 3). These experiments demonstrate that the isolated enzymes contain all the components required for electron transport and for Δp generation by polysulfide respiration.

Each of the three enzymes involved in polysulfide respiration consists of two hydrophilic and one hydrophobic membrane subunit (Fig. 2). The hydrophobic subunit of polysulfide reductase probably carries methyl-menaquinone (MM_b). The hydrophobic subunits of hydrogenase and formate dehydrogenase are

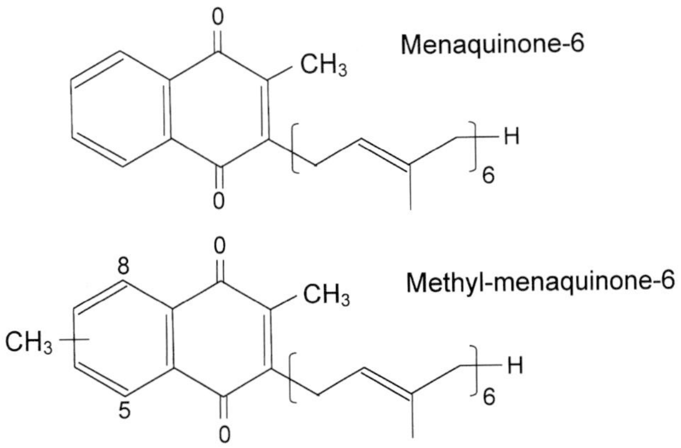

Menaquinone-6

Methyl-menaquinone-6

Fig. 1. Structures of the quinones of *W. succinogenes* (Collins and Fernandez, 1984) The second methyl group is bound either in position 5 or 8.

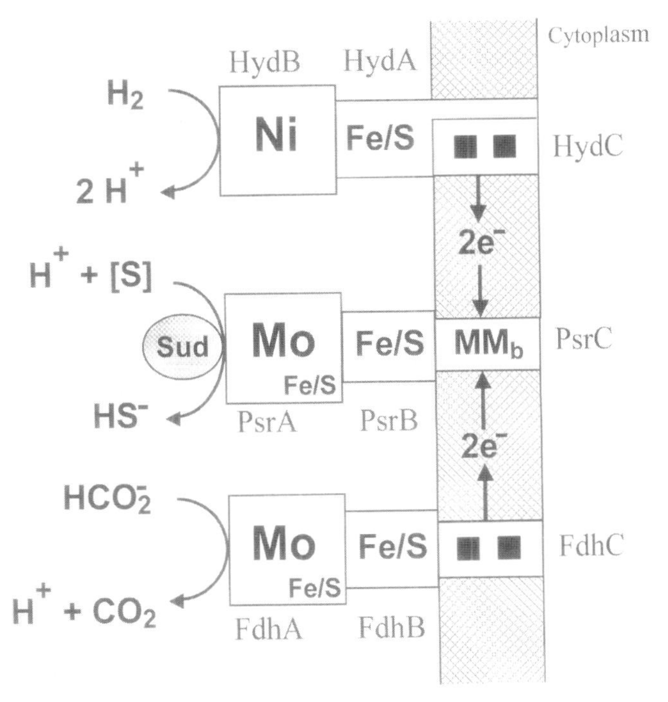

Fig. 2. Composition of the electron transport chain catalyzing polysulfide respiration with H_2 or formate in *W. succinogenes*. HydA/B/C, hydrogenase; PsrA/B/C, polysulfide reductase; FdhA/B/C, formate dehydrogenase; Ni, nickel-iron center; Fe/S, iron-sulfur centers; Mo, molybdenum ion bound to molybdopterin guanine dinucleotide; the dark squares designate heme *b* groups.

related di-heme cytochromes *b* (Biel et al., 2002). The larger hydrophilic catalytic subunits of the enzymes carry the substrate sites and are oriented towards the periplasmic side of the membrane. The smaller hydrophilic subunits carry three (HydA) or four iron-sulfur centers (PsrB and FdhB) which probably mediate electron transfer from the catalytic to the membrane-integrated subunit or vice versa (PsrB). The genes encoding the subunits of each enzyme are located in operons, the sequences of which were determined. Deletion mutants were constructed which lack one or more genes of each operon. The mutants can be complemented by genomic integration of the respective genes. This allows the construction of variants of the enzymes by site-directed mutagenesis.

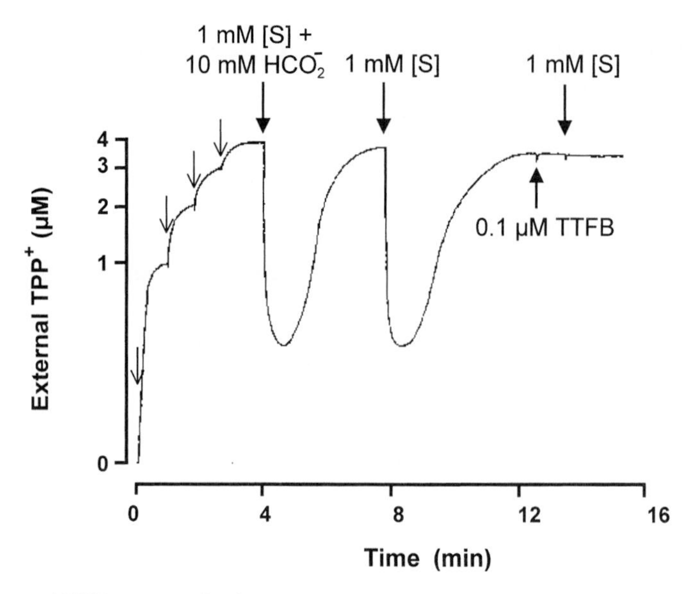

Fig. 3. Recording of the external TPP^+ concentration in a suspension of proteoliposomes catalyzing polysulfide reduction by formate. Proteoliposomes (3.0 g phospholipide · l^{-1}) containing formate dehydrogenase (62 nmol · g^{-1} phospholipid), polysulfide reductase (30 nmol · g^{-1} phospholipid) and methyl-menaquinone (10 µmol · g^{-1} phospholipid) were suspended in an anoxic Tris chloride buffer (pH 8.0, 37 °C). Calibration of the TPP^+ electrode was done by four additions of 1 µM TPP^+ (thin arrows). Electron transport was started by the addition of polysulfide ([S]) and formate.

Table 3. Activities of polysulfide respiration in proteoliposomes containing different naphthoquinones (Dietrich and Klimmek, 2002). Polysulfide respiration with H_2 (H_2 → [S]) or with formate (HCO_2^- → [S]) was measured in proteoliposomes containing polysulfide reductase and either hydrogenase or formate dehydrogenase isolated from *W. succinogenes*. Fumarate respiration with formate (HCO_2^- → Fumarate) was measured in proteoliposomes containing fumarate reductase and formate dehydrogenase. The activities are given as substrate turnovers of polysulfide reductase or fumarate reductase at 37 °C.

Quinone	H_2 → [S]	HCO_2^- → [S] (s^{-1})	HCO_2^- → Fumarate
---	25	5	17
Methyl-menaquinone-6	370	175	35
Menaquinone-6	27	5	1490
Menaquinone-4	34	7	1455
Vitamin K_1	25	5	1180

1. Polysulfide Reductase

Isolated polysulfide reductase consists of the three subunits predicted by the *psrABC* operon, and contains a molybdenum ion coordinated by molybdopterin guanine nucleotide (MGD) (Krafft et al., 1992; Jankielewicz et al., 1994) (Fig. 2). These cofactors are likely to be bound to the catalytic subunit PsrA together with a [4Fe – 4S] iron-sulfur center which is predicted by the sequence of PsrA. Four additional [4Fe – 4S] centers are predicted by the sequence of PsrB. The isolated enzyme contains approximately 20 mol of free iron and sulfide, in agreement with the presence of five [4Fe – 4S] centers (Hedderich et al., 1999). The enzyme does not contain heme, flavin or heavy metal ions in addition to molybdenum. About one mol of a mixture of methyl-menaquinone and menaquinone was consistently found to be bound to the isolated enzyme; the data were corrected for the amount of quinone which is associated with the phospholipid present in the preparation. Because of its lipophilic nature, the quinone is likely to be bound to the membrane anchor PsrC of the enzyme.

The isolated enzyme catalyzes the reduction of polysulfide by BH_4^- (reaction *h*) and the oxidation of sulfide by dimethylnaphthoquinone (DMN) (reaction *i*) at

$$S_n^{2-} + BH_4^- \rightarrow S_{n-1}^{2-} + HS^- + BH_3 \quad (h)$$

$$(n+1)HS^- + nDMN + (n-1)H^+ \rightarrow S_{n+1}^{2-} + n\,DMNH_2 \quad (i)$$

commensurate turnover numbers ($1.7 \cdot 10^3\ s^{-1}$ and $1.1 \cdot 10^3\ s^{-1}$) (Hedderich et al., 1999). The apparent K_M for polysulfide sulfur (reaction *h*) is 50 μM and that for sulfide (reaction *i*) is 25 mM. PsrC is not required for both catalytic activities. A mutant lacking *psrC* still catalyzed both reactions. In the absence of the membrane anchor PsrC the enzyme was located in the periplasmic cell fraction (Krafft et al., 1995; Dietrich and Klimmek, 2002). Furthermore, nineteen mutants with an altered residue of PsrC had wild-type activities with respect to reaction (*h*) and (*i*), and PsrA was bound to the membrane (Dietrich and Klimmek, 2002). However, six of these mutants did not catalyze electron transport from H_2 or formate to polysulfide, suggesting that the oxidation of reduced methyl-menaquinone by polysulfide was impaired in these mutants (see Fig. 7). Therefore, the site of DMN reduction (reaction *i*) appears to be different from that of methyl-menaquinone interaction. Each of the four iron-sulfur centers of PsrB appears to be required for the activities of reaction (*h*) and (*i*) as well as for electron transport (reactions *e* and *f*). This is suggested by mutants in which each cysteine residue of the four cysteine clusters coordinating the four iron-sulfur centers was replaced by alanine (O Klimmek, unpublished results). PsrA of these mutants was bound to the membrane, suggesting that the structure of the enzyme was not drastically impaired. However, the mutants did not catalyze reactions (*h*) and (*i*), or electron transport from H_2 or formate to polysulfide, suggesting that each of the four iron-sulfur centers of PsrB was required for enzymic activity.

As seen from its amino acid sequence, the catalytic subunit PsrA of polysulfide reductase belongs to the DMSO reductase family of molybdo-oxidoreductases (Krafft et al., 1992; Kisker et al., 1997; see also Chapter 9 by McEwan et al., Vol. 15 of this series). The structure of several single-subunit enzymes of this family is known. As a rule, the molybdenum ion coordinated by two MGD molecules appears to be the electron donor to or acceptor from the respective substrate in these enzymes. A cavity extending from the surface of the protein to the molybdenum close to its center is seen in all structures. The substrates probably reach the active site near the molybdenum through this cavity, and the products are released to the surface through the cavity. With the assumption that PsrA resembles these enzymes, it is likely that the substrates, polysulfide and protons, reach the molybdenum through a similar cavity. Here a sulfur atom is reductively cleaved from the end of the sulfur chain of polysulfide according to reaction (*j*) (Fig. 4), and the

$$S_n^{2-} + 2e^- + H^+ \rightarrow S_{n-1}^{2-} + HS^- \quad (j)$$

products are released through the cavity. Since PsrA is oriented towards the periplasmic side of the membrane, the substrates are taken up from the periplasm and the products are released to the same side. The electrons (reaction *j*) are thought to be provided by the molybdenum ion in the reduced state (+4). In the oxidized state, molybdenum (+6) ion is thought to be reduced via the iron-sulfur centers of PsrA and PsrB by reduced methyl-menaquinone bound to PsrC.

2. Formate Dehydrogenase

The two *fdh* operons on the genome of *W. succinogenes* differ in their promoter regions, but are nearly identical in their gene sequences (Bokranz et al., 1991; Lenger et al., 1997). Deletion mutants lacking one of the operons still grow by anaerobic respiration with polysulfide or fumarate using formate as electron donor. The formate dehydrogenase serving in polysulfide respiration is also involved in fumarate respiration where it catalyzes the reduction by formate of menaquinone which is dissolved in the membrane (Biel et al., 2002; Kröger et al., 2002). Like *E. coli* formate dehydrogenase-N, the enzyme consists of three different subunits, whose sequences are similar to those of the *E. coli* enzyme. The structure of *E. coli* formate dehydrogenase-N was determined recently, and that of the *W. succinogenes* enzyme is thought to be similar (Jormakka et al., 2002). The two enzymes are anchored in the membrane by related di-heme cytochromes *b* which carry the site of quinone reduction. The two heme groups of the *E. coli* enzyme are nearly on top of each other when viewed along the membrane normal. The iron-sulfur subunit is situated

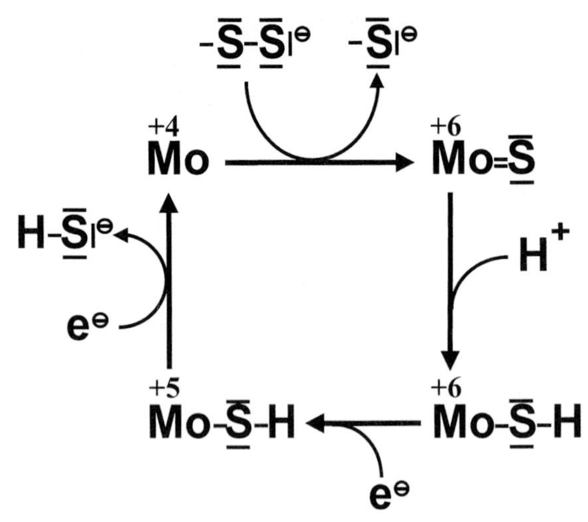

Fig. 4. Hypothetical mechanism of polysulfide reduction at the substrate site of polysulfide reductase. A sulfur atom is cleaved from the end of the polysulfide chain (-S-S|θ) and bound to the molybdenum ion (Mo) which is thereby oxidized. After the uptake of a proton and two electrons, HS^- is released and molybdenum ion returns to the reduced state again.

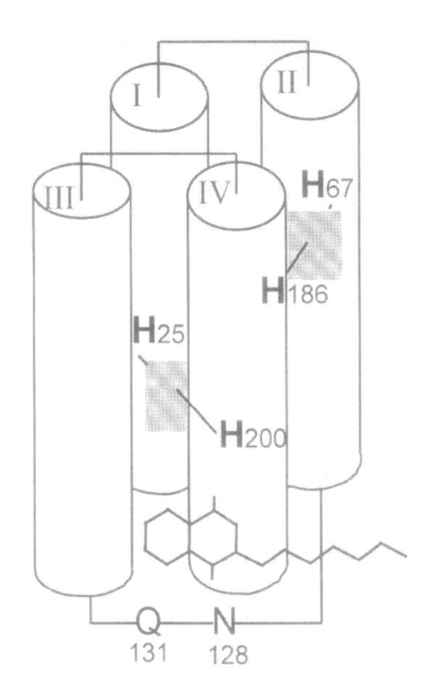

Fig. 5. Hypothetical arrangement of the four predicted membrane-spanning helices of *W. succinogenes* HydC (Biel et al., 2002). The scheme is based on the crystal structure of *E. coli* formate dehydrogenase-N (Jormakka et al., 2002). The shaded squares represent the heme *b* groups. A molecule of HQNO is shown bound to the site of quinone reduction which is confined by the axial ligand H200 of the distal heme group and by N128 and Q131 in the stretch connecting the hydrophobic parts of helices II and III.

between the cytochrome *b* and the catalytic subunit which carries a molybdenum ion coordinated by two MGD molecules and a [4Fe – 4S] center (Fig. 2). The prosthetic groups of the enzyme form a pathway for rapid electron transfer from the molybdenum ion to the distal heme group which is thought to be close to the site of quinone reduction (see Fig. 5). As in other molybdenum enzymes, the substrate formate reaches the active site near the molybdenum through a cavity extending from the molybdenum to the surface of the subunit. After oxidation of formate according to reaction (*k*),

$$HCO_2^- \rightarrow CO_2 + H^+ + 2e^- \qquad (k)$$

the products (CO_2 and H^+) leave the substrate site through the cavity. Since the catalytic subunit FdhA of formate dehydrogenase is exposed to the periplasmic side of the membrane, formate is taken up from the periplasm and the products are released to the same side of the membrane (Kröger et al., 1980). The electrons (reaction *k*) are thought to be passed to the cytochrome *b* subunit by the iron-sulfur centers of FdhA and FdhB.

3. Hydrogenase

Hydrogenase consists of the three subunits encoded by the *hydABC* operon (Droß et al., 1992). The enzyme is the same hydrogenase involved in fumarate respiration with H_2, where it catalyzes the reduction of menaquinone by H_2 (Kröger et al., 2002; Biel et al., 2002). The two hydrophilic subunits HydA and B of the enzyme (Fig. 2) are similar to those making up the periplasmic Ni-hydrogenases of two *Desulfovibrio* species, the structures of which are known (Volbeda et al., 1995; Higuchi et al., 1997). At the active site of these enzymes, H_2 is split to yield protons and electrons. The protons are released via a proton pathway extending from the active site to the surface of the larger catalytic subunit. The electrons are guided by three consecutive iron-sulfur centers to the binding site of the electron acceptor, a cytochrome *c*, on the surface of the smaller hydrophilic subunit. Since all the relevant residues are also conserved in HydA and B of *W. succinogenes*, it is likely that the catalytic mechanism also applies here. In the *W. succinogenes* enzyme the membrane-integrated di-heme cytochrome *b* subunit HydC serves as the acceptor of electrons from HydA to which HydC is tightly bound (Fig. 2). The midpoint potentials of the two heme groups of HydC were determined as -240 mV and -100 mV relative to the standard hydrogen electrode (Gross et al., 1998b). In contrast to the periplasmic hydrogenases, HydA of *W. succinogenes* carries a C-terminal hydrophobic helix which is integrated in the membrane (Gross et al., 1998a). The protons generated by H_2 oxidation are released into the periplasm, since the catalytic subunit HydB is oriented to the periplasmic side of the membrane.

The sequences of the di-heme cytochrome *b* subunits of hydrogenase (HydC) and of formate dehydrogenase (FdhC) of *W. succinogenes* are similar to that of formate dehydrogenase-N of *E. coli* (Jormakka et al., 2002). Notably, the axial ligands of the two heme groups in subunit FdnI of the *E. coli* enzyme are conserved in HydC and FdhC (Gross et al., 1998b; Biel et al., 2002). Therefore, the structures of the di-heme cytochrome *b* subunits of the *W. succinogenes* enzymes are thought to be similar to the known structure of *E. coli* FdnI. In Fig. 5 *W. succinogenes* HydC is drawn according to the structure of *E. coli* FdnI. The proximal heme group of FdnI is within electron transfer distance to an iron-sulfur center of FdnH and to the distal heme group. The proximal and the distal heme groups represent the upper and the lower shaded squares in Fig. 5. The site of quinone reduction on FdnI seems to be occupied by a molecule of HQNO which is located close to the distal heme group. HQNO is in close proximity to the axial heme ligand on helix IV and to three residues (N110, G112 and Q113 of FdnI) within the hydrophilic stretch connecting helices II and III. Another two residues (M172 and A173 of FdnI) at the cytoplasmic end of helix IV make contact with HQNO.

HQNO is known to inhibit menaquinone reduction by formate or H_2 in *W. succinogenes* (Kröger and Innerhofer, 1976; Unden and Kröger, 1982; Schröder et al., 1985). The midpoint potential of at least one heme *b* group in the membrane fraction of *W. succinogenes* was shifted from –190 mV to –230 mM upon the addition of HQNO (Kröger et al., 1979). Therefore, HQNO probably interacts at the site of quinone reduction of hydrogenase and of formate dehydrogenase of *W. succinogenes*.

To determine whether the site of quinone reduction on HydC is situated as in *E. coli* FdnI, mutants were constructed in which each of three residues (H122, N128 and Q131) of *W. succinogenes* HydC were replaced. These residues are predicted to be located in the loop connecting helix II and III (Fig. 5). Residues N110 and Q113 of FdnI which are in contact with HQNO, correspond to N128 and Q131 of HydC, and are also conserved in *W. succinogenes* FdhC. Mutants N128D and Q131L did not grow by anaerobic respiration with fumarate or polysulfide and H_2 as electron donor (not shown). The membrane fraction of mutants N128D and Q131L grown with formate and fumarate catalyzed benzyl viologen reduction by H_2 with wild-type specific activities, whereas the activities of DMN or polysulfide reduction by H_2 were drastically inhibited (Table 4). Replacement of H122, which is not conserved, did not affect the enzymic activities. The heme groups of HydC were present in the mutants, and their reduction by H_2 was not impaired by the mutations. This is seen from the amount of heme *b* reduced by H_2 after the addition of Triton X-100 and fumarate to the membrane fraction of the mutants.

Residues M203 and A204 at the cytoplasmic end of helix IV of HydC are conserved in FdhC and in *E. coli* FdnI where they make contact with HQNO (Jormakka et al., 2002). The corresponding *W. succinogenes* mutants M203I and A204F were found to have wild-type properties. In contrast, mutant Y202F showed negligible activities of DMN or polysulfide reduction and wild-type activity of benzyl viologen reduction by H_2. However, heme *b* was not reduced by H_2 in this mutant, although the HydC protein was detected by ELISA. This suggests that electron transfer from HydB to HydC was interrupted in mutant Y202F. This view is supported by the structure of *E. coli* formate dehydrogenase-N. The tyrosine residue corresponding to Y202 is seen to be bound to a conserved histidine residue in the C-terminal membrane helix of the iron-sulfur subunit. A mutant (H305M) with the corresponding histidine residue of *W. succinogenes* HydA replaced was found to have the same properties as Y202F (Gross et al., 1998b). This result indicates that Y202 of HydC and H305 of HydA serve the same function as the corresponding residues in *E. coli* formate dehydrogenase-N. Furthermore, tight binding of Y202 to H305 appears to be required for electron transfer from HydA to HydC which in turn is a prerequisite for quinone and polysulfide reduction.

The results of Table 4 suggest that the site of quinone reduction on HydC is located close to the cytoplasmic membrane surface as on *E. coli* FdhI. Furthermore, HydC and its site of quinone reduc-

Table 4. Enzymatic activities at 37 °C of *W. succinogenes hydC* mutants grown with formate and fumarate. The measurements were performed as described by Gross et al. (1998b).

Strain	$H_2 \rightarrow$ Polysulfide	$H_2 \rightarrow$ DMN	$H_2 \rightarrow$ Benzyl viologen	Heme *b* reduced by H_2
	(U · mg^{-1} cell protein)			(µmol · g^{-1} membrane protein)
Wild-type	1.4	4.5	2.1	0.35
H122A	1.8	4.5	2.3	0.33
N128D	≤0.05	0.25	2.5	0.29
Q131L	≤0.05	0.08	2.0	0.31
Y202F	≤0.05	≤0.01	2.2	≤0.02
M203 I	1.7	4.9	2.3	0.35
A204F	1.2	3.8	2.2	0.35

tion appear to be involved in electron transfer from hydrogenase to polysulfide reductase. The function of FdhC of *W. succinogenes* formate dehydrogenase in quinone reduction and in electron transfer to polysulfide reductase is thought to be equivalent to that of HydC.

C. Electron Transfer from Hydrogenase and Formate Dehydrogenase to Polysulfide Reductase

Proteoliposomes containing polysulfide reductase and either hydrogenase or formate dehydrogenase isolated from *W. succinogenes* do not catalyze polysulfide respiration unless methyl-menaquinone is present (Table 3). Menaquinone with a side chain consisting of six or four isoprene units, or vitamin K_1 served in reconstituting fumarate respiration, but did not replace methyl-menaquinone in polysulfide respiration. The low activities of polysulfide respiration observed without added methyl-menaquinone were probably due to the small amounts of this quinone associated with the enzyme preparations used. Maximum activity of polysulfide respiration required about 10 µmol methyl-menaquinone per gram of phospholipid (not shown).

Most of the methyl-menaquinone of *W. succinogenes* is probably dissolved in the lipid phase of the bacterial membrane. The redox potential of methyl-menaquinone dissolved in the membrane was estimated to be 170 mV more positive than that of polysulfide (Table 1) (Dietrich and Klimmek, 2002). Therefore, it is unlikely that electron transfer from the dehydrogenases to polysulfide reductase is mediated by diffusion of methyl-menaquinone in the membrane. The methyl-menaquinone involved in polysulfide respiration is thought to be bound to PsrC. In the reduced state, the bound quinone (MM_b) is thought to form the hydroquinone anion (MM_bH^-). This assumption is consistent with the finding that an arginine residue (R305) in one of the membrane helices of PsrC is essential for polysulfide respiration (see Fig. 7). Mutants with the arginine residue replaced by phenylalanine or lysine had negligible activities of polysulfide respiration with H_2 or formate, whereas the activities of polysulfide reductase (reactions *h* and *i*) were close to those of the wild-type strain, and PsrA was bound to the membrane of the mutants (Dietrich and Klimmek, 2002). The hydroquinone anion MM_bH^- is probably bound and stabilized by the positive charge of arginine. The redox potential of the MM_b/MM_bH^- couple is expected to be close to that of polysulfide or more negative.

Since electron transfer from the dehydrogenases to polysulfide reductase by diffusion of methyl-menaquinone is unlikely, two alternative mechanisms have to be considered. The dehydrogenases may either form a stable electron transport complex with polysulfide reductase, or electron transfer is achieved by diffusion and collision of the enzymes within the membrane. The latter mechanism is supported by experimental evidence (Fig. 6). When equimolar amounts of a dehydrogenase and polysulfide reductase are incorporated into liposomes containing methyl-menaquinone, the activity of electron transport increases in proportion to the protein / phospholipid ratio. The same linear relationship was observed when the membrane fraction of *W. succinogenes* was fused with increasing amounts of liposomes containing methyl-menaquinone (Hedderich et al., 1999). The linear relationship argues against the existence of stable electron transport complexes within the membrane, and cannot be explained by complex dissociation upon dilution with phospholipid. Complex dissociation would result in a hyperbolic rather than a linear relationship of electron transport activity to the protein/phospholipid ratio according to Ostwalds law of dilution. The linear relationship suggests that the activity of electron

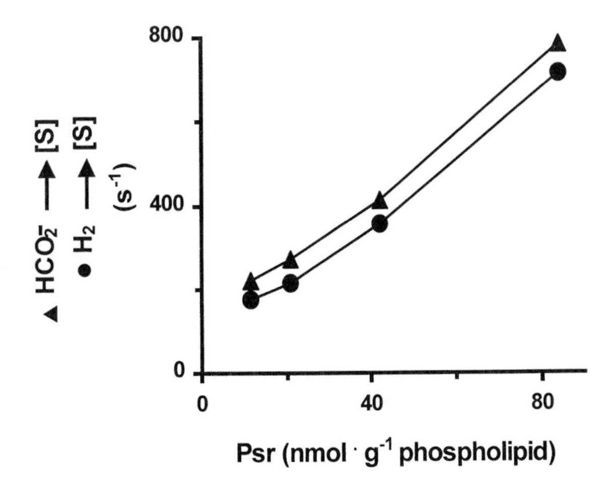

Fig. 6. Activity of polysulfide reduction by H_2 ($H_2 \rightarrow$ [S]) or formate ($HCO_2^- \rightarrow$ [S]) in proteoliposomes at 37°C. The proteoliposomes contained methyl-menaquinone (10 μmol · g^{-1} phospholipid) and approximately the same molar amount of hydrogenase or formate dehydrogenase as polysulfide reductase. The activities were determined as described by Dietrich and Klimmek (2002).

transport is limited by the diffusion of the enzymes within the membrane. The diffusion coefficient (D) evaluated from Fig. 6 according to Eq. (2) is approximately 10^{-8} $cm^2 \cdot s^{-1}$, in agreement with direct measurements of the diffusion

$$D = d^2/t \tag{2}$$

coefficients of membrane proteins of similar sizes (Chazotte and Hackenbrock, 1988). The average membrane surface occupied by an enzyme molecule (d^2) was calculated from the protein / phospholipid ratio, and the turnover numbers of the enzyme in electron transport was used as 1 / t. Thus the experiment of Fig. 6 suggests that the electron transfer from the dehydrogenases to polysulfide reductase is achieved by diffusion and collision of the enzyme molecules within the membrane. This view is supported by the finding that the activity of fumarate respiration is not decreased by enzyme dilution with phospholipid. In this case, the electron transfer from the dehydrogenases to fumarate reductase is mediated by diffusion of menaquinone. The diffusion of menaquinone is two orders of magnitude faster than that of the enzymes and, therefore, does not limit electron transport.

PsrC is predicted to form eight membrane helices. The nineteen residues of PsrC which were replaced by others are indicted in Fig. 7. Active polysulfide reductase was bound to the membrane in all mutants, but six of them had negligible electron transport activity with H_2 or formate and polysulfide. The corresponding residues (in bold type in Fig. 7) are either charged at neutral pH or are tyrosines, the phenolic hydroxyl groups of which appear to be essential for polysulfide respiration. Y23, Y159, and R305 may serve in binding MM_b and MM_bH^- to PsrC (Fig. 8). D218 and E225 are well suited for release of the proton formed by MM_bH^- oxidation into the periplasmic

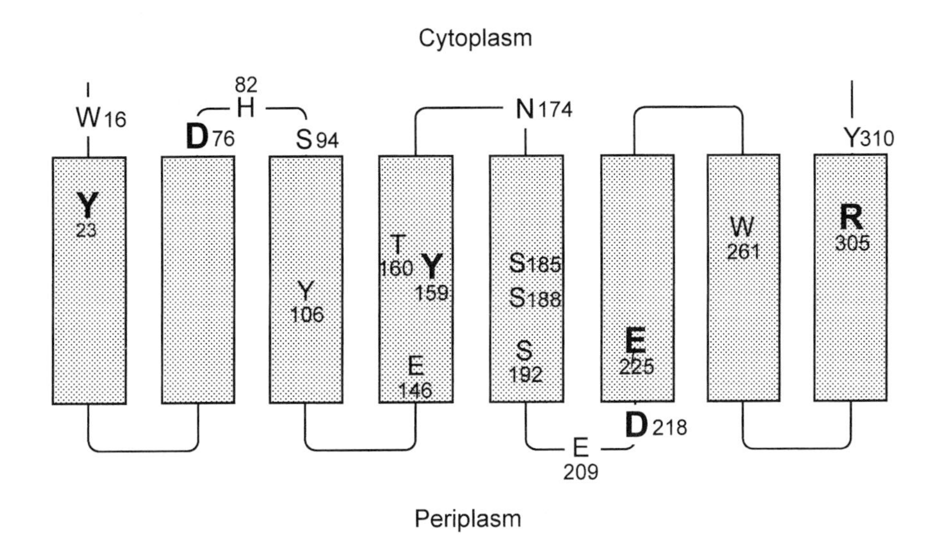

Fig. 7. Hypothetical topology of PsrC (Dietrich and Klimmek, 2002). Residues replaced in PsrC by site-directed mutagenesis are indicated. Residues in bold letters correspond to mutants with 5% or less of the wild-type specific activities of polysulfide respiration with H_2 or formate.

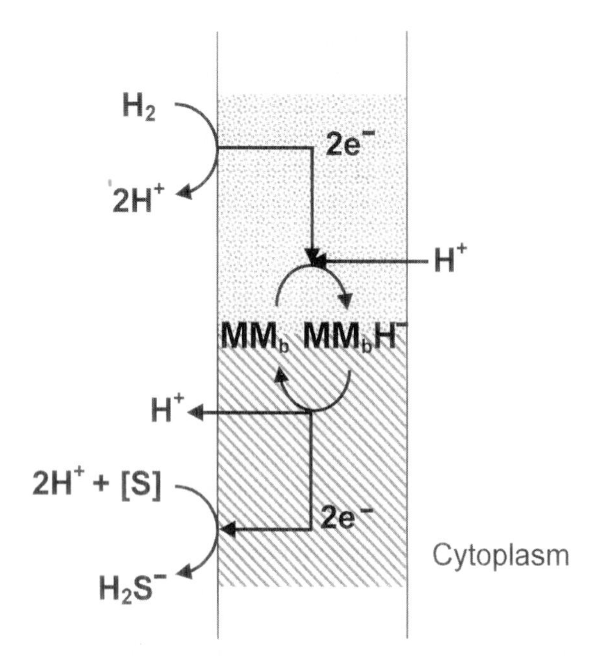

Fig. 8. Hypothetical mechanism of polysulfide respiration with H_2 (Dietrich and Klimmek, 2002). The reduction of methyl-menaquinone bound to PsrC (MM_b) is coupled to the uptake of a proton from the cytoplasmic side of the membrane. The oxidation of the hydroquinone anion of MM_b (MM_bH^-) by polysulfide is coupled to the release of a proton to the periplasmic side. The dotted and the striped areas designate HydC and PsrC, respectively.

aqueous phase.

The activity of polysulfide respiration in cells of *W. succinogenes* or in proteoliposomes is decreased to about 30% upon the addition of a protonophore (Dietrich and Klimmek, 2002). This effect can be explained if it is assumed that the activity is limited by the amount of MM_bH^- in the absence of a Δp, and that MM_bH^- dissociates from PsrC (C) according to reaction (*1*), where H_i^+ designates a proton

$$MM_bH^- + H_i^+ \leftrightarrows MMH_2 + C \qquad (1)$$

taken up from the cytoplasmic side of the membrane and MMH_2 reduced methyl-menaquinone dissolved in the lipid phase of the membrane. In the presence of a Δp (negative inside), MM_bH^- formation from MMH_2 is favored according to reaction (*1*). It is feasible that D76 which is located near the cytoplasmic end of a PsrC membrane helix (Fig. 7) is involved in the uptake and release of H_i^+. Replacement of D76 by aspargine or leucine resulted in mutants with negligible activities of polysulfide respiration, possibly because the formation of MM_bH^- from MMH_2 is prevented. The view that electron transport is limited by MM_bH^- in the absence of a Δp is supported by the finding that the activity of polysulfide reduction by H_2 is stimulated twofold by the incorporation of additional methyl-menaquinone into the membrane fraction of *W. succinogenes* (Dietrich and Klimmek, 2002). Considering its redox potential (Table 1), the methyl-menaquinone dissolved in the membrane should be fully reduced in the steady state of polysulfide respiration. Therefore, the amount of MM_bH^- is predicted to increase with the quinone content of the membrane according to reaction (*1*).

D. The Coupling Mechanism

The electron transfer from HydC of hydrogenase to methyl-menaquinone bound to PsrC (MM_b) at the moment of collision of the two enzymes is illustrated in Fig. 8. As depicted the Δp is thought to be generated by proton translocation across the membrane which is coupled to the redox reactions of MM_b. The same mechanism is thought to apply when formate dehydrogenase replaces hydrogenase. The proton consumed in MM_b reduction is taken up from the cytoplasmic side of the membrane, and the proton liberated by MM_bH^- oxidation is released on the periplasmic side. Proton uptake is envisaged to be accomplished by HydC and proton release by PsrC. D218 and E225 of PsrC (Fig. 7) possibly serve in the release of the proton formed by MM_bH^- oxidation to the periplasmic side. Mutants D218N and E225Q did not catalyze polysulfide respiration with H_2 or formate, whereas mutant E225D showed 25% of the wild-type specific activity. The H^+/e^- ratio of polysulfide respiration with H_2 (or formate) predicted by the mechanism (0.5) is consistent with the value given in Table 2. According to this mechanism, Δp is exclusively generated by MM_b reduction with H_2 (Fig. 8). In this process two protons are released on the periplasmic side from H_2 oxidation, and one proton disappears from the cytoplasmic side for MM_b reduction. This is consistent with the location of the site of quinone reduction on HydC close to the cytoplasmic membrane surface (Fig. 5). Furthermore, menaquinone or DMN reduction by H_2 catalyzed by hydrogenase in the membrane of *W. succinogenes* and in proteoliposomes was found to generate a Δp (Kröger et al., 2002; Biel et al., 2002), and the site of MM_b reduction is suggested to be close to that of menaquinone or DMN reduction (Table 4).

MM_bH^- oxidation by polysulfide dissipates part

of the Δp generated by MM_b reduction with H_2 according to the mechanism (Fig. 8), since one proton per two electrons disappears from the periplasmic (positive) side in this process. Hence, MM_bH^- oxidation by polysulfide is predicted to be driven by a Δp across the membrane. The view that Δp generation is coupled solely to MM_b reduction is consistent with the energetics of the two reactions (Dietrich and Klimmek, 2002). Only MM_b reduction is considered to be sufficiently exergonic to drive the translocation of a proton against a Δp of 0.17 Volt. In contrast, the oxidation of MM_bH^- might be slightly endergonic under standard conditions.

E. The Function of Sud Protein

Growth and survival by polysulfide respiration may be critical for bacteria living at low pH and/or low sulfide concentration, since the concentration of polysulfide may become very low under these conditions (reaction *c*). Therefore, it is not surprising that sulfur reducers synthesize a binding protein for polysulfide sulfur which is designed to allow rapid polysulfide respiration at low polysulfide concentration. The periplasmic Sud protein of *W. succinogenes* appears to serve as such a binding protein (Klimmek et al., 1998, 1999). Sud is formed by the bacterium when grown with polysulfide and is nearly absent upon growth with fumarate. The apparent K_M for polysulfide sulfur is 70 μM in polysulfide reduction by H_2 (reaction *e*) catalyzed by bacteria grown with fumarate, and is 10 μM with polysulfide-grown bacteria. A similar decrease of K_M was observed upon the addition of Sud protein to the bacterial membrane fraction catalyzing polysulfide reduction.

Sud consists of two identical subunits (14.3 kDa) and does not contain prosthetic groups or heavy metal ions. Sud catalyzes the transfer of sulfur from polysulfide to cyanide (reaction *m*) at a high turnover number (10^4 s^{-1} at 37°C).

$$S_n^{2-} + CN^- \rightarrow SCN^- + S_{n-1}^{2-} \qquad \text{(m)}$$

Sud binds up to 10 mol sulfur per mol subunit when incubated in a polysulfide solution. In solutions containing less than 0.1 mM polysulfide the dissociation constant of sulfur bound to Sud was 10 μM or less. Under these conditions, the decrease of K_M with Sud was observed in polysulfide reduction. At higher concentrations of polysulfide, the dissociation constant was 0.2 mM and no effect of Sud on the K_M for polysulfide was observed.

Sud contains a single cysteine residue which is essential for sulfur transferase activity (reaction *m*), for sulfur binding, and for the decrease of K_M for polysulfide in polysulfide reduction (Klimmek et al., 1999). Sulfur appears to be covalently bound to the cysteine residue. When subjected to MALDI mass spectrometry, Sud incubated with polysulfide was found to carry one or two sulfur atoms per monomer. No sulfur was bound to the monomer after treatment of Sud with cyanide. A Sud variant carrying a serine instead of the cysteine residue did not carry sulfur even upon incubation with polysulfide. To explain the effect of Sud on the apparent K_M for polysulfide in polysulfide reduction, it is assumed that sulfur is transferred from Sud to the active site of polysulfide reductase. From the concentration of Sud required for K_M decrease and from its concentration in the bacterial periplasm it is likely that Sud is bound to polysulfide reductase during sulfur transfer. The two enzymes occur in about equimolar amounts in cells of *W. succinogenes* grown with polysulfide.

The structure of the Sud dimer was determined by NMR spectroscopy (Lin et al., 2000; Löhr et al., 2000) (Fig. 9A). The two cysteine residues (C89), one above (left side) and one below the paper plane (right side), are highlighted. A view of the left monomer (Fig. 9A) from its left side and along the paper plane is presented in Fig. 9B. The monomer has an α/β topology with six α-helices packing against a central core of five parallel β-strands. The dimer interface consists of a four helix bundle around the rotational symmetry axis. There is no indication of cooperative interaction of the two cysteine residues which are 30 Å apart. The interaction of Sud with polysulfide reductase was investigated comparing the [^{1}H, ^{15}N] – TROSY spectra of Sud and of an equimolar mixture of Sud dimer and polysulfide reductase in the presence of polysulfide at low concentration (0.1 mM). The resonances of eight residues (D47, D49, M54, S66, R67, G115, D118 and K119) of Sud were found to be shifted by the presence of polysulfide reductase. These residues, which form a ring around the cysteine residue, are thought to make contact with polysulfide reductase. In the functional complex consisting of Sud and polysulfide reductase, the end of the sulfur chain bound to the cysteine residue of Sud may come close to the molybdenum where a sulfur atom is reductively cleaved according to the mechanism depicted in Fig. 4. This view was supported by the response of the EPR signal of the

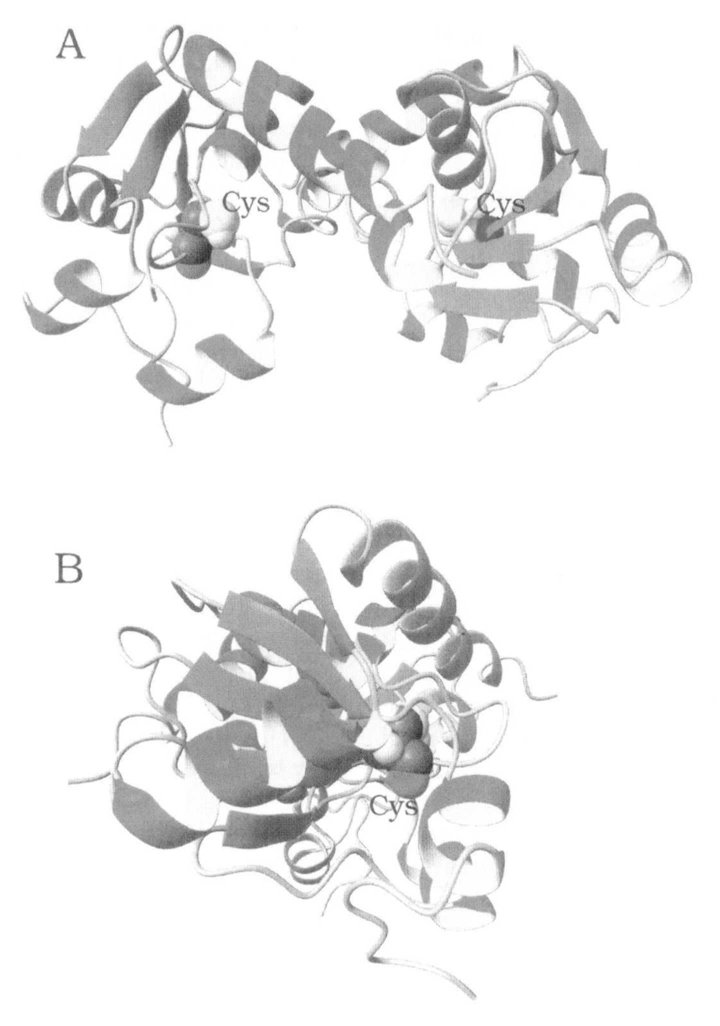

Fig. 9. Structure of Sud. Ribbon representations of Sud dimer drawn with program MOLMOL (Koradi et al., 1996). The A and B views are related by a 90° rotation around the symmetry axis. The cysteine residues were depicted using a CPK model (spheres with WdV radius) (See Color Plate 2 for details).

molybdenum (+5) ion in polysulfide reductase which indicated a slight change in the geometry of the ligand sphere of molybdenum when Sud was added in the presence of polysulfide at low concentration (Prisner et al., 2003). This effect of Sud was not observed in the absence of polysulfide or with polysulfide present at a high concentration.

Acknowledgments

The work was supported by grants from the Deutsche Forschungsgemeinschaft (SFB 472) and the Fonds der Chemischen Industrie. The authors are indebted to Mrs. Monica Sänger for editing the manuscript.

References

Biel S, Simon J, Gross R, Ruiz T, Ruitenberg M and Kröger A (2002) Reconstitution of coupled fumarate respiration in liposomes by incorporating the electron transport enzymes isolated from *Wolinella succinogenes*. Eur J Biochem 269: 1974–1983

Bokranz M, Gutmann M, Körtner C, Kojro E, Fahrenholz F, Lauterbach F and Kröger A (1991) Cloning and nucleotide sequence of the structural genes encoding the formate dehydrogenase of *Wolinella succinogenes*. Arch Microbiol 156: 119–128

Boulègue J (1978) Solubility of elemental sulfur in water at 298 K. Phosphorus Sulfur 5: 127–128

Chazotte B and Hackenbrock CR (1988) The multicollisional, obstructed, longs-range diffusional nature of mitochondrial electron transport. J Biol Chem 28: 14359–14367

Clark WM (1960) Oxidation-reduction potentials of organic

systems. Williams & Wilkins, Baltimore

Collins MD and Fernandez F (1984) Menaquinone-6 and thermoplasmaquinone-6 in *Wolinella succinogenes*. FEMS 22: 273–276

Dietrich W and Klimmek O (2002) The function of methyl-menaquinone-6 and PsrC in polysulfide respiration of *Wolinella succinogenes*. Eur J Biochem 296: 1086–1095

Droß F, Geisler V, Lenger R, Theis F, Krafft T, Fahrenholz F, Kojro E, Duchêne A, Tripier D, Juvenal K and Kröger A (1992) The quinone-reactive Ni/Fe-hydrogenase of *Wolinella succinogenes*. Eur J Biochem 206: 93–102

Fauque G, Klimmek O and Kröger A (1994) Sulfur reductases from mesophilic sulfur-reducing eubacteria. Meth Enzymol 243: 367–383

Giggenbach W (1972) Optical spectra and equilibrium distribution of polysulfide ions in aqueous solution at 20°. Inorg Chem 11: 1201–1207

Gross R, Simon J, Theis F and Kröger A (1998a) Two membrane anchors of *Wolinella succinogenes* hydrogenase and their function in fumarate and polysulfide respiration. Arch Microbiol 170: 50–58

Gross R, Simon J, Lancaster CRD and Kröger A (1998b) Identification of histidine residues in *Wolinella succinogenes* hydrogenase that are essential for menaquinone reduction by H_2. Mol Microbiol 30: 639–646

Hedderich R, Klimmek O, Kröger A, Dirmeier R, Keller M and Stetter KO (1999) Anaerobic respiration with sulfur and with organic disulfides. FEMS Microbiol Rev 22: 353–381

Higuchi Y., Tatsuhiko Y. and Yasuoka N. (1997) Unusual ligand structure in Ni-Fe active center and an additional Mg site in hydrogenase revealed by high resolution X-ray structure analysis. Structure 5: 1671–1680.

Jankielewicz A, Schmitz RA, Klimmek O and Kröger A (1994) Polysulfide reductase and formate dehydrogenase from *Wolinella succinogenes* contain molybdopterin guanine dinucleotide. Arch Microbiol 162: 238–242

Jormakka M, Tornroth S, Byrne B and Iwata S (2002) Molecular basis of proton motive force generation: Structure of formate dehydrogenase-N. Science 295: 1863–1868

Kisker C, Schindelin H and Rees DC (1997) Molybdenum-cofactor-containing enzymes: Structure and mechanism. Annu Rev Biochem 66: 233–267

Klimmek O (1996) Aufbau und Funktion der Polysulfid-Reduktase von *Wolinella succinogenes*. Doctoral (Ph D) Thesis, JW Goethe-Universität Frankfurt, FB Biologie

Klimmek O, Kröger, Steudel R and Holdt G (1991) Growth of *Wolinella succinogenes* with polysulphide as terminal acceptor of phosphorylative electron transport. Arch Microbiol 155: 177–182

Klimmek O, Kreis V, Klein C, Simon J, Wittershagen A and Kröger A (1998) The function of the periplasmic Sud protein in polysulfide respiration of *Wolinella succinogenes*. Eur J Biochem 253: 263–269

Klimmek O, Stein T, Pisa R, Simon J, Kröger A (1999) The single cysteine residue of the Sud protein is required for its function as a polysulfide-sulfur transferase in *Wolinella succinogenes*. Eur J Biochem 263: 79–84

Koradi R, Billeter M, Wüthrich K (1996) MOLMOL: A program for display and analysis of macromolecular structures. J Mol Graphics 14: 51–55

Krafft T, Bokranz M, Klimmek O, Schröder I, Fahrenholz F, Kojro E and Kröger A (1992) Cloning and nucleotide sequence of the *psrA* gene of *Wolinella succinogenes* polysulphide reductase. Eur J Biochem 206: 503–510

Krafft T, Gross R and Kröger A (1995) The function of *Wolinella succinogenes psr* genes in electron transport with polysulphide as the terminal electron acceptor. Eur J Biochem 230: 601–606

Kröger A and Innerhofer A (1976) The function of menaquinone, covalently bound FAD and iron-sulfur protein in the electron transport from formate to fumarate of *Vibrio succinogenes*. Eur J Biochem 69: 487–495

Kröger A, Winkler E, Innerhofer A, Hackenberg H and Shägger H (1979) The formate dehydrogenase involved in electron transport from formate to fumarate in *Vibrio succinogenes*. Eur J Biochem 94:465–475

Kröger A, Dorrer E and Winkler E (1980) The orientation of the substrate sites of formate dehydrogenase and fumarate reductase in the membrane of *Vibrio succinogenes*. Biochim Biophys Acta 589: 119–136

Kröger A, Biel S, Simon J, Gross R, Unden G and Lancaster CRD (2002) Fumarate respiration of *Wolinella succinogenes*: Enzymology, energetics, and coupling mechanism. Biochim Biophys Acta 1553: 23–38

Lenger R, Herrmann U, Gross R, Simon J and Kröger A (1997) Structure and function of a second gene cluster encoding the formate dehydrogenase of *Wolinella succinogenes*. Eur J Biochem 246: 646–651

Lin Y-J, Pfeiffer S, Löhr F, Klimmek O and Rüterjans H (2000) Backbone resonance assignment and secondary structure of the 30 kDa Sud dimer from *Wolinella succinogenes*. J Biomol NMR 18: 285–286

Löhr F, Pfeiffer S, Lin YJ, Hartleib J, Klimmek O and Rüterjans H (2000) HNCAN pulse sequence for sequential backbone resonance assignment across proline residues in perdenterated proteins. J Biomol NMR 18: 337–346

Macy JM, Schöder I, Thauer RK and Kröger A (1986) Growth of *Wolinella succinogenes* on H_2S plus fumarate and on formate plus sulfur as energy sources. Arch Microbiol 144: 147–150

Prisner T, Lyubenova S, Atabay Y, MacMillan F and Klimmek O (2003) Multifrequency cw-EPR investigation of the catalytic molybdenum cofactor of *Wolinella succinogenes* polysulfide reductase. J Biol Inorg Chem 8:419–426

Richardson DJ (2000) Bacterial respiration: A flexible process for a changing environment. Microbiology 146: 551–571

Rigaud JL, Pitard B and Levy D (1995) Reconstitution of membrane proteins into liposomes: Application to energy-transducing membrane proteins. Biochim Biophys Acta 1231: 223–246

Schauder R and Kröger A (1993) Bacterial sulphur respiration. Arch Microbiol 159: 491–497

Schauder R and Müller E (1993) Polysulfide as a possible substrate for sulfur-reducing bacteria. Arch Microbiol 160: 377–382

Schröder I, Roberton AM, Bokranz M, Unden G, Böcher R and Kröger A (1985) The membraneous nitrite reductase involved in the electron transport of *Wolinella succinogenes*. Arch Microbiol 140: 380–386

Schwarzenbach G and Fischer A (1960) Die Acidität der Sulfane und die Zusammensetzung wässeriger Polysulfidlösungen. Helv Chim Acta 43: 1365–1388

Simon J, Gross R, Klimmek O and Kröger A (2000) The Genus *Wolinella*. In: M. Dworkin et al (eds) The Prokaryotes: An evolving electronic resource for the microbiology community, 3rd edition, Springer-Verlag, New York

Thauer R, Jungermann K and Decker K (1977) Energy conservation in chemotrophic anaerobic bacteria. Bacteriol Rev 41: 100–180

Unden G and Bongaerts J (1997) Alternative respiratory pathways of *Escherichia coli*: energetics and transcriptional regulation in response to electron acceptors. Biochim Biophys Acta 1320: 217–234

Unden G and Kröger A (1982) Reconstitution in liposomes of the electron transport chain catalyzing fumarate reduction by formate. Biochim Biophys Acta 682: 258–263

Volbeda A, Charon MH, Piras C, Hatchikian EC, Frey M and Fontecilla-Camps JC (1995) Crystal structure of the nickel-iron hydrogenase from *Desulfovibrio gigas*. Nature 373: 580–587

Wloczyk C, Kröger A, Göbel T, Holdt G and Steudel R (1989) The electrochemical proton potential generated by the sulphur respiration of *Wolinella succinogenes*. Arch Microbiol 152: 600–605

Chapter 11

Hydrogen Respiration

Paulette M. Vignais*, John C. Willison and Annette Colbeau
UMR CNRS/CEA/UJF n° 5092, CEA/Grenoble, Laboratoire de Biochimie et Biophysique des Systèmes Intégrés, Département de Réponse et Dynamique Cellulaires, 17 Avenue des Martyrs, 38054 Grenoble cedex 9, France

Summary

Hydrogen respiration can be considered either as (i) the oxidation of H_2 to H^+, with the electrons released being channeled into a membrane-bound, respiratory electron transport chain or (ii) as the reduction of H^+ to H_2 in the terminal reaction of an anaerobic, low-potential electron transport system. In both cases, the redox reaction involving H_2 is catalyzed by a hydrogenase enzyme (H_2ase) and electron transport to or from H_2 is coupled to the vectorial translocation of H^+ across a membrane, leading to the conservation of energy in the form of a protonmotive force.

H_2ases are a diverse group of enzymes, and can be classified into 3 groups according to their metal content, i.e. [NiFe]-H_2ases, [Fe]-H_2ases and iron-sulfur cluster-free H_2ases. The [NiFe]-H_2ases can be subdivided into 4

*Author for correspondence, email: paulette.vignais@cea.fr

Davide Zannoni (ed): Respiration in Archaea and Bacteria. Vol 2: Diversity of Prokaryotic Respiratory Systems, pp. 233–260.

subgroups: (1) the uptake [NiFe]-H_2ases; (2) the cytoplasmic H_2 sensors and the cyanobacterial uptake [NiFe]-H_2ases; (3) the bidirectional cytoplasmic [NiFe]-H_2ases; (4) the H_2-evolving, energy-converting [NiFe]-H_2ases. Unlike the [NiFe]-H_2ases, the [Fe]-H_2ases form a homogeneous group and are primarily involved in H_2 evolution. The classification of H_2ases on the basis of phylogenetic analysis, the correlation of the different phylogenetic groupings with H_2ase function, the possible role of the [Fe]-H_2ases in the genesis of the eukaryotic cell, and the structural and functional relationships of H_2ases subunits with those of Complex I of the respiratory electron transport chain, are all discussed.

The diversity of H_2ases may seem surprising in view of the simplicity of the reaction catalyzed ($H_2 \leftrightarrow 2H^+ + 2e^-$) but can be understood in terms of the wide range of physiological roles that they play. In addition to the direct roles in respiration described above, work in recent years has revealed that H_2ases may be involved in regulating gene expression, e.g. by acting as H_2-sensors, or may interact directly with membrane-bound electron transport systems in order to maintain redox poise, particularly in some photosynthetic micro-organisms, such as cyanobacteria and green algae. In addition, some H_2-evolving H_2ases thought to be involved in purely fermentative processes have been shown to play a role in membrane-linked energy conservation, through the generation of a protonmotive force.

This diversity of metabolic roles is reflected in a diversity of regulatory mechanisms, and our current knowledge of H_2ases gene regulation is extensively described. Most studies of regulation have been carried out on members of the *Proteobacteria*, in which H_2ases genes are frequently clustered, but these studies are now being extended to members of the *Archaea*. The purpose of this chapter is to emphasize recent advances that have greatly increased our knowledge of H_2ases and H_2ases function, to underline the importance of H_2 metabolism in both prokaryotic and eukaryotic organisms, and to point out that, nevertheless, much remains to be discovered about the structure, function and regulation of this interesting class of enzymes.

I. Introduction

The title of this chapter, H_2 respiration, needs some explanation. Respiration was initially understood as the vital process that sustains life in the presence of O_2, as opposed to fermentation which sustains life in the absence of O_2. It is now known that respiration, which involves electron transfer between redox components of a respiratory chain located in a membrane and is coupled to vectorial proton transfer across the membrane, can also occur in the absence of O_2 in the presence of alternative electron acceptors such as nitrate, fumarate or sulfate. We can therefore distinguish *aerobic* respiration, which uses O_2 as terminal electron acceptor, and *anaerobic* respiration, in which other compounds are used, with the type of respiration being referred to more specifically as nitrate respiration, fumarate respiration or sulfate respiration, depending on the electron acceptor used. In all cases, respiration allows the recovery of the energy, released by nutrient oxidation, in the form of protonmotive force. The term H_2 respiration, on the other hand, may be understood in two different ways, depending on whether H_2 is the electron donor to the electron transport chain or the product of the terminal redox reaction. In the former case, H_2 is respired by microorganisms possessing a so-called uptake hydrogenase (H_2ase), with electrons from H_2 being transferred via a respiratory chain either to O_2 or to an alternative electron acceptor. In the latter case, protons are used as the final electron acceptors in the oxidation of low potential substrates, such as formate and carbon monoxide, under strictly anaerobic conditions, with H_2 evolution by H_2ase being coupled to energy conservation in the form of an electrochemical proton gradient. Strictly, this process, by analogy to other forms of anaerobic respiration, should be referred to as proton respiration (an awkward expression!).

In addition to the respiratory H_2ases, some H_2ases participate in purely fermentative processes and are therefore excluded from the present analysis. On the other hand, some of these 'fermentative' enzymes have recently been shown to be involved in membrane-linked energy conservation, and this novel

Abbreviations: Cyt – cytochrome; DMK – demethylmenaquinone; FAD – flavine adenine nucleotide; Fdh – formate dehydrogenase; [Fe]-H_2ases – iron-only containing hydrogenases; [FeS] – iron-sulfur cluster; FHL – formate hydrogenase; FMN – flavine mononucleotide; FNR – ferredoxin $NAD(P)^+$-oxidoreductase; H_2ase – uptake hydrogenase enzyme; [NiFe]-H_2ases – nickel-iron containing hydrogenases; MK – menaquinone; PS – photosynthetic system; SDH – succinate dehydrogenase; TCA cycle – tricarboxylic acid cycle; UQ – ubiquinone; UQH_2 – reduced ubiquinone or ubiquinol; Δp – proton motive force

aspect will be emphasized in the present chapter. This chapter will also cover recent findings concerning the role of soluble H_2ases in maintaining membrane redox poise in some photosynthetic microorganisms. H_2ases are widely distributed among procaryotic organisms belonging to the *Bacteria* and *Archaea* domains of life and are also found in some eukaryotes. The structure, classification and evolution of these enzymes has been extensively reviewed elsewhere (Vignais et al., 2001) and the present chapter will focus on those aspects directly related to respiration, namely the diversity of respiratory H_2ases, their physiological roles with respect to membrane-linked energy conservation, and their regulation.

II. Diversity of Hydrogenases

A. Classification of Hydrogenases

Three classes of H_2ase of distinct phylogenetic origin have been identified, namely the [NiFe]-H_2ases, the ('iron-only') [Fe]-H_2ases and the metal-free H_2ases (Vignais et al., 2001). Most of the known H_2ases are metalloenzymes. They are iron-sulfur proteins with two metal atoms at their active site, either a Ni and an Fe atom (in [NiFe]-H_2ases) (Volbeda et al., 1995; Higuchi et al., 1997) or two Fe atoms (in [Fe]-H_2ases) (Peters et al., 1998; Nicolet et al., 1999). Metal-free or rather [Fe-S] cluster-free (Lyon et al., 2004) H_2ases have been discovered in some methanogens (Zirngibl et al., 1990) and they function as H_2-forming methylenetetrahydromethanopterin dehydrogenases (Thauer et al., 1996).

The [NiFe]-H_2ases were first isolated as $\alpha\beta$-heterodimers with the large subunit (α-subunit) of ca 60 kDa hosting the Ni-containing active site and the small subunit (β-subunit) of ca 30 kDa, the Fe-S clusters (reviewed in Friedrich and Schwartz 1993; Vignais and Toussaint 1994; Vignais et al., 2001). Crystal structures have revealed the general fold and the nature of the Ni- and Fe-containing active site (Volbeda et al., 1995, 1996; Higuchi et al., 1997, 1999; Garcin et al., 1999). The extensive interaction of the two subunits through a large contact surface has been unveiled by the X-ray structure of the [NiFe]-H_2ase from *Desulfovibrio gigas* (Volbeda et al., 1995) and of *Desulfovibrio vulgaris* (Higuchi et al., 1997). Phylogenetic analyzes have shown that the two subunits of [NiFe]-H_2ases evolved conjointly (Vignais et al., 2001).

Unlike [NiFe]-H_2ases, many [Fe]-H_2ases are monomeric, formed by the catalytic subunit containing the so-called H-cluster of ca 350 residues at the active site (Adams, 1990), but dimeric, trimeric and tetrameric enzymes are also known (Atta and Meyer, 2000 and references therein; Vignais et al., 2001). The H-cluster comprises two Fe atoms bound to a [4Fe-4S] cluster via a bridging cysteine (Peters et al., 1998; Nicolet et al., 1999). The crystallographers (Peters 1999; Nicolet et al., 2000) have pointed out striking similarities between the structures of the Ni-containing active site and that containing Fe atoms only. The other Fe-S clusters are located in additional domains of the catalytic subunit or in additional subunits (Atta and Meyer 2000; Vignais et al., 2001).

The eukaryotic H_2ases whose genes have been cloned and sequenced appear to be of the monomeric [Fe]-H_2ase type. They are localized either in hydrogenosomes in anaerobic eukaryotes (Bui and Johnson, 1996; Hackstein et al., 1999; Horner et al., 2000, 2002; Voncken et al., 2002), or in chloroplasts of green algae (Florin et al., 2001, Happe and Kaminski 2002). Interestingly, the chloroplastic [Fe]-H_2ases consist of an H-cluster domain only and lack the additional Fe-S cluster in the N-terminal domain.

Although the Ni-containing and the Fe-only active sites display striking structural similarities, alignments of full [NiFe]- and [Fe]-H_2ase sequences (from ca one hundred enzymes in total) indicate that the [NiFe]- and the [Fe]-enzymes have a different evolutionary origin (Vignais et al., 2001).

B. The [NiFe]-Hydrogenases

The most numerous and best studied of H_2ases have been the [NiFe]-H_2ases of the domain *Bacteria* (Vignais et al., 2001). [NiFe]-H_2ases have also been isolated from *Archaea*, archaeal genes have been cloned, and recently several new [NiFe]-H_2ases revealed by archaeal genome sequencing have been identified and isolated (Künkel et al., 1998; Meuer et al., 1999; Tersteegen and Hedderich 1999; Sapra et al., 2000; Silva et al., 2000).

[NiFe]-H_2ases enable procaryotes to use H_2 as an electron and energy source either aerobically (Bowien and Schlegel, 1981; Vignais et al., 1981) or anaerobically, as in methanogens (Thauer et al., 1996; Thauer, 1998), sulfate reducers (Fauque et al., 1988), photosynthetic bacteria (Meyer et al., 1978; Vignais et al., 1985) and cyanobacteria (Houchins, 1984). Many organisms, e.g. *Escherichia coli*, sulfate reducers and methanogens, contain more than two different H_2ases, which are differentially expressed under various growth conditions. This variety is in keeping with the importance of H_2 in their metabolism and with the need to adapt quickly to changes in the environment.

Evolutionary trees have been derived from the full sequence alignments of the small and of the large subunits of [NiFe]-H_2ases. The classification of [NiFe]-H_2ases in the four groups described below, based on sequence similarities, is consistent with the functions of the enzymes (Vignais et al., 2001) (Fig. 1).

1. The Uptake [NiFe]-Hydrogenases (Group 1)

The *uptake* H_2ases, essentially found in *Proteobacteria*, link the oxidation of H_2 to the reduction of anaerobic electron acceptors such as NO_3^-, SO_4^{2-}, fumarate or CO_2 or, in the presence of oxygen, to aerobic respiration. The core H_2ase dimer is anchored to the membrane by the third H_2ase subunit, a diheme cytochrome *b* (Dross et al., 1992; Bernhard et al., 1997), which connects it to the quinone pool of the respiratory chain in the membrane, and by the hydrophobic C terminus of the small subunit (Cauvin et al., 1991; Gross et al., 1998) (Fig. 2). A mutant of *Rhodobacter capsulatus* lacking the third subunit, HupC, of the HupSLC uptake H_2ase (Fig. 3) was unable to grow autotrophically or to transfer electrons from H_2 to the respiratory chain, although H_2ase activity was present and measurable with artificial electron acceptors (Cauvin et al., 1991; Magnani et al., 2000). Electrons from H_2 are donated to the quinone pool (Henry and Vignais, 1983; Vignais et al., 1985; Komen et al., 1996) and the energy of H_2 oxidation is recovered by vectorial proton transfer at the level of the quinol oxidase (Komen et al., 1996), cytochrome bc_1 complex, cytochrome oxidase (Paul et al., 1979; Porte and Vignais, 1980) and fumarate reductase (Kröger et al., 2002)

The periplasmic H_2ases of sulfate reducing bacteria, such as *Desulfovibrio*, represent another type of uptake H_2ase able to interact with low-potential *c*-type cytochromes and a transmembrane redox protein complex encoded by the *hmc* operon (Rossi et al., 1993) and to participate in the creation of a proton gradient across the membrane for energy conservation (Fauque et al., 1988; Vignais et al., 2001). Deletion of the *hmc* operon of *D. vulgaris* (Hildenborough) impaired growth on hydrogen (but not that

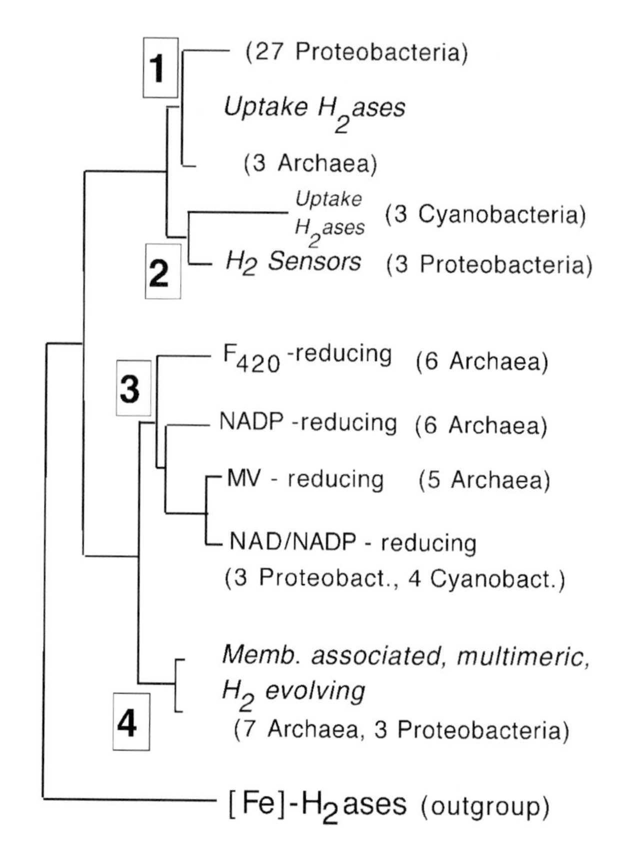

Fig. 1. Schematic representation of the phylogenetic tree of [NiFe]-H_2ases based on the complete sequences of the small and the large subunits (the same tree was obtained with each type of subunit) originally established by Vignais et al. (2001).

on lactate or pyruvate) confirming the importance of the Hmc complex in electron transport from H_2 in the periplasm to sulfate in the cytoplasm (Dolla et al., 2000). However, Hmc does not appear to be the only complex capable of coupling electron transport to proton pumping (Dolla et al., 2000) and some of the redox partners involved in the establishment of proton gradient across the membrane for energy conservation during H_2 oxidation have not yet been identified. (The products of homologs of *ech* genes (see below), recently discovered in the genome of *D. vulgaris* (cited in Pohorelic et al., 2002) might participate in the creation of a protonmotive force in the membrane). Other uptake H_2ases, e.g. H_2ase 2 of *E. coli* encoded by the *hybOABCDEFG* operon (Dubini et al, 2002) (Fig. 2), and H_2ase 1 of *Thiocapsa roseopersicina* encoded by the *hydSisp1isp2hydL* operon (Rákhely et al., 1998), have subunits that share significant degrees of identity with two subunits of the Hmc complex.

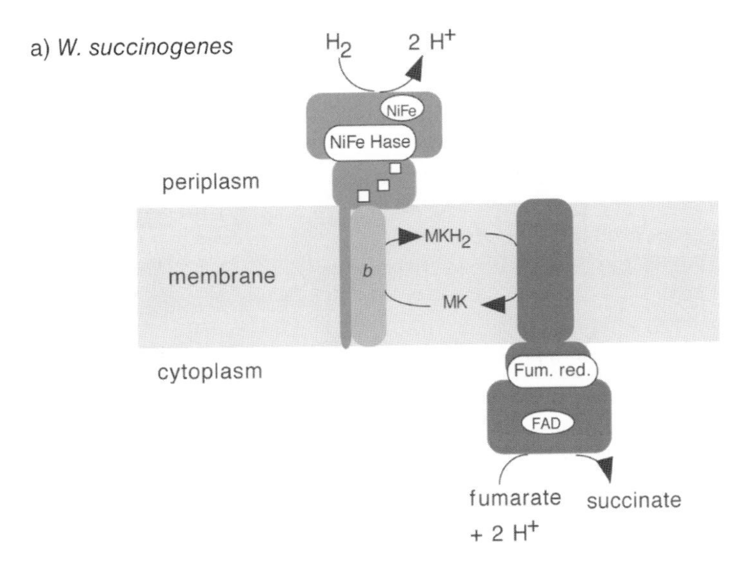

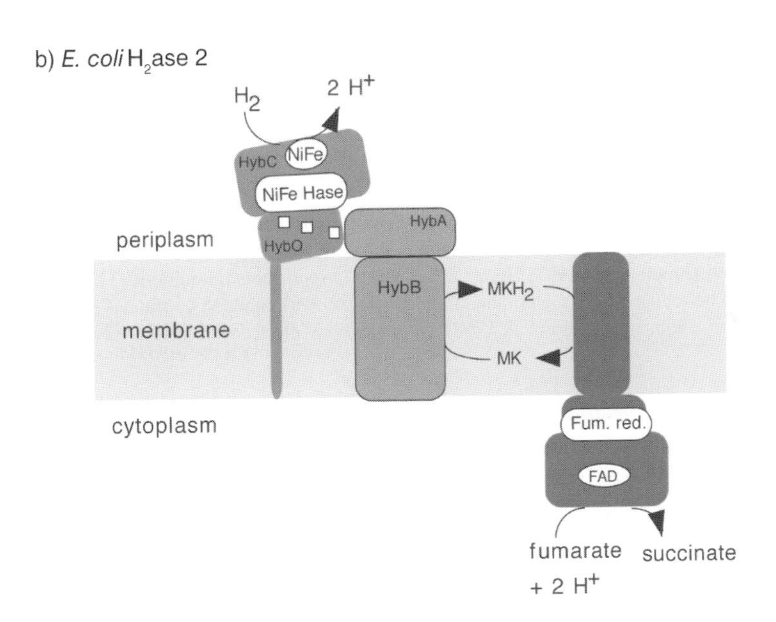

Fig. 2. Hypothetical mechanism of fumarate respiration with H_2 a) in *W. succinogenes* (adapted from Kröger et al., 2002), b) in *E. coli* (adapted from Dubini et al., 2002)

An uptake H_2ase has been identified in the methanogenic archaeon *Methanosarcina mazei* Gö1 (Ide et al., 1999). In this microorganism, the membrane-bound H_2ase encoded by *vhoGA* transfers electrons from H_2 to a cytochrome *b* (encoded by *vhoC*); the electrons are then channeled through methanophenazine to heterodisulfide reductase, which reduces the CoenzymeM-S-S-CoenzymeB heterodisulfide to produce CoenzymeB-SH, the reductant for the formation of methane from methyl-S-CoM. This electron transfer system is coupled to vectorial proton transfer, leading to the creation of a protonmotive force.

The small subunits of all the [NiFe]-H_2ases of group 1 bear at their N-terminus a long signal peptide (30–50 amino acid residues), which targets the fully folded heterodimer to the membrane and the periplasm (Berks et al., 2000; Vordouw 2000; Wu et al., 2000; Vignais et al., 2001; Sargent et al., 2002). The

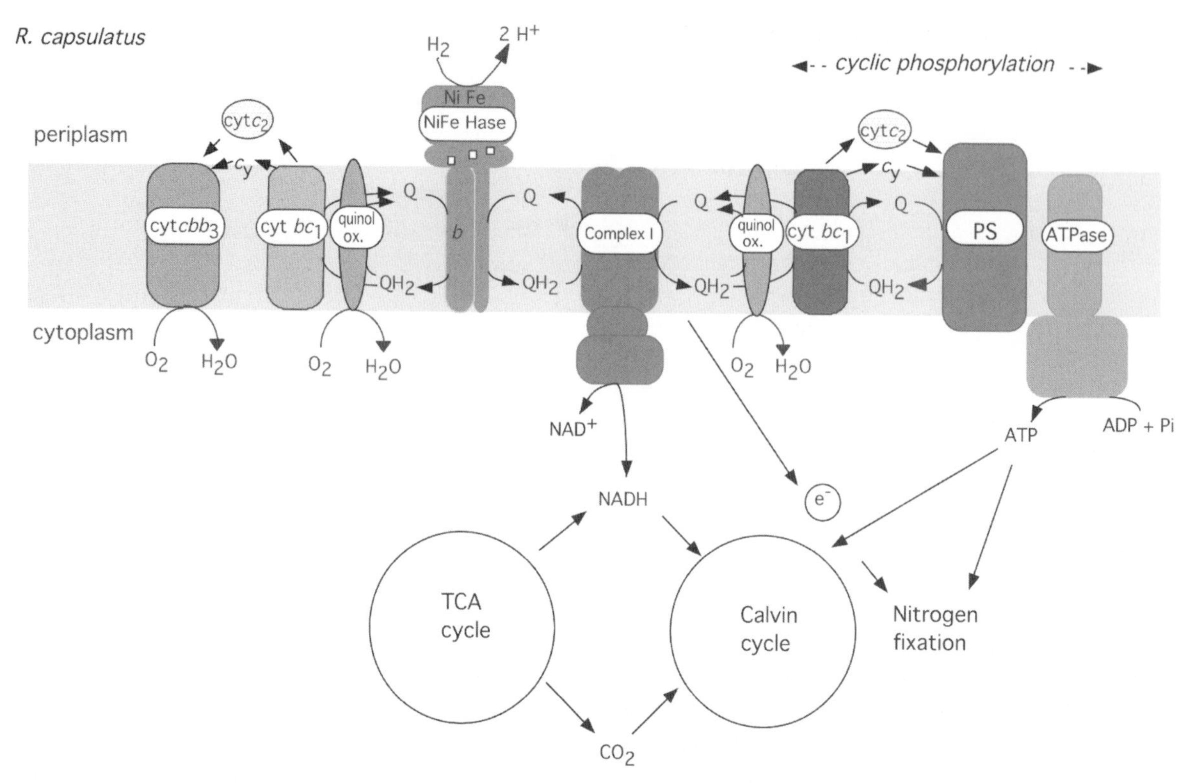

Fig. 3. Photosynthetic and respiratory electron transport chains of *Rb. capsulatus*. The scheme shows that Complex I catalyzes electron transport both in the downhill direction (respiratory chain) and in the uphill direction (reversed electron flow) to provide reductant for CO_2 fixation (Klemme, 1969; Dupuis et al, 1997; Herter et al 1998). Q represents ubiquinone. The electron donor to nitrogenase is ferredoxin I. It is thought to be reduced by a membrane complex (not shown) encoded by the *rnf* (for *Rhodobacter*-specific nitrogen fixation) genes (Schmehl et al, 1993, Jouanneau et al, 1998), which receives electrons either from NADH or the ubiquinone pool.

signal peptide contains a conserved (S/T)RRxFxK motif recognized by a specific protein translocation pathway known as the membrane targeting and translocation (Mtt) (Weiner et al., 1998) or twin-arginine translocation (Tat) (Sargent et al., 1998) pathway. Several H_2ases of this group, *E. coli* H_2ase 1 (Bogsch et al., 1998 ; Sargent et al., 1998), *E. coli* H_2ase 2 (Bogsch et al., 1998; Chanal et al., 1998; Sargent et al., 1998; Rodrigue et al., 1999), the membrane-bound H_2ase of *Wolinella succinogenes* (Gross et al., 1999) and of *Ralstonia eutropha* (formerly *Alcaligenes eutrophus*) (Bernhard et al., 2000) have been shown to be exported by this so-called hitchhiker type of cotranslocation of the two subunits.

2. The Cytoplasmic H_2 Sensors and the Cyanobacterial Uptake [NiFe]-Hydrogenases (Group 2)

In contrast to group 1, the small subunit of the enzymes of group 2 does not contain a signal peptide at its N-terminus. This indicates that these enzymes are not exported but remain in the cytoplasm. Indeed, two representatives of group 2, the *Rhodobacter* (*Rb.*) *capsulatus* HupUV H_2ase (Vignais et al., 2000) and the *R. eutropha* HoxBC H_2ase (Kleihues et al., 2000) have been shown to be soluble, cytoplasmic enzymes. These two H_2ases are not directly involved in energy transducing reactions. Instead, they are able to detect the presence of H_2 in the environment and to trigger a cascade of cellular reactions leading to the biosynthesis of respiratory [NiFe]-H_2ases (see below). Group 2 also includes the so-called uptake H_2ases (HupSL) of the cyanobacteria *Nostoc* (Oxelfelt et al., 1998) and *Anabaena variabilis* (Happe et al., 2000). These enzymes, which are induced under N_2 fixing conditions, are localized on the cytoplasmic side of either the cytoplasmic or thylakoid membrane (Appel and Schulz, 1998; Tamagnini et al., 2002).

3. The Bidirectional Cytoplasmic [NiFe]-Hydrogenases (Group 3)

In group 3, which is subdivided into four subgroups, the dimeric H_2ase moiety is associated with other subunits that are able to bind soluble cofactors, such as cofactor 420 (F_{420}, 8-hydroxy-5-deazaflavin), NAD or NADP (Fig. 1). They function reversibly and can thus reoxidize the cofactors under anaerobic conditions by using the protons of water as electron acceptors. Members of the first three subgroups belong to the archaeal domain of life. They are: the trimeric F_{420}-reducing H_2ases (group 3a), the tetrameric bifunctional H_2ases of hyperthermophiles, able to reduce S° to H_2S in vitro (Ma et al., 1993) and which use NADPH as electron donor (Ma et al., 1994) (group 3b) and the methyl-viologen reducing H_2ases (group 3c) (Fig. 1).

The H_2ases of Methanosarcinales belong to group 3a. They use H_2 or $F_{420}H_2$ as electron donors to membrane-bound electron transport systems. Electron transfer, which involves H_2ase, *b*-type cytochrome and $F_{420}H_2$ dehydrogenase, is coupled to proton translocation across the cytoplasmic membrane resulting in the formation of protonmotive force. The $F_{420}H_2$ dehydrogenase of *M. mazei*, encoded by the *fpo* genes, is a redox-driven proton pump sharing similarities with the proton-translocating NADH:quinone oxidoreductase of respiratory chains (reviewed in Deppenmeier et al., 1999; Bäumer et al., 2000).

The bacterial, multimeric, bidirectional, NAD(P)-linked H_2ases form group 3d. The first NAD-dependent, tetrameric [NiFe]-H_2ase was isolated from *R. eutropha* (*A. eutrophus*) (Schneider and Schlegel, 1976) in which it is encoded by a gene located on a megaplasmid (Tran-Betcke et al., 1990). Genes capable of encoding an enzyme homologous to the soluble NAD-linked H_2ase of *R. eutropha* were later discovered in cyanobacteria (Schmitz et al., 1995, 2002; Appel and Schulz, 1996; Kaneko et al., 1996) and in the photosynthetic bacterium *T. roseopersicina* (Rákhely et al., 2004). The function of this bidirectional H_2ase in cyanobacteria is now being actively studied (see below). The cyanobacterial enzyme is composed of two moieties: the diaphorase moiety, encoded by the *hoxE*, *hoxU* and *hoxF* genes, is homologous to three subunits of Complex I of the mitochondrial and bacterial respiratory chains and contains NAD(P), FMN and Fe-S binding sites (Table 1, Fig. 4); the [NiFe]-H_2ase moiety is encoded by the *hoxY* and *hoxH* genes (reviewed by Appel and Schulz, 1998; Vignais et al., 2001; Tamagnini et al., 2002).

4. The H_2-Evolving, Energy-Converting Hydrogenases (Group 4)

The enzymes of this group are multimeric (six subunits or more), function under strictly anaerobic conditions and reduce protons in order to dispose of excess reducing equivalents produced by the anaerobic oxidation of C_1 organic compounds. They are associated with the oxidation of low potential substrates such as carbon monoxide and formate (reviewed by Hedderich, 2004). The prototype of this group is *E. coli* H_2ase 3, which forms part of the formate hydrogen lyase complex (FLH-1) encoded by the *hyc* operon (Böhm et al., 1990; Sauter et al., 1992) (cf Fig. 8). Another member of this group is the CO-induced H_2ase of *Rhodospirillum rubrum* (Fox et al., 1996a,b). This enzyme is a component of the CO-oxidizing system that allows *R. rubrum* to grow in the dark with CO as sole energy source. CO-dehydrogenase and the H_2ase encoded by the *coo* operon oxidize CO to CO_2 with concomitant production of H_2. Since the CO dehydrogenase is a peripheral membrane protein, it has been proposed that the H_2ase component of the oxidizing system constitutes the energy coupling site (Fox et al., 1996a,b). *E. coli* H_2ase 3 and *R. rubrum* CooLH H_2ase are labile enzymes, and the exact number of subunits in each enzyme remains to be determined.

In *E. coli*, the recently discovered *hyf* operon is thought to encode a putative, ten-subunit H_2ase, designated H_2ase 4, which combines with formate dehydrogenase H (Fdh-H) to form a second formate hydrogenlyase (FHL-2). H_2ase 4 was suggested to be energy transducing since, in common with other group 4 H_2ases, some of its subunits show marked sequence similarities with subunits of Complex I, in particular two intrinsic membrane subunits which play a crucial role in proton translocation and energy coupling (Andrews et al., 1997) (see below). A recent report (Bagramyan et al, 2002) has suggested that, at slightly alkaline pH, H_2 production and Fdh-H activity in *E. coli* are dependent both on the F_0F_1 ATPase and FHL-2. Under these conditions, H_2 production would be driven by a proton gradient established by the F_0F_1 ATPase, as has been reported for fermentative gas (H_2 and H_2S) production in *Salmonella typhimurium* (Sasahara et al., 1997).

The majority of H_2ases that have been assigned to this group have been found in *Archaea* (Fig. 1), in-

able 1. Relationships between Complex I and NDH-1 subunits and subunits of selected [NiFe]-hydrogenases

	Bovine[1]	*Synechocystis*[2]		*E. coli*[3] or *Rb. capsulatus*[4]	*P. denitrificans*[5]	*E. coli*[6]	*M. barkeri*[7]	*R. rubrum*[8]
			HoxEFUYH					
	Complex I	NDH-1	H_2ase	NDH-1	NDH-1	Hyc H_2ase	Ech H_2ase	Coo H_2ase
	9 kDa							
Hydrophilic	24 kDa		HoxE	NuoE	Nqo2			
NADH-oxidizing	51 kDa		HoxF	NuoF	Nqo1			
module	75 kDa		HoxU[9]	NuoG	Nqo3			
				NuoCD (*E.c.*)[10]		HycE		
	30 kDa	NdhJ		NuoC (*R.c.*)	Nqo5	N-ter HycE		
subunits	49 kDa	NdhH		NuoD (*R.c.*)	Nqo4	C-ter HycE	EchE	CooH
of the	20 kDa (PSST)	NdhK		NuoB	Nqo6	HycG	EchC	CooL
connectig	23 kDa (TYKY)	NdhI		NuoI	Nqo9	HycF	EchF	CooX
module	39 kDa							
	18 kDa							
	13 kDa B							
	39 kDa							
Intrinsic-	NDI	NdhA		NuoH	Nqo8	HycD	EchB	CooK
membrane	ND2	NdhB		NuoN	Nqo14	HycC[11]	EchA[11]	N-ter CooM[11]
hydrophobic	ND3	NdhC		NuoA	Nqo7			
subunits	ND4	NdhD		NuoM	Nqo13	HycC[11]	EchA[11]	N-ter CooM[11]
	ND4L	NdhE		NuoK	Nqo11			
	ND5	NdhF		NuoL	Nqo12	HycC[11]	EchA[11]	N-ter CooM[11]
	ND6	NdhG		NuoJ	Nqo10			

Ref. [1] Fearnley and Walker, 1992 ; [2] Kaneko et al., 1996; [3]Weidner et al., 1993; [4] Dupuis et al., 1998; [5] Yagi, 1993; [6] Sauter et al., 1992; [7] Künkel et al., 1998; [8] Fox et al., 1996a,b; [9] Sequence similarities between HoxU and N-ter NuoG; [10] NuoC and NuoD are fused in *E. coli*, Blattner et al, 1997; [11]NuoL, NuoM and NuoN originated from gene duplication

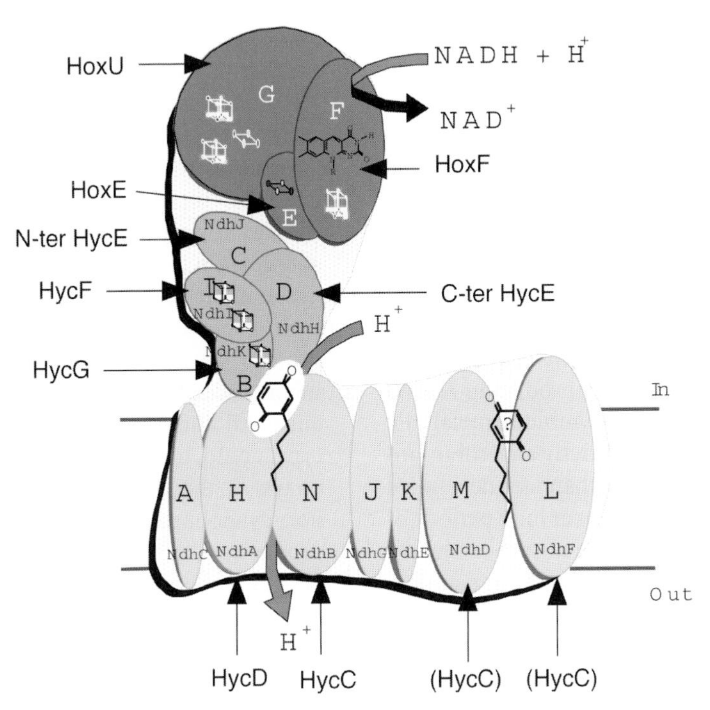

Fig. 4. Relationship of *Rb. capsulatus* Complex I subunits (Dupuis et al., 2001) to NDH-1 subunits of *Synechocystis* PCC 6803 (Kaneko et al., 1996) and to [NiFe]-H_2ases subunits (*Synechocystis* HoxEFUYH (Appel and Schulz, 1996; Kaneko et al., 1996; Schmitz et al., 2002) and *E. coli HycCDEFG* (Sauter et al., 1992)) (adapted from Sazanov et al., 2000 and Dupuis et al., 2001).

cluding *Methanosarcina barkeri* (Künkel et al., 1998), *Methanobacterium thermoautotrophicum* strain Marburg (now called *Methanobacter marburgensis)* (Tersteegen and Hedderich, 1999) and *Pyrococcus furiosus* (Sapra et al., 2000; Silva et al., 2000). The group 4 H_2ase from *M. barkeri*, encoded by the *ech* genes, has been purified from acetate-grown cells; it has been proposed to oxidize the carbonyl group of acetate to form CO_2 and H_2 (Künkel et al., 1998; Meuer et al., 1999). The methanogenic archaeon *M. thermoautotrophicum,* which is able to grow on CO_2/H_2 as carbon and energy source contains, in addition to F_{420}-reducing and F_{420}-non-reducing H_2ases, two gene groups designated 'energy converting H_2ase A' (*eha*) and 'energy converting H_2ase B' (*ehb*), which encode putative, multisubunit, membrane-bound H_2ases homologous to *E. coli* H_2ase 3 and *R. rubrum* CooHL H_2ase (Tersteegen and Hedderich, 1999). The hyperthermophilic archaeon *P. furiosus* contains two cytoplasmic H_2-evolving H_2ases (I and II) (Pedroni et al., 1995; Ma et al., 2000), members of group 3, and a membrane-bound H_2ase, member of group 4, encoded by a 14-gene operon (Schut et al., 2001) termed *mbh* (either *mbh1-14* (Sapra et al., 2000) or *mbhA-N* (Silva et al., 2000)). Four gene products of this operon share similarities with subunits of Complex I. These multimeric membrane-bound H_2ase complexes comprise transmembrane subunits homologous to Complex I subunits involved in proton pumping and energy coupling (Table 1) and appear to be able to couple the oxidation of a carbonyl group (originating from formate, acetate or carbon monoxide) and the reduction of protons to H_2 with energy conservation.

C. [Fe]-Hydrogenases

While [NiFe]-H_2ases tend to be involved in H_2 consumption, [Fe]-H_2ases are usually involved in H_2 production. However, the periplasmic [Fe]-H_2ase of *D. vulgaris* has been recently demonstrated to function as an uptake H_2ase (Pohorelic et al., 2002). This type of enzyme is found in anaerobic prokaryotes, such as clostridia and sulfate reducers (see reviews by Adams, 1990; Atta and Meyer, 2000; Vignais et al., 2001) and in eukaryotes (Horner et al., 2000, 2002; Florin et al., 2001, Wünschiers et al., 2001a, Happe and Kaminsky, 2002; Voncken et al., 2002). [Fe]-H_2ases are the only type of H_2ase to have been found in eukaryotes, and they are located exclusively in membrane-limited organelles, i.e. in chloroplasts or in hydrogenosomes (reviewed in Vignais et al., 2001 and Horner et al., 2002).

Two novel hypotheses for the origin of the eukaryotic cell (Martin and Müller, 1998; Moreira and López-García, 1998; López-García and Moreira, 1999) posit that a metabolic symbiosis (syntrophy) between a methanogenic archeon and a proteobacterium able to release H_2 in anaerobiosis was the first step in eukaryogenesis. The hydrogen hypothesis (Martin and Müller, 1998) suggests that an anaerobic, heterotrophic α-*Proteobacterium* producing H_2 and CO_2 as waste products formed a symbiotic metabolic association (syntrophy) with a strictly anaerobic, autotrophic archaeon, possibly a methanogen dependent on H_2. The intimate relationship over long periods of time allowed the symbiont and the host to co-evolve and become dependent on each other. In an anaerobic environment the symbiont was either lost, as in type I amitochondriate eukaryotes, or became a hydrogenosome, i.e. a hydrogen-generating and ATP-supplying organelle, as in type II amitochondriate eukaryotes (Müller, 1993). By further evolution, the host lost its autotrophic pathway and its dependence on H_2 and the endosymbiont adopted a more efficient aerobic respiration to become the ancestral mitochondrion. Thus, the eukaryotic cell would have arisen through the symbiotic association of two prokaryotes, an archaeon (the host) and a eubacterium (the symbiont), the common ancestor of hydrogenosomes and mitochondria. The syntrophy hypothesis for the origin of eukaryotes, proposed at the same time and independently by Moreira and López-García (1998), was based on similar metabolic consideration (interspecies hydrogen transfer) but the latter authors speculated that the organisms involved were δ-*Proteobacteria* (ancestral sulfate-reducing myxobacteria) and methanogenic *Archaea* (see also López-García and Moreira, 1999). (These authors suggested also that a second anaerobic symbiont was involved in the origin of mitochondria). Hydrogenosomes are either considered to be relics of ancestral endosymbiont and to share a common origin with mitochondria (Martin and Müller, 1998) or to have evolved several times as adaptations of mitochondria to anaerobic environments (Hackstein et al., 2001). With respect to the various theories on the origin of mitochondria and hydrogenosomes, which have been reviewed recently (Martin et al., 2001), it is interesting to mention that eukaryotic organelles contain only [Fe]-H_2ases, and that [Fe]-H_2ases have been found in δ-*Proteobacteria* (Malki et al., 1995; Pohorelic et al., 2002) but not in present-day α-*Proteobacteria*, which contain only [NiFe]-H_2ases (Vignais et al., 2001).

Phylogenetic analysis of eukaryotic [Fe]-H_2ases (Horner et al., 2000, 2002) suggests a polyphyletic origin of these enzymes, implying an acquisition by lateral gene-transfer from different prokaryotic sources. On the other hand, the [Fe]-H_2ases from the green algae emerge as a monophyletic group with hydrogenosomal [Fe]-H_2ases from microaerophilic protists (Horner et al., 2002). Although a vertical inheritance of orthologous enzymes from the putative ancestral endosymbiont of the δ-*Proteobacterium*-type could be envisaged for the hydrogenosomal enzymes, the plastidial [Fe]-H_2ases appear to have a non-cyanobacterial origin, since cyanobacteria, the progenitors of chloroplasts, contain only [NiFe]-H_2ases and no [Fe]-H_2ase (Vignais et al. 2001; Tamagnini et al., 2002).

D. Hydrogenases and Complex I

The proton-pumping NADH-ubiquinone oxidoreductase is the main entry site of reducing equivalents into the mitochondrial and the bacterial respiratory chains (Fearnly and Walker, 1992; Walker, 1992). The mitochondrial enzyme is also called Complex I, whereas the bacterial enzyme is more often referred to as type I NADH-dehydrogenase, or NDH-1. The bovine mitochondrial Complex I contains a total of 45 different subunits (Carroll et al., 2002), whereas NDH-1 from the bacteria *Paracoccus denitrificans* (Yagi, 1993; Yagi et al., 1998) and *Rb. capsulatus* (Dupuis et al., 1998) contains 14 subunits, and that from *E. coli* (Weidner et al., 1993; Blattner et al., 1997) 13 subunits, all of which have homologs in the bovine enzyme (Table 1). Both the mitochondrial and bacterial enzymes are L-shaped, with a membrane domain and a peripheral arm extending into the cytosol (Fig. 4) (Friedrich, 1998; Sazanov et al., 2000; Schultz and Chan, 2001). The hydrophilic NADH-oxidizing module, distal from the membrane, comprises three hydrophilic subunits containing FMN and five iron-sulfur clusters; a second hydrophilic module constituted of four subunits connects the NADH-oxidizing proteins to the membrane-bound hydrophobic subunits. The two extramembranous modules contain all the redox centers of the enzyme (Yagi et al., 1998 ; Dupuis et al., 2001; Friedrich, 2001; Schultz and Chan, 2001).

Sequence similarities between H_2ases and Complex I were first reported by Böhm et al. (1990) and by Pilkington et al. (1991) and have been emphasized in several subsequent reports (Albracht and de Jong, 1997; Friedrich and Weiss, 1997; Albracht and Hedderich, 2000; Friedrich and Scheide, 2000; Dupuis et al., 2001; Friedrich, 2001; Vignais et al., 2001 ; Yano and Ohnishi, 2001). To compare the subunits of H_2ases to those of Complex I, we have adopted the nomenclature used for *E. coli* and *Rb. capsulatus* Complex I (Table 1, Fig. 4). Subunits NuoE, NuoF, NuoI and the N-terminal Fe-S binding domain (ca 220 residues) of NuoG have homologous counterparts in accessory subunits and domains of soluble [NiFe]- and [Fe]-H_2ases. In addition, three subunits located within the connecting module of Complex I share similarities with subunits of [NiFe]-H_2ases, the NuoB subunit with the small H_2ase subunit, the NuoC and NuoD subunits (fused as a single NuoCD protein in *E. coli*) with the large H_2ase subunit. Furthermore, hydrophobic subunits of multimeric, membrane-bound [NiFe]-H_2ases belonging to the group 4, namely *E. coli* Hyc (Böhm et al., 1990; Sauter et al., 1992), *R. rubrum* Coo (Fox et al., 1996a,b), *M. barkeri* Ech (Künkel et al., 1998; Meuer et al., 1999), *M. marburgensis* Eha and Ehb (Tersteegen and Hedderich, 1999) and *P. furiosus* Mbh (Sapra et al., 2000; Silva et al. 2000), are also homologous to transmembrane subunits of Complex I (NuoH, NuoL, NuoM, NuoN). It should be noted that these group 4 H_2ases are also proton pumps. Thus, the presumed evolutionary links between H_2ases and Complex I concern not only the electron transferring subunits but also the proton pumping units, i.e. the coupling between electron transport and energy recovery by a chemiosmotic mechanism.

On the basis of the similarities between [NiFe]-H_2ases and the NuoB-NuoD dimer of the connecting module, Dupuis et al. (2001) have recently proposed that the [NiFe] active site of H_2ases was reorganized into a quinone-reduction site carried by the NuoB-NuoD dimer and a hydrophobic subunit such as NuoH (Fig. 4), and that NuoD might provide both the quinone gate and a potential proton channel entry for a minimal 'proton pumping' module, composed of subunits NuoB, NuoD, NuoI, NuoH and NuoL (Friedrich and Scheide, 2000; Dupuis et al., 2001; see also Kashani-Poor et al., 2001). Subunit NuoL, (or NuoM, or NuoN, which apparently evolved by triplication of an ancestral gene related to bacterial H^+/K^+ antiporters (Fearnley and Walker, 1992; Friedrich and Weiss, 1997)), would have provided the transmembrane channel required to complete the proton pump (Dupuis et al., 2001).

III. Physiological Roles of Respiratory Hydrogenases

A. Energy-Conservation via Electron Transfer Pathways

Electron transfer pathways in bacteria are initiated by a series of substrate-specific dehydrogenases feeding electrons into a common quinone pool, from which electrons are transferred via specific quinol oxidases to terminal reductases, e.g. in the absence of oxygen, TMAO or nitrate or fumarate reductase and, in the presence of oxygen, cytochrome oxidase. The cellular content of specific dehydrogenases, terminal reductases and also of the type of quinone (e.g. ubiquinone, UQ, menaquinone, MK, or demethylmenaquinone, DMK, in *E. coli* (Søballe and Poole, 1999), depends on the prevailing environmental conditions (Richardson, 2000) and it is not possible to give a general picture valid in all cases. The oversimplified scheme of Fig. 5 is meant to emphasize the role of the quinone pool in respiration.

This type of electron transfer chain is also found in photosynthetic organisms, which have the capacity to convert light energy into chemical energy (NAD(P)H and ATP). In purple photosynthetic bacteria, such as *Rb. capsulatus*, the photosynthetic electron transport chain comprises two multisubunit trans-membrane complexes, the photochemical reaction center (PS) and the cytochrome bc_1 complex, connected by the water-soluble cytochrome c_2 (and/or cytochrome c_y) and, in the membrane domain, by ubiquinone (UQ). Photosynthetic electron transport is cyclic in the light, PS accepts the electrons from Cyt c_2/Cyt c_y and donates them to the UQ pool, ubiquinol (UQH_2) in turn being oxidized by the cytochrome bc_1 complex which reduces cytochrome c_2 (Fig. 3) (see also Chapter 3 by Cooley et al., Vol. 15 of the series and Chapter 13 by Vermeglio et al., this volume). Cyclic electron transport generates a proton-motive force (Δp) across the membrane, which is consumed by ATP synthase to make ATP. Cyclic electron transport generates ATP but not NADH. In this system, NADH is formed in part by direct oxidation of low-potential donors and in part by reverse electron flow from UQH_2 to NAD^+, a reaction which consumes Δp (reviewed by Gest, 1972 ; and Knaff, 1978; see also Dupuis et al., 1997;

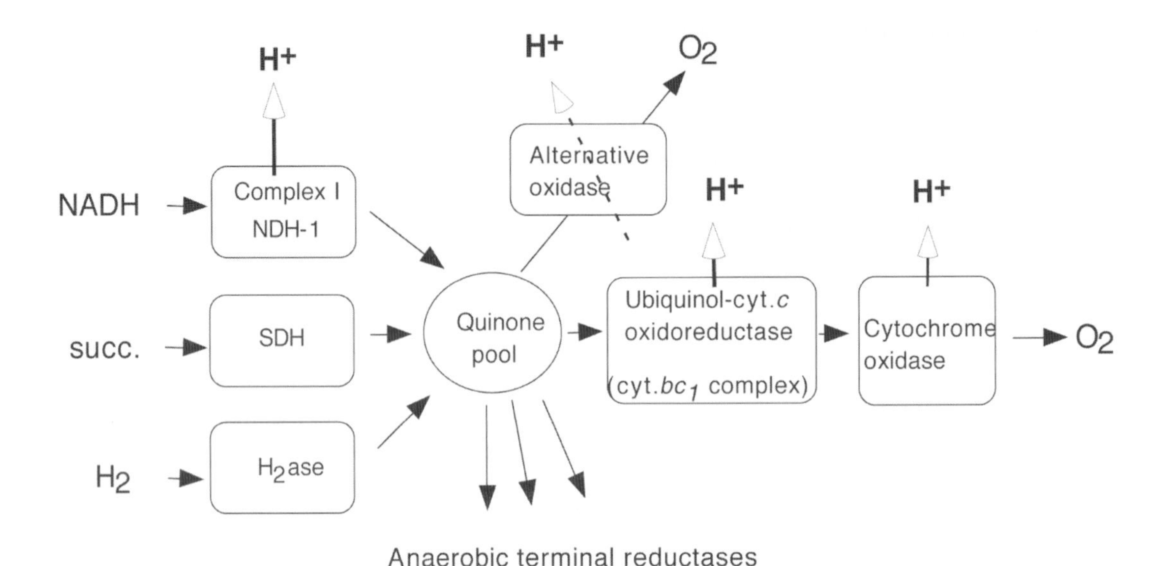

Fig. 5. Simplified and general scheme of electron pathways in bacteria. The energy coupling sites are indicated by arrows showing vectorial proton ejection. The proton pumping activity of *Rb. capsulatus* alternative oxidase (cytb_{260}) (Zannoni, 1995) is indicated by a dashed arrow since it is not a common case among the so called 'cyanide-insensitive oxidases'.

Herter et al., 1998). In the latter case, the donor can be of higher redox potential than NADH, provided that it can reduce quinones.

The photosynthetic apparatus of oxygenic photosynthetic prokaryotes (Cyanobacteria) comprises two photosystems, PSII and PSI, linked by a linear electron transport chain (Fig. 6) similar to those of higher plant and algal chloroplasts (Fig. 7). In cyanobacteria, photosynthetic electron transport takes place in the thylakoid membrane while respiratory electron transport occurs in both the thylakoid and the cytoplasmic membranes (Scherer, 1990; Schmetterer, 1994; Cooley and Vermaas, 2001; see also Chapter 12 by G. Schmetterer, this volume). The two photosystems operating in series are capable of oxidizing water to O_2 and generating ATP and low potential reductant for use in biosynthetic reactions. In *Synechocystis*, the proton-pumping complex, NDH-1, localized in the thylakoid membrane (Ohkawa et al., 2001), acts as a plastoquinone oxidoreductase, with NADH or NADPH as the reductant (Schemetterer, 1994) (Fig. 6). In the thylakoid membrane, this complex has been shown to participate in the cyclic electron flow around PSI (Mi et al., 1995).

The discovery of a plastid-encoded NAD(P)H-dehydrogenase (Ndh) complex homologous to Complex I and of a plastid terminal oxidase (PTOX) (Cournac et al., 2000), homologous to the plant mitochondrial alternative oxidase, has provided molecular support for the existence of a respiratory electron transport chain which interacts with the photosynthetic electron transport chain in the thylakoid membranes of chloroplasts. This respiratory activity in chloroplasts, called chlororespiration, participates in the nonphotochemical reduction and oxidation of plastoquinones (PQs). Whether this activity has or not a bioenergetic role in itself remains controversial. In addition, chlororespiration may be involved in the initiation of cyclic electron transfer reactions around PSI (see review by Peltier and Cournac, 2002).

B. Disposal of Excess Reducing Equivalents

Although a primary function of the respiratory system is to generate a protonmotive force utilized for ATP synthesis and a variety of other essential functions, the respiratory system is also necessary to regenerate NAD^+ from NADH and to eliminate excess reducing equivalents.

Growth of bacteria depends on dissimilatory and

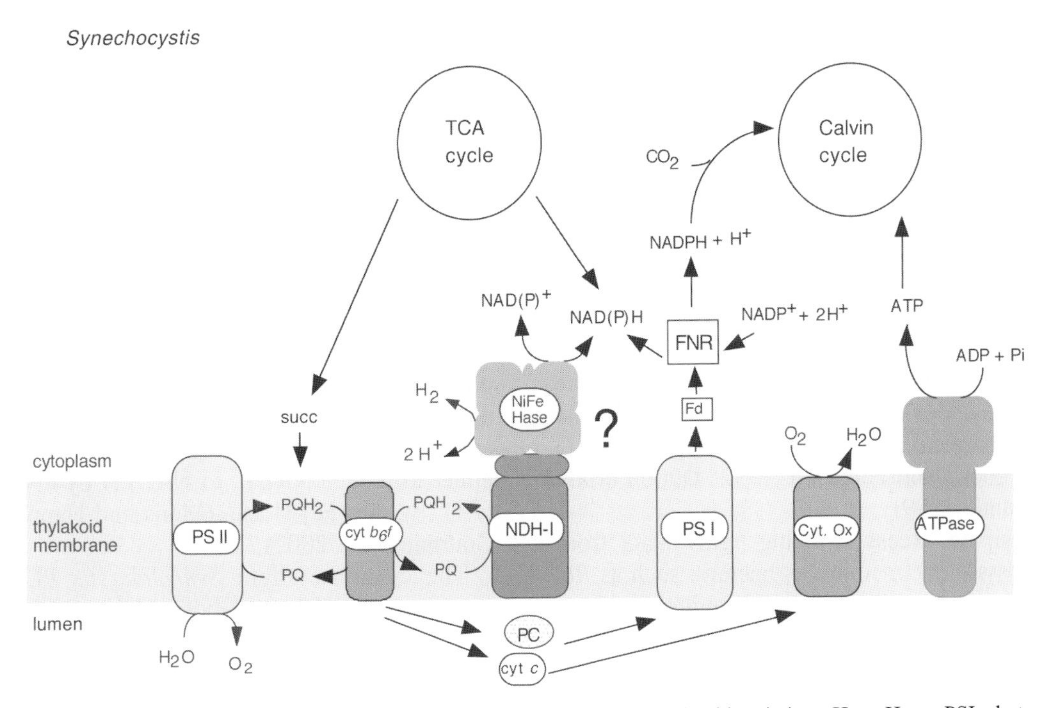

Fig. 6. Electron transfer pathways involved in H_2 production in *Synechocystis* PCC 6803. Abbreviations: Hase: H_2ase; PSI: photosystem I; PSII: photosystem II; Cyt: cytochrome; NDH-I: Complex I-like NAD(P)H dehydrogenase; PQ: plastoquinone; PC: plastocyanin; FNR: Ferredoxin-NADP$^+$ oxidoreductase; Fd: ferredoxin; ATPase: ATP synthase. The question mark means that a direct interaction between the bidirectional H_2ase and NDH-1 is hypothetical.

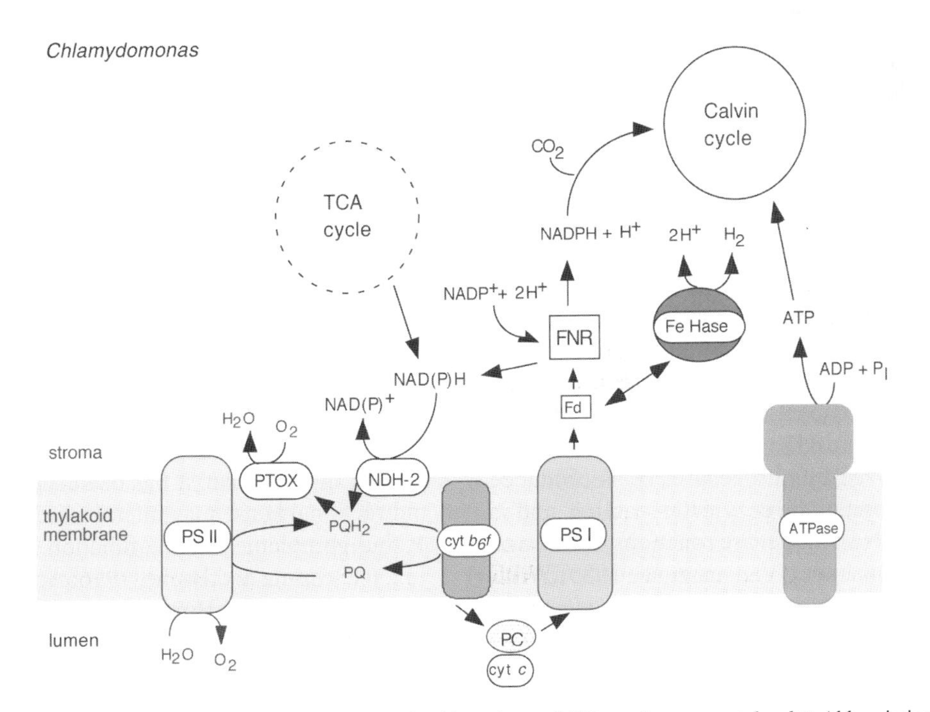

Fig. 7. Electron transfer pathways involved in H_2 production in chloroplasts of *Chlamydomonas reinhardtii.* Abbreviations (see legend to Fig. 6): NDH-2: type II NAD(P)H dehydrogenase; PTOX: plastid terminal oxidase. The TCA cycle takes place in the stroma. The ferredoxin functions as electron donor linking the H_2ase to the photosynthetic electron transport chain.

assimilatory processes. Oxidation of inorganic or organic substrates results in the formation of reducing power (NADH) and ATP which is used to drive assimilatory processes leading to the synthesis of cell materials. Growth rates depend on ATP content and the (photo)phosphorylation rate is regulated by redox balance. 'Over-reduction' or 'over-oxidation' of the redox components of the electron transport chain (including the quinone pool) leads to inhibition of phosphorylation (Bose and Gest, 1963). The requirement for a membrane redox poise close to the oxidation-reduction potential of the ubiquinone pool (Candela et al., 2001) can be explained by the involvement of a semiquinone intermediate in the Q cycle (Nicholls and Ferguson, 1992; Dutton et al., 1998; Brandt, 1999)

To dissipate excess reducing equivalents from the photosynthetic membrane, bacteria such as *Rb. capsulatus* can use an alternative quinol oxidase, which allows the cell to control the redox state of the Q pool and the rate of photophosphorylation activity (Zannoni and Marrs, 1981). This pathway drives a light-induced oxygen uptake when the Q-pool reduction level reaches 25% (Zannoni, 1995). Under respiratory steady-state conditions, continuous light does not perturb the redox state of the Q pool when the quinol oxidase pathway is functioning (Zannoni and Moore, 1990 ; see also Chapter 13 by Vermeglio et al., this volume).

During anaerobiosis, auxiliary oxidants serve to sustain the redox poise of the Q pool for photosynthetic electron transport (McEwan et al., 1985; Ferguson et al., 1987). In *Rb. capsulatus*, under anaerobic conditions in the light, excess reducing equivalents are transferred to NAD^+ by reverse electron flow through Complex I (Klemme 1969; Dupuis et al., 1997). Reducing equivalents stored in NADH can be dissipated by metabolic systems such as CO_2 fixation (Calvin cycle), nitrogen fixation or oxidation of auxiliary oxidants (Hillmer and Gest 1977; Tichi et al., 2001). In the case of nitrogen fixation, which is catalyzed by nitrogenase, H_2 is produced as an intrinsic part of the enzymatic reaction, and in the absence of N_2, nitrogenase functions as a H_2ase, reducing protons to H_2 (Vignais et al., 1985; Willison, 1993). Since nitrogenase is an ATP-dependent enzyme, this reaction dissipates energy as well as reducing equivalents.

H_2 production as a mechanism to dissipate reducing equivalents is also observed in some algae. It is the case, for example, when dark-adapted *Chlamydomonas reinhardtii* cells are illuminated (Cournac et al., 2002). In *Scenedesmus obliquus* (Florin et al., 2001; Wünschiers et al., 2001b) or *C. reinhardtii* (Happe and Kaminsky, 2002) the electrons provided by fermentative metabolism are transferred to PSI via the plastoquinone pool (probably through a type-2 NAD(P)H dehydrogenase, see Peltier and Cournac, 2002). In turn, PSI reduces a [2Fe-2S] ferredoxin, the physiological electron donor to [Fe]-H_2ase. Figure 7 shows how electrons originating at PSII upon photooxidation of water or at the plastoquinone pool upon oxidation of cellular endogenous substrates are transported via PSI to ferredoxin and then serve either to reduce $NADP^+$ to NADPH by FNR or H^+ to H_2 by the [Fe]-H_2ase. (Melis and Happe, 2001, Cournac et al., 2002).

The cyanobacterium *Synechocystis* PCC 6803 contains an NAD(P)-dependent bidirectional [NiFe]-H_2ase located in the cytoplasm. Significant H_2 production was observed when cells achieved anaerobiosis, the rate of H_2 production being higher in the presence of fermentative substrates such as glucose (Cournac et al., 2004). The transient H_2 burst observed upon illumination (Appel et al., 2000; Cournac et al., 2002, 2004) probably reflects the increase in NAD(P)H concentration in response to PSI activity. Appel et al. (2000) have proposed that the bidirectional H_2ase functions as an electron valve for the disposal of low-potential electrons generated at the onset of illumination. Fig. 6 shows that the production of H_2 depends on the supply of NAD(P)H. NAD(P)H can also donate electrons to the NDH-1 complex but the mechanism of this transfer is not clear. The [NiFe]-H_2ase is therefore in competition with NDH-1 for the NAD(P)H produced either by FNR or by glycolysis and the TCA cycle (Fig. 6). However, it is not yet clear how NAD(P)H can transfer electrons to NDH-1, since the *Synechocystis* NDH-1 complex lacks 3 of the usual 14 subunits, including the NAD(P)H-binding subunit. Since homologs of the missing subunits are present in the H_2ase, it has been suggested that the bidirectional H_2ase may interact directly with NDH-1, thus complementing its function, as illustrated in Fig. 6 (Schmitz and Bothe, 1996; Appel and Schulz, 1998; although see Boison et al., 1999).

IV. Regulation of Hydrogenase Gene Expression

A. Clustering of Hydrogenase Genes in Proteobacteria

The *Proteobacteria*, formerly known as 'purple bacteria and relatives,' comprise five subdivisions and form a large group of phenotypically diverse, Gram-negative bacteria, many of which are endowed with [NiFe]-H_2ases. The organization of genes encoding [NiFe]-H_2ases has been most extensively studied in this group. The genes are often found to be clustered, a feature that has facilitated their identification and the study of their products. The best studied of the H_2ase gene clusters, i.e. those from the α-*Proteobacteria, Rb. capsulatus, R. leguminosarum* and *B. japonicum*, the β-*Proteobacterium*, *R. eutropha*, and the γ-*Proteobacterium, E. coli*, are illustrated in Fig. 8.

The different operons have been termed variously, *hup* (for hydrogen uptake), *hox* (for hydrogen oxidation) or, in the case of *E. coli, hya, hyb, hyc* and *hyf,* (for H_2ases 1, 2, 3 and 4, respectively). In each structural operon coding for uptake H_2ase, the gene encoding the small subunit precedes that of the large subunit, followed by a third gene, named *hupC*, *hoxZ*, *hyaC* or *hybB*, encoding a cytochrome *b*, (the *E. coli* H_2ases 3 and 4 are part of formate hydrogen lyase complexes and evolve H_2). The structural genes are followed by genes necessary for the processing and maturation of the enzyme protein and by the *hyp* genes necessary for the insertion, at the active site, of the Ni and the Fe atom with its ligands, CO and CN. Altogether, there are 16 genes involved in the biosynthesis and the maturation of these H_2ases. The involvement of *hyp* genes in the assembly of the NiFe active center in *E. coli* H_2ase 3 has been intensively studied by the group of A. Böck, in Munich. Homologous *hyp* genes are present in all H_2ase gene clusters and have been found to function similarly in *E. coli* and in other proteobacterial organisms (see relevant references concerning the function of various accessory H_2ase genes in the reviews of Friedrich and Schwartz 1993; Vignais et al., 2001; Casalot and Rousset, 2001; Blokesch et al., 2002; Vignais and Colbeau, 2004).

B. Signaling and Transcription Control

Regulation of [NiFe]-H_2ase biosynthesis has mainly been studied in the organisms listed in Fig. 8. Some of these organisms are endowed with physiological attributes allowing their growth under very diverse environmental conditions: autotrophically or heterotrophically, in the light or in darkness, aerobically or anaerobically. The control of H_2ase synthesis represents a means to quickly and efficiently respond to changes in the environment and in particular to new energy demands. This control is exerted at the transcription level, as demonstrated in particular by the use of transcriptional reporter systems measuring levels of promoter activity of structural operons (Colbeau and Vignais, 1992). H_2ase synthesis responds to three main signals:

a. Molecular hydrogen, which is also the substrate, activates H_2ase gene expression in aerobic bacteria (e.g. *R. eutropha)*, in photosynthetic bacteria (e.g. *Rb. capsulatus*) and in free-living *Rhizobia* (e.g. *B. japonicum)*. In all these bacteria, the regulatory cascade responding to H_2 uses the same elements.

b. Molecular oxygen is an inhibitory signal for most of the H_2ases. In symbiotic *Rhizobia*, the regulation of H_2ase gene expression is linked to the expression of nitrogen fixation genes.

c. Metabolites functioning as electron donors or acceptors, such as formate, nitrate or sulfur, regulate the expression of H_2ases.

1. Response to H_2

The products of four genes function together as a *H_2-specific regulatory cascade*, which controls the transcription of the structural operon*s, hupSLC* in *Rb. capsulatus*, and *B. japonicum* and *hoxKGZ* in *R. eutropha*. The *H_2-specific regulatory system* comprises: (i) an *H_2-sensing H_2ase* (Vignais et al., 1997, 2000; Lenz and Friedrich, 1998) encoded by the *hupUV* genes in *B. japonicum* (Black et al., 1994) and *Rb. capsulatus* (Elsen et al., 1996) and by the homologous *hoxBC* genes in *R. eutropha* (Lenz et al., 1997) (Fig. 8). This H_2-sensing H_2ase, also called regulatory H_2ase (Kleihues et al., 2000), is the H_2 sensor for a two-component regulatory system which controls the transcription of the structural H_2ase operons. (ii) a *two-component regulatory system* which consists of a *protein histidine kinase*, termed HupT in *Rb. capsulatus* (Elsen et al., 1993, 1997) and in

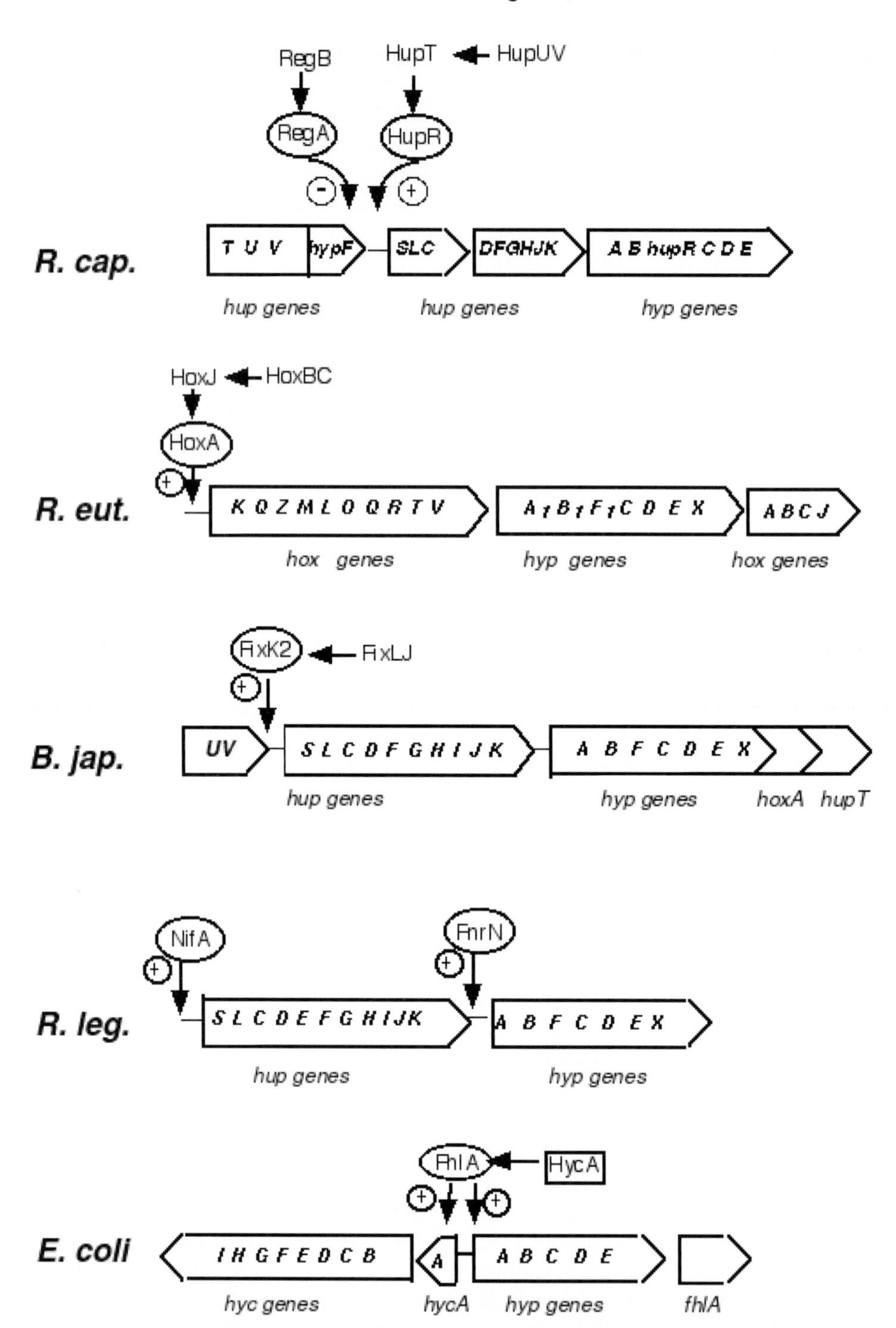

Fig. 8. Examples of [NiFe]-Hase gene clusters and transcriptional factors regulating gene expression in some *Proteobacteria*. In *Rb. capsulatus* (*R. cap.*), H_2ase expression is activated by an H_2-responding specific system (HupUV/HupT/HupR) and repressed by the global redox-sensing two-component regulatory system RegBA. In *R. eutropha* (*R. eut.*), the HoxBC/HoxJ/HoxA is homologous to HupUV/HupT/HupR from *Rb. capsulatus*. The same regulatory system is involved in free-living *B. japonicum* cells (not shown). In bacteroid cells, expression of H_2ase is linked to that of nitrogenase, by the intermediary of the two-component system FixLJ in *B. japonicum* (*B. jap.*) or the NifA regulator in *R. leguminosarum (R. leg.*). The scheme for *E. coli* concerns the control of the expression of H_2ase 3.

B. japonicum (Van Soom et al., 1999) and HoxJ in *R. eutropha* (Lenz et al., 1997; Lenz and Friedrich, 1998), and a *response regulator*, an NtrC-like transcription factor termed HupR (Richaud et al., 1991; Toussaint et al., 1997; Dischert et al., 1999) or HoxA (Zimmer et al., 1995; Durmowicz and Maier, 1997; Van Soom et al., 1997).

The NtrC superfamily of response regulators presents three domains, the N-terminal domain with the conserved phosphorylable aspartate residue, a central domain with Walker motifs for ATP binding, and a C-terminal domain with a helix-turn-helix motif. In *hupR* mutants (Richaud et al., 1991; Toussaint et al., 1997), expression of H_2ase no longer responded to the presence of H_2 and was even abolished. Expression of the response regulator HoxA gene is itself regulated in *R. eutropha* (Schwarz et al., 1999), while HupR in *Rb. capsulatus* is constitutively expressed (Dischert et al., 1999).

The gene coding for the histidine kinase (*hupT*) is localized upstream from the H_2ase structural genes in *Rb. capsulatus* (Fig. 8) and is cotranscribed with the *hupU* and *hupV* genes (Elsen et al., 1996). *HupT* mutants exhibit high H_2ase activities, even in absence of H_2, indicating that HupT negatively controls H_2ase gene expression (Elsen et al., 1993). This is the opposite phenotype to that shown by *hupR* mutants. However, the two types of mutant are similar in that expression of their uptake H_2ase does not respond to H_2, suggesting that both proteins are involved in the transduction of the H_2 signal. Indeed, it was later clearly demonstrated that the HupT and HupR proteins act as partners of a two-component regulatory system, which regulates the expression of the *hupSLC* H_2ase genes in *Rb. capsulatus* (Dischert et al, 1999).

The HupT histidine kinase is capable of autophosphorylation in the presence of [^{32}P]-γ-ATP (Elsen et al., 1997) and in the presence of HupT-P, the response regulator HupR becomes phosphorylated (Dischert et al., 1999). Footprinting experiments and in vivo experiments with plasmid-borne *hupS::lacZ* fusions have demonstrated that the transcription factor HupR binds to the *hupS* promoter and protects a palindromic sequence TTG-N_5-CAA localized at nt -162/-152 from the transcription start site (Toussaint et al., 1997; Dischert et al., 1999). Although HupR binds to an enhancer site of the *hupS* promoter and requires IHF to activate transcription (Toussaint et al., 1991), the *Rb. capsulatus hupSL* genes are transcribed by a σ^{70}-linked RNA polymerase (Dischert et al., 1999). This is a rather exceptional case, since enhancer-binding transcription factors usually activate σ^{54}-dependent RNA polymerase, as is the case for the activation of *hoxKG* gene transcription in *R. eutropha* (Römermann et al., 1989) or *hupSL* transcription in *B. japonicum* (Black and Maier, 1995).

A second peculiarity concerning the functioning of the two-component HupT/HupR and HoxJ/HoxA regulatory systems is that the response regulator (HupR or HoxA) activates the transcription of H_2ase genes in the nonphosphorylated form (Lenz and Friedrich, 1998; Dischert et al., 1999), in contrast to NtrC in enteric bacteria, which activates transcription under the phosphorylated form (Austin and Dixon, 1992). This was demonstrated in vivo by complementing a *hupR* null mutant of *Rb. capsulatus,* with mutated forms of the HupR protein in which the conserved phosphorylable Asp residue had been replaced by other aminoacids or had been deleted. In all cases, the mutant HupR proteins fully activated HupSL synthesis (Dischert et al., 1999). Similar types of experiment have demonstrated that HoxA in *R. eutropha* is active in the nonphosphorylated form (Lenz and Friedrich, 1998) and a similar conclusion was reached for *B. japonicum* HoxA (Van Soom et al., 1999). The relationships between the H_2 sensor HupUV, the HupT/HupR regulatory system and the promoter of the *hupSL* H_2ase are schematized in Fig.9. Although in *Rb. capsulatus, R. eutropha* and *B. japonicum*, the regulatory cascade is similar, *Rb. capsulatus* Hup(UV)$^-$ mutants have high level of H_2ase activity, while in *R. eutropha* and *B. japonicum*, mutants defective in the H_2 sensor are devoid of H_2ase activity.

2. Response to O_2 and Redox Regulation

Global regulatory proteins with cofactors that are able either to bind O_2 (heme and flavins) or to transfer electrons (heme, flavins and Fe-S clusters) (Dixon, 1998; reviewed by Bauer et al., 1999; Beinert and Kiley, 1999; Sawers, 1999) function in biological O_2-sensing pathways.

The *E. coli anaerobic global regulator Fnr* (for fumarate nitrate reduction) (Spiro and Guest, 1990; Kiley and Beinert, 1999) is a cytoplasmic, O_2-responsive regulator with a sensory and a regulatory DNA-binding domain. Fnr regulates transcription initiation in response to oxygen starvation. It activates the transcription of genes involved in anaerobic respiratory pathways, while repressing the expression

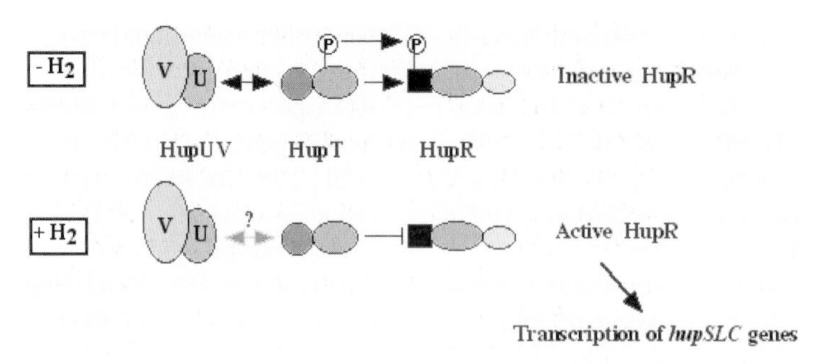

Fig. 9. Regulatory cascade in response to H_2 in *Rb. capsulatus.* In the absence of H_2, the heterodimer, HupUV, interacts with the histidine kinase, HupT (Elsen et al., 2003), which is able to autophosphorylate on the His_{217} residue. The transfer of the phosphate group inactivates the response regulator HupR. In the presence of H_2, sensed by HupUV, the interaction of HupUV with HupT is blocked or modified. As a result, the regulator HupR remains in an unphosphorylated, active form, which is able to activate the transcription of the *hupSLC* genes (Dischert et al, 1999). This scheme explains the opposing phenotypes observed in *hupT* and *hupUV* mutants on the one hand and in *hupR* mutants on the other.

of genes involved in aerobic energy generation (Iuchi and Lin, 1993). Upstream regions of Fnr-regulated genes are characterized by Fnr consensus sequences of dyad symmetry, TTGAT-N_4-ATCAA, to which the protein binds as a dimer. Fnr activity depends on the presence of a $[4Fe\text{-}4S]^{2+}$ cluster, which is labile in the presence of O_2, being converted rapidly to a more O_2-stable $[2Fe\text{-}2S]^{2+}$ form (Kiley and Beinert, 1999). That the O_2-sensing ability of Fnr is related to the O_2-lability of the $[4Fe\text{-}4S]^{2+}$ cluster was demonstrated using a mutant Fnr protein bearing a substitution of Leu-28 with His (Fnr-L28H), which stabilizes the $[4Fe\text{-}4S]^{2+}$ cluster in the presence of O_2. The mutant protein was functional under aerobic conditions (Bates et al., 2000). Fnr homologs are present in other *Proteobacteria* and have been shown to be required for gene expression under anaerobic or microaerobic conditions. They are classified into two types: Fnr-like (such as FixK1 in *B. japonicum* or FnrN in *R. leguminosarum*) and FixK-like (such as FixK2 in *B. japonicum*.) FixK-like proteins lack the redox sensitive cysteines and are activated by an associated, O_2-sensitive two-component system, FixLJ (Guttiérez et al., 1997) (Fig. 8).

Rhizobia can grow in symbiosis with leguminous plants, forming bacteroids that are located within specific root nodules. Under these conditions, expression of H_2ase is included in a complex network coordinating the process of N_2 fixation. In *B. japonicum*, the sensing of low O_2 concentrations and signal transduction are organized in two regulatory cascades involving the activators FixK2 and NifA for *hup* and *nif* genes, respectively. The transcription of H_2ase structural genes is activated by DNA binding of FixK2, whose expression depends on the FixL/FixJ two-component regulatory system (Durmowicz and Maier, 1998). The transmembrane sensor, FixL, has a heme-binding domain which belongs to the PAS superfamily (Taylor and Zhulin, 1999). The loss of O_2 from FixL induces a conformational change which activates its kinase activity (Gilles-Gonzales et al., 1991; Gong et al, 2000), triggers the phosphorylation of FixJ and initiates the regulatory cascade at reduced levels of O_2. *B. japonicum* contains two homologs of FixK but only FixK2 and FixJ mutants are devoid of H_2ase activity (Hup^-) and are also Nif^-. The H_2ase promoter presents two potential binding sites for FixK2: TTGA-C-GATCAA-G. In *B. japonicum*, besides the H_2ase genes, FixK2 regulates the expression of the *fixNOQP/ccoNOQP* operon encoding a terminal oxidase with a high affinity for O_2, *rpoN1, fixK1* and several genes involved in nitrate metabolism in response to low oxygen concentrations.

As in *B. japonicum*, H_2ase gene transcription in *R. leguminosarum* is co-regulated with that of nitrogenase genes. It is controlled by two global regulators, NifA and FnrN, in response to low oxygen concentrations in the nodules. However, in this bacterium, NifA directly activates H_2ase gene expression by binding to an UAS of the promoter region of the structural *hupSL* genes. This promoter also binds IHF and is transcribed by a σ^{54}-RNA polymerase. In addition, FnrN indirectly activates H_2ase gene expression by binding to the σ^{70}-dependent promoter of the *hyp* genes. There are two copies of the *fnrN* gene (Gutiérez et al., 1997) and only double mutants form ineffective nodules lacking both H_2ase and nitrogenase (Colombo et al., 2000).

Another redox-dependent sensor/regulator system is the ArcB/ArcA regulatory system of *E. coli* (Iuchi and Lin, 1993). ArcB is a transmembrane sensor protein, the histidine kinase activity of which is stimulated in absence of O_2. ArcB can then autophosphorylate and the phosphoryl group is transferred to a conserved aspartate of ArcA. Upon phosphorylation (anoxic conditions) ArcA is converted to Arc-phosphate (Arc-P), the active form which represses target genes of aerobic metabolism and activates genes of anaerobic metabolism. Quinones are redox signals for the Arc system. Oxidized forms of quinone electron carriers inhibit autophosphorylation of ArcB during aerobiosis, thus providing a link between the electron transport chain and gene expression (Georgellis et al., 2001). This two-component system is involved in the synthesis of H_2ase 1 and H_2ase 2 of *E. coli,* encoded by *hya* and *hyb* operons, respectively. However, while ArcA activates the expression of the *hya* genes it represses that of the *hyb* genes (Richard et al, 1999).

E. coli H_2ase 1 is encoded by the *hyaABCDEF* operon. Its physiological role is unclear, but it is generally assumed that it recycles the hydrogen produced by H_2ase 3. It is active under fermentative conditions (growth on glucose and formate) or during anaerobic respiration (growth on glycerol and fumarate). Anaerobic expression of the *hya* operon is independent of Fnr and is induced by two global regulators, ArcA and AppY. Depending on the growth conditions, the ArcA and AppY proteins act either cooperatively or independently. AppY is itself regulated by a two-component redox sensitive system, DpiAB, which functions in aerobiosis and represses ArcA synthesis in aerobiosis (Ingmer et al., 1998). Thus the *hya* operon is controlled by two two-component redox sensitive systems, one active under anaerobic conditions (ArcB/ArcA) and the other (AppY/DpiAB) under aerobic conditions (Richard et al., 1999). The transcriptional regulation of the *hybOABCDEFG* operon of *E. coli*, coding for H_2ase 2, is less well understood than that of the other *E. coli* H_2ases. Anaerobic induction is only slightly affected by *fnr* mutations and seems to be negatively regulated by ArcA but unaffected by AppY.

Oxidation of H_2 by uptake H_2ase produces electrons that can supply various energy-consuming processes, such as CO_2 reduction, nitrogen fixation and ATP synthesis (Fig. 3). In *Rb. capsulatus*, a global two-component signal transduction system, RegB/RegA (also called PrrB/PrrA in *Rb. sphaeroides*), which was originally found to regulate key operons involved in photosynthesis (reviewed by Bauer et al., 1999) and more recently to control respiratory electron transfer components (Swem et al., 2001), is implicated also in the control of CO_2 fixation (Qian and Tabita, 1996; Vichivanives et al., 2000) as well as nitrogen fixation and H_2 oxidation (Joshi and Tabita, 1996; Qian and Tabita, 1998; Elsen et al. 2000). In *Rb. capsulatus,* the RegA protein regulates negatively the expression of H_2ase by binding to *phupS,* the promoter of the structural H_2ase genes. RegB- and RegA-defective mutant strains have three to five times more activity than the wild type. Interestingly, this effect was observed under both aerobic and anaerobic growth conditions (Elsen et al., 2000). By foot-printing experiments, two binding sites for RegA were found, a high affinity site located between the binding sites for the HupR regulator and for RNA polymerase, and a low affinity site overlapping the binding site for the global regulator IHF (Elsen et al., 2000).

The effect of the Reg/Prr system is mediated by a regulatory cascade originating from the respiratory chain. This cascade has been elucidated in *Rb. sphaeroides* by the group of Kaplan (Eraso and Kaplan, 2002). The signal comes from the *cbb*$_3$ cytochrome *c* oxidase via its interaction with O_2 in conjunction with the redox protein Rdx. The *cbb*$_3$ cytochrome *c* oxidase, present in cells grown either in aerobiosis or anaerobiosis, shows properties of an O_2 sensor. It has a high affinity for O_2 but it is the rate or the volume of the electron flux through the *cbb*$_3$ Cyt *c* oxidase which is the signal rather than the binding of O_2 to the oxidase per se (Oh and Kaplan, 2000). The conformation of the *cbb*$_3$ oxidase changes with the volume of the electron flux; this long range conformational change in the *cbb3* oxidase is sensed and ultimately transponded by CcoQ subunit to the two-component PrrB/PrrA system by the intermediary of the membrane-localized PrrC polypeptide (reviewed by Oh and Kaplan, 2001; see also Chapter 13 by Kaplan and Oh, Vol. 15 of this series and Chapter 13 by Vermeglio et al., present volume). In *Rb. capsulatus*, signal transduction by RegB is mediated by a redox-active cysteine (Swem et al., 2003). Under oxidizing conditions, formation of an intermolecular disulfide bond converts the RegB kinase from an active dimer into an inactive tetramer state. In *B. japonicum,* the two-component system RegS/RegR is homologous to PrrB/PrrA or RegB/RegA. However, it is not known if this system regulates H_2ase synthesis. Under symbiotic conditions, H_2ase synthesis is regulated by the FixLJ-FixK2 system (see above) which is independent of RegS/RegR (Emmerich et al, 1999).

3. Response to Metabolites, Electron donors or Acceptors

a. Formate Regulation

E. coli H_2ase 3, encoded by the *hyc* operon, is the H_2ase component of formate hydrogen lyase (FHL-1) which evolves H_2 under fermenting growth conditions and is regulated by the intracellular concentration of formate. The genes coding for *E. coli* H_2ase 3 also belong to the formate regulon (Rossmann et al., 1991), which is regulated by the transcriptional regulator FhlA. FhlA shares homology with σ^{54}-dependent regulators of the NtrC family in its central and C terminal domains but differs in possessing an extended N terminal domain lacking the aspartate residue, which is the site of phosphorylation of response regulators. Thus, FhlA is not activated by phosphorylation but by binding an effector molecule, formate.

FhlA is an homotetramer which binds to and activates the *hyc, hyp, fhlF* and *hydN/hypF* promoters. The *hyc* and *hyp* operons are divergently transcribed. FhlA binds in the intergenic region and activates the *hyc* operon (Fig. 8). An additional binding site between the *hycA* (coding for a repressor) and the *hycB* genes is responsible of the activation of the *hyp* operon. Thus the regulator FhlA controls the expression of the structural and accessory genes of H_2ase. These promoters are of the σ^{54} type and are activated by the global regulator IHF (Rossman et al., 1991).

The H_2ase 4 operon (*hyf*) of *E. coli* is regulated by the σ^{54}-dependent activators FhlA (thus *hyf* expression resembles that of the *hyc* operon) and HyfR (related to FhlA). HyfR-dependent induction is dependent on low pH, anaerobiosis and postexponential growth and is weakly enhanced by formate. However, no *hyf*-related hydrogenase or formate dehydrogenase activity has so far been detected (Skibinski et al., 2002).

The particular case described above concerns fermentation rather than respiration. No evidence has yet been obtained for the regulation of uptake H_2ases by specific metabolic intermediates. Global regulatory networks have been discovered in some bacteria, that control both carbon and hydrogen metabolism, as well as nitrogen metabolism (e.g., Qian and Tabita, 1996).

b. Carbon Monoxide Regulation

In *R. rubrum*, CO induces the synthesis of CooLH H2ase. This synthesis is activated by CooA (CO-oxidation activator), a heme containing transcription factor, member of the (CAP/CRP)/FNR superfamily of regulatory proteins (Lanzilotta et al., 2000). CooA is unable to bind CO when the Fe heme is oxidized; upon reduction, there is an unusual switch of protein ligands to the six-coordinate heme and the reduced heme is able to bind CO. CO binding stabilizes a conformation of the dimeric protein that allows sequence-specific DNA binding and transcription is activated through contacts between CooA and RNA polymerase. Thus, the CooA dimer functions both as a redox sensor and as a specific CO sensor (Aono et al., 2000; reviewed by Roberts et al., 2001; Aono, 2003)

c. Nitrate Regulation

Expression of *hya* and *hyb* operons are strongly repressed in presence of nitrate, repression being mediated by the two-component systems NarQ/NarP and especially NarX/NarL. Several potential binding sites for these regulators are present on *phya* and p*hyb* (Richard et al., 1999). However, it has been suggested for H_2ase 1 that this repression could be indirectly linked to ArcA. Indeed, phosphorylation of ArcA by ArcB depends on the respiratory state of the cell: when nitrate is present, ArcA may be more activated than in its absence (Brønsted and Atlung, 1994).

Many H_2-oxidizing bacteria, such as *R. eutropha* (*A. eutrophus*) are also capable of nitrate respiration and denitrification and are able to grow anaerobically with H_2 as electron donor and nitrate as electron acceptor. However, no regulatory link has yet been established between H_2 uptake and the presence of nitrate or any of the denitrification pathway intermediates. In *R. eutropha* H16, H_2 oxidation and nitrate reduction are both under the control of the σ^{54} factor of RNA polymerase, and pleiotropic mutants deficient in both activities have been isolated (Römermann et al., 1988, 1989). The genes for uptake H_2ase are located on a 450-kb megaplasmid in this strain and the elimination of this plasmid resulted in the simultaneous loss of H_2 oxidation and nitrate reduction, suggesting that denitrification genes might also be plasmid-located. However, the loss of the ability to grow with nitrate was subsequently found to be due to the presence, on the mega-plasmid, of a gene for anaerobic ribonucleotide reductase, which is absolutely required for anaerobic growth (Siedow et al., 1999).

A regulatory interaction between H_2ase activity and the aerobic electron transport chain in *A. eutrophus* H16 was shown by the analysis of H_2ase-deficient mutants (Kömen et al., 1992). Mutants lacking the membrane-bound H_2ase contained increased amounts of the cytochrome *aa*$_3$-type cytochrome *c* oxidase and used this terminal oxidase preferentially for respiration, whereas mutants deficient in soluble H_2ase used predominantly the quinol oxidases for electron transfer. It would clearly be of interest to determine the effects of these mutations on the activities of nitrate reductase and other anaerobic respiratory enzymes.

d. Sulfur Regulation

The hyperthermophilic archaeon *P. furiosus* can grow on maltose either in the absence of elemental sulfur S^0 (it produces H_2 as an end product rather than H_2S) or in the presence of S^0. Recently, the effect of S^0 on the levels of gene expression in *P. furiosus* cells grown at 95 °C with maltose as the carbon source was investigated with the use of DNA microarrays (Schut et al., 2001). Eighteen ORFs that encode subunits associated with the three H_2ases characterized in *P. furiosus* (two cytoplasmic, H_2ases I and II, and one membrane-bound) were found to be strongly down-regulated by S^0 (an indication that these H_2ases are probably not directly involved in S^0 reduction). The nature of the enzyme system that reduces S^0 remains unknown as is unknown the mechanism by which S^0 affects H_2ase gene expression in *P. furiosus*.

V. Conclusions and Perspectives

H_2ase enzymes play many different roles in respiration. Chief among these is clearly the oxidation of H_2 or the reduction of protons, coupled to energy-conserving electron transfer chain reactions, which allow energy to be obtained either from H_2 or from the oxidation of lower potential substrates. These energy-conserving reactions are generally restricted to the prokaryotes, but are widely distributed among the bacterial and archaeal domains of life. In the last decade, additional roles have been revealed. Thus, the so-called H_2-sensor H_2ases are involved in regulating the biosynthesis of uptake [NiFe]-H_2ases in response to their substrate, H_2. These H_2ases form a separate class with some cyanobacterial H_2ases, which are apparently non-regulatory and function as uptake H_2ases. Other, bidirectional H_2ases may interact with respiratory electron transport chains and act as electron 'valves' to control the redox poise of the respiratory chain at the level of the quinone pool. This is essential to ensuring the correct functioning of the respiratory chain in the presence of excess reducing equivalents, particularly in photosynthetic microorganisms. It is interesting to note that a similar role has long been proposed for the nitrogenase of photosynthetic bacteria, which functions as a H_2-evolving H_2ase in the absence of N_2. An additional finding concerns some H_2ases which were originally thought to play a purely fermentative role, but which are now known to be involved in membrane-linked energy conservation through the generation of a transmembrane protonmotive force.

H_2ases are a structurally diverse, as well as a functionally diverse, group of enzymes and phylogenetic analysis has led to the identification of several phylogenetically distinct groups and subgroups which form the basis of a coherent system of classification. The large number of H_2ase gene sequences has been augmented by whole genome sequencing, which has revealed the presence of multiple H_2ases in several *Bacteria* and *Archaea*. Post-genomic analysis (transcriptome, proteome, metabolome) has and will be essential to elucidating the metabolic roles of these enzymes and the regulation of their biosynthesis and activity. Studies of regulation have essentially been restricted to the uptake [NiFe]-H_2ases of *Proteobacteria*, but have recently been extended to other types of H_2ase and other microorganisms, such as the hyperthermophilic archaeon, *P. furiosus*.

The phylogenetic diversity of H_2ases attests to the importance of H_2 metabolism in a wide range of environments, and suggests that H_2ases may have appeared very early in evolution. It has been proposed that H_2 metabolism may have been the driving force that led to the cellular symbiosis and fusion events that are thought to have been involved in the formation of the first eukaryotic cells. The present day H_2ases that are found in the organelles (hydrogenosomes or chloroplasts) of some unicellular eukaryotes may be relics of these evolutionary events or the results of more recent lateral gene transfer events. The continued study of these enzymes will assuredly lead to further progress in this field.

The current interest in H_2 as an alternative energy source to fossil fuels has led to a resurgence of interest in the biological production of H_2, and research into H_2ases will clearly play a major role in this area. Structural studies of H_2ases will be important in directing protein engineering, e.g. in decreasing the O_2-sensitivity of these enzymes. Studies of H_2 metabolism and regulation will also be important in engineering microorganism at the cellular level in order to maximize H_2 production. The isolation of novel H_2-producing organisms will also be a priority. Recent advances in molecular taxonomy (e.g. the analysis of environmental 16S rDNA sequences) have revealed that the number of existing bacterial species is at least two orders of magnitude higher than the number of isolated ones. Prokaryotic biodiversity is therefore much greater than previously thought, and whole phylogenetic groupings exist, e.g. the mesophilic *Crenarchaeota*, which have never been cultivated. Given the importance of H_2 metabolism among microorganisms generally, it can be anticipated that many of these so-far uncultivated species will contain H_2ases and that novel types of H_2ase and of H_2 metabolism remain to be discovered.

Acknowledgments

We are grateful to our colleagues and friends, Laurent Cournac, Alain Dupuis, Sylvie Elsen and André Vermeglio, for careful reading of the manuscript and for many helpful suggestions and comments, and Isabelle Prieur for the scheme of *Rb. capsulatus* Complex I used in Fig. 4. The expert assistance of Gérard Klein with the preparation of the figures is gratefully acknowledged.

References

Adams MWW (1990) The structure and mechanism of iron-hydrogenases. Biochim Biophys Acta 1020: 115–14

Albracht SPJ and de Jong AMPh (1997) Bovine-heart NADH-ubiquinone oxidoreductase is a monomer with 8 Fe-S clusters and 2 FMN groups. Biochim Biophys Acta 1318: 92–106

Albracht SPJ and Hedderich R (2000) Learning from hydrogenases: Location of a proton pump and of a second FMN in bovine NADH-ubiquinone oxidoreductase (Complex I). FEBS Lett 485: 1–6

Andrews SC, Berks BC, Mcclay J, Ambler A, Quail MA, Golby P and Guest JR (1997) A 12-cistron *Escherichia coli* operon (*hyf*) encoding a putative proton-translocating formate hydrogenlyase system. Microbiology 143: 3633–3647

Aono S, Honma Y, Ohkubo K, Tawara T, Kamiya T and Nakajima H (2000) CO sensing and regulation of gene expression by the transcriptional activator CooA. J Inorg Biochem 82: 51–56

Aono S (2003) Biochemical and biophysical properties of the CO-sensing transcriptional activator CooA. Acc Chem Res 36: 825–831

Appel J and Schulz R (1996) Sequence analysis of an operon of a NAD(P)-reducing nickel hydrogenases from the cyanobacterium *Synechocystis* sp. PCC 6803 gives additional evidence for direct coupling of the enzyme to NAD(P)H-dehydrogenase (complex I). Biochim Biophys Acta 1298: 141–147

Appel J and Schulz R (1998) Hydrogen metabolism in organisms with oxygenic photosynthesis: Hydrogenases as important regulatory devices for a proper redox poising? J Photochem Photobiol B-Biol 47: 1–11

Appel J, Phunpruch S, Steinmuller K, Schulz R (2000) The bidirectional hydrogenase of *Synechocystis* sp PCC 6803 works as an electron valve during photosynthesis. Arch Microbiol 173: 333–338

Atta M and Meyer J (2000) Characterization of the gene encoding the [Fe]-hydrogenase from *Megasphaera elsdenii*. Biochim Biophys Acta 1476: 368–371

Austin S and Dixon R (1992) The procaryotic enhancer binding protein NtrC has an ATPase activity which is phosphorylation and DNA dependent. EMBO J 11: 2219–2228

Bagramyan K, Mnatsakanyan N, Poladian A, Vassilian A and Trchounian A (2002) The roles of hydrogenases 3 and 4, and the F_0F_1-ATPase, in H_2 production by *Escherichia coli* at alkaline and acidic pH. FEBS Lett 516: 172–178

Bates DM, Popescu CV, Khoroshilova N, Vogt K, Beinert H, Munck E, Kiley PJ (2000) Substitution of leucine 28 with histidine in the *Escherichia coli* transcription factor FNR results in increased stability of the $[4Fe\text{-}4S]^{2+}$ cluster to oxygen J Biol Chem 275: 6234–6240

Bauer CE, Elsen S and Bird TH (1999) Mechanisms for redox control of gene expression. Annu Rev Microbiol 53: 495–523

Bäumer S, Ide T, Jacobi C, Johann A Gottschalk G and Deppenmeier U (2000) The $F_{420}H_2$ dehydrogenase from *Methanosarcina mazei* is a redox-driven proton pump closely related to NADH dehydrogenase. J Biol Chem 275: 17968–17973

Beinert H and Kiley PJ (1999) Fe-S proteins in sensing and regulatory functions. Curr Opin Chem Biol 3: 152–157.

Berks BC, Sargent F and Palmer T (2000) The Tat protein export pathway. Mol Microbiol 35: 260–274

Bernhard M, Benelli B, Hochkoeppler A, Zannoni D and Friedrich B (1997) Functional and structural role of the cytochrome *b* subunit of the membrane-bound hydrogenase complex of *Alcaligenes eutrophus* H16. Eur J Biochem 248: 179–186

Bernhard M, Friedrich B and Siddiqui RA (2000) *Ralstonia eutropha* TF93 is blocked in Tat-mediated protein export. J Bacteriol 182: 581–588

Black LK and Maier RJ (1995) IHF- and RpoN-dependent regulation of hydrogenase expression in *Bradyrhizobium japonicum*. Mol Microbiol 16: 405–413

Black LK, Fu C and Maier RJ (1994) Sequence and characterization of *hupU* and *hupV* genes of *Bradyrhizobium japonicum* encoding a possible nickel-sensing complex involved in hydrogenase expression. J Bacteriol 176: 7101–7106

Blattner FR, Plunkett IIIG, Bloch CA, Perna NT, Burland V, Riley M, Collado-Vides J, Glasner JD, Rode CK, Mayhew GF, Gregor J, Davis NW, Kirkpatrick HA, Goeden MA, Rose DJ, Mau B and Shao Y (1997) The complete genome sequence of *Escherichia coli* K-12. Science 277: 1453–1474

Blokesch M, Paschos A, Theodoratou E, Bauer A, Hube M, Huth S and Bock A (2002) Metal insertion into NiFe-hydrogenases.

Biochem Soc Trans 30: 674–680
Bogsch E, Sargent F, Stanley NR, Berks BC, Robinson C and Palmer T (1998) An essential component of a novel bacterial protein export system with homologues in plastids and mitochondria. J Biol Chem 273: 18003–18006
Böhm R, Sauter M and Böck A (1990) Nucleotide sequence and expression of an operon in *Escherichia coli* coding for formate hydrogenlyase components. Mol Microbiol 4: 231–243
Boison G, Bothe H, Hansel A and Lindblad P (1999) Evidence against a common use of the diaphorase subunits by the bidirectional hydrogenase and by the respiratory complex I in cyanobacteria. FEMS Microbiol Lett 174: 159–165
Bose SK and Gest H (1963) Bacterial photophosphorylation: Regulation by redox balance. Proc Natl Acad Sci USA 49: 337–345
Bowien B and Schlegel HG (1981) Physiology and biochemistry of aerobic hydrogen-oxidizing bacteria. Annu Rev Microbiol 350: 405–452
Brandt U (1999) Proton translocation in the respiratory chain involving ubiquinone — a hypothetical semiquinone switch mechanism for complex I. Biofactors 9: 95–101
Brøndsted L and Atlung T (1994) Anaerobic regulation of the hydrogenase 1 (*hya*) operon of *Escherichia coli*. J Bacteriol 176: 5423–5428
Bui ETN and Johnson PJ (1996) Identification and characterization of [Fe]-hydrogenases in the hydrogenosome of *Trichomonas vaginalis*. Mol Biochem Parasitol 76: 305–310
Candela M, Zaccherini E and Zannoni D (2001) Respiratory electron transport and light-induced energy transduction in membranes from the aerobic photosynthetic bacterium *Roseobacter denitrificans*. Arch Microbiol 175: 168–177
Carroll J, Shannon RJ, Fearnley IM, Walker JE and Hirst J (2002) Definition of the nuclear encoded protein composition of bovine heart mitochondrial complex I. Identification of two new subunits. J Biol Chem 277: 50311–50317
Casalot L and Rousset M (2001) Maturation of the [NiFe] hydrogenase. Trends Microbiol. 9: 228–236
Cauvin B, Colbeau A and Vignais PM (1991) The hydrogenase structural operon in *Rhodobacter capsulatus* contains a third gene, *hupM*, necessary for the formation of a physiologically competent hydrogenase. Mol Microbiol 5: 2519–2527
Chanal A, Santini C-L and Wu L-F (1998) Potential receptor function of three homologous components, TatA, TatB and TatE, of the twin-arginine signal sequence-dependent metalloenzyme translocation pathway in *Escherichia coli*. Mol Microbiol 30: 674–676
Colbeau A and Vignais PM (1992) Use of *hupS::lacZ* gene fusion to study regulation of hydrogenase expression in *Rhodobacter capsulatus*: stimulation by H_2. J Bacteriol 174: 4258–4264
Colombo MV, Gutiérrez D, Palacios JM, Imperial J and Ruiz-Argüeso (2000) A novel autoregulation mechanism of *fnr* expression in *Rhizobium leguminosarum* bv viciae. Mol Microbiol 36: 477–486
Cooley JW and Vermaas WFJ (2001) Succinate dehydrogenase and other respiratory pathways in thylakoid membranes of *Synechocystis* sp strain PCC 6803: Capacity comparisons and physiological function. J Bacteriol 183: 4251–4258
Cournac L, Redding K, Ravenel J, Rumeau D, Josse EM, Kuntz M and Peltier G (2000) Electron flow between photosystem II and oxygen in chloroplasts of photosystem I-deficient algae is mediated by a quinol oxidase involved in chlororespiration. J Biol Chem 275: 17256–17262
Cournac L, Mus F, Bernard B, Guedeney G, Vignais P and Peltier G (2002) Limiting steps of hydrogen production in *Chlamydomonas reinhardtii* and *Synechocystis* PCC 6803 as analysed by light-induced gas exchange transients. Int J Hydrogen Energy 27: 1229–1237
Cournac C, Guedeney G, Peltier G and Vignais PM (2004) Sustained photoevolution of molecular hydrogen in a mutant of *Synechocystis* sp. strain PCC 6803 deficient in the type I NADPH-dehydrogenase complex. J Bacteriol 186: 1737–1746
Deppenmeier U, Lienard T and Gottschalk G (1999) Novel reactions involved in energy conservation by methanogenic archaea. FEBS Lett. 457: 291–297
Dischert W, Vignais PM and Colbeau A (1999) The synthesis of *Rhodobacter capsulatus* HupSL hydrogenase is regulated by the two-component HupT/HupR system. Mol Microbiol 34: 995–1006
Dixon R (1998) The oxygen-responsive NIFL-NIFA complex: A novel two-component regulatory system controlling nitrogenase synthesis in gamma- proteobacteria. Arch Microbiol 169: 371–380
Dolla A, Pohorelic BK, Voordouw JK and Voordouw G (2000) Deletion of the *hmc* operon of *Desulfovibrio vulgaris* subsp. *vulgaris* Hildenborough hampers hydrogen metabolism and low-redox-potential niche establishment. Arch Microbiol 174: 143–151
Dross F, Geisler V, Lenger R, Theis F, Krafft T, Fahrenholz F, Kojro E, Duchêne A, Tripier D, Juvenal K and Kröger A (1992) The quinone-reactive Ni/Fe-hydrogenase of *Wolinella succinogenes*. Eur J Biochem 206: 93–102 [Erratum (1993) 214: 949–950]
Dubini A, Pye RL, Jack RL, Palmer T and Sargent F (2002) How bacteria get energy from hydrogen: A genetic analysis of periplasmic hydrogen oxidation in *Escherichia coli*. Int J Hydrogen Energy 27: 1413-1420
Dupuis A, Peinnequin A, Darrouzet E, Lunardi J (1997) Genetic disruption of the respiratory NADH-ubiquinone reductase of *Rhodobacter capsulatus* leads to an unexpected photosynthesis-negative phenotype. FEMS Microbiol Lett 149: 107–114
Dupuis A, Chevalet M, Darrouzet E, Duborjal H, Lunardi J and Issartel JP (1998) The complex I from *Rhodobacter capsulatus*. Biochim Biophys Acta 1364: 147–165
Dupuis A, Prieur I and Lunardi J (2001) Towards a characterization of the connecting module of Complex I. J Bioenerg Biomemb 33: 159–168
Durmowicz MC and Maier RJ (1997) Roles of HoxX and HoxA in biosynthesis of hydrogenase in *Bradyrhizobium japonicum*. J Bacteriol 179: 3676–3682
Durmowicz MC and Maier RJ. (1998) The FixK2 protein is involved in regulation of symbiotic hydrogenase expression in *Bradyrhizobium japonicum* J Bacteriol 180: 3253–3256
Dutton PL, Moser CC, Sled VD, Daldal F and Ohnishi T (1998) A reductant-induced oxidation mechanism for complex I. Biochim Biophys Acta 1364: 245–257
Elsen S, Richaud P, Colbeau A and Vignais PM (1993) Sequence analysis and interposon mutagenesis of the *hupT* gene, which encodes a sensor protein involved in repression of hydrogenase synthesis in *Rhodobacter capsulatus*. J Bacteriol: 175: 7404–7012
Elsen S, Colbeau A, Chabert J and Vignais PM (1996) The *hupTUV* operon is involved in negative control of hydrogenase synthesis in *Rhodobacter capsulatus*. J Bacteriol 178: 5174–5181

Elsen S, Colbeau A and Vignais PM (1997) Purification and in vitro phosphorylation of HupT, a regulatory protein controlling hydrogenase gene expression in *Rhodobacter capsulatus* J Bacteriol 179: 968–971

Elsen S, Dischert W, Colbeau A and Bauer CE (2000) Expression of uptake hydrogenase and molybdenum nitrogenase in *Rhodobacter capsulatus* is coregulated by the RegB-RegA two-component regulatory system. J Bacteriol 182: 2831–2837

Elsen S, Duche O and Colbeau A (2003) Interaction between the H2 sensor HupUV and the histidine kinase HupT controls hydrogenase synthesis in *Rhodobacter capsulatus*. J Bacteriol 185: 7111–7119

Emmerich R, Panglungtshang K, Strehler P, Hennecke H and Fischer H (1999) Phosphorylation, dephosphorylation and DNA-binding of the *Bradyrhizobium japonicum* RegSR two-component regulatory proteins. Eur J Biochem 263: 455–463

Eraso JM and Kaplan S (2002) Redox flow as an instrument of gene regulation. Methods Enzymol 348: 216–229

Fauque G, Peck HD, Jr, Moura JJG, Huynh BH, Berlier Y, DerVertanian DV, Teixeira M, Przybyla AE, Lespinat PA, Moura I, and Le Gall J (1988) The three classes of hydrogenases from sulfate-reducing bacteria of the genus *Desulfovibrio*. FEMS Microbiol Rev 54: 299–344

Fearnley IM and Walker JE (1992) Conservation of sequences of subunits of mitochondrial complex I and their relationships with other proteins. Biochim Biophys Acta 1140: 105–134

Ferguson SJ, Jackson JB and McEwan AG (1987) Anaerobic respiration in the Rhodospirillaceae: Characterisation of pathways and evaluation of roles in redox balancing during photosynthesis. FEMS Microbiology Reviews 46: 117–143

Florin L, Tsokoglou A and Happe T (2001) A novel type of iron hydrogenase in the green alga *Scenedesmus obliquus* is linked to the photosynthetic electron transport chain. J Biol Chem 276: 6125–6132

Fox J D, He Y, Shelver D, Roberts GP and Ludden PW (1996a) Characterization of the region encoding the CO-induced hydrogenase of *Rhodospirillum rubrum*. J Bacteriol 178: 6200–6208

Fox JD, Kerby RL, Roberts GP and Ludden PW (1996b) Characterization of the CO-induced, CO-tolerant hydrogenase from *Rhodospirillum rubrum* and the gene encoding the large subunit of the enzyme. J Bacteriol 178: 1515–1524

Friedrich B and Schwartz E (1993) Molecular biology of hydrogen utilization in aerobic chemolithotrophs. Annu Rev Microbiol 47: 351–383

Friedrich T (1998) The NADH:ubiquinone oxidoreductase (complex I) from *Escherichia coli*. Biochim Biophys Acta 1364: 134–146

Friedrich T (2001) Complex I: A chimaera of a redox and conformation-driven proton pump? J Bioenerg. Biomemb. 33: 169–177

Friedrich T and Scheide D (2000) The respiratory complex I of bacteria, archaea and eukarya and its module common with membrane-bound multisubunit hydrogenases. FEBS Lett 479: 1–5

Friedrich T and Weiss H (1997) Modular evolution of the respiratory NADH-ubiquinone oxidoreductase and the origin of its modules. J Theor Biol 187: 529–540

Garcin E, Vernede X, Hatchikian EC, Volbeda A, Frey M and Fontecilla-Camps JC (1999) The crystal structure of a reduced [NiFeSe] hydrogenase provides an image of the activated catalytic center. Structure Fold Des 7: 557–566

Georgellis D, Kwon O and Lin EC (2001) Quinones as the redox signal for the arc two-component system of bacteria. Science 292: 2314–2316

Gest H (1972) Energy conversion and generation of reducing power in bacterial photosynthesis. Adv Microb Physiol 7: 243–282.

Gilles-Gonzales MA, Ditta GS and Helinski DR (1991) A haemoprotein with kinase activity encoded by the oxygen sensor of *Rhizobium meliloti*. Nature 350: 170–172

Gong W, Hao B and Chan MK (2000) New mechanistic insights from structural studies of the oxygen-sensing domain of *Bradyrhizobium japonicum* FixL. Biochemistry 39: 3955–3962

Gross R, Simon J, Theis F and Kröger A (1998) Two membrane anchors of *Wolinella succinogenes* hydrogenase and their function in fumarate and polysulfide respiration. Arch Microbiol 170: 50–58

Gross R, Simon J and Kröger A (1999) The role of the twin-arginine motif in the signal peptide encoded by the *hydA* gene of the hydrogenase from *Wolinella succinogenes*. Arch. Microbiol. 172: 227–232

Guttiérez D, Hernando Y, Palacios J-M, Imperial J and Ruiz-Argüeso T (1997) FnrN controls symbiotic nitrogen fixation and hydrogenase activities in *Rhizobium leguminosarum* Biovar viciae UPM791. J Bacteriol 179: 5264–5270

Hackstein JHP, Akhmanova A, Boxma B, Harhangi HR, and Voncken FGJ (1999) Hydrogenosomes: Eukaryotic adaptations to anaerobic environments. Trends Microbiol 7: 441–447

Hackstein JHP, Akhmanova A, Voncken F et al. (2001) Hydrogenosomes: Convergent adaptations of mitochondria to anaerobic environments. Zoology 104: 290–302

Happe T and Kaminski A (2002) Differential regulation of the Fe-hydrogenase during anaerobic adaptation in the green alga *Chlamydomonas reinhardtii*. Eur J Biochem 269: 1022–1032

Happe T, Schütz K and Böhme H (2000) Transcriptional and mutational analysis of the uptake hydrogenase of the filamentous cyanobacterium *Anabaena variabilis* ATCC 29413. J Bacteriol 182: 1624–1631

Hedderich R (2004) Energy-converting [NiFe] hydrogenases from archaea and extremophiles: Ancestors of complex I? J Bioenerg Biomemb 36: 65–75

Henry M-F and Vignais PM (1983) Electron pathways from H_2 to nitrate in *Paracoccus denitrificans*: Effects of inhibitors of the UQ-cytochrome *b* region. Arch Microbiol 136: 64–68

Herter SM, Kortluke CM and Drews G (1998) Complex I of *Rhodobacter capsulatus* and its role in reverted electron transport. Arch Microbiol 169: 98–105

Higuchi Y, Yagi T and Yasuoka N (1997) Unusual ligand structure in Ni-Fe active center and an additional Mg site in hydrogenase revealed by high resolution X-ray structure analysis. Structure 5: 1671–1680

Higuchi Y, Ogata H, Miki K, Yasuoka N, and Yagi T (1999) Removal of the bridging ligand atom at the Ni-Fe active site of [NiFe] hydrogenase upon reduction with H_2, as revealed by X-ray structure analysis at 1.4 Å resolution. Structure 7: 549–556

Hillmer P and Gest H (1977) H_2 metabolism in the photosynthetic bacterium *Rhodopseudomonas capsulata*: H_2 production by growing cultures. J Bacteriol 129: 724–731

Horner DS, Foster PG and Embley TM (2000) Iron hydrogenases

and the evolution of anaerobic eukaryotes. Mol Biol Evol 17: 1695–1709

Horner DS, Heil B, Happe T and Embley TM (2002) Iron hydrogenases — ancient enzymes in modern eukaryotes. Trends Biochem Sci 27: 148–153

Houchins JP (1984) The physiology and biochemistry of hydrogen metabolism in cyanobacteria. Biochim Biophys Acta 768: 227–255

Ide T, Bäumer S and Deppenmeier U (1999) Energy conservation by the H_2:heterodisulfide oxidoreductase from *Methanosarcina mazei* Gö1: Identification of two proton-translocating segments. J Bacteriol 181: 4076–4080

Ingmer H, Miller CA and Cohen SN (1998) Destabilised inheritance of pSC101 and other *Escherichia coli* plasmids by DpiA, a novel two-component system regulator. Mol Microbiol 29: 49–59

Iuchi S and Lin ECC (1993) Adaptation of *Escherichia coli* to redox environments by gene expression. Mol Microbiol 9: 9–15

Jouanneau Y, Jeong HS, Hugo N, Meyer C and Willison JC (1998) Overexpression in *Escherichia coli* of the *rnf* genes from *Rhodobacter capsulatus*. Characterization of two membrane-bound iron-sulfur proteins. Eur J Biochem 251: 54–64

Joshi HM and Tabita FR (1996) A global two component signal transduction system that integrates the control of photosynthesis, carbon dioxide assimilation, and nitrogen fixation. Proc Natl Acad Sci USA 93: 14515–14520

Kaneko T, Sato S, Kotani H, Tanaka A, Asamizu E, Nakamura Y, Miyajima N, Hirosawa M, Sugiura M, Sasamoto S, Kimura T, Hosouchi T, Matsuno A, Muraki A, Nakazaki N, Naruo K, Okumura S, Shimpo S, Takeuchi C, Wada T, Watanabe A, Yamada M, Yasuda M, and Tabata S. (1996) Sequence analysis of the genome of the unicellular cyanobacterium *Synechocystis* sp. strain PCC6803. II. Sequence determination of the entire genome and assignment of potential protein-coding regions. DNA Res 3: 109–136 and 3: 185–209.

Kashani-Poor N, Zwicker K, Kerscher S and Brandt U (2001) A central functional role for the 49-kDa subunit within the catalytic core of mitochondrial complex I. J Biol Chem 276: 24082–24087

Kiley PJ and Beinert H (1999) Oxygen sensing by the global regulator, FNR: the role of the iron-sulfur cluster. FEMS Microbiol Rev 22: 341–352

Kleihues L, Lenz O, Bernhard M, Buhrke T and Friedrich B (2000) The H_2 sensor of *Ralstonia eutropha* is a member of the subclass of regulatory [NiFe] hydrogenases. J Bacteriol 182: 2716–2724

Klemme JH (1969) Studies on the mechanism of NAD-photoreduction by chromatophores of the facultative phototroph, *Rhodopseudomonas capsulata*. Z Naturforsch 24b: 67–76

Knaff DB (1978) Reducing potentials and the pathway of NAD^+ reduction. In: Clayton RK and Sistrom WR (eds) The Photosynthetic Bacteria, pp 629–640. Plenum, New York

Kömen R, Schmidt K and Friedrich B (1992) hydrogenase mutants of *Alcaligenes eutrophus* H16 show alterations in the electron transport system. FEMS Microbiol Lett 96: 173–178

Kömen R, Schmidt K and Zannoni D (1996) Hydrogen oxidation by membranes from autotrophically grown *Alcaligenes eutrophus* H16: Role of the cyanide-resistant pathway in energy transduction. Arch Microbiol 165: 418–420

Kröger A, Biel S, Simon J, Gross R, Unden G and Lancaster CRD (2002) Fumarate respiration of *Wollinella succinogenes*: Enzymology, energetics and coupling mechanism. Biochim Biophys Acta 1553: 23–38

Künkel A, Vorholt JA, Thauer RK and Hedderich R (1998) An *Escherichia coli* hydrogenase-3-type hydrogenase in methanogenic archaea. Eur J Biochem 252: 467–476

Lanzilotta WN, Schuller DJ, Thorsteinsson MV, Kerby RL, Roberts GP and Poulos TL (2000) Structure of the CO sensing transcription activator CooA. Nat Struct Biol 7: 876–880

Lenz O and Friedrich B (1998) A novel multicomponent regulatory system mediates H_2 sensing in *Alcaligenes eutrophus*. Proc Natl Acad Sci USA 95: 12474–12479

Lenz O, Strack A, Tran-Betcke A and Friedrich B (1997) A hydrogen-sensing system in transcriptional regulation of hydrogenase gene expression in *Alcaligenes* species. J Bacteriol 179: 1655–1663

López-García P and Moreira D (1999) Metabolic symbiosis at the origin of eukaryotes. Trends Biochem Sci 24: 88–93

Lyon EJ, Shima S, Buurman G, Chowdhuri S, Batschauer A, Steinbach K and Thauer RK (2004). UV-A/blue-light inactivation of the 'metal-free' hydrogenase (Hmd) from methanogenic archaea. Eur J Biochem 271: 195–204

Ma K, Schicho RN, Kelly RM and Adams MWW (1993) hydrogenase of the hyperthermophile *Pyrococcus furiosus* is an elemental sulfur reductase or sulfhydrogenase: Evidence for a sulfur-reducing hydrogenase ancestor. Proc Natl Acad Sci USA 90: 5341–5344

Ma K, Zhou ZH and Adams MWW (1994) Hydrogen production from pyruvate by enzymes purified from the hyperthermophilic archaeon, *Pyrococcus furiosus*: A key role for NADPH. FEMS Microbiol Lett 122: 245–250

Ma K, Weiss R and Adams MWW (2000) Characterization of hydrogenase II from the hyperthermophilic archaeon *Pyrococcus furiosus* and assessment of its role in sulfur reduction. J Bacteriol 182: 1864–1871

Magnani P, Doussière J and Lissolo T (2000) Diphenylene iodonium as an inhibitor for the hydrogenase complex of *Rhodobacter capsulatus*: Evidence for two distinct electron donor sites. Biochim Biophys Acta 1459: 169–178

Malki S, Saimmaime I, De Luca G, Rousset M, Dermoun Z and Belaich J-P (1995) Characterization of an operon encoding an NADP-reducing hydrogenase in *Desulfovibrio fructosovorans*. J Bacteriol 177: 2628–2636

Martin W and Müller M (1998) The hydrogen hypothesis for the first eukaryote. Nature 392: 37–41

Martin W, Hoffmeister M, Rotte C and Henze K (2001) An overview of endosymbiotic models for the origins of eukaryotes, their ATP-producing organelles (mitochondria and hydrogenosomes), and their heterotrophic lifestyle. Biol Chem 382: 1521–1539

McEwan AG, Cotton NPJ, Ferguson SJ and Jackson JB (1985) The role of auxiliary oxidants in the maintenance of a balanced redox poise for photosynthesis in bacteria. Biochim Biophys Acta 810: 140–147

Melis A and Happe T (2001) Hydrogen Production. Green algae as a source of energy. Plant Physiol 127: 740–748

Meuer J, Bartoschek S, Koch J, Künkel A and Hedderich R (1999) Purification and catalytic properties of Ech hydrogenase from *Methanosarcina barkeri*. Eur J Biochem 265: 325–335

Meyer J, Kelley B and Vignais PM (1978) Nitrogen fixation and hydrogen metabolism in photosynthetic bacteria. Biochimie 60: 245–260

Mi H, Endo T, Schreiber U, Ogawa T, and Asada K. (1995) Thylakoid membrane-bound, NADPH-specific pyridine nucleotide dehydrogenase complex mediates cyclic electron transport in the cyanobacterium *Synechocystis* sp. PCC 6803. Plant Cell Physiol 36: 661–668

Moreira D and López-García P (1998) Symbiosis between methanogenic archaea and delta-proteobacteria as the origin of eukaryotes: the syntrophic hypothesis. J Mol Evol 47: 517–530

Müller M (1993) The hydrogenosome. J Gen Microbiol 139: 2879–2889

Nicholls DG and Ferguson SJ (1992) Bioenergetics 2. Academic Press, London

Nicolet Y, Piras C, Legrand P, Hatchikian CE and Fontecilla-Camps JC (1999) *Desulfovibrio desulfuricans* iron hydrogenase: The structure shows unusual coordination to an active site Fe binuclear center. Structure Fold Des 7: 13–23

Nicolet Y, Lemon BJ, Fontecilla-Camps JC and Peters JW (2000) A novel FeS cluster in Fe-only hydrogenases. Trends Biochem Sci 25: 138–43

Oh J and Kaplan S (2000) Redox signaling: Globalization of gene expression. EMBO J 19: 4237–4247

Oh J and Kaplan S (2001) Generalized approach to the regulation and integration of gene expression. Mol Microbiol 39: 1116–1123

Ohkawa H, Sonoda M, Shibata M and Ogawa T (2001) Localization of NAD(P)H dehydrogenase in the cyanobacterium *Synechocystis* sp strain PCC 6803. J Bacteriol 183: 4938–4939

Oxelfelt F, Tamagnini P and Lindblad P (1998) Hydrogen uptake in *Nostoc* sp. strain PCC 73102. Cloning and characterization of a *hupSL* homolog. Arch Microbiol 169: 267–274

Paul F, Colbeau A and Vignais PM (1979) Phosphorylation coupled to H_2 oxidation by chromatophores from *Rhodopseudomonas capsulata*. FEBS Lett 106: 29–33

Pedroni P, Della Volpe A, Galli G, Mura GM, Pratesi C and Grandi G (1995) Characterization of the locus encoding the [Ni-Fe] sulfhydrogenase from the archaeon *Pyrococcus furiosus*: Evidence for a relationship to bacterial sulfite reductases. Microbiology 141: 449–458

Peltier G and Cournac L (2002) Chlororespiration. Annu Rev Plant Physiol Plant Mol Biol 53: 523- 550

Peters JW (1999) Structure and mechanism of iron-only hydrogenases. Curr Opin Struct Biol 9: 670–676

Peters JW, Lanzilotta WN, Lemon BJ and Seefeldt LC (1998) X-ray crystal structure of the Fe-only hydrogenase (CpI) from *Clostridium pasteurianum* to 1.8 Ångstrom resolution. Science 282: 1853–1858

Pilkington SJ, Skehel JM, Gennis RB, and Walker JE (1991) Relationship between mitochondrial NADH-ubiquinone reductase and a bacterial NAD-reducing hydrogenase. Biochemistry 30: 2166–2175

Pohorelic BK, Voordouw JK, Lojou E, Dolla A, Harder J, and Voordouw G (2002) Effects of deletion of genes encoding Fe-only hydrogenase of *Desulfovibrio vulgaris* Hildenborough on hydrogen and lactate metabolism. J Bacteriol 184: 679–86

Porte F and Vignais PM (1980) Electron transport chain and energy transduction in *Paracoccus denitrificans* under autotrophic growth conditions. Arch Microbiol 127: 1–10

Qian Y and Tabita FR (1996) A global signal transduction system regulates aerobic and anaerobic CO_2 fixation in *Rhodobacter sphaeroides*. J Bacteriol 178: 12–18

Qian T and Tabita FR (1998) Expression of *glnB* and *glnB*-like gene (*glnK*) in a ribulose bisphosphate carboxylase/oxygenase-deficient mutant of *Rhodobacter sphaeroides*. J Bacteriol 180: 4644–4649

Rákhely G, Colbeau A, Garin J, Vignais PM and Kovács KL (1998) Unusual organization of the genes coding for HydSL, the stable [NiFe] hydrogenase in the photosynthetic bacterium *Thiocapsa roseopersicina* BBS. J Bacteriol 180: 1460–1465

Rákhely G, Kovács ÁT, Maróti G, Fodor BD, Csanádi G, Latinovics D and Kovács KL (2004) Cyanobacterial-type, heteropentameric, NAD^+- reducing NiFe hydrogenase in the purple sulfur photosynthetic bacterium, *Thiocapsa roseopersicina*. Appl Environ Microbiol 70: 722–728

Richard DJ, Sawers G, Sargent F, McWalter L and Boxer D H (1999) Transcriptional regulation in response to oxygen and nitrate of the operons encoding the [NiFe] hydrogenases 1 and 2 of *Escherichia coli*. Microbiology 145: 2903–12

Richardson DJ (2000) Bacterial respiration: A flexible process for a changing environment. Microbiology 146: 551–571

Richaud P, Colbeau A, Toussaint B and Vignais PM (1991) Identification and sequence analysis of the *hupR*$_1$ gene, which encodes a response regulator of the NtrC family required for hydrogenase expression in *Rhodobacter capsulatus*. J Bacteriol 173: 5928–5932

Roberts GP, Thorsteinsson MV, Kerby RL, Lanzilotta WN and Poulos T (2001) CooA: A heme-containing regulatory protein that serves as a specific sensor of both carbon monoxide and redox state. Prog Nucleic Acid Res Mol Biol. 67: 35–63

Rodrigue A, Chanal A, Beck K, Müller M and Wu L-F (1999) Co-translocation of a periplasmic enzyme complex by a hitchhiker mechanism through the bacterial Tat pathway. J Biol Chem 274: 13223–13228

Römermann D, Lohmeyer M, Friedrich CG and Friedrich B (1988) Pleiotropic mutants from *Alcaligenes eutrophus* defective in the metabolism of hydrogen, nitrate, urea, and fumarate. Arch Microbiol 149: 471–475

Römermann D, Warrelmann J, Bender RA and Friedrich B (1989) An *rpoN*-like gene of *Alcaligenes eutrophus* controls expression of diverse metabolic pathways, including hydrogen oxidation. J Bacteriol 171: 1093–1099

Rossi M, Pollock WB, Reij M., Keon RG, Fu R and Voordouw G (1993) The *hmc* operon of *Desulfovibrio vulgaris* subsp. *vulgaris* Hildenborough encodes a potential transmembrane redox protein complex. J Bacteriol 175: 4699–4711

Rossmann R, Sawers G and Böck A (1991) Mechanism of regulation of the formate-hydrogenlyase pathway by oxygen, nitrate, and pH: Definition of the formate regulon. Mol Microbiol 5: 2807–2814

Sapra R, Verhagen MF, Adams MW (2000) Purification and characterization of a membrane-bound hydrogenase from the hyperthermophilic archaeon *Pyrococcus furiosus*. J Bacteriol 182: 3423–3428

Sargent F, Bogsch EG, Stanley NR, Wexler M, Robinson C, Berks BC and Palmer T (1998) Overlapping functions of components of a bacterial Sec-independent protein export pathways. EMBO J 17: 3640–3650

Sargent F, Berks BC and Palmer T (2002) Assembly of membrane-bound respiratory complexes by the Tat protein transport

system. Arch Microbiol 178: 77–84
Sasahara KC, Heinzinger NK and Barrett EL (1997) Hydrogen sulfide production and fermentative gas production by *Salmonella typhimurium* require F_0F_1 ATP synthase activity. J Bacteriol 179: 6736–6740
Sauter M, Böhm R and Böck A (1992) Mutational analysis of the operon (*hyc*) determining hydrogenase 3 formation in *Escherichia coli*. Mol Microbiol 6: 1523–1532
Sawers G (1999) The aerobic/anaerobic interface. Curr Opin Microbiol 2: 181–187
Sazanov LA, Peak-Chew SY, Fearnley IM and Walker JE (2000) Resolution of the membrane domain of bovine complex I into subcomplexes: Implications for the structural organization of the enzyme. Biochemistry 39: 7229–7235
Scherer S (1990) Do photosynthetic and respiratory electron transport chains share redox proteins. Trends Biochem Sci 15: 458–462
Schmehl M, Jahn A, Meyer zu Vilsendorf A, Hennecke S, Masepohl B, Schuppler M, Marxer M, Oelze J and Klipp W (1993) Identification of a new class of nitrogen fixation genes in *Rhodobacter capsulatus*: A putative membrane complex involved in electron transport to nitrogenase. Mol Gen Genet 241: 602–15
Schmetterer G (1994) Cyanobacterial respiration. In : Bryant DA (ed) The Molecular Biology of Cyanobacteria, pp. 409–435. Kluwer Academic Publishers, Dordrecht
Schmitz O and Bothe H (1996) The diaphorase subunit HoxU of the bidirectional hydrogenase as electron transferring protein in cyanobacterial respiration? Naturwissenschaften 83: 525–527
Schmitz O, Boison G, Hilscher R, Hundeshagen B, Zimmer W, Lottspeich F and Bothe H (1995) Molecular biological analysis of a bidirectional hydrogenase from cyanobacteria. Eur J Biochem 233: 266–276
Schmitz O, Boison G, Salzmann H, Bothe H, Schutz K, Wang SH and Happe T (2002) HoxE—a subunit specific for the pentameric bidirectional hydrogenase complex (HoxEFUYH) of cyanobacteria. Biochim Biophys Acta 1554: 66-74
Schneider K and Schlegel HG (1976) Purification and properties of soluble hydrogenase from *Alcaligenes eutrophus H 16*. Biochim Biophys Acta 452: 66–80
Schultz BE and Chan SI (2001) Structures and proton-pumping strategies of mitochondrial respiratory enzymes. Annu Rev Biophys Biomol Struct 30: 23–65
Schut GJ, Zhou J and Adams MWW (2001) DNA microarray analysis of the hyperthermophilic archaeon *Pyrococcus furiosus*: Evidence for a new type of sulfur-reducing enzyme complex. J Bacteriol 183: 7027–7036.
Schwartz E, Buhrke T, Gerischer U and Friedrich B (1999) Positive transcriptional feedback controls hydrogenase expression in *Alcaligenes eutrophus* H16. J Bacteriol 181: 5684–5692
Siedow A, Cramm R, Siddiqui RA and Friedrich B (1999) A megaplasmid-borne anaerobic ribonucleotide reductase in *Alcaligenes eutrophus* H16. J Bacteriol 181: 4919–4928
Silva PJ, van den Ban ECD, Wassink H, Haaker H, de Castro B, Robb FT, and Hagen WR (2000) Enzymes of hydrogen metabolism in *Pyrococcus furiosus*. Eur J Biochem 267: 6541–6551
Skibinski DAG, Golby P, Chang YS, Sargent F, Hoffman R, Harper R, Guest JR, Attwood MM, Berks BC and Andrews SC (2002) Regulation of the hydrogenase-4 operon of *Escherichia coli* by the σ^{54}-dependent transcriptional activators FhlA and HyfR. J Bacteriol 184:6642–6653
Søballe B and Poole RK (1999) Microbial ubiquinones: Multiple roles in respiration, gene regulation and oxidative stress management. Microbiology 145: 1817–1830
Spiro S and Guest JR (1990) FNR and its role in oxygen-regulated gene expression in *Escherichia coli*. FEMS Microbiol Rev 75: 399–428
Swem LR, Elsen S, Bird TH, Swem DL, Koch HG, Myllykallio H, Daldal F and Bauer CE (2001) The RegB/RegA two-component regulatory system controls synthesis of photosynthesis and respiratory electron transfer components in *Rhodobacter capsulatus*. J Mol Biol 309: 121–138
Swem LR, Kraft BJ, Swem DL, Setterdahl AT, Masuda S, Knaff DB, Zaleski JM and Bauer CE (2003) Signal transduction by the global regulator RegB is mediated by a redox-active cysteine. EMBO J 22: 4699–4708
Tamagnini P, Axelsson R, Lindberg P, Oxelfelt F, Wünschiers R and Lindblad P (2002) hydrogenases and hydrogen metabolism of Cyanobacteria. Microbiol Mol Biol Rev 66: 1–20
Taylor BL and Zhulin I (1999) PAS domains: internal sensors of oxygen, redox potential, and light. Microbiol. Mol Biol Rev 63: 479–506
Tersteegen A and Hedderich R (1999) *Methanobacterium thermoautotrophicum* encodes two multisubunit membrane-bound [NiFe] hydrogenases. Transcription of the operons and sequence analysis of the deduced proteins. Eur J Biochem 264: 930–943
Thauer RK (1998) Biochemistry of methanogenesis: A tribute to Marjory Stephenson. 1998 Marjory Stephenson Prize Lecture. Microbiology 144: 2377–2406
Thauer RK, Klein AR and Hartmann GC (1996) Reactions with molecular hydrogen in microorganisms: Evidence for a purely organic hydrogenation catalyst. Chem Rev 96: 3031–3042
Tichi MA, Meijer WG and Tabita FR (2001) Complex I and its involvement in redox homeostasis and carbon and nitrogen metabolism in *Rhodobacter capsulatus*. J Bacteriol 183: 7285–7294
Toussaint B, Bosc C, Richaud P Colbeau A and Vignais PM (1991) A mutation in a *Rhodobacter capsulatus* gene encoding an integration host factor-like protein impairs in vivo hydrogenase expression. Proc Natl Acad Sci USA 88: 10749–10753
Toussaint B, de Sury d'Aspremont R, Delic-Attree I, Berchet V, Elsen S, Colbeau A, Dischert W, Lazzaroni Y, and Vignais PM (1997) The *Rhodobacter capsulatus hupSLC* promoter: Identification of *cis*-regulatory elements and of *trans*-activating factors involved in H_2 activation of *hupSLC* transcription. Mol Microbiol 26: 927–937
Tran-Betcke A, Warnecke U, Boecker C, Zaborosch C and Friedrich B (1990) Cloning and nucleotide sequences of the genes for the subunits of NAD-reducing hydrogenase of *Alcaligenes eutrophus* H16. J Bacteriol 172: 2920–2929
Van Soom C, de Wilde C and Vanderleyden J (1997) Hox A is a transcriptional regulator for expression of the *hup* structural genes in free-living *Bradyrhizobium japonicum*. Mol Microbiol 23: 967–977
Van Soom C, Lerouge I, Vanderleyden J, Ruiz-Argüeso T and Palacios JM (1999) Identification and characterization of *hupT*, a gene involved in negative regulation of hydrogen oxidation in *Bradyrhizobium japonicum*. J. Bacteriol 181: 5085–5089
Vichivanives P, Bird TH, Bauer CE and Tabita FR (2000) Multiple regulators and their interaction in vivo and in vitro with

the *cbb* regulons of *Rhodobacter capsulatus* J Mol Biol 300: 1079–1099

Vignais PM and Colbeau A (2004) Molecular biology of microbial hydrogenases. Curr Issues Mol Biol 6: 159–188

Vignais PM and Toussaint B (1994) Molecular biology of membrane bound H_2 uptake hydrogenases. Arch Microbiol 161: 1–10 [Erratum Arch Microbiol 16: 196]

Vignais PM, Henry M-F, Sim E and Kell DB (1981) The electron transport system and hydrogenase of *Paracoccus denitrificans*. Current Topics in Bioenergetics 12: 115–196

Vignais PM, Colbeau A, Willison JC and Jouanneau Y (1985) hydrogenase, nitrogenase and hydrogen metabolism in the photosynthetic bacteria. Adv Microbial Physiol 26: 155–234

Vignais PM, Dimon B, Zorin NA, Colbeau A and Elsen S (1997) HupUV proteins of *Rhodobacter capsulatus* can bind H_2: Evidence from the H-D exchange reaction. J Bacteriol 179: 290–292

Vignais PM, Dimon B, Zorin NA, Tomiyama M and Colbeau A (2000) Characterization of the hydrogen-deuterium exchange activities of the energy-transducing HupSL hydrogenase and the H_2-signaling HupUV hydrogenase in *Rhodobacter capsulatus*. J Bacteriol 182: 5997–6004

Vignais PM, Billoud B and Meyer J (2001) Classification and phylogeny of hydrogenases. FEMS Microbiol Rev 25: 455–501

Volbeda A, Charon MH, Piras C, Hatchikian EC, Frey M and Fontecilla-Camps JC (1995) Crystal structure of the nickel-iron hydrogenase from *Desulfovibrio gigas*. Nature 373: 580–587

Volbeda A, Garcin E, Piras C, de Lacey AL, Fernandez VM, Hatchikian EC, Frey M and Fontecilla-Camps JC (1996) Structure of the [NiFe] hydrogenase active site: Evidence for biologically uncommon Fe ligands. J Am Chem Soc 118: 12989–12996

Voncken FGJ, Boxma B, van Hoek AH, Akhmanova AS, Vogels GD, Huynen M, Veenhuis M and Hackstein JH (2002) A hydrogenosomal [Fe]- hydrogenase from the anaerobic chytrid Neocallimastix sp. L2. Gene 284: 103–112

Voordouw G (2000) A universal system for the transport of redox proteins: Early roots and latest developments. Biophys Chem 86: 131–140

Walker JE (1992) The NADH:ubiquinone oxidoreductase (complex I) of respiratory chains. Q Rev Biophys 25: 253–324

Weidner U, Geier S, Ptock A, Friedrich T, Leif H and Weiss H (1993) The gene locus of the proton-translocating NADH-ubiquinone oxidoreductase in *Escherichia coli*: Organization of the 14 genes and relationship between the derived proteins and subunits of mitochondrial complex I. J Mol Biol 233: 109–122

Weiner JH, Bilous PT, Shaw GM, Lubitz SP, Frost L, Thomas GH, Cole JA and Turner RJ (1998) A novel and ubiquitous system for membrane targeting and secretion of cofactor-containing proteins. Cell 93: 93–101

Willison JC (1993) Biochemical genetics revisited: The use of mutants to study carbon and nitrogen metabolism in the photosynthetic bacteria. FEMS Microbiol Rev 104: 1–38

Wu L-F, Chanal A and Rodrigue A (2000) Membrane targeting and translocation of bacterial hydrogenases. Arch Microbiol 173: 319–324

Wünschiers R, Stangier K, Senger H and Schulz R (2001a) Molecular evidence for a Fe-hydrogenase in the green alga *Scenedesmus obliquus*. Curr Microbiol 42: 353-360

Wünschiers R, Senger H and Schulz R (2001b) Electron pathways involved in H_2-metabolism in the green alga *Scenedesmus obliquus*. Biochim Biophys Acta 1503: 271–278

Yagi T (1993) The bacterial energy-transducing NADH-quinone oxidoreductases. Biochim Biophys Acta 1141: 1–17

Yagi T, Yano T, Di Bernardo S, and Matsuno-Yagi A (1998) Procaryotic complex I (NDH-1), an overview. Biochim Biophys Acta 1364: 125–133

Yano T and T Ohnishi (2001) The origin of cluster N2 of the energy-transducing NADH-quinone oxidoreductase: Comparisons of phylogenetically related enzymes. J Bioenerg Biomemb 33: 213–222

Zannoni D and Marrs B (1981) Redox chain and energy transduction in chromatophores from *Rhodopseudomonas capsulata* cells grown anaerobically in the dark on glucose and dimethylsulfoxide. Biochim Biophys Acta 637: 96–106

Zannoni D and Moore AL (1990) Measurement of the redox state of the ubiquinone pool in *Rhodobacter capsulatus* membrane fragments. FEBS Lett 271: 123–127

Zannoni D (1995) Aerobic and anaerobic transport chains in anoxygenic phototrophic bacteria. In: Blankenship RE, Madigan MT and Bauer C (eds) Anoxygenic Photosynthetic Bacteria, pp 949–971. Kluwer Academic Publishers, Dordrecht

Zirngibl C, Hedderich R and Thauer RK (1990) N^5,N^{10}-Methylenetetrahydromethanopterin dehydrogenase from *Methanobacterium thermoautotrophicum* has hydrogenase activity. FEBS Lett 261: 112–116

Zimmer D, Schwartz E, Tran-Betcke A, Gewinner P and Friedrich B (1995) Temperature tolerance of hydrogenase expression in *Alcaligenes eutrophus* is conferred by a single amino acid exchange in the transcriptional activator HoxA. J Bacteriol 177: 2373–2380

Chapter 12

Cyanobacterial Respiration

Georg Schmetterer and Dietmar Pils
Institute of Physical Chemistry, University of Vienna, UZA2, Althanstrasse 14, A-1090 Vienna, Austria

---

*Author for correspondence, email: georg.schmetterer@univie.ac.at

Davide Zannoni (ed): Respiration in Archaea and Bacteria. Vol 2: Diversity of Prokaryotic Respiratory Systems, pp. 261–278.

Summary

Cyanobacterial respiration has some unique features. Cyanobacteria contain two independent respiratory chains, one in the cytoplasmic membrane and one in the intracytoplasmic membranes (also called thylakoids). The latter is intimately linked with the photosynthetic electron transport chain and some components, e.g. the quinone pool, the cytochrome b_6f complex, and cytochrome c_6, are shared by both photosynthesis and respiration. However, all components of the respiratory electron transport chain might have some—possibly regulatory—function in photosynthesis. In recent years the total genomic sequences of several cyanobacteria have been invaluable tools for understanding cyanobacterial respiration. Especially in *Synechocystis* sp. strain PCC 6803 the majority of the components of the respiratory chains were finally identified. Respiratory electron transport is highly branched, with several routes of electron input and several routes to the final electron acceptor, which is always dioxygen. The single component common to all respiratory electron pathways is the quinone pool. Furthermore, total genomic sequencing has shown that different cyanobacteria may contain different electron transport components. Several strains, especially (but not exclusively) those that are capable of cell differentiation, contain more than one copy of genes for some respiratory electron transport chain components. The different functions of these genes and their gene products have only partially been elucidated. Several components of the respiratory chains are subject to complex gene regulations, about which information is only starting to appear. Some proteins, e.g. cytochrome c_M, and the dehydrogenase DrgA, are likely to be involved in respiratory electron transport, but their positions in the electron transport chain (i.e. their in vivo electron donors and acceptors) have not yet been fully clarified. The electron transport chain of the cytoplasmic membrane is not yet well characterized in any strain.

I. Introduction

In recent years considerable progress has been made in the understanding of cyanobacterial respiration and a number of reviews have appeared on the topic (e.g. Schmetterer, 1994; Vermaas, 2001). This review stresses the fact that despite much work many questions are still open, not the least being that almost certainly the components of the cyanobacterial respiratory chain(s) are not yet known in their entirety. The availability of a steadily growing number of genomic sequences of different cyanobacteria has had a profound impact on the study of cyanobacteria, and, specifically, their respiratory chains. One important lesson learned was that a detailed picture of respiration in cyanobacteria as a whole cannot be given, because not all respiratory components are present in all strains and for many pertinent genes the number of homologues per chromosome differ from strain to strain. Therefore, this review attempts to list all components that are currently known to be part of cyanobacterial respiratory chain(s). Not all components are present in all cyanobacteria, and only for *Synechocystis* sp. PCC 6803, the first totally sequenced cyanobacterium (Kaneko et al., 1996), an attempt will be made to define the function of each component identified so far.

In order to keep the list of references to a manageable size, the majority of references presented in this review are from 1994 or younger. For older references the reader is referred to an earlier review on the same topic (Schmetterer, 1994).

A. Cyanobacteria

Cyanobacteria are prokaryotes capable of oxygenic photosynthesis. They are very widely distributed in many different habitats and play a huge role in total primary production of organic material (Partensky et al., 1999). All cyanobacteria are also capable of aerobic respiration. This is an important property, since cyanobacteria are thus the only cells, in which the two most important bioenergetic processes, aerobic respiration and oxygenic photosynthesis occur in the same compartment. Much attention has been placed on the question of the interaction of photosynthesis and respiration in cyanobacteria. Since the overall reactions of these two processes formally are the reverse of each other, their simultaneous occurrence in the same compartment should lead to futile cycles.

Abbreviations: ARTO – alternate respiratory terminal oxidase(s); CM – cytoplasmic membrane; Cox – cytochrome *c* oxidase; FNR – ferredoxin-NADP$^+$ oxidoreductase; HQNO – 2-heptyl-4-hydroxy-quinoline-N-oxide; ICM – intracytoplasmic membranes or thylakoids; PCP – pentachlorophenol; Qox – quinol oxidase; tsp – transcriptional starting point

Despite great progress in the understanding of the interaction between respiration and photosynthesis in cyanobacteria, it seems likely that not all the components of the respiratory chain(s) have been identified, at least not with a clear idea of their function.

With one single exception (*Gloeobacter* sp. that has no ICM) all cyanobacteria have three distinct lipid bilayer membranes: 1. the outer membrane that is part of the gram-negative cell wall and located outside of the peptidoglycane layer is not bioenergetically active and thus not of immediate relevance for respiratory processes. 2. the cytoplasmic membrane (CM) that encloses the cytoplasm, and 3. the network of intracellular membranes (ICM, also known as cyanobacterial thylakoids) that enclose a further compartment, the intrathylakoid lumen, which is distinct from the cytoplasm and may have a physical connection to the periplasm. Purification techniques have shown that CM and ICM are clearly distinct in their chemical composition (Omata and Murata, 1983), but the question whether there is physical contact between the two membranes has not been solved yet. It seems now certain that both the CM and the ICM carry respiratory electron transport chains that may, however, consist of different components. Photosynthetic electron transport is confined to the ICM and the CM contains no or at most very little chlorophyll. A general scheme of cyanobacterial membranes, the cellular compartments formed by them and their bioenergtic significance is presentd in Fig. 1.

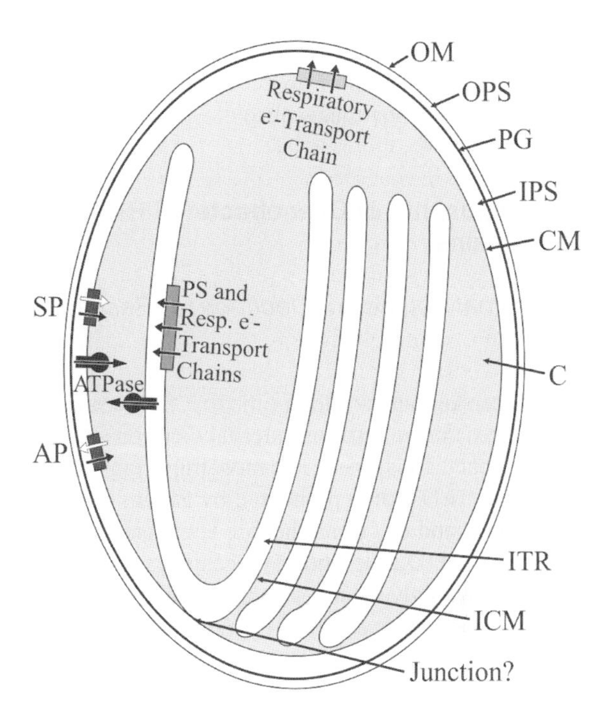

Fig. 1. A schematic model of the bioenergetics of a 'typical' cyanobacterial cell. OM, outer membrane; OPS, outer periplasmic space; PG, peptidoglycan cell wall; IPS, inner periplasmic space; CM, cytoplasmic membrane; C, cytoplasm; ICM, intracellular membranes or thylakoids; ITR, intrathylakoid space; PS, photosynthetic; SP, active transport by symporters; AP, active transport by antiporters.

B. Specific Properties of Cyanobacterial Respiration

Several properties of cyanobacterial respiration are unique. As mentioned above, there are two bioenergetically active membranes, and each membrane apparently contains a different respiratory chain. In the ICM respiratory and photosynthetic electron transport chains share several components so that both processes must be considered to be part of a single large multi-enzyme bioenergetic complex.

Cyanobacterial respiration is also characterized by some components that are not usually part of respiratory electron transport chains, among them NADPH, the cytochrome b_6f complex, and plastocyanin. NADH, not NADPH, is generally a donor to the electron transport chain. Cyanobacteria are the only organisms to use the cytochrome b_6f complex and the peripheral electron carrier plastocyanin that are components of photosynthetic electron transport in chloroplasts (and cyanobacteria) in respiratory electron transport.

Some filamentous cyanobacteria are capable of cell differentiation, such as heterocysts (specialized for nitrogen fixation), akinetes (spores), or hormogonia (highly motile, important for the infection process establishing symbioses with other organisms). Especially in heterocysts, where the extremely oxygen sensitive nitrogenase is localized, respiration apparently plays an important role in maintaining the necessary microaerobic environment.

Cyanobacteria were the first organisms on Earth that produced O_2 in massive amounts (see also Castresana, Chapter 1, Vol. 15), and thus may have been the first organisms that used the available O_2 to generate ATP by aerobic respiration in the dark. Indeed all components of cyanobacterial respiration apparently have at least some function in photosynthesis. This reasoning led Broda (1975) to the formulation of his 'conversion hypothesis,' which states essentially that all aerobic respirers evolved from photosynthetic organisms (specifically cyanobacteria) by loss of the photosystems. Until now no evidence has turned up that makes this hypothesis obsolete.

II. Components of Cyanobacterial Respiratory Chains

A. External Sources of Electrons for Respiration

It has been known for a long time that the number of substances that are used as external electron sources in cyanobacteria is small. The most important one is, of course, H_2O that is processed by Photosystem II to yield O_2 and electrons that are transferred to the quinone pool. During photosynthesis, carbohydrate (in the form of glycogen) is accumulated that is used during dark periods as a primary source of electrons for respiration. Although, in nature, glycogen is most likely the major electron source for respiratory electron transport, it seems that there is no study that describes quantitatively respiratory activity at the expense of glycogen, despite the existence of a useful method to quantify glycogen (Ernst et al., 1984). Several cyanobacteria can use exogenous sugars and a few related substances (such as glycerol in *Synechococcus* sp. PCC 7002) that are taken up and then oxidized to yield NADH and/or NADPH (a list is available in Rippka et al., 1979). Exogenous H_2 can be used as a primary electron donor through the action of one of the hydrogenases (see below).

B. Intracellular Primary Electron Donors

1. NADPH

An unusual property of cyanobacterial respiratory chains is that NADPH is an important primary reductant. In the dark NADPH is probably generated primarily by the oxidation of glucose-6-phosphate in the oxidative pentose phosphate cycle. In the light, NADPH is produced by ferredoxin-$NADP^+$ oxidoreductase.

2. NADH

Many years ago Biggins (1969) studied the relative importance of NADH and NADPH as primary electron donors for respiration. He noticed that upon transfer of *Synechococcus* sp. PCC6301 from aerobiosis to anaerobiosis the amount of NADPH increased but the amount of NADH remained constant and concluded that NADPH, not NADH, was the primary donor. Although several primary dehydrogenases have been identified that oxidize NADH in vitro, the in vivo function of NADH in respiration remains largely unknown. Even the major route(s) of formation of NADH from NAD^+ are not known with certainty. Notably, in the sequenced genomes of cyanobacteria, pyridin nucleotide transhydrogenases that use NADPH as an electron donor to generate NADH have been identified.

3. Succinate

An important result from the first available complete sequence of a cyanobacterium [*Synechocystis* sp. PCC6803, Kaneko et al., (1995)] was the identification of succinate dehydrogenases that have been shown to be important for respiration (Cooley et al., 2000). The source of succinate is not known with certainty, but Cooley et al. (2000) have obtained evidence that despite the absence of a recognizable 2-oxoglutarate dehydrogenase in cyanobacterial genomes, and thus the existence of an incomplete citric acid cycle, 2-oxoglutarate can be converted into succinate. Cooley and Vermaas (2001) have presented evidence that the levels of NAD(P)H and succinate are interrelated, especially a mutant lacking type I NAD(P)H dehydrogenase having only about 2% of the succinate levels of the wild type. They therefore considered that a considerable part of the oxidized NAD(P) formed by NAD(P)H dehydrogenases may be used for the cycle producing succinate, specifically for conversion of isocitrate to 2-oxoglutarate and malate to oxaloacetate.

4. Molecular Hydrogen (H_2)

Hydrogen is a possible electron donor to the respiratory chain. Besides being taken up from the environment, H_2 can also be generated intracellularly by a reversible hydrogenase or—in heterocysts—by nitrogenase.

C. Primary Oxidoreductases

1. Type I NAD(P)H-Dehydrogenase

Cyanobacteria contain many genes that have high sequence similarity with the NADH dehydrogenase of mitochondria (Table 1). Generally, the arrangement of the genes in the different genomes is very similar. In *Synechococcus* sp. PCC 7002 the arrangement is highly similar to the situation in *Synechocystis* sp. PCC 6803 (Nomura, 2001): there are *ndhAIGE* and *ndhCKJ* gene clusters, solitary *ndhB*, *ndhH* and *ndhL* genes, and several *ndhD* (possibly four) and *ndhF* (possibly five) genes. The assignment of a gene to *ndhD* or *ndhF* must be viewed with caution, since they are highly related to each other and—to a smaller degree—also to *ndhB*. The nomenclature used is that of Nomura (2001) who performed detailed sequence comparisons. Detailed functional studies that would allow the unequivocal assignment of these genes are not yet available. Table 1 shows alternate names assigned to *ndhD* and *ndhF* genes in different organisms. The gene arrangement of the putative *ndhD* and *ndhF* genes in *Synechococcus* sp. PCC 7002 is also exactly as in *Synechocystis* sp. PCC 6803 with the exception of the small ORF (116 AA) between *ndhF2* and *nhdF5* tentatively named *ndhM* in Table 1 that is missing in *Synechococcus* sp. PCC 7002. This ORF has sequence similarity to the *mnhC* gene of *Staphylococcus aureus* (Kuroda et al., 2001) and the *mrpC* gene of *Bacillus firmus* (Ito et al., 2001), both of which are involved in Na^+/H^+ antiport. It is striking that in *Synechococcus* sp. PCC 7002 the gene for the Na^+/H^+ antiporter is located nearby. Thus these *ndh* genes may be involved in ion transport.

Important questions are still open concerning cyanobacterial type I NAD(P)H dehydrogenases. One is the in vivo electron donor to the enzyme. Homologues of the diaphorase subunits of type I NADH dehydrogenase (*nuoEFG* in *E. coli*), i.e. those subunits that react directly with NADH, are not present in any cyanobacterium tested so far. It has therefore been discussed (Boison et al., 1999) whether the HoxEFU diaphorase subunits of the bidirectional hydrogenase might substitute for the 'missing' diaphorase of type I NADH dehydrogenase, but the evidence is neither compelling for nor against this proposal and more work on this question is necessary. Another unsolved question is the location of the cyanobacterial type I NAD(P)H dehydrogenase, for which conflicting reports have been published. Berger et al. (1991) found it in both the CM and the ICM of *Synechocystis* sp. PCC6803, Ogawa (1992) only in the ICM of the same strain, while Howitt et al. (1993) located the enzyme in *Anabaena* sp. PCC7120 only in the CM.

Several mutants of different subunits of type I dehydrogenase have been isolated, of which mutant M55 of *Synechocystis* sp. PCC6803 that lacks the *ndhB* gene is the most interesting (Ogawa, 1991). Respiration in this strain was reduced to about 7% of wild type activity, the best indication available that type I NAD(P)H dehydrogenase is important for respiration in cyanobacteria. Interestingly, mutant M55 also showed a severe reduction in uptake of inorganic carbon, which made this strain high (3%) CO_2-requiring. The reason for this link between inorganic carbon uptake and respiration has not yet been explained. Similarly, an *ndhB* mutant in *Synechococcus* sp. PCC7942 was high-CO_2-requiring, but its respiratory capacity was only reduced by 50–70% (Marco et al., 1993). In *Synechococcus* sp. PCC7002 an attempted *ndhB* mutant did not segregate to homozygocity (Nomura, 2001). The *ndhL* gene is only known from cyanobacteria. Its mutation in *Synechocystis* sp. PCC6803 also led to a defect in inorganic carbon uptake (Ogawa, 1992). Another interesting NAD(P)H dehydrogenase mutant was constructed by Schluchter et al. (1993). Despite the occurrence of about four different genes for *ndhF* (see above) in this strain, mutation in one such gene, the *ndhF1* gene, displayed several striking phenotypes: respiratory activity was lowered to about 20–25% of the wild type, the capacity for photoheterotrophic growth with glycerol as the substrate was lost, and cyclic electron transport around Photosystem I was partially inhibited (Yu et al., 1993).

2. Type II NAD(P)H Dehydrogenase

Many bacteria contain a type II NAD(P)H dehydrogenase that consists—in contrast to the many subunits of type I—only of a single subunit and has no iron-sulfur clusters. Several cyanobacteria have been found to contain a number of homologues to this enzyme. *Synechocystis* sp. PCC6803 contains three such genes designated *ndbA* (slr0851), *ndbB* (slr1743) and *ndbC* (sll1484) (Kaneko et al., 1996), *Anabaena* sp. PCC7120 contains at least six, all1127 (most similar to *ndbA*), alr4094 (most similar to *ndbB*), alr5211 (most similar to *ndbC*), all2964, all1126, and all1553 (Kaneko et al., 2001), *Thermosynechococcus elongatus* BP-1 contains only one (tlr1136, most similar to *ndhB*), and *Synechococcus* sp. PCC7002 contains two that display high similarities with *ndhA* and *ndhB*, respectively (Nomura, 2001). Only the *Synechocystis* enzymes are characterized in some detail (Howitt et al., 1999). *ndbA* and *ndbB* have no discernible transmembrane stretches and may be

Table 1. NAD(P)H dehydrogenases in Cyanobacteria

Gene Name	Corresponding ORF in *Synechocystis* sp. PCC 6803	*Anabaena (Nostoc)* sp. PCC 7120	*Thermosynechococcus elongatus* BP-1
Type I			
ndhA	sll0519	alr0223	tlr0667
ndhI	sll0520	alr0224	tlr0668
ndhG	sll0521	alr0225	tlr0669
ndhE	sll0522	alr0226	tlr0670
ndhB	sll0223	all4883	tll0045
ndhC	slr1279	all3842	tlr1429
ndhK	slr1280	all3841	tlr0705
ndhJ	slr1281	all3840	tlr1430
ndhL	ssr1386	asr4809	tsr0706
ndhD1	slr0331	alr3957	tlr0719
ndhF1	slr0844	alr3956	tlr0720
ndhD2	slr1291	alr0348	–
ndhD2b	–	alr5050	–
ndhD3	sll1733	alr4157	tlr0905
ndhF3	sll1732	alr4156	tlr0904
ndhD4	sll0027	alr0870	tlr2125
ndhF4	sll0026	alr0869	tlr2124
ndhF2	slr2009 (*ndhD6*)	–	tll1819 (*ndhD2*)
ndhM (?)	slr2008	all1843	tll1818
ndhF5	slr2007 (*ndhD5*)	all1842	tll1817 (*ndhD5*)
ndhH	slr0261	alr3355	tlr1288
Type II			
ndbA	slr0851	all1127	–
		all1126	–
		all1553	–
		all2964 (?)	–
ndbB	slr1743	alr4094	–
		alr5211	–
DrgA			
drgA	slr1719	all1864	tll2201

cytoplasmic proteins, while *ndbC* has one possible transmembrane region. Mutants were constructed that lack one each, or two or all three of the genes, whose phenotype with respect to growth rates or respiratory activities were not dramatically different from each other or the wild type. Under the conditions tested the level of expression was found to be quite low. A curious phenotype was observed, when any or all of the *ndb* genes were deactivated in a strain carrying a PS I background. PS I-less strains are very sensitive to light and only grow at low light intensities (5 $\mu E \cdot m^{-2} \cdot s^{-1}$), while all *ndb*-PS I double mutants are able to grow at 50 $\mu E \cdot m^{-2} \cdot s^{-1}$. The authors therefore considered the possibility that the type II NAD(P)H dehydrogenases are important for regulation of electron transport and may have no bioenergetic function. On the other hand, it was demonstrated that the *ndbB* gene from *Synechocystis* complemented an *E. coli* strain that lacked all NADH dehydrogenases for growth on minimal medium with mannitol, suggesting that it can function as an orthodox NADH dehydrogenase.

3. Other NAD(P)H Dehydrogenase(s)

Two other enzymes with NAD(P)H dehydrogenase activities have been identified in cyanobacteria, the DrgA protein and Ferredoxin-NADP$^+$ Oxidoreductase (FNR). The *drgA* gene was originally discovered as a gene whose inactivation conferred resistance to herbicides such as dinoseb or metronidazole in *Synechocystis* sp. PCC6803 (Elanskaya et al., 1998). Matsuo et al. (1998) showed that the DrgA protein was an NAD(P)H-quinone oxidoreductase. It is probably a soluble protein and its function in the electron transport chain is still uncertain. Ferredoxin-NADP$^+$ oxidoreductase (FNR), whose main function is undoubtedly the synthesis of NADPH in non-cyclic photosynthetic electron transport, has been suggested to play a role as an NAD(P)H dehydrogenase with quinones as electron and proton acceptors (see Schmetterer, 1994). It is still not known whether this activity of FNR contributes in a significant way to the total in vivo NAD(P)H dehydrogenase activity. Howitt et al. (1999) have performed native gel electrophoresis of NAD(P)H dehydrogenases of *Synechocystis* sp. PCC6803 and found evidence for at least two more such enzymes, about which no details are available.

4. Succinate Dehydrogenases

Synechocystis sp. PCC6803 contains two succinate-quinone oxidoreductases that contribute significantly to the total respiratory electron transport activity. In *E. coli*, succinate dehydrogenase consists of four subunits encoded by the *sdhCDAB* operon. Cyanobacteria contain homologues to *sdhA* (binding FAD) and *sdhB* (binding the FeS cluster), but not to *sdhC* or *sdhD*, which encode the subunits binding to heme b_{556}. However, in none of the cyanobacteria investigated so far the *sdhA* and *sdhB* genes are located close to each other on the chromosome (*Synechocystis* sp. PCC6803: *sdhA*, slr1233 and *sdhB1*, sll0823 and *sdhB2* sll1625; *Anabaena* sp. PCC 7120: *sdhA*, all2970 and *sdhB* all0945, *Thermosynechococcus elongatus* BP–1: *sdhA*, tlr1377 and *sdhB* tlr1754, *Gloeobacter violaceus* PCC 7421: *sdhA*, glr2988 and *sdhB*, glr2944). Cooley et al. (2000) constructed mutants in each of the two *sdhB* genes of *Synechocystis* sp. PCC6803 and also a double mutant lacking both genes. They clearly showed that these proteins form part of a functional succinate dehydrogenase that is lacking in the double mutant. No information is available as to other subunits this enzyme(s) might contain. Cooley and Vermaas (2001) have shown that succinate dehydrogenase may be a major electron donor to the respiratory chain in *Synechocystis* sp. PCC 6803. In wild type cells, respiration was approximately 50% inhibited by malonate, an inhibitor of succinate dehydrogenase and in a mutant lacking succinate dehydrogenase, the residual respiratory oxygen uptake was only about 25% of the wild type.

5. Hydrogenases

At least three enzymes exist in cyanobacteria that have hydrogenase activity:

1) Nitrogenase, which produces hydrogen;

2) Uptake hydrogenase, which oxidizes hydrogen and transfers electron into the quinone pool. This enzyme can be considered part of a respiratory electron transport branch;

3) Bidirectional hydrogenase, whose biological function is not clear yet.

[A recent review on cyanobacterial hydrogenases has appeared (Tamagnini et al., 2002) and therefore these enzymes will not be treated further here].

D. Quinone Pool

Stoichiometrically, the quinones represent by far the most abundant components of the cyanobacterial respiratory electron transport chains. Cyanobacterial membranes contain large amounts of plastoquinone-9 and less phylloquinone, with their relative amounts varying in the ICM and CM. The quinone pool of the ICM (thylakoids) is also used by the photosynthetic electron transport chain, and it is of central importance in defining the redox state of the cell. Given their importance in bioenergetic processes, comparatively little is known about the biosynthesis of the quinones. Using the complete genomic sequence of *Synechocystis* sp. PCC6803, Johnson et al. (2000) have proposed a hypothetical pathway for the biosynthesis of phylloquinone from chorismate, based on the assumption that this pathway consists of homologues of eight genes that have been identified in *E. coli* and *Bacillus subtilis*. They showed that inactivation of the *menA* or *menB* genes (1,4-dihydroxy-2-naphthoic acid phytyl transferase and 1,4-dihydroxy-2-naphthoate synthase, respectively) leads to cells that lack phylloquinone, but have almost normal photosynthetic electron transport rates. Unfortunately, no data on respiratory electron transport rates are presented, but the ability of the mutants to grow photoheterotrophically at rates similar to wild type cells could indicate that electron transport from NAD(P)H to cytochrome b_6f may be largely unaltered in the mutants. Thus, phylloquinone may not be essential for respiration.

E. Cytochrome b_6f

Bioenergetically, the cytochrome b_6f complex is the most important component of cyanobacterial electron transport, since its redox reaction (from plastoquinol to either cytochrome c_6 or plastocyanin) is coupled to the chemiosmotic translocation of protons. In the ICM, it is used in both respiratory and photosynthetic electron transport processes. Despite earlier reports indicating that cyanobacteria contain a mitochondrial-type cytochrome bc_1 complex, the complete genomic sequences have shown that this is not the case. The use of cytochrome b_6f in respiratory electron transport is unique to cyanobacteria.

Cyanobacterial cytochrome b_6f consists of four subunits, cytochrome b_6, cytochrome f (a c-type cytochrome, which is not related to the cytochrome c of the cytochrome bc_1 complex), the so-called subunit IV, and the Rieske protein. The active protein may be a dimer (Poggese et al., 1997). In the homologous cytochrome bc_1 complex, cytochrome b_6 and subunit IV are fused to one protein. The three first named subunits are present as single copy genes, while the number of genes for the Rieske protein varies. Thus, *Synechocystis* sp. PCC6803 contains three (*sll1182*, *sll1316*, *sll1185*), *Anabaena* sp, PCC7120 four (*all2453*, *all1512*, *all4511*, *all0606*) and *Thermosynechococcus elongatus* BP-1 one gene (*tlr0959*) coding for putative Rieske Fe-S proteins. The significance of the different Rieske proteins has not been elucidated yet.

Several reports have provided evidence for the existence of a chlorophyll *a* molecule as a component of cytochrome b_6f in cyanobacteria (see Poggese et al., 1997, and the literature cited there). These authors also provide evidence that the cytochrome b_6 subunit is the actual binding site for the chlorophyll *a* molecule. While it is unclear, what the possible function of this chlorophyll component might be, it appears nevertheless remarkable that a chlorophyll-containing protein complex is an important component of a respiratory chain.

Attempts to construct mutants lacking the cytochrome b_6f complex have not been successful. This probably implies that cytochrome b_6f is an essential component in all bioenergetic regimes that effect growth. However, Lee et al. (2001) succeeded in constructing site-specific mutations in the cytochrome b_6f complex of *Synechococcus* sp. PCC7002. Especially striking was a D148G substitution in *Synechococcus* sp. PCC7002 cytochrome b_6. In contrast to cytochrome bc_1, the cytochrome b_6f complex is insensitive to the inhibitor myxothiazol. The above mentioned substitution generated a functional, but myxothiazol-sensitive cytochrome b_6f complex. On the other hand, in *Rhodobacter capsulatus*, which contains a cytochrome bc_1, the mutation G152D of the corresponding amino acid yielded a strain that was insensitive to myxothiazol (see also Cooley et al., Chapter 3, Vol. 15).

A cytochrome b_6f complex with reduced activity was generated by Tichy and Vermaas (1999). The *ccsB* gene of *Synechocystis* sp. PCC 6803, which is involved in the maturation of cytochrome f, could not be deleted completely from the genome. However, a *ccsB* gene lacking the first 24 amino acids could be constructed. This strain was partially deficient in maturation of cytochrome f and accumulated the cytochrome *f* apoprotein. Mature cytochrome *f* was produced to a level of about 20% of the wild type, and the mutant strain grew only under anaerobic conditions.

F. Peripheral Intermediate Electron Carriers

Cyanobacteria contain two types of peripheral intermediate electron carriers, cytochrome *c* and plastocyanin, the former containing a heme-iron cofactor, the latter containing copper. The main function of these soluble proteins in cyanobacteria is the transfer of electrons from the cytochrome b_6f complex to Photosystem I. However, they also are important parts of the respiratory chains. With respect to the expression of the peripheral intermediate electron carriers in response to the copper concentration of the medium, Sandmann (1986) found that all known cyanobacteria belong to one of three groups:

1) Strains containing only cytochrome c_6 (gene for plastocyanin absent);

2) Strains forming cytochrome c_6 in low, and plastocyanin in high, (about 1 μM) Cu^{2+}. Cells grown in intermediate Cu^{2+} concentrations (about 300 nM) contain both proteins;

3) Strains forming cytochrome c_6 constitutively and plastocyanin only in high Cu^{2+} concentration.

Of the several strains that were thought to belong to group 1) only a single strain has remained that does not seem to contain a plastocyanin gene (*Synechococcus* sp. strain PCC 7002).

1. Cytochromes c

There are three types of soluble cytochrome *c*'s in cyanobacteria, namely: cytochrome c_6 (also called cytochrome c_{553}), cytochrome c_M, and cytochrome c_{550}. The latter cytochrome has an unusually low midpoint redox potential (about –260 mV), is a component of Photosystem II and thus it is not involved in respiration. Cytochrome c_6 is highly related to mitochondrial cytochrome *c* and several cytochrome *c*'s from many bacterial species, while cytochrome c_M seems to occur only in cyanobacteria.

a. Cytochrome c_6

All cyanobacteria contain at least one copy of a gene for cytochrome c_6. It constitutes a branching point in the electron transport chains of the cyanobacterial ICM. It accepts electrons from the cytochrome b_6f complex and transfers them either to Photosystem I or to cytochrome *c* oxidase. In strains where the expression of cytochrome c_6 depends on the copper concentration (such as *Synechocystis* sp. PCC 6803) its function in the ICM can apparently be taken over completely by plastocyanin, since mutants show very little phenotype. In contrast, the respiratory electron transport chain in the CM is fully dependent on the presence of cytochrome c_6. This has been demonstrated by Pils and Schmetterer (2001) using the uptake of 3-O-methyl-glucose by *Synechocystis* sp. PCC 6803 cells as a probe. They showed that the energization of the CM (necessary for the active uptake of the non-metabolizable glucose-analog) is lost, when a mutant containing ARTO (stands for Alternate Respiratory Terminal Oxidase) as the only respiratory terminal oxidase (see below) also lacks the *petJ* gene (encoding cytochrome c_6) and therefore concluded that cytochrome c_6 is an essential part of the electron transport chain ending in ARTO, which is likely a component of the CM. Very little is known about the mechanisms that target proteins into either the CM or the ICM in cyanobacteria. This is also true for soluble proteins that are fully transported through the membranes, into the intrathylakoid lumen (transport through the ICM) or into the periplasmic space (transport through the CM). Therefore, while at least in *Synechocystis* sp. PCC 6803 it is quite likely that cytochrome c_6 is present in both cellular spaces, what property of the *petJ* gene is responsible for this dual targeting remains unknown.

The total genomic sequence of *Anabaena* sp. PCC 7120 (Kaneko et al., 2001) has revealed the presence of three copies of genes that may code for cytochrome c_6's. No data are available what the function of these genes and the proteins they may encode is, but one might speculate that the multitude of genes is related to the fact that this strain develops heterocysts upon removal of combined nitrogen.

An unusual property has been observed for cytochrome c_6 of *Synechocystis* sp. PCC 7002. This strain contains a single gene *petJ* coding for cytochrome c_6. In contrast to all other cyanobacteria, this gene cannot be removed from the genome and is therefore essential under both autotrophic and heterotrophic growth conditions (Nomura, 2001). This is clearly due to the absence of a plastocyanin gene in this strain, which in other cyanobacteria can take over the functions of cytochrome c_6 (see below). Curiously, even the attempted substitution of the *Synechococcus* sp. PCC 7002 *petJ* gene with the *petJ* gene or the *petE* gene (encoding plastocyanin) from *Synechocystis* sp,. PCC 6803, was not possible (Nomura 2001). In view of the high similarity between the cytochrome c_6's from the two organisms (45% identity, 68% similarity) it is unclear what specific property of the cytochrome c_6 from strain PCC 7002 makes it unique.

b) Cytochrome c_M

Cytochrome c_M was discovered by sequencing a region of *Synechocystis* sp. PCC 6803 (Malakhov et al., 1994). Despite earlier reports that cytochrome c_M might occur only in a few strains, a homolog has been identified in each cyanobacterium whose total sequence was determined (*Synechocystis* sp. PCC 6803, *Anabaena* sp. PCC7120, *Thermosynechococcus elongatus* BP-1, *Gloeobacter violaceus* PCC 7421, *Prochlorococcus marinus* SS120, *Prochlorococcus marinus* MED4, *Prochlorococcus marinus* MIT9313, *Synechococcus* sp. WH 8102, *Synechococcus* sp. PCC 7002). It is present in very small amounts under normal growth conditions, but induced under stress such as low temperature (22 °C) and high light (2.10^3 E m^{-2} s^{-1}), under which conditions the expression of the other two peripheral intermediate electron carriers (cytochrome c_6 and plastocyanin) is down-regulated (Malakhov et al., 1999). This has led to the idea that cytochrome c_M might substitute for the other two proteins under stress. Indeed, Shuvalov et al. (2001) have shown that it is redox active in vivo and light absorbed by Photosystem I led to the oxidation of reduced cytochrome c_M, suggesting that cytochrome c_M might donate electrons to Photosystem I. However, Molina-Heredia et al. (2002) found very low electron transport rates in vitro from purified reduced cytochrome c_M to isolated PS I particles so that this reaction may not be of importance in the whole cell. Furthermore, the redox midpoint potential of cytochrome c_M from *Synechocystis* PCC 6803 has been determined independently by two groups (Cho et al., 2000; Molina-Heredia et al., 2002) to be +150 mV at pH 7, which is very different from the redox midpoint potentials of cytochrome c_6 and plastocyanin from the same strain (about +350 mV), so that a similar redox function also appears unlikely. Mutants lacking the *cytM* gene (encoding cytochrome c_M) have little phenotype. However, Pils et al. (1997) found that a *Synechocystis* PCC6803 double mutant lacking both the *petE* gene (encoding plastocyanin) and the *cytM* gene will not segregate to homozygocity, so that in a *cytM* mutant plastocyanin appears to be essential. Manna and Vermaas (1997) found that in a Photosystem I minus strain cytochrome c_M could not be removed from the genome and thus is essential in this mutant. Since Photosystem I minus strains need cytochrome *c* oxidase as the terminal electron transport component, they suggested that cytochrome c_M might be a part of the electron transport branch ending in cytochrome *c* oxidase. While this idea is compatible with the results of Pils et al. (1997), the position of cytochrome c_M in this branch, taking into account its rather low midpoint redox potential, remains unclear. In summary, although cytochrome c_M may be ubiquitous among cyanobacteria, further work is warranted for understanding its real function.

2. Plastocyanin

Plastocyanin is a soluble protein containing one atom of copper per polypeptide. It is only expressed in media containing high amounts of copper (about 0.3 μM or higher). Its primary function in cyanobacteria is the electron transfer from the cytochrome b_6f complex to Photosystem I, as in chloroplasts. However, Pils et al. (1997) have also obtained evidence for the involvement of plastocyanin in the respiratory chain in *Synechocystis* sp. PCC 6803. Using a number of mutants lacking different respiratory terminal oxidases and peripheral soluble intermediate electron carriers they found that the ability of this strain to grow chemoheterotrophically is strictly coupled to the functional presence of cytochrome *c* oxidase. Interestingly, a mutant containing cytochrome *c* oxidase but lacking the *petJ* gene encoding cytochrome c_6 (the normal electron donor to cytochrome *c* oxidase) is still able to grow chemoheterotrophically. Thus, there must be an alternative pathway for electrons to cytochrome *c* oxidase and plastocyanin may thus be not only an alternative electron donor to Photosystem I but also to cytochrome *c* oxidase.

Mutants lacking the *petE* gene encoding plastocyanin are easy to obtain and have little phenotype. Clarke and Campbell (1996) reported an enhanced sensitivity to chilling in a *petE* mutant of *Synechococcus* sp. PCC 7942. Despite attempts in several laboratories it has not been possible to construct a double mutant lacking both the genes for cytochrome c_6 and plastocyanin. Thus one must conclude that one of these peripheral intermediate electron carriers must be present to achieve transfer of electrons to Photosystem I (for phototrophic growth) or to cytochrome *c* oxidase (for chemoheterotrophic growth). Two reports have claimed that both proteins might be dispensible in *Synechococcus* sp. strain PCC 7942 (Laudenbach et al., 1990) or in *Synechocystis* sp. PCC 6803 (Zhang et al., 1994). In the former case a mutant lacking the *petJ* gene was constructed and no gene for plastocyanin was found at the time and still normal photosynthetic and respiratory electron transport capacities were found in the mutant. However, it is now known that this strain does contain a *petE* gene, so that the alternative peripheral intermediate electron carrier most likely is plastocyanin in this strain, as in other cyanobacteria. In the case of *Synechocystis* sp. strain PCC 6803 — a strain belonging to Sandmann group 2 (see above) — a mutant lacking the *petJ* gene (and thus cytochrome c_6) was constructed and grown under very low copper concentrations (about 20 nM) so that no plastocyanin was detectable. This result was thought to be evidence for the existence of a third electron carrier between cytochrome b_6f and Photosystem I. However, Pils et al. (unpublished results) have found that the repression of the *petJ* gene (in high copper) or *petE* gene (in low copper) is very effective but not fully complete so that wild type cells grown in high copper still contain tiny amounts of cytochrome c_6 that may be sufficient for function. Therefore, currently there is no compelling evidence for the existence of any other peripheral intermediate electron carrier besides cytochrome c_6 and plastocyanin (except in one case, see below). A possible function of plastocyanin as a donor to cytochrome *c* oxidase has been described above (see Cytochrome c_6).

3. Blue Copper Protein BcpA

A new potential peripheral intermediate electron carrier was discovered in *Synechococcus* sp. PCC 7002 (Nomura, 2001). It is a copper protein and related to plastocyanin, the main similarities being around the copper binding site. *bcpA* genes are also found in the genomes of *Anabaena* sp. PCC 7120 (*all1020*) (Kaneko et al., 2001) and *Nostoc punctiforme* (NCIB accession number ZP 00111122), but absent in the other completely sequenced cyanobacterial genomes currently available. It is expressed and translated in *Synechocystis* sp. PCC 7002. A mutant lacking the *bcpA* gene can be constructed, but has no known phenotype (Nomura, 2001). Thus it must remain uncertain what the function of BcpA might be and whether it is associated with photosynthetic and/or respiratory electron transport.

G. Respiratory Terminal Oxidases

Cyanobacterial respiratory terminal oxidases are the key enzymes of respiration, since they may be the only components of the respiratory chain in the ICM that are not directly involved in photosynthesis. All cyanobacterial respiratory terminal oxidases belong to only two protein families, the relatives of the heme copper oxidases (Cox and ARTO) and the quinol oxidases of the cytochrome *bd* type (Qox). The best characterized members of the first group are mitochondrial cytochrome *c* oxidase from mammals, from the second group the cytochrome *bd* from *Escherichia coli*. The heme copper oxidases of cyanobacteria are of two types, the genuine cytochrome *c* oxidases (Cox) and a group of related enzymes, whose function is not well understood, and which we therefore — for the time being — have called ARTO. All known cyanobacterial respiratory terminal oxidases are sensitive to KCN. This and the complete genomic sequences show definitely — despite earlier reports to the contrary — the absence of cyanide insensitive terminal respiratory oxidases in cyanobacteria, and specifically, of alternative terminal oxidase as found in mitochondria of plants and some other eukaryotes (Vanlerberghe and McIntosh, 1997).

An important result of the sequencing of cyanobacterial genomes is that even closely related strains may have different sets of respiratory terminal oxidases (Table 2). This makes it very difficult to assign a specific function to any of the terminal respiratory oxidases in cyanobacteria, since results obtained in one strain cannot necessarily be applied to another one.

Unfortunately, there is currently no method to assay the in vivo contribution of the different respiratory terminal oxidases to total respiration in any strain. A number of mutants exist in *Synechocystis* sp. PCC6803 (Schmetterer et al., 1994; Howitt and Vermaas, 1998; Pils and Schmetterer 2001), *Anabaena* sp. PCC7120 (Valladares et al., 2003), *Anabaena variabilis* ATCC29413 (Schmetterer et al., 2001), and *Synechococcus* sp. PCC7002 (Nomura, 2001) that lack one or more respiratory terminal oxidases. In

Table 2. Respiratory terminal oxidases in Cyanobacteria

Strain	Cox	ARTO	Qox
Anabaena sp. PCC7120	2	1	1
Synechocystis sp, PCC6803	1	1	1
Thermosynechococcus elongatus BP-1	1	0	1
Gloeobacter violaceus PCC7421	1	1(?)	1
Synechococcus sp. WH8102	1	1	0
Prochlorococcus marinus SS120	1	0	0
Prochlorococcus marinus MED4	1	0	0
Prochlorococcus marinus MIT9313	1	0	0
Synechococcus sp. PCC7002	1	1	0
Nostoc punctiforme	2	2	0
Anabaena variabilis ATCC29413	2	1	1

The numbers given represent the number of gene sets encoding the corresponding respiratory terminal oxidases. The last three named strains are not yet sequenced completely so that the actual number of respiratory terminal oxidases in these strains may be higher. Concerning the question mark (?) for the ARTO of *Gloeobacter violaceus* PCC7421, see text.

Synechocystis sp. PCC 6803 (and only in this strain) all possible combinations of mutants are available and a striking result is that removal of one or more respiratory terminal oxidase(s) influences the total respiratory activity (as determined by the uptake of O_2) in a currently unpredictable way. Especially the respiratory rates of all those mutants that contain only one respiratory terminal oxidase do by far not add up to the respiratory rate of the wild type. This implies that removal of any one of the terminal respiratory oxidases changes the contribution (through different expression or activity or degree of saturation with substrates) of all the other respiratory terminal oxidases to total respiration. Mutants lacking all known respiratory terminal oxidases are available currently only from *Synechocystis* sp. PCC6803 (Howitt and Vermaas, 1998) and *Synechococcus* sp. PCC 7002 (Nomura, 2001). The former shows no respiratory activity, demonstrating that all respiratory terminal oxidases are knocked out and thus known, the latter displays some residual uptake of O_2 and may therefore contain another respiratory terminal oxidase.

1. Heme Copper Family

a) Cytochrome c Oxidases

All cyanobacteria contain at least one set of three genes (called *coxBAC* or *ctaCDE*) coding for a cytochrome *c* oxidase (Cox). The three subunits correspond to the three mitochondrially encoded subunits I,II, and III of cytochrome *c* oxidase from eukaryotes. The total genomic sequences of cyanobacteria have produced no evidence for genes with sequence similarity to those encoded in the nucleus of eukaroytes. Cyanobacterial Cox can be conveniently assayed in vitro using isolated membranes and horse heart cytochrome *c* as the electron donor. In contrast, ARTO (see below) shows no detectable horse heart cytochrome *c* oxidase activity, despite the high sequence similarity of the corresponding three subunits. This property has been used by Pils and collaborators (unpublished) to show that of the at least three sets of *cox* genes identified in *Anabaena variabilis* ATCC 29413 two encode genuine cytochrome c oxidases, since a double mutant lacking Cox1 and Cox2 in this strain has no horse heart cytochrome c oxidase activity.

In *Synechocystis* sp. PCC 6803, cytochrome *c* oxidase is probably located only in the ICM. This may not be a universal feature, however, since isolated CM from other strains (e.g. *Plectonema boryanum* PCC 6306 (Peschek et al., 1988), *Synechococcus* sp. PCC7942 (Schmetterer, unpublished)) has been found to contain in vitro horse heart cytochrome *c* oxidase activity.

The genes encoding Cox constitute an operon, as has been demonstrated for the *coxBAC* genes of *Synechocystis* sp. PCC 6803 (Howitt and Vermaas, 1998), the *cox1BAC* (Schmetterer et al., 2001) and *cox2BAC* (Pils and collaborators, unpublished) genes of *Anabaena variabilis* ATCC29413 and the *cox2BAC* genes of *Anabaena* sp. PCC7120 (Valladares et al., 2003). However, the primary transcript could be processed in *Anabaena* strains, since a strong signal in Northern blots was obtained for an mRNA containing only the first gene of the operon, the *coxB* gene. Whether this is of functional significance in the cells cannot be stated currently, especially as it is expected that the three subunits are synthesized in equal amounts in vivo. Cytochrome *c* oxidase is subject to gene regulation in response to the growth medium in *Anabaena* strains. In *A. variabilis* ATCC29413 the *cox1BAC* genes are upregulated by fructose (Schmetterer et al., 2001), but this regulation occurs only with one of the two transcriptional start sites upstream of *coxB* (tsp2), while the other one (tsp1) is constitutive. In *Anabaena* sp. PCC7120 the *cox2BAC* genes (and also the *cox3BAC* genes encoding the ARTO, see below) are upregulated when the cells are grown in the absence of combined nitrogen and thus contain heterocysts (Valladares et al., 2003). The *cox2BAC* genes are either preferentially or possibly even exclusively expressed in heterocysts.

Mutants lacking cytochrome *c* oxidase(s) have been constructed in strains *Synechocystis* sp. PCC 6803 (Schmetterer et al. 1994, Pils et al. 1997), *Anabaena* sp. PCC 7120 (Valladares et al., 2003), *Anabaena variabilis* ATCC 29413 (Schmetterer et al., 2001; Pils and collaborators, unpublished), and *Synechococcus* sp. PCC7002 (Nomura, 2001). These mutants have allowed to determine that cytochrome *c* oxidase is the most important respiratory terminal oxidase with respect to bioenergetic capacity. When the genes encoding cytochrome *c* oxidase were inactivated in *Synechocystis* sp. PCC 6803 (*coxBAC*) or in *Anabaena variabilis* (*cox1BAC*), the resulting mutants lost the capacity for heterotrophic growth. This is especially striking for the latter strain, since it contains at least one other cytochrome *c* oxidase (*cox2BAC*). In *Synechocystis* sp. PCC6803, the loss of Cox also led to the loss of the so-called salt respiration, a short term enhancement of respiratory activity upon the addition of NaCl, even if the other two respiratory terminal oxidases were still present. In *Anabaena* sp. PCC7120 the *cox2* locus is involved in the protection of nitrogenase in heterocysts from the damage by oxygen. Curiously, this protection can also be performed by the *cox3* locus (coding for ARTO, see below), but not by the *cox1* locus that encodes a second cytochrome *c* oxidase enzyme. One possible explanation is that Cox2 and the ARTO, but not Cox1 are localized in the heterocysts.

An important finding was the observation that cytochrome *c* oxidase cannot be inactivated in a *Synechocystis* sp. PCC 6803 lacking Photosystem I and vice versa (Manna and Vermaas, 1997). Therefore, Cox is a necessary electron sink for a Photosystem I minus strain (Vermaas et al., 1994). One may speculate that this might have been the original function of cytochrome *c* oxidase, namely regulating electron flow between Photosystem II and Photosystem I.

b) Alternate Respiratory Terminal Oxidases (ARTO)

The second type of enzymes of the heme copper family was originally discovered by sequencing the genome of *Synechocystis* sp. PCC6803 (Kaneko et al., 1996), and later called ARTO (Pils et al., 1997). The three putative genes encoding ARTO (proposed nomenclature: *ctaD*, *ctaC*, *ctaE*) have high sequence similarity to subunits I, II, and III, respectively, of mitochondrial cytochrome *c* oxidases, but are easily discernible from the genuine cytochrome *c* oxidases by two highly characteristic sequence differences, as first pointed out by Howitt and Vermaas (1998): the Mg^{2+} binding motif in cytochrome *c* oxidase subunit I (HisAsp around amino acid 360 to 400) is lacking and replaced by GlyAsn or AsnAsn, and the Cu_A binding motif of cytochrome *c* oxidase subunit II (Cys-X-Glu-X-Cys around amino acid 190 to 250) is not or only partially present in the corresponding ARTO subunit. An enzyme related to ARTO is the cytochrome *bo* quinol oxidase from *Escherichia coli*, but currently there are no data that show a quinol oxidase activity for cyanobacterial ARTO.

As shown in Table 2, not all cyanobacteria contain an ARTO, and the gene arrangement is not the same in those strains that do. The most common gene arrangement is *ctaCDE*. However, in *Synechocystis* sp. PCC6803 the *ctaC* gene is located far away from the *ctaDE* genes, and in *Gloeobacter violaceus* PCC7421 there is apparently no *ctaC* gene. Especially the latter observation raises the question, whether a CoxB protein (subunit II of cytochrome *c* oxidase) is able to form a functional ARTO with the CtaD (subunit I) and CtaE (subunit III) proteins from the same strain.

In no cyanobacterium has this been tested so far.

The enzymatic reaction catalyzed by ARTO is still unknown. While Pils and Schmetterer (2001) have shown that a *Synechocystis* sp. PCC6803 mutant that contains ARTO as the only terminal respiratory oxidase does consume O_2 (albeit at a very low rate), practically no information is available about the electron donor(s). However, Pils and Schmetterer (2001) were able to demonstrate that the ARTO is probably a bioenergetically active enzyme. They used the uptake of 3-O-methylglucose, a non-metabolizable analog of glucose that is actively transported through the CM, to monitor the energization of the CM and found that despite its low total respiratory activity (about 1% of the wild type), a mutant having ARTO as the only respiratory terminal oxidase is able to support active transport across the CM. This activity depends critically on the presence of cytochrome c_6, so that an involvement of this peripheral intermediate electron carrier in the respiratory branch ending in ARTO had to be postulated. This does not necessarily imply that ARTO is a cytochrome *c* oxidase, especially as the in vitro cytochrome *c* oxidase assay gives no detectable activity with isolated membranes from a mutant containing only ARTO. In view of the low respiratory activity of this mutant and the rather high active transport capacity of the strain, Pils and Schmetterer (2001) suggested that ARTO might be the terminal respiratory oxidase of the respiratory chain in the CM. Independent evidence for this came from a proteomics project designed to identify proteins in the purified CM of *Synechocystis* sp. PCC6803 (Huang et al., 2002): the only peptide from a respiratory terminal oxidase found was subunit II of ARTO. The activity of ARTO in the wild type is not necessarily low, however. In a mutant containing both Qox and ARTO (that had about 1/2 the respiratory activity of the wild type), the specific Qox inhibitors HQNO (2-heptyl-4-hydroxy-quinoline-N-oxide) and PCP (pentachlorophenol) inhibited respiration by only 72% and 83%, respectively, so that appreciable contributions of ARTO remained.

In *Synechococcus* sp. PCC7002 a mutant lacking Cox almost has the same (low) respiratory rate as a mutant lacking both Cox and ARTO, so that in this strain ARTO also appears to have a low contribution to total respiratory activity.

In *Anabaena* sp. PCC 7120 a different aspect of ARTO activity was discovered (Valladares et al., 2003). This filamentous strain differentiates about one in every 10–20 cells in a semi-regular fashion, when the cells are grown in a medium lacking combined nitrogen. The resulting differentiated cells, called heterocysts, are specialized for nitrogen fixation. The key enzyme of nitrogen fixation, nitrogenase, is confined to the heterocysts, and unusually sensitive to O_2. Respiration, as a process removing O_2 from the cytoplasm, apparently plays a decisive role in the protection of nitrogenase. Like Cox2 (see above), ARTO was found to be preferentially or possibly completely expressed in heterocysts. Mutants lacking both ARTO and Cox2 (but not the single mutants) were unable grow on N_2 and fixed very little N_2, implying that ARTO alone is sufficient to remove O_2 from the cytoplasm of heterocysts to allow diazotrophic growth. No information is available about the possible (heme?) cofactors of ARTO and nothing is known about the intracellular location of ARTO in heterocysts of *Anabaena* sp. PCC7120.

2. Cytochrome bd Type Quinol Oxidases

Cyanobacterial genes with high sequence similarity to the two subunits of *E. coli* cytochrome *bd* (Qox) were first detected in the process of the determination of the total genomic sequence of *Synechocystis* sp. PCC6803 (Kaneko et al., 1996). All cyanobacteria that have such an enzyme contain the two subunits encoded by the *cydA* and *cydB* genes in adjacent positions on the chromosome, as in *E. coli*. In *Synechocystis* sp. PCC6803 Howitt and Vermaas (1998) constructed a mutant that lacks both cytochrome *c* oxidase and ARTO so that the *cydAB* genes code for the only functional respiratory terminal oxidase. By showing that this strain respired almost at the same rate of the wild type and additional removal of the *cydAB* genes led to a non-respiring strain, they produced conclusive evidence that the *cydAB* genes code for a respiratory terminal oxidase. Whether it is a cytochrome *bd* type enzyme cannot be stated with certainty, however, since no conclusive evidence for the existence of heme *d* exists in the cyanobacterial literature.

Recently a good indication of the function of cyanobacterial Qox was obtained (Berry et al., 2002). While it has not yet been demonstrated directly that the enzymatic reaction of Qox is a quinol-dioxygen oxidoreductase (as in *E. coli*), these authors provide strong evidence for this by showing that it catalyses the oxidation of the quinone pool without using the cytochrome b_6f complex. The function of the Qox is therefore probably the prevention of the overreduction of the quinone pool. Since the photosystems are confined to the ICM and are not present in the CM, Qox is probably located in the ICM, although its presence in the CM cannot be completely excluded (see Vermaas, 2001). In the dark, Qox apparently contributes considerably to total respiration. Using a *Synechocystis* sp. PCC 6803 mutant, in which the Qox was the only remaining respiratory terminal oxidase (Howitt and Vermaas, 1998), Pils and Schmetterer (2001) identified HQNO and PCP as specific inhibitors of *Synechocystis* Qox. Thus in cells with more than one respiratory terminal oxidase the HQNO- or PCP- sensitive portion of total respiratory activity could be taken to represent the electron flux through the electron transport branch ending in Qox. In wild type *Synechocystis* sp. PCC 6803 this contribution of Qox is about 40–50%.

Curiously, the *Synechocystis* sp. PCC 6803 mutant containing only Qox as the respiratory terminal oxidase does not display an enhancement of respiration upon the addition of glucose to the external medium, in contrast to all other mutants of this strain (except the one completely devoid of respiration, of course). This seems to imply that the respiratory branch ending in Qox could be linked to a different electron transport branch at the reducing end of the respiratory chain than the branches ending in the other two respiratory terminal oxidases. How this might work, is currently unknown, since all respiratory branches contain the quinone pool as intermediate electron buffer.

H. Terminal Electron Acceptor

Only one terminal electron acceptor of the cyanobacterial respiratory electron transport chains is known, namely O_2. Neither physiologic experiments nor the total genomic sequences have yielded evidence for the existence of anaerobic respiration in cyanobacteria.

I. The Respiratory Chains of Synechocystis sp. PCC 6803

Figure 2 shows a model for electron transport in *Synechocystis* sp. PCC 6803. Due to the early avail-

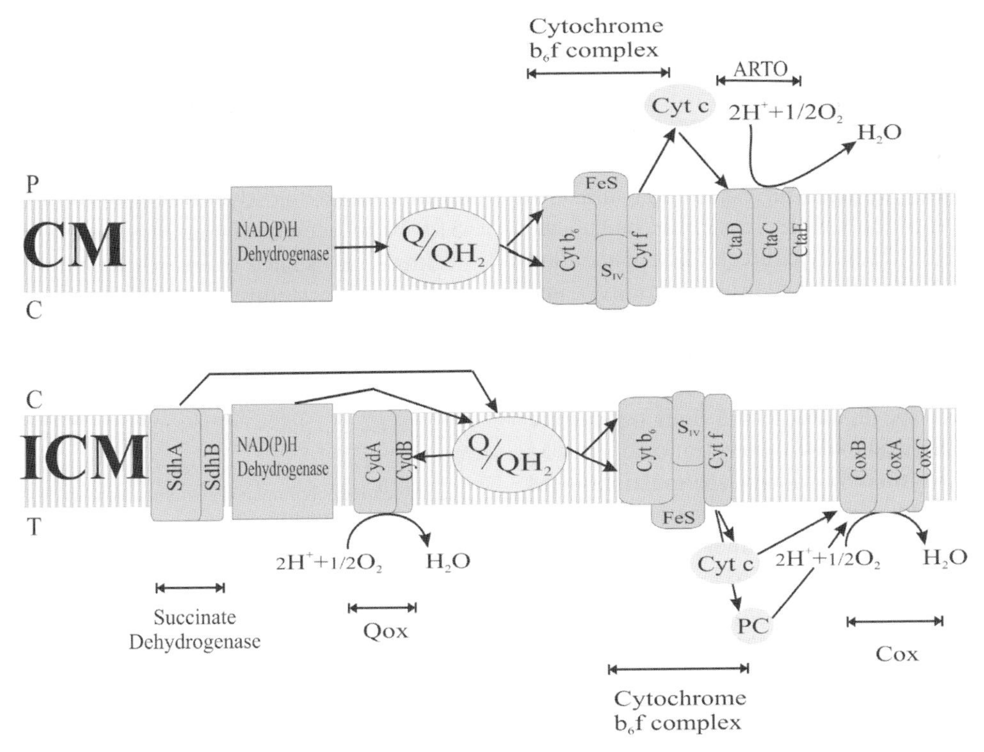

Fig. 2. Model for the respiratory electron transport chains in *Synechocystis* sp. PCC 6803. For clarity, Photosystem II (donating electrons directly into the quinone pool of the ICM) and Photosystem I (accepting electrons from either cytochrome c_6 or plastocyanin in the ICM) have been omitted. There is also an electron transport from Photosystem I back into the quinone pool (cyclic photosynthetic electron transport), whose components have also been omitted. P, inner periplasmic space; C, cytoplasm; T, intrathylakoid space; CM, cytoplasmic membrane; ICM, intracytoplasmic membranes or thylakoids; Q, oxidized quinones; QH_2 , reduced quinones.

ability of the total genomic sequence this is the only strain, in which such a model can be constructed with some confidence. An application of this scheme to other cyanobacteria should be made only with great caution. Especially in strains capable of cell differentiation (e.g. heterocysts) the situation is certainly much more complicated. However, even in *Synechocystis* there remain several uncertainties. Cytochrome c_M is not shown, because both its cellular location and its position in the electron transport chains is unclear. Similarly, the DrgA dehydrogenase and NADH dehydrogenase of type II have been deliberately omitted. Furthermore, the type I NAD(P)H dehydrogenase is not shown with the details of its subunits, especially since the subunit responsible for NAD(P)H oxidation is not yet known. The electron transport chain of the CM is less well characterized than that in the ICM.

III. Interaction between Respiration and Photosynthesis

The question of interaction between respiration and photosynthesis in cyanobacteria is an old one and indeed one of the main reasons that cyanobacterial respiration has originally been studied. The interaction is very different for the two different respiratory chains in the CM and the ICM. Since the CM contains no photosystems, its respiratory chain does not share components with photosynthesis and the interaction is of necessity indirect. The energization of the CM is a very important process, however, and must be possible in order to allow the cells controlled transport of relevant substances into or out of the cell. Since it was possible to construct a mutant strain of *Synechocystis* sp. PCC 6803 that does not respire (Howitt and Vermaas, 1998), energization of the CM is evidently possible by primary energization of the ICM (by photosynthesis) and subsequent transfer of energy to the CM. Biochemically, this process involves probably diffusion of ATP generated by an ATP synthase in the ICM to an ATPase located in the CM. Curiously, cyanobacteria contain only one set of F_oF_1 ATPase genes, so that either this ATPase is present in both membranes or a completely different ATPase is active in one of them. In the first case, which on current evidence seems to be the more likely one, the important question remains how this enzyme is targeted during biosynthesis into both membranes, while most membrane proteins occur only in CM or ICM.

The interaction between respiratory and photosynthetic electron transport chains in the ICM is characterized by their sharing of common components, of which the quinone pool is the most important. The construction of mutants that lack different components of this large bioenergetic complex has clearly demonstrated that electrons can enter from either the 'respiratory' side (i.e. the primary dehydrogenases) or the 'photosynthetic' side (i.e. Photosystem II) and end up either in the 'respiratory' oxygen-consuming terminal oxidases or Photosystem I. Unfortunately it is not yet known well to what extent the different branches of electron transport contribute to total electron flow under natural conditions in wild type cells. Besides some trivial statements, such as the inactivity of the two photosystems in darkness, this is largely due to the fact that no in vivo assay exists that allows to follow the redox reactions of one respiratory branch independent of the others that are also present.

Acknowledgments

We are grateful to Profs S. Tabata and D. Bryant for access to unpublished results. Work in the authors' laboratory was partially supported by HFSP (Human Frontiers for Science Program, project No. RG 51/97) and Acciones Integradas Austria-España. The technical assistance of Mr. O. Kuntner and Mr. C. Kolowrat is gratefully acknowledged.

References

Berger S, Ellersiek U, Steinmüller K (1991) Cyanobacteria contain a mitochondrial complex-I homologous NADH-dehydrogenase. FEBS Lett 286: 129–132

Berry S, Schneider D, Vermaas WFJ and Rögner M (2002) Electron transport routes in whole cells of *Synechocystis* sp. strain PCC 6803: The role of the cytochrome *bd*-type oxidase. Biochemistry 41: 3422–3429

Biggins J (1969) Respiration in blue-green algae. J Bacteriol 99: 570–575

Boison G, Bothe H, Hansel A and Lindblad P (1999) Evidence against a common use of the diaphorase subunits by the bidirectional hydrogenase and by the respiratory complex I in cyanobacteria. FEMS Microbiol Lett 174: 159–165

Broda E (1975) The Evolution of the Bioenergetic Processes. Pergamon Press, Oxford

Cho YS, Pakrasi HB and Whitmarsh J (2000) Cytochrome c_M from *Synechocystis* 6803. Detection in cells, expression in *Escherichia coli*, purification and physical characterization. Eur J Biochem 267: 1068–1074

Clarke AK and Campbell D (1996) Inactivation of the *petE* gene for plastocyanin lowers photosynthetic capacity and exacerbates chilling-induced photoinhibition in the cyanobacterium *Synechococcus*. Plant Physiol 112: 1551–1561

Cooley JW and Vermaas WJF (2001) Succinate dehydrogenase and other respiratory pathways in thylakoid membranes of *Synechocystis* sp. strain PCC 6803: Capacity comparisons and physiological function. J Bacteriol 183: 4251– 4258

Cooley JW, Howitt CA and Vermaas WFJ (2000) Succinate:quinol oxidoreductase in the cyanobacterium *Synechocystis* sp. strain PCC 6803: Presence and function in metabolism and electron transport. J Bacteriol 182: 714–722

Elanskaya IV, Chesnavichene EA, Vernotte C and Astier C (1998) Resistance to nitrophenolic herbicides and metronidazole in the cyanobacterium *Synechocystis* sp. PCC 6803 as a result of the inactivation of a nitroreductase-like protein encoded by *drgA* gene. FEBS Lett 428: 188–192

Ernst A, Kirschenlohr H, Diez J and Böger P (1984) Glycogen content and the stability of nitrogenase activity in *Anabaena variabilis*. Arch Microbiol 140: 120–125

Howitt C, Smith GD and Day DA (1993) Cyanide-insensitive oxygen uptake and pyridine nucleotide dehydrogenases in the cyanobacterium *Anabaena* PCC 7120. Biochim Biophys Acta 1141: 313–320

Howitt CA, Udall PK and Vermaas WFJ (1999) Type 2 NADH dehydrogenases in the cyanobacterium *Synechocystis* sp. strain PCC 6803 are involved in regulation rather than respiration. J Bacteriol 181: 3994–4003

Howitt CA and Vermaas WFJ (1998) Quinol and cytochrome oxidases in the cyanobacterium *Synechocystis* sp. PCC 6803. Biochemistry 37: 17944–17951

Huang F, Parmryd I, Persson A, Nilsson F, Pakrasi H, Andersson B and Norling B (2002) Proteomics of *Synechocystis* 6803: Identification of plasma membrane proteins. In: 5th European Workshop on the Molecular Biology of Cyanobacteria, Stockholm, Sweden, June 9–12, 2002, abstract no. 135

Ito M, Guffanti AA and Krulwich TA (2001) Mrp-dependent Na^+/H^+ antiporters of *Bacillus* exhibit characteristics that are unanticipated for completely secondary active transporters. FEBS Lett 496: 117–120

Johnson TW, Shen G, Zybailov B, Kolling D, Reategui R, Beauparlant S, Vassiliev IR, Bryant DA, Jones AD, Golbeck JH and Chitnis PR (2000) Recruitment of a foreign quinone into the A_1 site of Photosystem I. J Biol Chem 275: 8523–8530

Kaneko T, Nakamura Y, Wolk CP, Kuritz T, Sasamoto S, Watanabe A, Iriguchi M, Ishikawa A, Kawashima K, Kimura T, Kishida Y, Kohara M, Matsumoto M, Matsuno A, Muraki A, Nakazaki N, Shimpo S, Sugimoto M, Takazawa M, Yamada M, Yasuda M, and Tabata S (2001) Complete Genomic Sequence of the Filamentous Nitrogen-fixing Cyanobacterium *Anabaena* sp. Strain PCC 7120. DNA Res 8: 205–213

Kaneko T, Sato S, Kotani H, Tanaka A, Asamizu E, Nakamura Y, Miyajima N, Hirosawa M, Sugiura M, Sasamoto S, Kimura T, Hosouchi T, Matsuno A, Muraki A, Nakazaki N, Naruo K, Okumura S, Shimpo S, Takeuchi C, Wada T, Watanabe A, Yamada M, Yasuda M, Tabata S (1996) Sequence Analysis of the Genome of the Unicellular Cyanobacterium *Synechocystis* sp. Strain PCC6803. II. Sequence Determination of the Entire Genome and Assignment of Potential Protein-coding Regions. DNA Res 3: 109–136

Kuroda M, Ohta T, Uchiyama I, Baba T, Yuzawa H, Kobayashi I, Cui L, Oguchi A, Aoki K, Nagai Y, Lian J, Ito T, Kanamori M, Matsumaru H, Maruyama A, Murakami H, Hosoyama A, Mizutani-Ui Y, Kobayashi N, Tanaka T, Sawano T, Inoue R, Kaito C, Sekimizu K, Hirakawa H, Kuhara S, Goto S, Yabuzaki J, Kanehisa M, Yamashita A, Oshima K, Furuya K, Yoshino C, Shiba T, Hattori M, Ogasawara N, Hayashi H and Hiramatsu K (2001) Whole genome sequencing of meticillin-resistant *Staphylococcus aureus*. Lancet 357: 1225–1240

Laudenbach DE, Herbert SK, McDowell C, Fork DC and Grossman AR (1990) Cytochrome *c*-553 is not required for photosynthesis activity in the cyanobacterium *Synechococcus*. Plant Cell 2: 913–924

Lee TX, Metzger SU, Cho YS, Whitmarsh J and Kallas T (2001) Modification of inhibitor binding sites in the cytochrome *bf* complex by direct mutagenesis of cytochrome b_6 in *Synechococcus* sp. PCC 7002. Biochim Biophys Acta 1504: 235–247

Malakhov MP, Wada H, Los DA, Semenenko VE and Murata N (1994) A new type of cytochrome c from *Synechocystis* PCC6803. J Plant Physiol 144: 259–264

Malakhov MP, Malakhova OA and Murata N (1999) Balanced regulation of expression of the gene for cytochrome c_M and that of genes for plastocyanin and cytochrome c_6 in *Synechocystis*. FEBS Lett 444: 281–284

Manna P and Vermaas W (1997) Lumenal proteins involved in respiratory electron transport in the cyanobacterium *Synechocystis* sp. PCC6803. Plant Mol Biol 35: 407–416

Marco E, Ohad N, Schwarz R, Lieman-Hurwitz J, Gabay C and Kaplan A (1993) High CO_2 concentration alleviates the block in photosynthetic electron transport in an ndhB-inactivated mutant of *Synechococcus* sp. PCC 7942. Plant Physiol 101: 1047–1053

Matsuo M, Endo T and Asada K (1998) Isolation of a novel NAD(P)H-quinone oxidoreductase from the cyanobacterium *Synechocystis* PCC6803. Plant Cell Physiol 39: 751–755

Molina-Heredia FP, Balme A, Hervás M, Navarro JA and De la Rosa MA (2002) A comparative structural and functional analysis of cytochrome c_M, cytochrome c_6 and plastocyanin from the cyanobacterium *Synechocystis* sp. PCC 6803. FEBS Lett 517: 50–54

Nomura CT (2001) Electron transport proteins of *Synechococcus* sp. PCC 7002. PhD Thesis, The Pennsylvania State University, State College, PA

Ogawa T (1991) A gene homologous to the subunit-2 gene of NADH dehydrogenase is essential to inorganic carbon transport of *Synechocystis* PCC6803. Proc Natl Acad Sci USA 88: 4275–4279

Ogawa T (1992) Identification and characterization of the ictA/ndhL gene product essential to inorganic carbon transport of *Synechocystis* PCC6803. Plant Physiol 99: 1604–1608

Omata T and Murata N (1983) Isolation and characterization of the cytoplasmic membranes from the blue-green alga (cyanobacterium) *Anacystis nidulans*. Plant Cell Physiol 24: 1101–1112

Partensky F, Hess WR and Vaulot D (1999) *Prochlorococcus*, a marine photosynthetic prokaryote of global significance. Microbiol Mol Biol Rev 63: 106–127

Peschek GA, Molitor V, Trnka M, Wastyn M and Erber W (1988)

Characterization of cytochrome-*c* oxidase in isolated and purified plasma and thylakoid membranes from cyanobacteria. Meth Enyzmol 167: 437–449

Pils D, Gregor W and Schmetterer G (1997) Evidence for in vivo activity of three distinct respiratory terminal oxidases in the cyanobacterium *Synechocystis* sp. strain PCC6803. FEMS Microbiol Lett 152: 83–88

Pils D and Schmetterer G (2001) Characterization of three bioenergetically active respiratory terminal oxidases in the cyanobacterium *Synechocystis* sp. strain PCC 6803. FEMS Microbiol Lett 203: 217–222

Poggese C, Polverino de Laureto P, Giacometi GM, Rigoni F and Barbato R (1997) Cytochrome b_6/f complex from the cyanobacterium *Synechocystis* 6803: Evidence of dimeric organization and identification of chlorophyll-binding subunit. FEBS Lett 414: 585–589

Rippka R, Deruelles J, Waterbury JB, Herdman M and Stanier RY (1979) Generic assignments, strain histories and properties of pure cultures of cyanobacteria. J Gen Microbiol 111: 1–61

Sandmann G (1986) Formation of plastocyanin and cytochrome *c-553* in different species of blue-green algae. Arch Microbiol 145: 76–79

Schluchter WM, Zhao J and Bryant DA (1993) Isolation and characterization of the *ndhF* gene of *Synechococcus* sp. strain PCC 7002 and initial characterization of an interposon mutant. J Bacteriol 175: 3343–3352

Schmetterer G (1994) Cyanobacterial respiration. In: DA Bryant (ed) The Molecular Biology of Cyanobacteria, pp 409–435. Kluwer Academic Publishers, Dordrecht

Schmetterer G, Alge D and Gregor W (1994) Deletion of cytochrome *c* oxidase genes from the cyanobacterium *Synechocystis* sp. PCC6803: Evidence for alternative respiratory pathways. Photosynth Res 42: 43–50

Schmetterer G, Valladares A, Pils D, Steinbach S, Pacher M, Muro-Pastor AM, Flores E and Herrero A (2001) The *coxBAC* operon encodes a cytochrome *c* oxidase required for heterotrophic growth in the cyanobacterium *Anabaena variabilis* strain ATCC 29413. J Bacteriol 183: 6429–6434

Shuvalov VA, Allakhverdiev SI, Sakamoto A, Malakhov M and Murata N (2001) Optical study of cytochrome c_M formation in *Synechocystis*. IUBMB Life 51: 93–97

Tamagnini P, Axelsson R, Lindberg P, Oxelfelt F, Wünschiers R and Lindblad P (2002) Hydrogenases and hydrogen metabolism of cyanobacteria. Microbiol Mol Biol Rev 66: 1–20

Tichy M and Vermaas W (1999) Accumulation of pre-apocytochrome *f* in a *Synechocystis* sp. PCC 6803 mutant impaired in cytochrome *c* maturation. J Biol Chem 274: 32396–32401

Valladares A, Herrero A, Pils D, Schmetterer G and Flores E (2003) Cytochrome *c* oxidase genes required for nitrogenase activity and diazotrophic growth in *Anabaena* sp. PCC 7120. Mol Microbiol 47: 1239–1249

Vanlerberghe GC and McIntosh L (1997) Alternative oxidase: From gene to function. Annu Rev Plant Physiol Plant Mol Biol 48: 703–734

Vermaas WFJ (2001) Photosynthesis and Respiration in Cyanobacteria. In: Encyclopedia of Life Sciences. Nature Publishing Group

Vermaas WFJ, Shen G and Styring S (1994) Electrons generated by Photosystem II are utilized by an oxidase in the absence of Photosystem I in the cyanobacterium *Synechocystis* sp. PCC 6803. FEBS Lett 337: 103–108

Yu L, Zhao J, Mühlenhoff U, Bryant DA and Golbeck JH (1993) PsaE is required for in vivo cyclic electron flow around Photosystem I in the cyanobacterium *Synechococcus* sp. PCC 7002. Plant Physiol 103: 171–180

Zhang L, Pakrasi HB and Whitmarsh J (1994) Photoautotrophic growth of the cyanobacterium *Synechocystis* sp. PCC 6803 in the absence of cytochrome c_{553} and plastocyanin. J Biol Chem 269: 5036–5042

Chapter 13

Interactions Between Photosynthesis and Respiration in Facultative Anoxygenic Phototrophs

André Vermeglio
CEA-DSV-DEVM-LBC, Cadarache, F-13108, Saint Paul-lez-Durance, Cedex, France

Roberto Borghese and Davide Zannoni*
Department of Biology, University of Bologna, I-40126, Bologna, Italy

Summary

The respiratory and photosynthetic electron transport chains of the two facultative phototrophs *Rhodobacter (Rb.) sphaeroides* and *Rb. capsulatus* are arranged in such a way to be spatially segregated in separate regions of the internal membrane system (CM and ICM). The CM part contains the majority of the oxidative redox components which are therefore in redox non-equilibrium with most of the photochemical RCs; conversely, the major part of the photosynthetic carriers (including RCs, Cyt c_2 or Cyt c_y and Cyt bc_1 complex) are located in the ICM part of the membrane. This spatial level of organization is paralleled by an arrangement of these photosynthetic elements in supramolecular complexes in order to allow a fast and efficient cyclic electron transfer by limiting the diffusion of the reactants. However, these two levels of arrangement are not present in all types of photosynthetic bacteria. Indeed, species like *Blastochloris viridis* or *Rubrivivax gelatinosus,* contain a large excess of RCs over the Cyt bc_1 complexes so that the formation of supercomplexes is stoichiometrically hindered. Further, in obligate aerobic phototrophs such as for example *Roseobacter denitrificans*,

Author for correspondence, email: davide.zannoni@unibo.it

Davide Zannoni (ed): Respiration in Archaea and Bacteria. Vol 2: Diversity of Prokaryotic Respiratory Systems, pp. 279–295.

the Qa is fully reduced under anaerobic conditions and this might be due to the lack of both quinol oxidase and ICM system.

The expression of photosynthetic and respiratory components is controlled by the oxygen tension and by the redox state of the system. This genetic coordination mechanism does not necessarily require a direct interaction of the two sets of components in line with their different spatial membrane location. The signals to which the system responds originate from either specific respiratory components, e.g. cbb_3 oxidase, as in the case of oxygen sensing, or from redox carriers involved in both oxidative and photosynthetic ET, as for redox sensing. Although the genetic control of the supramolecular arrangement of the ETCs is, at present, largely undefined, the working scheme presented here, suggests a tentative framework of genetic regulatory connections in *Rb. capsulatus* and/or *Rb. sphaeroides*

I. Introduction

A. Background: What Is Important to Know About Physiology, Ecology, and Biochemistry of Facultative Photosynthetic Bacteria

Facultative photosynthetic bacteria are anoxygenic phototrophs, i.e. do not generate oxygen during their photosynthetic growth as Cyanobacteria do (see Chapter 12 by G. Schmetterer, Vol. 2), which are also capable to obtain energy from aerobic and anaerobic metabolism in darkness (Prince, 1990; Zannoni, 1995). In general, their photosynthetic apparatus (bacteriochlorophylls and carotenoids of light-harvesting systems, photochemical reaction center) is strongly repressed by molecular oxygen (Drews and Golecki, 1995); an exception to this rule, is a group of bacteria, the so called 'aerobic anoxygenic phototrophs' primarily isolated from marine environments (genera: *Erythrobacter* and *Roseobacter*), which are unable to synthesize their photosynthetic apparatus without the presence of molecular oxygen (Shimada, 1995; Yurkov and Beatty, 1998). The metabolic options available to typical facultative phototrophic bacteria, e.g. genera *Rhodobacter*, *Rhodoferax*, *Rhodospirillum*, *Rhodopseudomonas*, put them in a position to survive in quite different habitats and many species can use a variety of carbon sources such as organic acids or fatty acids but also CO_2 as sole carbon source and H_2 as sole electron donor. Facultative phototrophs can be found in natural habitats just below the oxic-anoxic interface of lakes and the capacity to grow in the dark by respiration is generally considered a mechanism for temporary survival in transiently oxygenated environments or as a way for maintaining the energy/redox balance (Madigan, 1988). This latter requirement is however the 'pre-requisite' for growth in the case of aerobic anoxygenic phototrophs (Yurkov and Beatty, 1998).

On a physiological/ecological point of view (Trüper and Pfennig, 1982), all anoxygenic phototrophs can be divided into four subgroups: (a) Purple sulfur (*Chromatiaceae*, *Ectothiorhodospiraceae*); (b) Purple nonsulfur (*Rhodospirillaceae*); (c) Green sulfur (*Chlorobiaceae*), and (d) Green gliding bacteria (*Chloroflexaceae*). These subgroups do not include strictly anaerobic spore-forming phototrophs of the genus *Heliobacterium* (Gest and Favinger, 1983) which are allocated apart (order *Clostridiales*) due to their physiological peculiarities, e.g. presence of bacteriochlorophyll *g* (Brochmann and Lipinski, 1983).

On a phylogenetic basis (16S rRNA analyses) purple bacteria and their relatives are grouped in a new class, the *Proteobacteria* (Stackebrandt et al., 1988) which is formed by several subclasses, namely: α subclass, containing most of the species of the *Rhodospirillaceae* family along with strictly aerobic species of the genera *Erythrobacter*; β subclass, including nonsulfur purple genera such as *Rhodocyclus* and *Rubrivivax*; γ subclass, which includes purple sulfur bacteria (*Chromatiaceae* and *Ectothiorhodospiraceae*). Green sulfur (*Chlorobiaceae*) and nonsulfur bacteria (*Chloroflexaceae*), resulted to be phylogenetically unrelated, form two distinct classes (Woese, 1987).

Before going into details about the metabolic

Abbreviations: CM – cytoplasmic membrane; COX – cytochrome *c* oxidase; Cyt – cytochrome; DMS – dimethylsulfide; DMSO – dimethylsulfoxide; ET – electron transport; ETCs – electron transport chains; HiPIP – High potential Iron sulfur Protein; ICM – intracytoplasmic membrane; LH – light harvesting complex; Q pool – ubiquinone pool; Qa – quinone primary acceptor; Qb – quinone secondary acceptor; QOX – quinol oxidase; RC – photochemical reaction center; TMPD – tetramethyl *p*-phenylenediamine

options characterizing anoxygenic facultative phototrophs (see below), the reader should be aware of a few basic notions on growth capacity of the above reported bacterial genera. Photosynthetic bacteria which are able to grow both in the light and in the dark under aerobic conditions belong to α and β subclasses while *Chromatiaceae*, *Chlorobiaceae*, and *Ectothiorhodospiraceae* are rather strict light-dependent anaerobes although some species of green sulfur bacteria may grow chemotrophically at very low oxygen tension.

As mentioned above, facultative phototrophs are endowed with several metabolic options; this peculiarity makes this group one of the most flexible of the microbial world. Species such as *Rhodobacter* (*Rb.*) *capsulatus* and *Rb. sphaeroides* may grow aerobically and photosynthetically using either organic or inorganic substrates but also anaerobically with trimethylamine-N-oxide (TMAO) or dimethyl sulfoxide (DMSO) as electron acceptors. The question whether anaerobic reduction of DMSO or TMAO is a true fermentative process requiring accessory oxidants or another example of anaerobic respiration is still a matter for debate (Madigan and Gest, 1978; McEwan et al., 1985; Jones et al., 1990; Zannoni, 1995) (see also Chapter 9 by McEwan et al., Vol. 1, for details on the molecular structure and function of DMSO and TMSO reductases). Furthermore, some strains of the species *Rhodopseudomonas* (*Rps.*) *palustris*, *Roseobacter* (*Rsb.*) *denitrificans* and *Rb. sphaeroides* can reduce nitrate (NO_3^-) into dinitrogen (N_2) via nitrite (NO_2^-), and in some cases also nitric oxide (NO) and nitrous oxide (N_2O) (Ferguson et al., 1987; Richardson et al., 1991) (see also Chapter 8 by Ferguson and Richardson, Vol. 2). These energy generating processes are catalyzed by electron transfer chains (ETCs) which redox components are inserted or associated with the cytoplasmic membrane (CM). Apparently not all the above summarized metabolic options are activated or can be available simultaneously in a single species; however, in some cases, e.g. aerobic conditions in the light, the electrochemical proton gradient ($\Delta\mu_{H^+}$) generated by a given electron transport chain may affect directly or indirectly other energy redox transducing apparatuses.

Photosynthetic bacteria convert the light-energy into chemical energy at the level of the photochemical reaction center (RC). This charge separation is followed by a cyclic electron transfer which involves quinone molecules, the cytochrome (Cyt) bc_1 complex and the soluble Cyt c_2 (Prince, 1990) or the membrane-bound Cyt c_y in the case of *Rhodobacter capsulatus* (Jenney and Daldal, 1993). Under dark aerobic conditions, these bacteria use a respiratory chain related to that present in mitochondria. The last step of O_2 reduction into H_2O is catalyzed by a Cyt *c* oxidase and also by the so-called alternative oxidase which is branched directly after the quinone pool (Zannoni et al., 1978). The photosynthetic and respiratory processes involve transmembrane complexes located on the cytoplasmic membrane (RC, Cyt bc_1, dehydrogenase, quinol oxidase, cbb_3 and/or aa_3 oxidases) and electron carriers (Cyt c_2, HiPIP, Cyt c_y) and enzymes in the periplasmic space (nitrate, nitrite, N_2O, DMSO reductases). In Fig. 1 a block scheme of the different electron transport pathways operating in *Rhodobacter sphaeroides*, is shown (see also Chapters 2,3,5,6,9,13 of Vol.1 for molecular details).

Although this metabolic flexibility allows the bacteria to prosper in various environments and to respond to changes of their surroundings, the assembly of the different bioenergetic pathways is highly regulated to prevent the unnecessary biosynthesis of specific enzymes and electron transfer components (as described in Section V). Short term regulations between the different bioenergetic chains are also used by the bacteria for an efficient and maximal utilization of the available energy. For example, photosynthesis strongly slows down respiration, denitrification or the reduction of DMSO or TMAO (Nakamura, 1937; Satoh, 1977; McEwan et al., 1982; Cotton et al., 1983; Sabaty et al., 1993). Under dark conditions the bacteria utilize the available electron acceptor which possesses the highest mid point potential to obtain the maximum of energy. Oxygen (O_2/H_2O, $E_{m7} = +820$ mV) is reduced in preference to nitrate (NO_3^-/NO_2^-, $E_{m7} = +420$ mV) or DMSO (DMSO/DMS, $E_{m7} = +160$ mV). The free energy recovered by the bacteria at the expense of NADH oxidation and reduction of oxygen, nitrate or DMSO is equal to 226, 146 and 92 kJ/mol, respectively. However, the most favorable bioenergetic process is photosynthesis where the light energy is utilized by the cells to form ATP and reducing power. The short term regulations between the different bioenergetic chains are illustrated in Fig. 2 in the case of *Rhodobacter sphaeroides* sp. *denitrificans,* a strain which possesses the three bioenergetic pathways. Continuous illumination partially inhibits both respiration and denitrification activities and this last bioenergetic process only occurs when all the oxygen has been consumed by the cells. Light

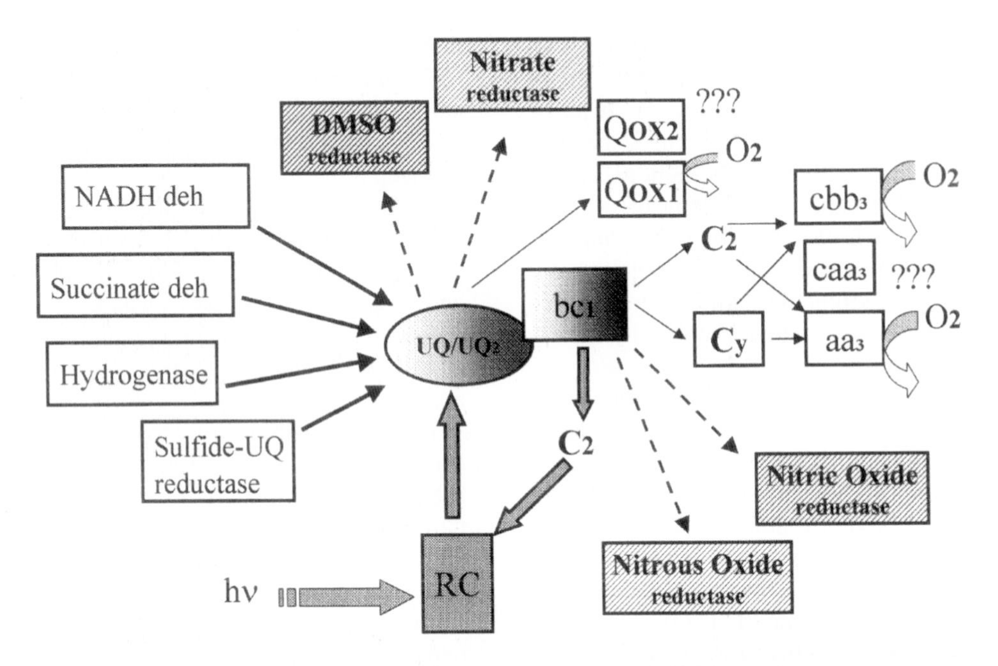

Fig. 1. Block-scheme illustrating the complex network of electron transport chains in *Rb. sphaeroides*. Symbols used: thick-black arrows indicate the influx of reducing equivalents while thin-black arrows symbolize the output of electrons leading to the membrane-bound oxidases. Hatched arrows indicate output of electrons involved in anaerobic respiration. Gray-colored redox components and arrows are those specifically involved in photocyclic electron transfer while those shaded off are shared by photosynthesis and respiration. Abbreviations: RC, photochemical reaction center; hv, radiant energy; C_2, soluble Cyt c_2; Cy, membrane-attached Cyt c_y; QOX1, functional ubiquinol oxidase (*qxtAB* operon); QOX2, not functional ubiquinol oxidase (*qoxBA* operon). See text and Chapters 3 and 13 by Cooley et al. and by Oh and Kaplan (Vol. 1), respectively, and Chapters 8 by Ferguson and Richardson and 9 by Vignais et al.(Vol. 2), respectively, for further details.

inhibits each step of the denitrification process, i. e. reduction of nitrite into N_2O or reduction of N_2O into N_2 (Sabaty et al., 1993).

The mode of interaction between the different bioenergetic chains of photosynthetic bacteria has been comprehensively studied. The interactions occur by two non exclusive mechanisms. The first mechanism is an indirect effect of the proton motive force produced by one bioenergetic process on trans-membrane complexes involved in the other chains like, for example, the NADH dehydrogenase, the Cyt bc_1 complexes or the oxidases (Cotton et al., 1983). This explanation is an extension of the respiratory control in mitochondria proposed by Mitchell (Nicholls and Ferguson, 2002) where the proton motive force induced by respiration exerts a back pressure on the proton translocating complexes. The second mechanism supposes a direct interaction between electron carriers common to different bioenergetic chains (Zannoni et al., 1978; Verméglio and Carrier, 1984). Since several electron carriers, the Cyt bc_1, the ubiquinone molecules and Cyt c, are engaged

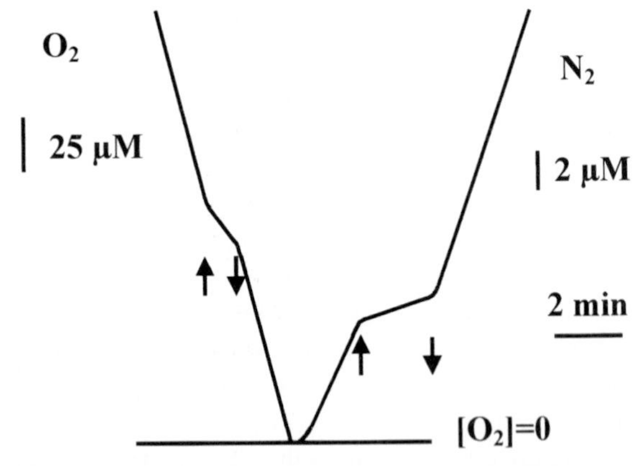

Fig. 2. Inhibition by light of aerobic respiration and denitrification. Cells of *Rb. sphaeroides* sp. forma *denitrificans* were placed in the presence of 2 mM nitrate. The concentrations of dissolved gases (O_2 and N_2) in the suspension were measured with a mass spectrometer. Illumination of the suspension (↑) inhibits the O_2 consumption. Production of N_2 from reduction of nitrate (or denitrification) only occurs when all the O_2 has been consumed ($[O_2] = 0$) by the cells. Denitrification is also inhibited by continuous illumination. The upwards arrows indicate the switching on of illumination while the downwards arrows correspond to its cessation. Figure adapted from Sabaty et al. (1993)

in different bioenergetic processes, the activity of a given bioenergetic chain affects the redox state of these components and consequently influences the functioning of the others chains.

The main objective of this Chapter is to outline all different levels of interaction (structural, functional, energetic and genetic) between photosynthetic and respiratory redox components, when simultaneously present, in the cytoplasmic membrane system (CM + ICM) of facultative photosynthetic bacteria. Rather than describing in details the data available for the numerous bacterial species studied, representative cases will be illustrated to provide a general picture. For further discussion along the same research lines we refer the reader to the books edited by C. Anthony (1988), A.J.B. Zehnder (1988) and by R.E. Blankenship et al. (1995).

II. Modes of Interaction Between Photosynthetic and Respiratory Activities

Most of the anoxygenic purple phototrophs contain a rather complex membrane system which is formed by an externally located membrane surrounding the cell, named cytoplasmic membrane or CM, and by internally located membranes with various arrangements, named intracytoplasmic membranes or ICM (Oelze and Drews, 1972; Collins and Remsen 1991; Drews and Golecki, 1995). A consistent amount of data tend to indicate that the major components of the ICM are the membrane-bound complexes forming the photosynthetic apparatus, e.g. photochemical reaction center (RC) or light-harvesting systems (LH), while respiratory enzymes, e.g. cytochrome oxidases, are the major components of the CM. This latter conclusion is derived from the observation that the synthesis of the photosynthetic apparatus is associated to the formation of large amounts of intracytoplasmic membranes. Nevertheless, the carotenoid band-shift, a specific indicator of the membrane potential due to a Stark effect on the light-harvesting carotenoid molecules, is identical when induced either by the photosynthetic or the respiratory activities (Chance and Smith, 1955; Wraight et al., 1978). This demonstrates that the membrane potential is delocalized over the entire internal membrane. Therefore, the electron components of the different bioenergetic chains experience the same membrane potential. The influence of the proton motive force generated during photosynthetic activity on the inhibition of the respiratory activity has been demonstrated by determining the effect of uncouplers. The addition of uncouplers prevents the inhibition of the respiratory activity by continuous illumination (Ramirez and Smith, 1968; Cotton et al., 1983; Richaud et al., 1986). The light-induced proton motive force inhibits the respiratory chain essentially at the complex I level. Indeed, this complex has been shown to reverse electron flow from the quinol pool to NAD^+ in the presence of a light-induced membrane potential (Klemme, 1969; Vignais et al., 1985; Knaff and Kämpf, 1987; Dupuis et al. 1997, 1998; Herter et al., 1998). On the other hand, the proton motive force appears to exert no control on the operation of the Cyt bc_1 or the terminal oxidases involved in the respiration or denitrification. This can be inferred from the observation that neither the reduction of nitrite, N_2O or O_2 can be inhibited by continuous illumination in the presence of an artificial electron donor like TMPD although a large membrane potential is induced under these particular conditions (Sabaty et al., 1993).

Besides this indirect interaction, direct connection between the different bioenergetic chains has been demonstrated by several approaches. The involvement of the Q-pool, acting as a branching point between respiratory and photosynthetic chains, has been clearly established by the stimulation of the ubiquinol oxidase activity in the presence of light and exogenous electron donors, e.g. horse-heart Cyt *c* or TMPD (Zannoni et al., 1978), in both chromatophores and hybrid-membranes generated from ET-mutants (Zannoni et al., 1986). This interaction between the respiratory and the photosynthetic chains at the level of the Q-pool has important implications for the operation of the photosynthetic apparatus. Under anaerobic conditions and in the presence of reduced carbon substrates, such as butyrate or succinate, the photochemical activity is partially inactivated due to the reduction of a fraction of the primary electron acceptor (Qa) of the RC (McEwan et al., 1985). This is caused by the entry of electrons via the dehydrogenases which are not removed by the respiratory activities. Under these reducing conditions, continuous illumination could reactivate all the RCs (Verméglio and Joliot, unpublished). One possible interpretation is that the photochemistry of the fraction of active RCs generates a membrane potential which induces a reverse electron flow from the quinol pool to NAD^+ (Klemme, 1969; Vignais et al., 1985; Knaff and Kämpf, 1987; Dupuis et al. 1997, 1998; Herter et al., 1998). This would partially reoxidize the Q-pool and the RCs' electron acceptors and fully reactivate their photochemical activity. Reoxidation of the quinol pool is also obtained by oxygenation of the bacterial suspension or by addition of alternative electron acceptors such as nitrate or DMSO (McEwan et al., 1985; Takamiya et al., 1988). This type of interaction is particularly crucial in the case of aerobic photosynthetic bacteria, e.g. *Roseobacter* genus. These bacteria can grow photosynthetically only under aerobic condition or in the presence of alternative electron acceptors (Takamiya et al., 1988; Shimada, 1995; Yurkov and Beatty,1998). Under anaerobic conditions, Qa is totally reduced and no photochemistry occurs. This particular behavior could be due to the mid-point potential ($E_{m,7.0}$) of the primary quinone acceptor which has been reported to be higher than that of anaerobic *Rhodobacter* species (Takamiya et al., 1987; Kramer et al., 1997; Yurkov et al., 1998), although this latter hypothesis has been recently challenged by Schwarze et al. (2000). Further, studies on the respiratory apparatus of *Rsb. denitrificans* have shown that the most striking difference between this photosynthetic-aerobe and typical purple non-sulfur bacteria is the lack of a quinol oxidase pathway (Candela et al., 2001). In this respect, it is important to remark that under respiratory steady-state conditions, continuous illumination does not perturb the redox state of the Q-pool when the quinol oxidase pathway is functioning (Zannoni and Moore, 1990). Indeed, as shown in Fig. 3, the major variation of the Q-pool redox state, as detected by a Q-electrode in membranes from *Rb. capsulatus* MR126, is seen when the COX is blocked by cyanide or after the onset of anaerobiosis (Zannoni and Moore, 1990). The lack of a Q-oxidase in *Rsb. denitrificans* has therefore a great impact on the rate of photophosphorylation because, as mentioned above, photochemistry does not occur when the Q-pool, in equilibrium with the secondary electron acceptor Qb, is fully reduced (Crofts, 1983). In fact, membrane vesicles (chromatophores) isolated from a typical representative of purple non-sulfur bacteria, *Rb. capsulatus*, and from *Rsb. denitrificans*, perform light-induced ATP-synthesis only under oxic conditions (Candela et al., 2001). On the other hand, it has also been demonstrated that the rate of photophosphorylation strongly depend on the ambient redox potential (Loach, 1966; Baccarini-Melandri et al., 1979). It is therefore evident that light-induced phosphorylation is not affected by oxygen *per se*, but requires a suitable redox poise ($\geq$+80 mV $\leq$+140 mV) which is close to the $E_{m7,0}$ of the Q-pool (+90 mV). In

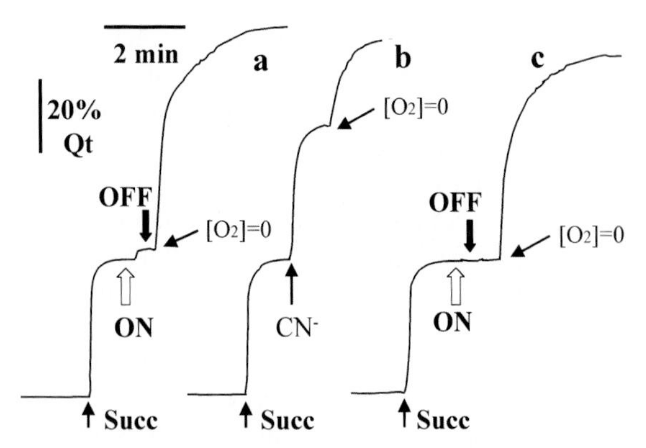

Fig. 3. Redox level of the Q-pool in membranes from wild type (strain MR126) (traces a and b) and mutant (strain MT-GS18) (c) cells of *Rb. capsulatus*. Strain MT-GS18 is a double mutant (c_2^-, c_1^-) unable to grow photosynthetically. Respiration was initiated by the addition of 5 mM succinate (Succ). Cyanide (CN-) addition was as indicated (trace b) at a concentration of 50 μM which is required to inhibit the Cyt *c* oxidase of *Rb. capsulatus* (*cbb*$_3$ type). In a and c, illumination was switched ON and OFF, as indicated. $[O_2] = 0$ symbolizes the onset of anaerobiosis. The vertical bar indicates the 20% of the total ubiquinone (Qt) which can be reduced by succinate under anaerobic conditions. This figure is redrawn and modified from Zannoni and Moore (1990).

the case of *Rsb. denitrificans*, a very active respiratory pathway leading to a COX of *aa*$_3$ type (Candela et al., 2001) would maintain the optimal redox poise to keep the Q-pool partially oxidized also under continuous illumination. Whatever the molecular explanation could be, it is apparent that respiration is required to allow photochemical activity. Another demonstration of the direct interaction between photosynthetic and respiratory chains in facultative phototrophs has been provided by analysis of the respiratory activity after flash excitation with a very sensitive O_2 electrode (Fig. 4) (Verméglio and Carrier, 1984; Richaud et al., 1986). Following a series of short saturating flashes, the respiratory activity is inhibited after each flash but stimulation is observed after an even number of flashes. Addition of uncouplers does not affect this oscillatory pattern. The lack of oscillation pattern for a mutant of *Rb. capsulatus* deficient in COX implies that this phenomenon is due to the modulation of the activity of this enzyme and not to the alternative oxidase (Richaud et al., 1986). After each flash, an electron is diverted from the respiratory chain to the photoxidized RC at the level of Cyt c_2. Because of the gating mechanisms at the level of the secondary electron acceptor, Qb (Verméglio, 1977; Wraight, 1977), respiration is restored only after an even number of flashes by formation of a QH_2 molecule.

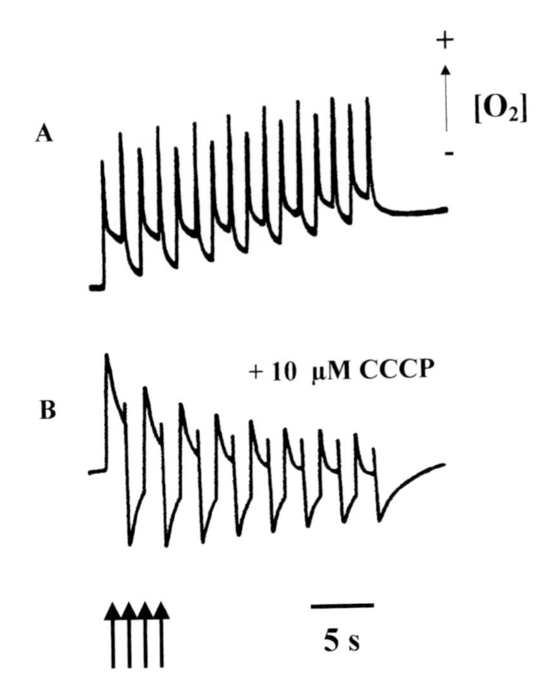

Fig. 4. Modulation of the respiratory activity by saturating flashes fired every second for a suspension of *Rb. sphaeroides*. Each flash induces an inhibition of the respiratory activity (this corresponds to an increase in O_2 concentration). The respiratory activity is restored after even flashes. Part A: No addition. Part B: in the presence of 10 μM of the uncoupler CCCP. Figure adapted from Richaud et al. (1986)

III. Modes of Interaction Between Aerobic and Anaerobic Respiratory Activities

As already stated, photosynthetic activity inhibits each single step of the reduction of nitrate into dinitrogen (Sabaty et al., 1993). These inhibitions are not prevented in the presence of uncouplers and therefore do not require the formation of a transmembrane potential. It has therefore been postulated that this is due to diversion of electrons from the denitrification chain to the photosynthetic chain at the level of Cyt bc_1 and Cyt c_2, two components which are in common to both pathways (Richardson et al., 1991; Sabaty et al., 1993, 1994)

The presence of oxygen, completely inhibits the electron transfer to alternative acceptors like nitrate or TMAO (McEwan et al., 1985; King et al., 1987; Sabaty et al., 1993). Contrary to the light-inhibition of aerobic respiration, inhibition of denitrification or TMAO respiration by oxygen is not suppressed by the addition of uncouplers (McEwan et al., 1985; King et al., 1987). Therefore this inhibition is not a consequence of the thermodynamic control of the transmembrane potential resulting from the respiratory activity on different elements of the denitrification or TMAO reduction chains. This is rather due to a direct change in the redox state of the electron transfer components. Possible candidates are the Cyt c_2, Cyt bc_1 complex in the case of the inhibition of both nitrite and N_2O reduction by oxygen. For nitrate and TMAO reduction, the inhibition by oxygen is certainly mediated via the redox state of the Q-pool. One possibility is that the reduction of nitrate or TMAO is attainable only when the Q-pool is largely reduced. The partial oxidation of the Q-pool by the respiratory activity may prevent the nitrate or TMAO reduction. Modulation of the activity of oxido-reducing enzymes by the redox state of the Q-pool has already been demonstrated in the case of the alternative oxidase or QOX (Zannoni and Moore, 1990). This oxidase is not functional until the quinone pool is at least 25% reduced. On the other hand, the rate of the COX is proportional to the reduction level of the quinone pool (Zannoni and Moore, 1990). This implies a preferential interaction, i.e. a lower k_m, between quinol molecules and Cyt bc_1 than between quinols and the alternative oxidase. Such behavior is illustrated in Fig. 5 where cells of a mutant of *Rhodobacter capsulatus* deficient in the cytochrome oxidase have been placed in the presence of the exogeneous electron donor TMPD. Upon illumination the quinone pool is reduced at the expense of TMPD photooxidation. This stimulates the oxygen uptake by the alternative oxidase. Upon cessation of illumination the oxygen uptake is strongly inhibited and is only restored when all the photooxidized TMPD is re-reduced via the Cyt bc_1 complex.

IV. Heterogeneity and Lack of Thermodynamic Equilibrium between Redox Components

The results described above demonstrate a direct interaction between different components of the photosynthetic and respiratory chains. Since these chains are localized in different parts of the internal membrane, this implies diffusion of some of electron carriers like the quinone in the lipid phase and the Cyt c_2 in the periplasmic space. Several observations indicate however that the diffusion of these elements does not occur on the entire internal membrane and that the different bioenergetic chains are not necessarily in thermodynamic equilibrium. Two distinct functional pools of Cyt c_2 have been demonstrated in

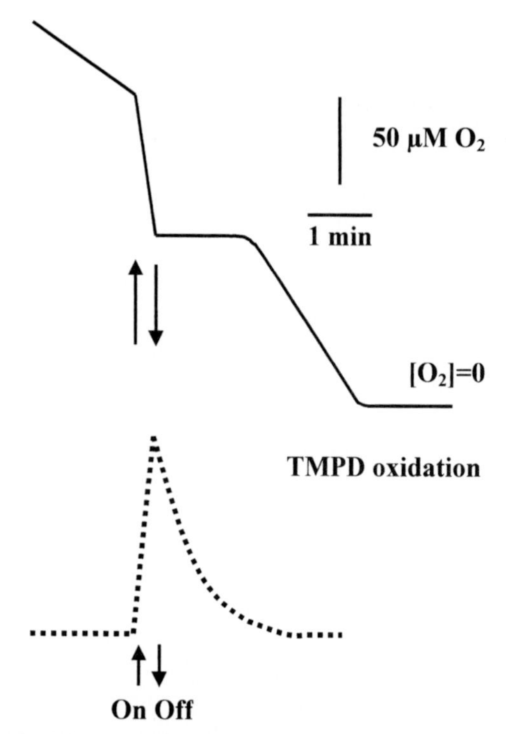

Fig. 5. Light stimulation of oxygen uptake by the quinol oxidase in the presence of 1 mM TMPD for a bacteria suspension of *Rb. capsulatus* deleted in the Cyt oxidase. The upwards arrows indicate the switching on of illumination while the downwards arrows correspond to its cessation. After illumination the respiratory activity is strongly inhibited. Reactivation of respiration only occurs when all the photooxidized TMPD has been rereduced via the Cyt bc_1 complex. Figure adapted from Richaud et al. (1986).

Rhodobacter sphaeroides and *Rhodobacter capsulatus*. This has clearly been shown for cells grown in the presence of nitrate or under dark aerobic condition, which presents two kinetically distinct phases (a fast and a slow phase) for the photooxidation of Cyt c_2 by continuous illumination at high pH or in the presence of glycerol (Matsuura et al., 1988; Verméglio et al., 1993). Cells grown under anaerobic photosynthetic condition exhibit only the fast photooxidation phase. This clearly establishes the occurrence of two distinct functional pools of Cyt c_2. The first pool of Cyt c_2, which is rapidly photooxidized during the fast phase or by the two first flashes of a series, corresponds to 1 Cyt c_2 connected to 2 RCs and 1 Cyt c_1. The second pool of Cyt c_2 is only slowly photooxidized following continuous illumination and requires a large number of excitations when cells are subjected to a series of flashes. This is linked to a small amount of RCs connected to this pool of Cyt c_2 (the ratio could be as high as 7 Cyt c_2 per RC in the case of dark semiaerobic grown cells). Growth of bacteria under different light intensities and dark aerobic conditions shows that the ratio between the fast and the slow phases correlates with the amount of ICM (Verméglio et al., 1993). Moreover, the second pool of Cyt c_2 is specifically oxidized by the respiratory or denitrification activities (Sabaty et al., 1993; Verméglio et al., 1993).

In summary, two pools of Cyt c_2 are not in thermodynamic equilibrium. The first pool is involved in an efficient cyclic photoinduced electron transfer. This Cyt c_2 is intimately connected to 2 RCs and 1 Cyt bc_1 complex in the intracytoplasmic part of the membrane and it is hardly oxidized by respiratory activities. The second pool interacts with a small number of RCs present in the cytoplasmic part of the membrane and with the elements of the respiratory chains localized in this membrane region. This allows a direct interaction between the two bioenergetic chains.

Lack of thermodynamic equilibrium has also been observed between the different photosynthetic chains in the case of *Rhodobacter sphaeroides*. In this species, the apparent equilibrium constant between Cyt c_2, Cyt bc_1 and the RC, measured under low intensity illumination, is much less than expected from the mid-point potentials of these electron carriers (Joliot et al., 1989). One possible explanation of this behavior is to suppose that a rapid thermodynamic equilibrium is achieved at a local level within a domain, or supercomplex, containing a small number of electron carriers whereas equilibration at a macro level between these domains is a much slower process (Lavergne et al., 1989). In other words, fast electron transfer only occurs in the supercomplex, i. e. the diffusion of Cyt c_2 is confined to a small domain included only 2 RCs and 1 Cyt bc_1. Simulation of the data suggests that each supercomplex contains 2 RCs, 1 Cyt c_2 and 1 Cyt bc_1 complex (Joliot et al., 1989; Verméglio and Joliot, 1999). The stoichiometry determined for these small domains is fully consistent with the one measured for the fast photooxidation phase of Cyt *c*.

A second explanation of the low equilibrium constant has been proposed by Crofts and coworkers (Crofts et al., 1998). In their model they assume free diffusion of the Cyt c_2 in the aqueous phase but heterogeneity in the stoichiometry of this electron transfer component and the membrane complexes. A further argument was taken from the observation that in mutant strains in which the stoichiometric ratios of the components of the supercomplexes have been changed, the components of the photosynthetic chains present the behavior expected for the diffusional model (Crofts et al., 1998; Witthuln et al., 1996). Moreover, the addition of a subsaturating concentration of myxothiazol, a specific inhibitor of the Cyt bc_1 complex, decreases the rate of electron transfer of re-reduction of Cyt c_2 but not its amount in isolated chromatophores (Fernàndez-Velasco and Crofts, 1991). However this behavior is not observed in intact cells where inhibition of part of the Cyt bc_1 complex does not affect the rate of electron transfer for the uninhibited complexes (Fernàndez-Velasco and Crofts, 1991; Joliot et al., 1996). This implies a restricted diffusion of Cyt c_2 at least in intact cells but its molecular explanation is still a matter for debate (Crofts, 2000a,b; Verméglio and Joliot, 2000).

A strong argument in favor of a supramolecular organization of the photosynthetic chains in *Rhodobacter capsulatus* came from the discovery of a membrane-bound cytochrome, named Cyt c_y (Jenney et al., 1986; Jenney and Daldal, 1993). In this species, in addition to Cyt c_2, the Cyt c_y is competent for the photoinduced electron transfer (Myllykallio et al., 1999). The membrane attachment of this cytochrome certainly limits its rapid diffusion and therefore impedes possible interactions with a large number of different redox partners. Consequently one expects that thermodynamic equilibrium between the RCs interacting with either a Cyt c_y or with a Cyt c_2 will be slow or even inoperative. This lack of equilibrium has been experimentally demonstrated using a mutant strain lacking Cyt c_2. In this mutant, only ~30% of the RCs had their photooxidized primary donor rapidly re-reduced by Cyt c_y. Although this cytochrome is re-reduced by the Cyt bc_1 complex in the millisecond time range, it remained unable to convey electrons to the large amount of unconnected RCs which remained photooxidized for several seconds although the two types of RCs are localized in the same membrane region (Verméglio et al., 1998; Myllykallio et al., 1999). These experiments not only demonstrate that the thermodynamic equilibrium is not reached between the different photosynthetic electron transport chains but also that RCs, Cyt c_y and the Cyt bc_1 complex are in very close association.

A c_y-type cytochrome has also been evidenced in *Rhodobacter sphaeroides* (Myllykallio et al., 1999). Contrary to the Cyt c_y of *Rhodobacter capsulatus*, the *Rhodobacter sphaeroides* Cyt c_y is not functional in the photoinduced electron transfer (Myllykallio et al., 1999). This has been clearly demonstrated by the non-photosynthetic growth of a mutant deleted in Cyt c_2 (Donohue et al., 1988). However, genetic introduction of Cyt c_y from *Rhodobacter capsulatus* in a Cyt c_2 *minus* mutant of *Rhodobacter sphaeroides* readily restores the photosynthetic activity (Jenney and Daldal, 1993; Jenney et al., 1996). The different behavior between the Cyt c_y of *Rhodobacter capsulatus* and of *Rhodobacter sphaeroides* is probably linked to variations in the precise location of charged residues, important for the interactions of the heme subdomain with its redox partners, but also it might depend on the different length of their membrane-anchors (Myllykallio et al., 1999). Cytochrome c_y is also involved in the respiratory activity of both *Rhodobacter sphaeroides* and *Rhodobacter capsulatus*. Indeed, this cytochrome mediates, in addition to the soluble Cyt c_2, electron transfer between the Cyt bc_1 complex and the cbb_3-type Cyt oxidase in *Rhodobacter capsulatus* (Jenney and Daldal, 1993; Gray et al., 1994) and between this former complex and the two types of cytochrome oxidases (cbb_3 and aa_3) present in *Rhodobacter sphaeroides* (Daldal et al., 2001). Consequently, the respiratory chains embracing a Cyt c_y could be organized in supramolecular complexes and operate independently of the redox state of electron transfer components of the photosynthetic chains or other respiratory pathways. Supramolecular organization of part of the respiratory chains have been evidenced in different bacterial species (mainly *Archaea*, see Chapter 1 by G. Schäfer, Vol. II) and more recently in mitochondria (Schägger, 2001; Schägger and Pfeiffer, 2001).

V. Genetic Regulation

The physiological relationships between photosynthesis and respiration, as described in the preceding paragraphs, are superimposed on the regulation at the genetic level. Oxygen is the key element in the coordinated regulation of photosynthetic and respiratory activities since it directly or indirectly participates in determining the levels of expression of all components involved. There are five regulatory elements, identified up to now in *Rb. capsulatus*, that are involved in mediating the response to varying oxygen levels, namely: RegA, HvrA, CtrJ, AerR, and FnrL. This paragraph will be focused on the regulatory aspects which are common to photosynthesis and respiration, and will touch only marginally the ones that are specific to photosynthesis (for details on this latter aspect see Pemberton et al., 1998; Gregor and Klug, 1999).

A. Regulatory Elements: RegA, HvrA, CtrJ, AerR, and FnrL

RegA (PrrA in *Rb. sphaeroides*) is the first element identified that acts as an activator of reaction center's (RC's) gene expression (*puf* and *puh*) in response to low oxygen (Sganga and Bauer, 1992). This protein is the effector component of a regulatory couple in which RegB (PrrB in *Rb. sphaeroides*) is the sensor partner (Mosley et al., 1994). The sensor-effector couple is functioning according to the classical model in which the sensor (RegB) detects changes in the cellular environment, i.e. variations in oxygen tension, and phosphorylates the effector (RegA); the latter, in its phosphorylated form, regulates the expression of a number of genes. The genes initially identified as responding to RegA under photosynthetic-anaerobic conditions were all components of the photosynthetic machinery: *puf* and *puh* operons, coding for the RCs and LHI subunits, and *puc* operon representing LHII subunits (Sganga and Bauer, 1992). Recently it has been shown that several other genes coding either for elements shared by photosynthetic and respiratory electrons transport chains (ETC) or coding for components unique to the respiratory apparatus are regulated by RegA (Swem et al., 2001; Swem and Bauer, 2002). To the first group of these genes belongs the *petABC* operon, (bc_1 complex) which expression is increased by RegA under anaerobic and semiaerobic conditions, and shows only a slight increase under high oxygen tension. The other two elements shared by the redox pathways in *Rb. capsulatus* are Cyt c_2 (*cytA*) and Cyt c_y (*cytY*). Cyt c_2 has a regulation profile similar to bc_1 with RegA that has a greater impact with low or no oxygen and a little effect with high O_2. Cyt c_y, on the contrary, is poorly affected by RegA under anaerobic conditions and significantly induced only under semiaerobic conditions. To the second group of RegA regulated elements belong the two terminal oxidases of the respiratory ETC. The cbb_3, Cyt *c* oxidase (*ccoNOQP*), is increased under aerobic and semiaerobic conditions but is limited in the absence of oxygen. The quinol oxidase (*cydAB*), on the contrary, is induced by RegA under all growth conditions (Swem et al., 2001). In addition to all the above described effects, RegA exerts its action, in response to changes in oxygen tension, on other basic metabolic activities such as: DMSO reduction (Kappler et al., 2002), CO_2 assimilation (Vichivanives et al., 2000), H_2 oxidation and N_2 fixation (Elsen et al., 2000).

The transducing events that connect the sensing of changes in oxygen levels to the phosphorylation/dephosphorylation of the regulatory couple RegB/RegA have not been yet identified in *Rb. capsulatus*. There are, however, a few considerations that can be made indicating that the cbb_3 COX could be the element, in the transducing chain, capable of sensing the level of O_2 via the electron flow going through it. First, it is likely to presume that a redox complex which is both expressed under all conditions and plays a fundamental part of the ETC might function as oxygen sensor. Second, Cyt *c* oxidase deficient mutants show an increased expression of RC's *puf* and *puh* genes under anaerobic and semiaerobic conditions, as if the lack of a functional cbb_3 oxidase could be sensed as a signal of anaerobiosis even under aerobic conditions (Buggy and Bauer, 1995). Third, in *Rb. sphaeroides*, a close relative of *Rb. capsulatus*, the role of cbb_3 in sensing the oxygen availability and transducing the signal to the genes regulated by the PrrB/PrrA couple (RegB/RegA in *Rb. capsulatus*) has clearly been demonstrated (O'Gara et al., 1998). In *Rb. sphaeroides* it has been shown that the regulatory relay from O_2 sensing to PS gene expression includes PrrC, a protein anchored in the membrane, that transmits the signal originated from cbb_3 to PrrB (Eraso and Kaplan, 2000; see also Chapter 13 by Oh and Kaplan, Vol. I). SenC, which is the equivalent to PrrC in *Rb. capsulatus*, appears to have a different and less clear role in this regulatory cascade (Buggy and Bauer, 1995).

RegA is the element that directly interacts with photosynthetic and respiratory genes that are controlled by the above described regulatory chain. According to this scheme, RegA needs to be phosphorylated in the presence of low to no oxygen in order to function as an activator of gene expression. As illustrated in Fig.6, RegA can also behave as a repressor of cbb_3 oxidase (*ccoNOQP*) transcription under anaerobic conditions (Swem et al., 2001). The ability of phosphorylated RegA to exert an inhibitory role on transcription has been described for the *senC-regA-hvrA* operon (Du et al., 1999), showing that RegA auto-regulates its own expression. In addition to its activator and/or inhibitory role under anaerobic or semiaerobic conditions, that is in the phosphorylated form, RegA participates in the aerobic regulation of the terminal oxidases (*cooNOQP* and *cydAB*) (Swem and Bauer, 2002). Under these growth conditions RegA is supposed to be present in the non phosphorylated form, which has been shown to have a 16 times lower affinity for its

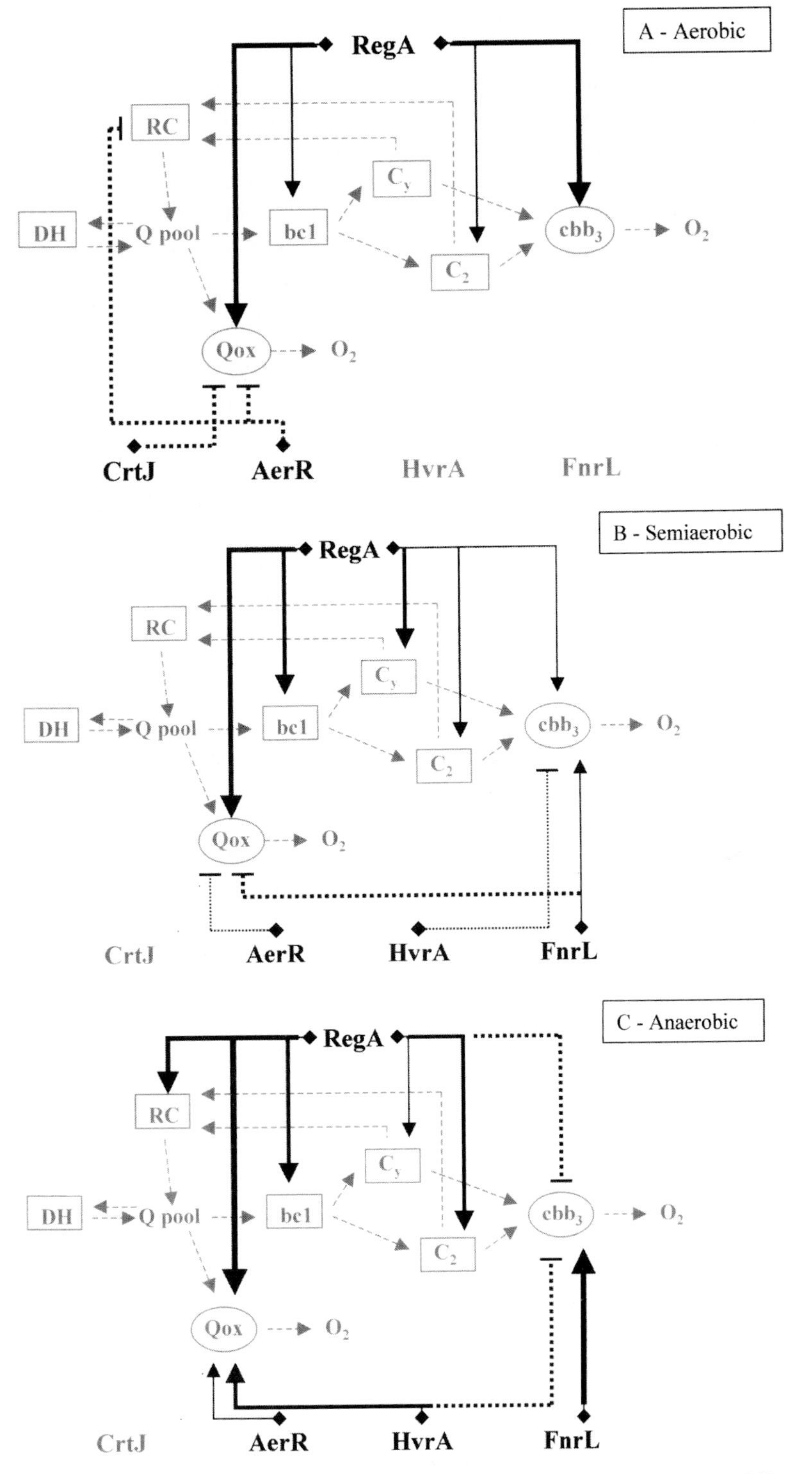

Fig. 6. Regulatory connections between respiratory and photosynthetic electron transport chains under different growth conditions. A, aerobic; B, semiaerobic; C, anaerobic. Redox components of the ETCs are gray. Squares indicate elements shared by both ETCs or specific to photosynthesis; circles indicate elements specific to respiration. Dashed gray arrows indicate electron transfer pathways between redox components. Continuous black arrows indicate genetic induction; dotted black lines indicate genetic repression. The thickness of the arrows indicate the relative strength of regulation. Standard abbreviations were used except for DH which stands for: NADH dehydrogenase. See text for further details.

target promoters than the phosphorylated counterpart (Bird et al., 1999).

HvrA is known to be part of a cluster of genes, including RegA, RegB and SenC, involved in controlling the expression of the photosynthetic apparatus of *Rb. capsulatus* (Buggy et al., 1994). HvrA has been shown to increase the expression of RC genes, under anaerobic conditions, when light intensity is lowered (Buggy et al., 1994). In addition, HvrA is involved in regulating N_2 fixation (Kern et al., 1998). A recent report has demonstrated that HvrA participates in the anaerobic regulation of both terminal oxidases: stimulates the expression of QOX in the absence of oxygen and limits the expression of COX under anaerobic and semiaerobic conditions (Swem and Bauer, 2002).

It has been hypothesized that HvrA, or an upstream regulatory element, may bind a chromophore and transduce its light-absorbing properties into regulation of *puf* and *puh* operons (Buggy et al., 1994). However, in the light of recent data about the effect of HvrA on the expression of the terminal oxidases of the ETC (Swem and Bauer, 2002), we are tempted to suggest that this regulatory element participates in the coordinated regulation of photosynthetic and respiratory functions by sensing the reduction level of the Q pool. In this respect, the Arc system in *E. coli* has been shown to regulate a number of gene involved in respiratory or fermentative metabolism by directly sensing the oxidation-reduction equilibrium of the Q pool (Georgellis et al., 2001). In *Rb. capsulatus* the redox state of the Q pool, which is likely to depend on the light intensity as in the close relative *Rb. sphaeroides* (Oh and Kaplan, 2000), and on the respiratory electron transport rate (Zannoni and Moore, 1990), could be sensed by a relay-regulatory system which includes HvrA, that would use this signal to optimize the expression of RC and terminal oxidases.

Two regulatory elements, **CrtJ** and **AerR**, act as aerobic repressors of genes involved in photosynthetic processes and genes specific for aerobic respiration. These repressors differ in the way they are controlled by oxygen. CrtJ responds directly to O_2 by the formation of an intramolecular disulfide bond that alters its binding to target promoters (Masuda et al., 2002), but its transcription is not regulated by aerobiosis or anaerobiosis (Dong et al., 2002). On the other hand, AerR is more expressed under anaerobic photosynthetic conditions and this transcriptional regulation does not depend on RegA, CrtJ or AerR itself, suggesting that there might exist an additional unidentified oxygen-responding regulator (Dong et al., 2002). The two repressors act on different sets of genes, with a subclass controlled by both regulatory elements as is the case of *puc* operon (LHII) on which they show cooperativity (Dong et al., 2002).

FNR is a global regulator that responds to shifts from aerobic to anaerobic conditions inducing or repressing anaerobically different sets of genes involved in the tricarboxylic acid cycle, in anaerobic respiratory pathways, and in aerobic respiration of *E. coli* (Bauer et al., 1999). The *E. coli* FNR homologue in *Rb. capsulatus,* FnrL, behaves similarly in that it acts under anaerobic conditions as an inducer of the cbb_3 oxidase and as a repressor of the quinol oxidase (Swem and Bauer, 2002). FnrL is required for anaerobic-dark growth in the presence of DMSO but it does not seem to be involved in regulation of photosynthetic activity in *Rb. capsulatus* (Zeilstra-Ryalls et al., 1997).

B. The Coordinated Regulation Framework: A Proposal

A synthetic description of the regulatory system used by *Rb. capsulatus* to coordinate the expression of photosynthetic and respiratory ETCs while adapting to growth mode changes, is here proposed according to the present knowledge, albeit largely incomplete, of the genetic and functional networks that connect photosynthesis and respiration. The regulatory scheme we propose is by no means exhaustive and is just intended as a preliminary proposal.

The level of oxygen and the redox state of the quinone pool represent the environmental and the physiological indicators that are sensed by the cell in order to regulate the relative expression of the two ETCs components. As reported in the preceding paragraph, the data gathered on the genetic regulation of the ETCs' elements, either shared by the two ATP generating systems or specific to the respiratory chain, concern mainly the terminal oxidases COX and QOX and only marginally the bc_1 complex, Cyt c_2 and Cyt c_y (Swem et al., 2001; Swem and Bauer, 2002). In our proposal, the environmental level of oxygen is sensed by the cbb_3 Cyt *c* oxidase, possibly via the electron flow going though it, and the signal is relayed to the RegA/RegB couple by SenC. RegA responds to this signal by regulating the expression of the ETCs elements shared by photosynthesis and respiration and the terminal oxidases. In this scheme COX would establish a regulatory loop with RegA, in which the first component affects the activity of the second as a function of the oxygen level, and the second component participates in modulating the expression of the first in a way that depends on the growth conditions. Ultimately, is the level of O_2 that sets the equilibrium position of the loop. The other regulatory elements, CrtJ, AerR, HvrA and FnrL, respond to aerobiosis/anaerobiosis transitions in different ways.

The quinone pool plays a fundamental role in that it connects all electron transport chains (Fig. 1) and its correct functioning, which depends on maintaining an optimal redox poise, is of paramount importance under all growth modes. In the regulatory scheme we propose, in order to maintain a correct interaction between photosynthesis and respiration, HvrA would have to respond to the redox status of the Q-pool, under low to no oxygen, and translate it into the regulation of the expression of the two terminal oxidases. The redox level of the Q-pool is directly modulated by the oxidative action of QOX, which, in turn, is controlled by the former through HvrA. The interplay between HvrA and QOX, mediated by the redox level of the quinone pool, defines another regulatory loop in which each component regulates and is regulated by the other. The 'oxygen' loop and the 'Q-pool' loop are considered to be integrated by the RegA/RegB couple which responds to both signals (Buggy and Bauer, 1995; Du et al., 1999) and, at the same time, exerts its regulatory action on all ETCs components involved (Fig. 6).

Under aerobic growth conditions the presence of abundant oxygen is sensed by the cbb_3 cyt oxidase, and this signal is relayed to the RegA/RegB couple. Under these conditions RegA, in its dephosphorylated form, regulates positively the expression of several components of the ETC, having a strong effect on the expression of the terminal oxidases (COX and QOX), and a milder effect on both Cyt c_2 and bc_1 complex (Swem et al., 2001) (see Fig. 6A). The aerobic repressors CrtJ and AerR counteract the action of RegA by limiting the expression of QOX (Swem and Bauer, 2002). AerR works also as a repressor of RC's genes expression (Dong et al., 2002). Based on the relative contribution of RegA on the expression of the operons involved, it is likely to conclude that the level of the cbb_3 Cyt oxidase (cooNOQP) mainly depends on the presence of RegA (Swem et al., 2001; Swem and Bauer, 2002). The expression of QOX would also depend mostly on the action of RegA (Swem et al., 2001) but, in addition, on that of CrtJ and AerR through the use of different mechanisms (Dong et al., 2002; Masuda et al., 2002; Swem and Bauer, 2002). The interplay of these regulatory effects generates a fine tuning of the QOX expression as a function of the cell redox state.

As the oxygen concentration decreases, the cell reacts adjusting the level of the ETC's components to the new growth conditions. Under semiaerobiosis, most of the redox elements, except the bc_1 complex, reach their maximum level of expression through the positive action of RegA (Fig. 6B) (Swem et al., 2001). With low to no oxygen, the cbb_3 oxidase shows a complex regulation pattern that includes HvrA and FnrL, in addition to RegA (Fig. 6B,C) (Swem and Bauer, 2002). Since the effect of RegA on the expression of cbb_3 is less pronounced under anaerobic than aerobic conditions, it might be suggested that, as oxygen decreases, the proportion of phosphorylated RegA increases and, in this form, the regulatory element would have less affinity for the cbb_3 promoter, thereby lowering its positive action. The high level of cbb_3 expression, under semiaerobic conditions, also depends on FnrL, which has a regulatory role comparable to RegA (Swem and Bauer, 2002). Although these regulators show a mild effect when lacking singularly, a dramatic decrease of cbb_3 expression is seen when both are deleted (Swem and Bauer, 2002). The third element participating in the regulation of the Cyt *c* oxidase is HvrA which has a negative function (Swem and Bauer, 2002). As suggested above, HvrA might respond, anaerobically or semiaerobically, to changes in the redox equilibrium of the Q pool. If this is true, it could be envisaged a regulatory scheme in which RegA and FnrL induce the expression of cbb_3 oxidase depending on the relative oxygen tension while HvrA may contrast their action by sensing the redox level of the Q pool.

VI. Conclusions

The picture we can propose presently for the interaction between the photosynthetic and respiratory chains in *Rhodobacter sphaeroides* and *Rhodobacter capsulatus* could be summarized as follows. The organization of the bioenergetic chains satisfies two apparently contradictory requirements, namely: 1) the photosynthetic apparatus has to interact with the respiratory chains (aerobic and anaerobic) for a better use of the available energy; 2) this latter interaction must preserve the RC from an excess of reducing equivalents. Indeed if full thermodynamic equilibrium is established between the respiratory and photosynthetic chains, the primary electron acceptor should be fully reduced due to its much higher mid point potential compared to the $NADH/NAD^+$ couple. Cells face these contradictory requirements by arranging the bioenergetic chains at two different levels in *Rhodobacter sphaeroides* and *Rb. capsulatus*. A first level of organization is the spatial segregation of the photosynthetic and respiratory chains in two separate regions of the internal membrane. The cytoplasmic part contains the greater part of the components of the respiratory chains which can go into redox equilibration with a small part of the RCs. The major part of the photosynthetic chains (including RCs, Cyt c_2 or Cyt c_y, and Cyt bc_1 complex) are localized in the intracytoplasmic part of the membrane. The second level of organization is the arrangement of these photosynthetic elements in supramolecular complexes. This tight connection allows a fast and efficient cyclic electron transfer by limiting the diffusion of the reactants.

These two levels of arrangement are not present in all types of photosynthetic bacteria. Species like *Blastochloris viridis* or *Rubrivivax gelatinosus,* possess a large excess of RCs over Cyt bc_1 complexes so that the formation of supercomplexes is stoichiometrically hindered. For species like *Rhodopseudomonas viridis* or *Thiocapsa pfennigii*, the quinone pool can however be maintained partially oxidized under anaerobic condition due to the lack of thermodynamic equilibration at the Q-pool level between the photosynthetic and respiratory chains due to the spatial isolation of these chains. Indeed, the high ordering of the photosynthetic apparatus in the intracytoplasmic membrane of these species certainly renders difficult the long range diffusion of quinone/quinol molecules. *Rubrivivax gelatinosus* possess a very low amount of intracytoplasmic membranes and no supramolecular organization of the photosynthetic apparatus. The lack of redox equilibration at the quinone level in this species is therefore due to a mechanism which remains to be revealed. In contrast, the lack of intracytoplasmic membranes in aerobic photosynthetic bacteria, e.g. *Roseobacter denitrificans*, could partially explain why its primary electron acceptor is fully reduced under anaerobic condition.

A further level of interaction between photosynthesis and respiration is based on the capability to coordinate the expression of photosynthetic and respiratory components in response to the oxygen tension and to the redox state of the system. Notably, the genetic coordination mechanism between photosynthetic and respiratory chains does not necessarily require a direct interaction of the two sets of components to exerts its sensing and regulatory functions. The signals to which the system responds may be originated from components that are specific for respiration, as in the case of oxygen sensing, or derive from components that may be either part of respiratory or photosynthetic ETCs, as for redox sensing. The picture that has been presented here, which is far from being exhaustive, tend to describe the framework of genetic regulatory connections existing in *Rb. capsulatus* and/or *Rb. sphaeroides* and gives support to the notion of a tight functional relationship between light- and dark-metabolic pathways. On the other hand, the way the supramolecular organization of the ETCs may be genetically controlled remains, at present, a relatively unexplored area.

Acknowledgments

DZ, RB and AV were supported by M.I.U.R. of Italy (PRIN2001, 2003) and Commissariat à l'Energie Atomique de France, respectively.

References

Anthony C (ed) (1988) Bacterial Energy Transduction. Academic Press, London

Baccarini-Melandri A, Melandri BA and Hauska G (1979) The stimulation of photophosphorylation and ATPase by artificial redox mediators in chromatophores of *Rhodopseudomonas capsulata* at different redox potentials. J Bioenerg Biomembr 11: 1–16

Bauer CE, Elsen S and Bird TH (1999) Mechanisms for redox control of gene expression. Annu Rev Microbiol 53: 495–523

Bird TH, Du S and Bauer CE (1999) Autophosphorylation, phosphotransfer, and DNA-binding properties of the RegB/RegA two-component regulatory system in *Rhodobacter capsulatus*. J Biol Chem 274: 16343–16348

Blankenship RE, Madigan MT and Bauer CE (eds) (1995) Anoxygenic Photosynthetic Bacteria. Kluwer Academic Publishers, Dordrecht

Brockmann H Jr and Lipinski A (1983) Bacteriochlorophyll *g*: A new bacteriochlorophyll from *Heliobacterium chlorum*. Arch Microbiol 136: 17–19

Buggy J and Bauer CE (1995) Cloning and characterization of senC, a gene involved in both aerobic respiration and photosynthesis gene expression in *Rhodobacter capsulatus*. J Bacteriol 177: 6958–6965

Buggy JJ, Sganga MW and Bauer CE (1994) Characterization of

a light-responding trans-activator responsible for differentially controlling reaction center and light-harvesting-I gene expression in *Rhodobacter capsulatus*. J Bacteriol 176: 6936–6943

Candela M, Zaccherini E and Zannoni D (2001) Respiratory electron transport and light-induced energy transduction in membranes from the aerobic photosynthetic bacterium *Roseobacter denitrificans*. Arch Microbiol 175: 169–177

Chance B and Smith L (1955) Respiratory systems of *Rhodospirillum rubrum*. Nature 175: 803–806

Collins MPL and Remsen CC (1991) The purple phototrophic bacteria. In: Stolz JF (ed) Structure of Phototrophic Prokaryotes, pp 49–97. CRC Press, Boca Raton

Cotton NJP, Clark AJ and Jackson JB (1983) Interaction between the respiratory and photosynthetic electron transport chains of intact cells of *Rhodopseudomonas capsulata* mediated by membrane potential. Eur J Biochem 130: 581–587

Crofts AR (1983) The mechanism of ubiquinol:cytochrome *c* oxido-reductases of mitochondria and of *Rhodopseudomonas sphaeroides*. In: Martonosi A (ed) The enzymes of Biological Membranes, Vol 4, pp 347–382. Plenum Press, New York

Crofts AR (2000a) Photosynthesis in *Rhodobacter sphaeroides*. Trends Microbiol 8: 105–106

Crofts AR (2000b) Response from Crofts. Trends Microbiol 8: 107–108

Crofts AR, Guergova-Kuras M and Hong S (1998) Chromatophore heterogeneity explains effects previously attributed to supercomplexes. Photosynth Res 55: 357–362

Daldal F, Mandaci S, Winterstein C, Myllkallio H, Duyck K and Zannoni D (2001) Mobile cytochrome c_2 and membrane-anchored cytochrome c_y are both efficient electron donors to the cbb_3- and aa_3-type cytochrome *c* oxidases during respiratory growth of *Rhodobacter sphaeroides*. J Bacteriol 183: 2013–2024

Dong C, Elsen S, Swem LR and Bauer CE (2002) AerR, a second aerobic repressor of photosynthesis gene expression in *Rhodobacter capsulatus*. J Bacteriol 184: 2805–2814

Donohue JT, McEwan AG, Van Doren S, Crofts AR and Kaplan S (1988) Phenotypic and genetic characterization of cytochrome c_2 deficient mutants of *Rhodobacter sphaeroides*. Biochemistry 27: 1918–1925

Drews G and Golecki J (1995) Structure, molecular organization, and biosynthesis of membranes of purple bacteria. In: Blankenship RE, Madigan MT and Bauer CE (eds) Anoxygenic Photosynthetic Bacteria, pp 231–257. Kluwer Academic Publishers, Dordrecht

Du S, Kouadio JL and Bauer CE (1999) Regulated expression of a highly conserved regulatory gene cluster is necessary for controlling photosynthesis gene expression in response to anaerobiosis in *Rhodobacter capsulatus*. J Bacteriol 181: 4334–4341

Dupuis A, Peinnequin A, Darrouzzet E and Lunardi J (1997) Genetic disruption of the respiratory NADH-ubiquinone reductase of *Rhodobacter capsulatus* leads to an unexpected photosynthesis-negative phenotype. FEMS Microbiol Lett. 148: 107–114

Dupuis A, Chevallet M, Darrouzzet E, Duborjal H, Lunardi J and Issartel JP (1998) The complex I from *Rhodobacter capsulatus*. Biochim Biophys Acta 1364: 147–165

Elsen S, Dischert W, Colbeau A and Bauer CE (2000) Expression of uptake hydrogenase and molybdenum nitrogenase in *Rhodobacter capsulatus* is coregulated by the RegB-RegA two component regulatory system. J Bacteriol 182: 2831–2837

Eraso JM and Kaplan S (2000) From redox flow to gene regulation: Role of the PrrC protein of *Rhodobacter sphaeroides* 2.4.1. Biochemistry 39: 2052–2062

Ferguson ST, Jackson JB and Mc Ewan AG (1987) Anaerobic respiration in the *Rhodospirillacea*e. FEMS Microbiol Rev 46: 117–143

Fernàndez-Velasco J and Crofts AR (1991) Complex or supercomplexes: Inhibitor titration show that electron transfer in chromatophores from *Rhodobacter sphaeroides* involves a dimeric UQH_2: cytochrome c_2 oxidase, and is delocalized. Biochem Soc Trans 19: 588–593

Georgellis D, Kwon O and Lin ECC (2001) Quinones as the redox signal for the Arc two-component system of bacteria. Science 292: 2314–2316

Gest H and Favinger JL (1983) *Heliobacterium chlorum* gen. nov. sp. nov., and anoxygenic brownish-green photosynthetic bacterium containing a new form of bacteriochlorophyll. Arch Microbiol 136: 11–16

Gray KA, Grooms M, Myllykallio H, Moomaw C, Slaughter C and Daldal F (1994) *Rhodobacter capsulatus* contains a novel *cb*-type cytochrome *c* oxidase without a Cu_A centre. Biochemistry 33: 3120–3127

Gregor J and Klug G (1999) Regulation of bacterial photosynthesis genes by oxygen and light. FEMS Microbiol Lett 179: 1–9

Herter SM, Kortlüle CM and Drews G (1998) Complex I of *Rhodobacter capsulatus* and its role in reverted electron transport. Arch Microbiol 169: 98–105

Jenney FE and Daldal F (1993) A novel membrane-associated *c*-type cytochrome, Cyt c_y, can mediate the photosynthetic growth of *Rhodobacter capsulatus* and *Rhodobacter sphaeroides*. EMBO J 12: 1283–1292

Jenney FE, Prince RC and Daldal F (1996) The membrane-bound cytochrome c_y of *Rhodobacter capsulatus* is an electron donor to the photosynthetic reaction center of *Rhodobacter sphaeroides*. Biochim Biophys Acta 1273: 159–164.

Joliot P, Verméglio A and Joliot A (1989) Evidence for supercomplexes between reaction centers, cytochrome c_2 and cytochrome bc_1 complex in *Rhodobacter sphaeroides* whole cells. Biochim Biophys Acta 975: 336–345

Joliot P, Verméglio A and Joliot A (1996) Supramolecular organization of the photosynthetic chain in chromatophores and cells of *Rhodobacter sphaeroides*. Photosynth Res 48: 291–299

Jones MR, Richardson DJ, McEwan AG, Ferguson SJ and Jackson JB (1990) In vivo redox poising of the cyclic electron transport system of *Rhodobacter capsulatus* and the effects of the auxiliary oxidants, nitrate, nitrous oxide and trimethylamine *N*-oxide, as revealed by multiple short flash excitation. Biochim Biophys Acta 1017: 209–216

Kappler U, Huston WM and McEwan AG (2002) Control of dimethylsulfoxide reductase expression in *Rhodobacter capsulatus*: The role of carbon metabolites and the response regulators DorR and RegA. Microbiology 148: 605–614

Kern M, Kamp P-B, Paschen A, Masepohl B and Klipp W (1998) Evidence for a regulatory link of nitrogen fixation and photosynthesis in *Rhodobacter capsulatus* via HvrA. J Bacteriol 180: 1965–1969

King GF, Richardson DJ, Jackson JB and Fergusson SJ (1987) Dimehylsulfoxide and trimethylmamine-*N*-oxide as bacterial electron transport acceptors: Use of nuclear magnetic resonance to assay and characterise the reductase system in *Rhodobacter capsulatus*. Arch Microbiol 149: 47–51

Klemme JH (1969) Studies on the mechanism of NAD-photoreduction by chromatophores of the facultative phototroph, *Rhodopseudomonas capsulata*. Z Naturforsch 24: 67–76

Knaff D and Kämpf C (1987) Substrate oxidation and NAD^+ reduction by phototrophic bacteria. In: Amesz J (ed) Photosynthesis. New comprehensive Biochemistry. Vol 15, pp 199–211. Elsevier, Amsterdam

Kramer DM, Kanazawa A and Fleishman D (1997) Oxygen dependence of photosynthetic electron transfer in a bacteriochlorophyll-containing *Rhizobium*. FEBS Lett 417: 275–278

Lavergne J, Joliot P and Verméglio A (1989) Partial equilibration of photosynthetic electron carriers under weak illumination: A theoretical and experimental study. Biochim Biophys Acta 975: 346–354

Loach (1966) Primary oxidation-reduction changes during photosynthesis in *Rhodospirillum rubrum*. Biochemistry 5: 592–600

Madigan MT (1988) Microbiology, Physiology, and Ecology of Phototrophic Bacteria. In: Zehnder JB (ed) Biology of Anaerobic Microorganisms, pp 39–111. John Wiley & Sons, New York

Madigan MT and Gest H (1978) Growth of a photosynthetic bacterium anaerobically in darkness, supported by oxidant-dependent sugar fermentation. Arch Microbiol 117: 119–122

Masuda S, Dong C, Swem D, Setterdahl AT, Knaff DB and Bauer CE (2002) Repression of photosynthesis gene expression by formation of a disulfide bond in CrtJ. Proc Natl Acad Sci USA 99: 7078–7083

Matsuura K, Mori M and Satoh T (1988) Heterogeneous pools of cytochrome c_2 in photodenitrifying cells of *Rhodobacter sphaeroides* forma sp. *denitrificans*. J Biochem 104: 1016–1020

McEwan AG, George CL, Ferguson SJ and Jackson JB (1982) A nitrate reductase activity in *Rhodopseudomonas capsulata* linked to electron transfer and generation of a membrane potential. FEBS Lett. 150: 277–280

McEwan AG, Cotton NPJ, Ferguson SJ and Jackson JB (1985) The role of auxillary oxidants in the maintenance of balanced redox poise for photosynthesis in bacteria. Biochim Biophys Acta 810: 140–147

Mosley CS, Suzuki JY and Bauer CE (1994) Identification and molecular genetic characterization of a sensor kinase responsible for coordinately regulating light harvesting and reaction center gene expression in response to anaerobiosis. J Bacteriol 176: 7566–7573

Myllykallio H, Zannoni D and Daldal F (1999) The membrane-attached electron carrier cytochrome c_y from *Rhodobacter sphaeroides* is functional in respiratory but not in photosynthetic electron transfer. Proc Natl Acad Sci USA 96: 4348–4353

Nakamura H (1937) Über die photosynthese bei der schelfreien purpurbakterie *Rhodobazillus palustris*. Beiträge zur stoffwechselphysiologie der purpurbakterie. Acta Phytochim 9: 189–234.

Nicholls DG and Ferguson SJ (2002) Bioenergetics 3. Academic Press, London and New York

Oelze J and Drews G (1972) Membranes of photosynthetic bacteria. Biochim Biophys Acta 265: 209–239

O'Gara JP, Eraso JM and Kaplan S (1998) A redox-responsive pathway for aerobic regulation of photosynthesis gene expression in *Rhodobacter sphaeroides* 2.4.1. J Bacteriol 180: 4044–4050

Pemberton JM, Horne IM and McEwan AG (1998) Regulation of photosynthetic gene expression in purple bacteria. Microbiology 144: 267–278

Prince RC (1990) Bacterial photosynthesis: From photons to Δp. In: Krulwich TA (ed) Bacterial Energetics of the Bacteria, Vol 12, pp 111–150. Academic Press, New York

Ramirez J and Smith L (1968) Synthesis of adenosine triphosphate in intact cells of *Rhodospirillum rubrum* and *Rhodopseudomonas sphaeroides* on oxygenation or illumination. Biochim Biophys Acta 153: 466–475

Richardson DJ, Bell LC, McEwan AG, Jackson JB and Ferguson SJ (1991) Cytochrome c_2 is essential for electron transfer to nitrous oxide reductase from physiological substrates in *Rhodobacter capsulatus* and act as an electron donor to the reductase in vivo. Correlation with photoinhibition studies. Eur J Biochem 199: 677–683

Richaud P, Marrs BL and Verméglio A (1986) Two modes of interaction between photosynthetic and respiratory electron chains in whole cells of *Rhodopseudomonas capsulata*. Biochim Biophys Acta 850: 256–263

Sabaty M, Gans P and Verméglio A (1993) Inhibition of nitrate reduction by light and oxygen in *Rhodobacter sphaeroides* forma sp. *denitrificans*. Arch Microbiol 159: 153–159

Sabaty M, Jappé J, Olive J and Verméglio A (1994) Organization of electron transfer components in *Rhodobacter sphaeroides* forma sp. *denitrificans*. Biochim Biophys Acta 1187: 313–323

Satoh T (1977) Light-activated, -inhibited and -independent denitrification by a denitrifying phototrophic bacterium. Arch Microbiol 115: 293–298

Schägger H (2001) Respiratory chain supercomplexes. IUBMB Life 52: 119–128

Schägger H and Pfeiffer K (2001) The ratio of oxidative phosphorylation complexes I-V in bovine heart mitochondria and the composition of the respiratory chain supercomplexes. J Biol Chem 276: 37861–37867

Schwarze C, Carluccio AV, Venturoli G and Labahn A (2000) Photo-induced cyclic electron transfer involving cytochrome bc_1 complex and reaction center in the obligate aerobic phototroph R*oseobacter denitrificans*. Eur J Biochem 267: 422–433

Shimada K (1995) Aerobic anoxygenic phototrophs. In: Blankenship RE, Madigan MT and Bauer CE (eds) Anoxygenic Photosynthetic Bacteria, pp 105–122, Kluwer Academic Publishers, Dordrecht

Sganga MW and Bauer CE (1992) Regulatory factors controlling photosynthetic reaction center and light-harvesting gene expression in *Rhodobacter capsulatus*. Cell 68: 945–954

Stackebrandt E, Murray RGE and Truper HG (1988) *Proteobacteria* classis nov., a name for the phylogenetic taxon that includes the 'purple bacteria and their relatives.' Intern J Syst Bacteriol 38: 321–325

Swem DL and Bauer CE (2002) Coordination of ubiquinol oxidase and cytochrome cbb_3 oxidase expression by multiple regulators in *Rhodobacter capsulatus*. J Bacteriol 184: 2815–2820

Swem LR, Elsen B, Bird TH, Swem DL, Koch HG, Myllykallio H, Daldal F and Bauer CE (2001) The RegB/RegA two-component regulatory system controls synthesis of photosynthesis and respiratory electron transfer components in *Rhodobacter capsulatus*. J Mol Biol 309: 121–138

Takamiya K, Iba K and Okamura K (1987) Reaction center complex from an aerobic photosynthetic bacterium *Erythrobacter* species OCh 114. Biochim Biophys Acta 890: 127–133

Takamiya K, Arata H, Shioi Y and Doi M (1988) Restoration of the optimal redox state for the photosynthetic electron transfer system by auxiliary oxidants in an aerobic photosynthetic bacterium *Erythrobacter* sp. OCh 114. Biochim Biophys Acta 935: 26–33

Truper HG and Pfennig N (1982) Characterization and identification of anoxygenic phototrophic bacteria In: Starr MP, Stolp H, Truper HG, Balows A and Shlegel HG (eds) The Prokaryotes, pp 299–312. Springer-Verlag, New York

Verméglio A (1977) Secondary electron transfer in reaction centers of *Rhodopseudomonas sphaeroides*: Out-of-phase periodicity of two for the formation of ubisemiquinone and fully reduced ubiquinone. Biochim Biophys Acta 459: 516–524

Verméglio A and Carrier JM (1984) Photoinhibition by flash and continuous light of oxygen uptake by intact photosynthetic bacteria. Biochim Biophys Acta 764: 233–238

Verméglio A and Joliot P (1999) The photosynthetic apparatus of *Rhodobacter sphaeroides*. Trends Microbiol 7: 435–440

Verméglio A and Joliot P (2000) Response from Verméglio and Joliot. Trends Microbiol 8: 106–107

Verméglio A, Joliot P and Joliot A (1993) The rate of cytochrome c_2 photooxidation reflects the subcellular distribution of reaction centers in *Rhodobacter sphaeroides* Ga cells. Biochim Biophys Acta 1183: 352–360

Verméglio A, Joliot A and Joliot P (1998) Supramolecular organization of the photosynthetic chain in mutants of *Rhodobacter capsulatus* deleted in cytochrome c_2. Photosynthesis Res 56: 329–337

Vichivanives P, Bird TH, Bauer CE and Tabita RF (2000) Multiple regulators and their interactions in vivo and in vitro with *cbb* regulons of *Rhodobacter capsulatus*. J Mol Biol 300: 1079–1099

Vignais PM, Colbeau A, Willison JC and Jouanneau Y (1985) Hydrogenase, nitrogenase and hydrogen metabolism in the photosynthetic bacteria. Adv in Microbial Physiol 26: 155–234

Woese CR (1987) Bacterial evolution. Microbiol Rev 51: 221–271

Witthuln VC, Goa J, Hong S, Halls S, Rott MA, Wraight CA, Crofts AR and Donohue TJ (1996) The reactions of the iso-cytochrome c_2 in the photosynthetic chain of *Rhodobacter sphaeroides*. Biochemistry 36: 903–911

Wraight CA (1977) Electron acceptors of photosynthetic bacterial reaction centers. Direct observation of oscillatory behavior suggesting two closely equivalent ubiquinones. Biochim Biophys Acta 459: 525–531

Wraight CA, Cogdell RJ and Chance B (1978) Ion transport and electrochemical gradients in photosynthetic bacteria. In: Clayton RK and Sistrom RW (eds) The Photosynthetic Bacteria, pp 471–502. Plenum Press, New York

Yurkov VV and Beatty JT (1998) Aerobic anoxygenic phototrophic bacteria. Microbiol Mol Biol Rev 62: 695–724

Yurkov V, Menin L, Schoepp B and Verméglio A (1998) Purification and characterization of reaction centers from obligate aerobic phototrophic bacteria *Erythrobacter litoralis*, *Erythromonas ursincola* and *Sandaracinobacter sibiricus*. Photosynthesis Res 57: 129–138

Zannoni D (1995) Aerobic and anaerobic electron transport chains in anoxygenic phototrophic bacteria. In: Blankenship RE, Madigan MT and Bauer CE (eds) Anoxygenic Photosynthetic Bacteria, pp 449–971. Kluwer Academic Publishers, Dordrecht

Zannoni D and Moore AL (1990) Measurement of the redox state of the ubiquinone pool in *Rhodobacter capsulatus* membrane fragments. FEBS Lett 271: 123–127

Zannoni D, Jasper P and Marrs BL (1978) Light-induced absorption changes in intact cells of *Rhodopseudomonas sphaeroides*. Evidence for interaction between photosynthetic and respiratory electron transfer chains. Biochim Biophys Acta 191: 625–631

Zannoni D, Peterson S and Marrs BL (1986) Recovery of the alternative oxidase dependent electron flow by fusion of membrane vesicles from *Rhodobacter capsulatus* mutant strains. Arch Microbiol 144: 375–380

Zehnder AJB (ed) 1988) Biology of Anaerobic Microorganisms. John Wiley & Sons, New York

Zeillstra-Ryallis J, Gabbert K, Mouncey NJ, Kaplan S and Kranz RG (1997) Analysis of the fnrL gene and its function in *Rhodobacter capsulatus*. J Bacteriol 179:7264–7273

Subject Index

A

B

C

D

H

I

K

L

M

N

O

P

Q

R

S

T

U

V

Species Index

Gene and Gene Product Index

O

P

Q

R

S

Y

Advances in Photosynthesis

Series editor: Govindjee, University of Illinois, Urbana, Illinois, U.S.A.

1. D.A. Bryant (ed.): *The Molecular Biology of Cyanobacteria.* 1994
ISBN Hb: 0-7923-3222-9; Pb: 0-7923-3273-3
2. R.E. Blankenship, M.T. Madigan and C.E. Bauer (eds.): *Anoxygenic Photosynthetic Bacteria.* 1995 ISBN Hb: 0-7923-3681-X; Pb: 0-7923-3682-8
3. J. Amesz and A.J. Hoff (eds.): *Biophysical Techniques in Photosynthesis.* 1996
ISBN 0-7923-3642-9
4. D.R. Ort and C.F. Yocum (eds.): *Oxygenic Photosynthesis: The Light Reactions.* 1996 ISBN Hb: 0-7923-3683-6; Pb: 0-7923-3684-4
5. N.R. Baker (ed.): *Photosynthesis and the Environment.* 1996
ISBN 0-7923-4316-6
6. P.-A. Siegenthaler and N. Murata (eds.): *Lipids in Photosynthesis: Structure, Function and Genetics.* 1998 ISBN 0-7923-5173-8
7. J.-D. Rochaix, M. Goldschmidt-Clermont and S. Merchant (eds.): *The Molecular Biology of Chloroplasts and Mitochondria in Chlamydomonas.* 1998
ISBN 0-7923-5174-6
8. H.A. Frank, A.J. Young, G. Britton and R.J. Cogdell (eds.): *The Photochemistry of Carotenoids.* 1999 ISBN 0-7923-5942-9
9. R.C. Leegood, T.D. Sharkey and S. von Caemmerer (eds.): *Photosynthesis: Physiology and Metabolism.* 2000 ISBN 0-7923-6143-1
10. B. Ke: *Photosynthesis: Photobiochemistry and Photobiophysics.* 2001
ISBN 0-7923-6334-5
11. E.-M. Aro and B. Andersson (eds.): *Regulation of Photosynthesis.* 2001
ISBN 0-7923-6332-9
12. C.H. Foyer and G. Noctor (eds.): *Photosynthetic Nitrogen Assimilation and Associated Carbon and Respiratory Metabolism.* 2002 ISBN 0-7923-6336-1
13. B.R. Green and W.W. Parson (eds.): *Light-Harvesting Antennas in Photosynthesis.* 2003 ISBN 0-7923-6335-3
14. A.W.D. Larkum, S.E. Douglas and J.A. Raven (eds.): *Photosynthesis in Algae.* 2003 ISBN 0-7923-6333-7
15. D. Zannoni (ed.): *Respiration in Archaea and Bacteria.* Diversity of Prokaryotic Electron Transport Carriers. 2004 ISBN 1-4020-2001-5
16. D. Zannoni (ed.): *Respiration in Archaea and Bacteria.* Diversity of Prokaryotic Respiratory Systems. 2004 ISBN 1-4020-2002-3
17. D. Day, A.H. Millar and J. Whelan (eds.): *Plant Mitochondria.* From Genome to Function. 2004 ISBN 1-4020-2399-5
18. *Forthcoming.*

19. G. Papageorgiou and Govindjee (eds.): *Chlorophyll a Fluorescence*. A Signature of Photosynthesis. 2004 ISBN 1-4020-3217-X

For further information about the series and how to order please visit our Website http://www.springeronline.com